Nebojša Nakićenović · Arnulf Grübler (Eds.)

Diffusion of Technologies and Social Behavior

With 107 Figures and 56 Tables

Springer-Verlag Berlin Heidelberg GmbH

Dr. Nebojša Nakićenović
Dr. Arnulf Grübler

International Institute for Applied Systems Analysis (IIASA)
A-2361 Laxenburg, Austria

ISBN 978-3-662-02702-8 ISBN 978-3-662-02700-4 (eBook)
DOI 10.1007/978-3-662-02700-4

Originally published by Springer-Verlag Berlin Heidelberg New York in 1991
Softcover reprint of the hardcover 1st edition 1991

Typesetting: Camera ready by author;

42/3020-543210 - Printed on acid-free paper

Preface

Wee felt it before in sense; but now wee know it by science.
Edward Misselden (1623)

The collective effort reported in this volume is the outcome of the diffusion of the idea of diffusion as a fundamental process in society. The considerable number of disciplines represented here indicates the weight of the problem area. The editors are to be congratulated for their initiative in drawing together present thinking at a vivid meeting, now also in print. An old-timer in the business has not much to add. But maybe some things, bearing in mind that a Preface is a celebration and not a review.

As always with ideas it is hard to identify those who first gave shape to the idea of diffusion. In a general sense it is probably an observation as old as human self-reflection that groups of populations exchange ideas and copy habits and implements from each other. Sometimes it has even been recommended, as a Chinese proverb suggested millenia ago, "If you want to become a good farmer, look at your neighbor".

A scholarly interest in the matter emerged toward the end of the last century. Friedrich Ratzel's *Anthropogeographie II* (1891) is one of the early landmarks. Ratzel spent much time studying archeological and ethnographic artifacts and pondered about the causes of their geographical distribution before they ended up on the shelves of museums. The dominating view at the time was that societies inevitably progressed, from simple to complex stages, almost like growing organisms. According to this belief, tools and habits of similar shape found among different populations indicated that they were at the same stage of a unilinear evolution.

Ratzel disagreed with this view, "Life is movement" and "Man is restless," he exclaimed. But more than that, he also found order in the changes and movements. He recognized centers of creation each exhibiting a radial spread. Since Ratzel was also the father of the concept of *Lebensraum* (later to become so disreputable) he could easily have said that life comprises multiplication and expansion. There are close relationships between diffusion and invasion.

Ratzel entertained quite realistic ideas about social responses to innovations.

> Some things are readily adopted and thus spread fast. It is naturally those which are necessary or comply with the inclinations of people, whereas what is difficult and toilsome is dropped and forgotten. Narcotic stimulants have fast conquered large areas. Consider how widely tobacco, betel, and spirits have spread. Then the diffusion of effective weapons is surprising. They compete with spirits not only in today's Africa but made their way equally fast into old America.

In some respects restless man does not seem to have changed preferences during this century. However, qualitatively and quantitatively we now live in a different world, reshaped by wave after wave of innovations.

At the same time as Ratzel carried out his studies on human cultures with biological analogies, the French scholar Gabriel Tarde saw things from a social psychology vantage point. In *Les lois de l'imitation* (1890) he saw society as the sum total of invention and subsequent imitation. And he pointed out that only a tiny fraction of all inventions were actually imitated and became collective property. In this respect things are not different today. Research into the mortality of inventions would be an interesting complement to studies of diffusion. In technology, at least, it would not be impossible to do this.

At the turn of the century the time was not yet ripe for quantitative answers to questions concerning diffusion. Nor were efforts at predicting the agenda. It did not occur to scholars that contemporary observations could reveal something of general interest with reference to their problem. Gradually, however, the mood changed. In his book *Cultural Change* (1928) F.S. Chapin made frequent use of the "curve of cumulative growth", borrowed from demography and epidemiology. He may not have been the first to do so, but since the 1920s the S-curve, with its many empirical and mathematical varieties, has been the favored device for describing and interpreting innovation diffusion.

Work in the spatial tradition of Ratzel and his generation has been more restricted. The reason is perhaps that it is more time-consuming to pin down events place by place than it is to take time-series data from statistical publications. In addition the mathematical treatment of space and time phenomena is a lot more cumbersome. Many would probably argue that space is less relevant in our present era of high physical mobility and easy global communication than it was in former times. However, to my mind there is more to space than mere distance. Only maps can show where items under study are missing. This might also reveal something about both causes and consequences. In the higher topology of networks we still have, in all likelihood, centers of creation and structured channels of spread. The "niche" is also a spatial concept as it refers to society's ability to provide "room" for the specific kind of "choreography" every new device requires. But to give empirical substance to such dimensions is clearly very difficult.

One can think of many reasons why it is important to understand the forces behind diffusion. One obvious reason is the use that people in public policy and industry can make of improved methods of prediction. A more purely academic reason is that the spread and nature of innovations can cast light on the structures and workings of cultures and societies.

A further reason – I believe the most important in our present historical situation – is the increasing necessity for global and national policy makers to concentrate on such innovations that put the world on the road toward *sustainable development*. This task must be tackled not only by UN resolutions and national legislations. Real implementation entails widespread dissemination of new values and new tools and materials. Careful selection of normative centers and suitable designs for inviting niches are required in order to enlist the needed support of market forces.

Rigid planning is out these days. The time has come for the art of caring about the content of innovations and guiding their diffusion.

Torsten Hägerstrand
Lund, Sweden
November 22, 1990

Introduction

The adoption of innovations is at the core of the dynamic processes underlying social, economic, and technological change. Diffusion phenomena occur at different hierarchical levels of society and on different temporal and spatial scales; they are all-embracing, starting at the micro level of products and processes and continuing up to large pervasive systems and infrastructures, sometimes enveloping certain forms of social and political organization. Consequently, the timing, duration, and patterns of innovation diffusion vary greatly. A deeper analysis of diffusion processes often reveals increasing levels of complexity and a multitude of interdependencies and nonlinear feedback mechanisms.

Considerable progress has been made in understanding the nature of diffusion processes and their underlying driving forces both theoretically and empirically. Many studies have been published during the last few decades that deal with technological and social diffusion processes. New theories of innovation diffusion have been proposed and numerous case studies have been analyzed. More recently, diffusion analyses have been applied to aid strategic business decisions and planning activities. It is therefore fair to say that both theories and applications have established a new body of knowledge that is being used more widely in a variety of ways: to explain the process of social and economic change and restructuring, to integrate technological change into economic theory, and as a policy tool. In view of the importance of this research area an international, interdisciplinary conference on *Diffusion of Technologies and Social Behavior* was held at the International Institute for Applied Systems Analysis (IIASA), Laxenburg,

Austria, in June 1989, to review the results of diffusion research, to discuss the value and usefulness of accomplished work, and to identify future research directions.

The aim of the conference was to stimulate future research and to enhance the integration of various diffusion research disciplines. The range of disciplines represented at the conference included physics, engineering, economics, political science, sociology, management science, history, and geography. In addition to researchers, the 91 conference participants from 16 countries included practitioners engaged in public and private policy making and marketing. The individual sessions were devoted to a review of diffusion theories, including spatial and temporal approaches, economic theory, and sociology; an overview of empirical case studies of both product and process innovations; and applications of diffusion studies for strategic policy formulation. The conference concluded with a summary session addressing the major open questions of diffusion research and possible future research strategies. The contributions to this volume are representative of the conference presentations[1]. The Conference Program and List of Participants are included at the end of the book.

Chapter 1 by Jesse Ausubel gives a comprehensive summary of both the individual contributions to the conference and the discussions. It provides an excellent insight into the complexity of diffusion phenomena. On a general note Ausubel raises the issue of what constitutes "diffusion-oriented policies" and notes the importance of heterogeneity between technological options, economic agents, and their expectations of the future which often appear as a prerequisite for evolution. He stresses the need for multiple perspectives, for the exploration of several paths in both the analyses and policies that deal with diffusion phenomena, for more cross-cultural comparisons, and he lists some of the issues that remain on the research agenda. Ausubel raises the question of whether by "looking at the patterns of the patterns" some meaningful taxonomy or classification can be attempted in order to identify and eventually explain differences and similarities of diffusion processes.

In Chapter 2, George Modelski and Gardner Perry III present an interesting answer to two vital but difficult questions: How much democracy is there in the world? Is there a regular pattern to the global spread of democracy? These are very timely questions in view of the encouraging process

[1] A smaller selection of conference papers entitled "From Democracy to Chain Saws: New Perspectives on Innovation Diffusion" is to be found in a Special Issue of *Technological Forecasting & Social Change* **39**(1–2), March–April, 1991.

of liberalization and spread of democracy in Eastern Europe and certainly very apt for a conference held in Laxenburg, Austria, the crossroads between East and West. The authors estimate that in 1986 approximately 40 percent of the world population was "democratic," with a projected increase to 50 percent shortly after the year 2000. By adopting a long-term perspective, democracy is analyzed as a superior "technology of collective choice," the spread of which is seen as a pervasive diffusion process. Pronounced discontinuities with peaks and setbacks are "strewn along the way," but the overall upward direction of the process is clearly demonstrated. However, it may take another 100 years or so before the majority (90 percent) of the world population will live under a democratic social and political order.

William Peirce (Chapter 3) analyzes diffusion processes from a regional perspective focusing on the "single Europe" and the possible effect of international integration on promoting or retarding innovativeness. He argues that a single market and the removal of national barriers within the Community would yield benefits from economies of scale and promote competitiveness, while some institutional rigidities would tend to exacerbate inefficient policies, abet the reformation, and augment the strength of existing distributional coalitions. The author argues that governments can remove some barriers by improving the general environment for diffusion and by taking specific measures to encourage technological change. However, he emphasizes that it is not clear that they have much direct influence on adoption decisions and that it is generally difficult for any government to accelerate diffusion processes. This finding may be the reason why some studies show similar diffusion rates in dissimilar countries and periods of time.

Harold Linstone demonstrates in Chapter 4 the limitations of reductionist approaches in systems research and the gap that exists between models and real-world decisions. He argues for the augmentation of what he calls the "technical" perspective, by explicitly introducing organizational and personal perspectives into the analysis. The chapter includes a number of case studies ranging from strategic planning to risk evaluation and management, energy forecasting, trade deficits, and cross-cultural analyses, where a multiple-perspective approach has been used. Guidelines are provided on how to balance the description and integrate different perspectives on a particular issue. He also identifies a growing mismatch between the rates of technological and socio-institutional change and arrives at the intriguing conclusion that "even the most advanced countries continue confronting 21st-century technology with 19th century institutions."

Linstone refers to information technology as a unique diffusion factor, the degree and speed of which will affect the diffusion of all other technologies. This idea is extended in Chapter 5 by Dirk-Jan Kamann and Peter Nijkamp stressing in particular the importance of information exchange (external and internal learning) and of the role of informational and social-spatial networks in the diffusion process. The authors analyze the intricate relationships in the "triangle" of technology, economy, and space within a given socio-cultural setting. They are critical in determining the development trajectories that in turn shape new conditions for technological progress and spatial dynamics. The latter are path-dependent and marked by nonlinear evolutionary patterns of growth and decline that sometimes seem to accompany each other even within the same region or sector. Spatial differentiation and diversity in the selection environment are the cause of social-spatial networks which serve as potential channels of diffusion between actors. The authors conclude that network cooperation between actors creates a synergetic surplus which is largely appropriated by those actors who dominate other actors or even entire network segments.

The next two chapters give an overview of the evolving perspectives and models developed to analyze diffusion processes from two different disciplines – marketing (Chapter 6) and economics (Chapter 7). Both address many common concepts and issues, e.g., "externalities" such as market intervention, or more generally, the role of institutions and policies in diffusion processes.

Chapter 6 by Vijay Mahajan, Eitan Muller, and Frank M. Bass reviews the evolution of marketing diffusion models since the pioneering work of Bass in 1969. A common characteristic of all models is that they analyze diffusion after the turbulent (cf. Chapter 9) selection phase of new ideas and innovations. The authors analyze the salient characteristics and the applications of the main models, and investigate ways to make them theoretically more sound and practically more effective. Eleven areas are considered to be particularly suitable for extensions and refinements. The authors also highlight the importance of interaction and experience sharing between firms and diffusion analysts to further validate diffusion models and enhance their usefulness.

In Chapter 7 Giovanni Dosi gives an overview of innovation diffusion research from the perspective of evolutionary economics. The discussion focuses first on the economics of R&D, highlighting in particular the dichotomy between the partly public good nature of information and the "tacitness" of cognitive structures, skills, and capabilities guiding inventive activities.

Departing from a number of "stylized facts" concerning diffusion the author then gives an overview of the hypotheses underlying diffusion models in economics. These are related, *inter alia*, to the questions of the heterogeneity of potential adopters, to their behavior and decision procedures, and to the nature of the dynamics of technological and economic change. A taxonomy of diffusion models is developed based on a two-dimensional classification (equilibrium versus disequilibrium dynamics and optimizing versus institutionalized behavior of the adopters). In particular the chapter highlights the merits of evolutionary approaches based on principles of self-organization (cf. Chapter 8) for modeling diffusion processes involving various forms of increasing returns, circular feedbacks between innovations, or diffusion environments where "rational behavior" is not only empirically unlikely but also theoretically impossible to define. The chapter concludes by identifying a number of broader macro issues and challenges for diffusion research which may bring together economics, history, social disciplines, and policy analysis.

The next three chapters describe the dynamics of selection and diffusion mechanisms from an evolutionary perspective and demonstrate that innovation adoption processes are governed by interdependence (e.g., between demand and supply), heterogeneity in expectations and appropriability conditions, and a multitude of nonlinear feedback mechanisms.

Gerald Silverberg (Chapter 8) emphasizes an important dimension of the evolutionary nature of diffusion phenomena – the path-dependency and self-organization of collective behavior in the innovation adoption process. He shows that a unique "optimal" payback period for innovation adoption may not exist independent of market structure and pricing strategies, and he demonstrates the basic nature of collective behavior that may lock an industry into alternative combinations of locally stable payback periods and consequently different adoption patterns among economic actors. He bases these conclusions on simulation "experiments" with an evolutionary model of market competition and incremental technical change. The model illustrates that the dynamic appropriability of an innovative strategy is a function of the rates of learning, both internal and public. He shows that S-shaped diffusion paths at the macro level result from diversity in technological design, individual behavior, and uncertainty about and interdependence between the actions of economic agents and their effect at the micro level.

In Chapter 9 Sergei Glaziev and Yuri Kaniovski present a stochastic model of the uncertainty and random fluctuations prevailing in the early phase of diffusion processes. It has been frequently argued that one of the conditions for "take-off" of diffusion is the emergence of standardization and a dominant design. Thus, it is particularly important to elucidate the conditions and dynamics underlying the selection mechanisms in the early diffusion phases. The authors treat two sources of uncertainty in this early phase concerning the characteristics of technologies and the market. The model clearly illustrates that adopters can influence but not predetermine the final outcome of the selection process under conditions of uncertainty. Yet, it is exactly this highly turbulent environment and diversity of expectations that shape and "lock in" the future characteristics of technologies and strategies of adopters.

Ove Granstrand presents a diffusion model in Chapter 10 where the organizational units representing users and producers of technology are explicitly coupled in order to incorporate the interdependence between the demand and supply sides of diffusion processes. The model considers both diffusion (entry) and substitution (exit) processes. Interesting insights are provided into the complex dynamics of (unstructured) subpopulations in terms of the interactions between buyer and seller populations that are subjected to a stream of innovations. At the macro level, conclusions are similar to those that emerge from the evolutionary modeling approaches: diffusion is a result of complex, nonlinear dynamics and the interaction of a diverse number of subpopulations of potential adopters.

The remaining chapters are concerned with specific case studies that illustrate innovation diffusion from the macro down to the micro level. They illustrate the pervasiveness and universality of diffusion phenomena underlying the processes of techno-economic and social change.

The chapters by Gerhard Rosegger and Sergei Glaziev analyze diffusion in the context of technology transfer, the former from the perspective of a single economic sector and the latter from the perspective of a whole national economy. Both papers illustrate that the interrelatedness between technological, institutional, and organizational settings account for many of the differences in the rates of technology diffusion and changes in competitiveness.

Gerhard Rosegger (Chapter 11) examines interfirm cooperation in the global automobile industry and the function of such partnerships in promoting the transfer of technology. He shows that the mature American automotive industry used partnerships with foreign rivals as a strategic instrument for the diffusion of innovations originally introduced into imported vehicles. It is demonstrated that this particular catch-up strategy on the product side did not succeed in closing the performance gap (between the American and the European and Japanese products). This illustrates that "corporate culture" and organizational styles (e.g., attitudes, routines, and standard operating practices) may require channels other than those through which individual, capital-embodied techniques are diffused, and that the "Japanese approach to manufacturing" cannot be copied but must be learned in context. The short-run benefits of promoting diffusion through cooperative agreements may be overshadowed by the longer-run loss of innovative capabilities or, as Rosegger provokingly states, there is a danger of "forgetting by not doing."

Sergei Glaziev (Chapter 12) introduces the notion of interrelatedness of whole clusters of technological, infrastructural, and organizational innovations. This aspect of interrelatedness, which has become familiar under the concept of socioeconomic paradigms is demonstrated using a principal component approach to describe technological change and diffusion in a number of market and centrally planned economies. The author illustrates that during the post–World War II period there was a simultaneous development of successive generations of technology clusters in the USSR. In view of the current restructuring efforts of *perestroika*, the author argues that an opportunity window exists in such transitional periods representing an "advantage for the backward." By a careful assessment of the emerging new directions, technologies, and growth sectors, optimal strategies for closing the technology and productivity gap could be developed by "surpassing without overtaking." Appropriate institutional and organizational reforms, however, are a precondition for redirecting investments and the future orientation of the economy as a whole toward newly emerging key industries.

The chapters by Charlie Karlsson and by Maryellen Kelley and Harvey Brooks analyze innovation diffusion in manufacturing industries. Both examine the diffusion of technology from the perspective of the structural characteristics of adopting firms. In particular, they focus on the size of the firm, external linkages, and information channels.

Charlie Karlsson (Chapter 13) analyzes the characteristics of firms in the engineering industry that were early adopters of information technology in two peripheral regions of Sweden. In addition to the influence of firm size on adoption date, he also studies other pertinent characteristics. The results indicate the importance of direct and indirect information channels of external and internal learning and appropriability conditions for the early adoption of technologies. More generally, the chapter also presents conclusions regarding regional technology and innovation policies stressing that both information channels and skill levels of the labor force can be promoted by appropriate government policies.

In Chapter 14, Maryellen Kelley and Harvey Brooks analyze the factors that influence the adoption of programmable automation by American manufacturers based on a recent survey of more than one thousand establishments. They show that there is a wide range of "incentive structures" between metal-working and machinery manufacturing. Differences in firm size appears to be the most important factor in explaining the range of adoption rates. Other important factors include external linkages and the network participation of firms. In fact, well-connected small firms tend to demonstrate higher adoption rates. Conversely isolated, small firms tend to rely on traditional techniques. One way of promoting diffusion would be to strengthen the linkages between economic agents by appropriate institutional arrangements that enhance external learning and learning by doing.

The next five chapters analyze more traditional economic and technological factors such as cost and performance in determining diffusion rates. Whereas the previous two chapters have shown that the social and institutional dimensions of innovation diffusion are difficult to treat in terms of "standard" diffusion models, the next five chapters show that it is often difficult to apply simple univariate analytical tools to more traditional diffusion measures such as market shares.

John Tilton (Chapter 15) analyzes the process of material substitution from the perspective of technology diffusion and from the more traditional perspective that emphasizes the role of relative material prices. He stresses that the two approaches are not mutually exclusive, but that the technological change view provides a better understanding of the dynamic processes that operate over longer time scales. He describes the American beverage container industry, where very dynamic and complex temporal patterns in material substitution for beer and soft-drink containers have resulted in six different container types competing simultaneously for market shares. This also illustrates the inadequacy of simple S-shaped models to describe multiple innovation diffusion processes. An important finding is that while a price rise over a limited range may have little or no effect on demand, once the price increases beyond some threshold level it may stimulate the introduction of a new technology that shifts the demand curve downward. This indicates that material substitution should be most pervasive in those applications experiencing rapid technology change and that significant changes in the prices of competing materials would enhance the substitution process.

Dominique Foray and Arnulf Grübler (Chapter 16) deal with another important conceptual and methodological area of diffusion research: the definition of the object diffusing, its interaction with the technological environment in which it is embedded, and the transformation it undergoes during its diffusion. To this end, the authors introduce the concepts of "technological neighborhood" and "distance" derived from a morphological analysis of the entire technological space for a particular function. This then serves as a methodology for defining competing technological routes, illustrated for the case of non-ferrous casting processes in the FRG and France. The authors illustrate the critical importance of small initial market niches providing the ground for experimentation and learning within the industry. Although initially inferior to its competitors, this allows a new technology to escape from a "lock-in" situation, become competitive, and diffuse into a larger market. The authors conclude by presenting an evolutionary tree developed to describe the changing nature and characteristics of technologies during diffusion.

Robert Ayres and Ike Ezekoye (Chapter 17) describe the complex interplay of competition and complementarity in the adoption of tetraethyl lead as an antiknock additive in gasoline. They show that it is difficult to apply the standard "life cycle" model of technological evolution in the context of complex systems that often portray nonlinear dynamic behavior and self-reinforcing "lock-in" mechanisms. In particular, the authors investigate possible explanations for the failure of antiknock additives to displace cracking as a means of raising gasoline octane (or vice versa). They demonstrate that technological substitution in the petroleum-cracking processes followed similar patterns as shown by Tilton (Chapter 15) for material substitution: simultaneous competition and a sequence of process replacements. However, petroleum cracking and production of additives have not been substituted for each other but have instead maintained a complementary character. The economies of adoption apparently favor this complementarity rather than the replacement of additives by further petroleum refining. The authors note another important difference between the two technological routes: while cracking evolved rapidly, tetraethyl lead additives remained static until regulation forced a change.

Arnulf Grübler (Chapter 18) introduces a further degree of complexity in describing temporal and spatial patterns of diffusion processes. Using several case studies ranging from merchant marine propulsion systems to raw steel production processes he demonstrates that observed diffusion processes often cannot be described in terms of a single (simple) diffusion curve. Also the importance of any individual driving force such as relative costs, prices, or technical change is not only different in the various phases of the diffusion process but also in successive technological generations. Therefore, diffusion phenomena and their driving forces need to be analyzed in a larger context, e.g., as elements of so-called "techno-economic paradigms" that constitute a hierarchy of cross-enhancing and related diffusion clusters. He therefore proposes a classification scheme for assessing the importance of diffusion processes: the longer a diffusion process lasts the more pervasive it would be in terms of its social and economic impacts. He concludes that the transition from one "cluster" of innovation diffusion to the next generates pronounced discontinuities in the longer-term evolution of our socioeconomic systems.

Chapter 19 by Nebojša Nakićenović discusses the linkages between individual diffusion processes that constitute a whole "cluster" associated with a given techno-economic regime. Based on the characteristic time constants of diffusion processes the chapter demonstrates that the time span between the start of diffusion in leading and lagging countries tends to decrease as the whole cluster matures. However, the leaders achieve higher adoption levels than the followers roughly in proportion to the time lag in introduction. This catch-up effect and its relationship to the adoption level are illustrated on the basis of the evolution of transport infrastructures and systems in several countries. A comparison of the evolution of transport systems in the United States and the Soviet Union indicates that the catch-up effect also occurs over longer periods, spanning three successive generations of diffusion clusters. These conclusions about the focusing of the diffusion clusters toward saturation are, of course, only tentative. It is suggested that a taxonomy and classification of hierarchies within each cluster could be a possible route toward determining the driving forces behind the clustering effect observed in the samples presented.

The following two chapters by Theodore Modis and Alain Debecker and by Paul Diederen and colleagues deal with diffusion phenomena in a rapidly expanding part of the economy: the service sector. Chapter 20 by Theodore Modis and Alain Debecker analyze an often neglected but important "opposite" to diffusion: the mortality of products. Using experience drawn from a very dynamic and rapidly expanding industry they report that the life cycles of computers are shortening considerably: 1 to 2 years for some models. Their modeling approach to determine product life spans (as opposed to sales life cycles only) is particularly appealing in its duality of growth and mortality formulations in the form of a convolution function. One of the most interesting findings emerging from their study is the difference in computer mortality: models phase out at the same time as their technological generation becomes outdated. This "accelerating" mortality of the later models of a particular generation provides an interesting symmetry to a similar phenomenon in the clustering of diffusion processes discussed in the previous Chapter.

Paul Diederen together with René Kemp, Joan Muysken, and Rombout de Wit report in Chapter 21 on the results of a larger study of technological change: employment and skill formation in Dutch banking. First an overview is given of new information technologies applied to banking and other related technological and organizational innovations. A model is then introduced that analyzes the diffusion of these new technologies based on the notion that technological change involves a learning process so that the availability of a more profitable technique does not automatically imply that it will be used immediately. Diffusion takes time because different techniques will be employed at each moment to produce a certain mix of services. These are captured in the model by the concepts of technological distance, efficiency barrier, competitive power, and pressure. The case study illustrates that technological change is a result and consequence of learning and adjustment processes. For example, the change in the handling of accounts from the traditional product-oriented approach to "custom banking" has repercussions for the whole nature of the business and especially on the skill requirements of human resources.

The next two chapters illustrate the social dimension of diffusion phenomena. Chapter 22 by Jonny Hjelm focuses on the change of skill requirements in Swedish forestry as a consequence of the diffusion of the motor-driven chain saw. While this had far-reaching consequences and fundamentally transformed the nature of the whole industry, the salient point is that it was the forest workers themselves who originally financed and bought chain saws to increase productivity. In fact, the diffusion of the chain saw was only one component of the overall process of structural change in the industry; another important accompanying institutional change was the introduction of the piece-rate wage system. Under this wage system any savings resulting from productivity increases due to innovation diffusion tended to increase the intensity of work. Hjelm's analysis deals explicitly with the social context into which any diffusion case is embedded, an aspect too frequently ignored in many diffusion studies. It stresses the collective nature of the process of technological change and the uncertainty and interdependence of individual adoption decisions.

Chapter 23 by Emilio Casetti and Cindy Fan returns to the origins of diffusion studies: the spread of epidemics. Their analysis illustrates the importance of spatial differentiation and contexts in diffusion phenomena and deals with the spatial dimension of the spread of AIDS at the regional level (Ohio, USA). In contrast to the other chapters, the importance of the social and behavioral changes that are necessary to slow down the spread of AIDS is stressed: a "diffusion" process that is not based on voluntary adoption decisions. The authors use a spatial polynomial expansion method, focusing upon variation in the temporal growth patterns of AIDS in response to population densities. The results from this regional analysis illustrate the importance of social and infrastructural networks in the spatial spread of the disease. On an optimistic note, the authors observe that the rate of increase in AIDS cases has started to slow down, pointing to the fact that increasing awareness and modification of individual behavior are crucial factors in controlling the spread of this epidemic.

The concluding chapter is by Cesare Marchetti, one of the pioneers of diffusion research. It is a very personal contribution; it annotates his professional preoccupation with diffusion phenomena for almost 20 years. It demonstrates the pervasiveness and universality of the phenomena that other contributions in this book have tried to embrace from a wide range of disciplinary, modeling, and empirical perspectives. Ranging from the process of technological and economic change to transformations in the social fabric and cultural traits, Marchetti describes diffusion phenomena as fractal in nature "branching out into the universe." Perhaps it is true that our understanding of the deeper underlying mechanisms of the process of technological, economic, and social change is still fragmented and insufficient. But as Marchetti says, the phenomenon is of true universal importance and holds the key to striking "a deeper level of truth."

Acknowledgments

The conference that led to this publication was jointly sponsored by the International Institute for Applied Systems Analysis (IIASA), the Swedish National Board for Technological Development, and the Swedish Council for Planning and Coordination of Research. We would like to extend particular thanks to the conference co-organizers Stören Wibe (University of Umeå) who actively participated in shaping the event and aided in obtaining financial support for the conference, and Robert U. Ayres (Carnegie-Mellon University) who was instrumental in organizing the conference, solicited the participation of many key participants, and provided essential intellectual input during the sessons and discussions. Special thanks are also due to the IIASA conference staff and to all of the participants for making it such a successful event.

For their advice during the preparation of the manuscripts, we would particularly like to thank Jesse Ausubel, Dominique Foray, Hans-Dieter Haustein, and Stören Wibe. We are also deeply indebted to Wendy Caron, Lourdes Cornelio, Eryl Maedel, Susan Riley, and Lilo Roggenland for their valuable help and assistance in the preparation of the manuscript. Needless to say, any remaining errors and shortcomings are ours.

Many other institutions and individuals have been helpful and supported this effort, and we cannot mention them all here, but we would like to thank IIASA's Director, Peter E. de Jánosi, for making the publication of this book possible.

Contents

Chapter 1

Rat Race Dynamics and Crazy Companies: The Diffusion of Technologies and Social Behavior

Jesse H. Ausubel

1.1 Introduction

How and why do technologies spread when and where they do? What are the implications and consequences for the structure, wealth, and management of human organizations? These expansive questions were the subject of the presentations and discussions of the International Conference on Diffusion of Technologies and Social Behavior, summarized in this chapter. The chapter is organized under the following headings: empirical regularities; theoretical issues; predictability; roles of time and space; definition of niche and innovation; selection dynamics; role of marketing; social aspects of diffusion; globalization of diffusion processes; and applications of diffusion. While the chapter treats some questions for policy in both the public and private sectors, it emphasizes research needs and opportunities in the diffusion field.

The conference represented a convergence and a maturation of studies of diffusion. A great range of disciplines was represented from both the social sciences and the natural sciences. There were geographers, historians, economists, sociologists, psychologists, and political scientists. There were physicists and mathematicians. Along with researchers, there were also practicing engineers and managers. The conference was made more special by the participation of several of the modern pioneers of the exploration of diffusion, including Torsten Hägerstrand, Harold Linstone, Cesare Marchetti, and Robert Pry, people who have facilitated diffusion research over the years and provided many of the ideas on which the conference was built.

The first point to address is why the group came together. The answer is the importance of diffusion as a key process in social and economic change, made powerfully evident by the growing and widespread recognition of regularities of diffusion processes.

1.2 Empirical Regularities

In a sense, the conference, like diffusion research itself, had an empirical origin and a phenomenological orientation. Each discipline, each group of researchers, discovered, somewhat independently, diffusion phenomena. One of the most satisfying aspects of the conference was the presentation of data on newly charted diffusion processes. There were examples of resins and plastics from Vladimir Falzman from the Soviet Union and examples in transport by Veniamin Livshits, also from the Soviet Union. There were two examples on AIDS. There were examples from Oskar Ullman from the Federal Republic of Germany in the solar energy area. There were examples on automated banking from the Netherlands (Paul Diederen and René Kemp), electronic mail from Sweden (Tomas Åstebro), chain-saws from Sweden (Johnny Hjelm), and Catholic saints from Italy (Marchetti). George Modelski presented the spread of democracy as a global diffusion process. The multitude of examples is most important. One of the significant features of the conference was the recognition that there is in fact now a large library of cases of diffusion, perhaps 3,000 cases that are well-documented and quantified.

One of the major tasks for diffusion research for the next years is the meta-analysis of ensembles of diffusion processes analyzed in the various disciplines. Can one undertake some meaningful taxonomy or classification of the many examples? There might be various criteria, for example, the time constant of diffusion processes (the so-called "Δt"). Other facets to examine might include relationships of clusters of technologies, relationships of levels in the system, and pervasiveness of phenomena, as Roberto Vacca suggested at one point. The job is to look at the patterns of the patterns: to compare countries, to compare industries, and so forth, according to their characteristic forms and configurations of technological innovation and diffusion. The objective would be to identify and eventually explain differences, similarities, and congruences of diffusion processes and their causes.

Whatever discipline one comes from, we all now have a rich empirical library on which to draw. There is, however, a need to make this library and its raw data more accessible.

1.3 Theoretical Issues

While there is excitement and satisfaction with the construction of the empirical base, there is considerable questioning of the adequacy of the theories resting upon the data. In each field, theoretical models have been developed. For example, in economics new mathematical approaches to treat the complex dynamics of diffusion and selection processes and the collective behavior of economic agents were presented by Giovanni Dosi and Gerald Silverberg. They also referred back to earlier contributions in economics by Josef Schumpeter, Edwin Mansfield and others. In geography, theories and models of Hägerstrand and others were mentioned. Vijay Mahajan presented an overview of diffusion models from the perspective of marketing and management science. New models were shared at the conference as well, by Heinz-Dieter Haustein and in a paper by Ove Granstrand. But, one senses that there is considerable dissatisfaction about the theoretical base. We do not feel able to explain the phenomena well. We do not feel we understand mechanisms.

At the same time, there seems to be an acceptance of a vocabulary for talking about the phenomena. The vocabulary is largely derived from the field of biology. There is not too much debate about the usefulness of the biological metaphor. Harvey Brooks presented the metaphor clearly and succinctly, describing principal features of the evolutionary process as generally understood in life sciences and extending them to the seemingly inanimate world of technological objects. Other biologically-derived ideas were presented as well, for example by Michael Sonis who suggested examining the usefulness of the competitive exclusion principal and principals of collective behavior, not based on optimization. There were other suggestions of this kind, but certainly no field represented seemed to be satisfied with its theoretical base.

At a general theoretical level, I was reminded of the statement by the late Elliot Montroll, a well-known American physicist, from 1978, that "Evolution is a sequence of replacements" (Montroll, 1978). That statement is, for the most part, consistent with research presented in the conference. There may be some examples to contest, but Montroll's assertion is the kind of theoretical hypothesis that might be used generally in diffusion studies. It also reminds researchers of the limits of focusing on only a single diffusion process in which a new technology replaces an old one. As Arnulf Grübler emphasized, evolution is composed of a series of interlaced and multiple diffusion processes, characterized by various driving forces and adoption environments.

Among the most promising directions for the search for theory appears to be the field of communications, as well as biology. In communication, the aspect stressed was networks. Networks were raised in different forms by several speakers, for example, Kirk-Jan Kamann and Peter Nijkamp from the Netherlands. Gerhard Rosegger, talked about a particular type of network, the curious mix of formal and informal alliances that give shape to the global automobile industry. Hägerstrand mentioned the need to research the architecture of social communication and relayed the wonderful quote, "First I make friends, then I make business." There were also the AIDS examples (Emilio Casetti and Cindy Fan), which stressed very much the importance of networks. Here is an area in which the existence of networks, the revealing of social networks over the last few years as the epidemic has spread, has been readily apparent. Illumination may come from understanding communication in more detail, as manifested in spatial and other characteristics of networks.

The question of networks is intimately linked to the question of people's behavioral rules. A relevant insight comes from *The Book of The Courtier*, by Baldesar Castiglione, one of the first advisors to policy makers (Castiglione, 1528). Castiglione did not emphasize analytic processes in explaining diffusion. "Usage is more powerful than reason in introducing new things among us and in blocking out old things and anyone who tries to judge of perfection in such matters is often deceived." That was published in 1528. Perhaps, as Marchetti would say, the rules of the game do not change very much.

The need to understand the filtering and acceptance of messages evokes a remark about causality. There was discussion at the conference, and I will return to this later, about the meaning of space or location. Åke Anderson raised the issue of space. Students of philosophy may remember that the 18th century Scottish philosopher and historian David Hume had three principles of causation: first, temporal precedence of cause over effect; second, spatial contiguity; and third, constant conjunction. In the context of diffusion, it may be useful to revisit ideas about causation from philosophy and other fields. What are the underlying principles and requirements for the reproduction and spread of technology?

As mentioned already, another field which merits a careful *diffused* look is biology. It is clear that there are exciting developments going on in the life sciences that bear both directly and indirectly on diffusion. Within the discipline, there is a highly developed vocabulary of *passive* and *facilitated* diffusion, active transport, and membrane inhibitors. There is also potentially relevant work with regard to the role of messenger chemicals (of course, the messenger RNA). In neurosciences, there have been fascinating discoveries about so-called cell adhesion molecules that also transfer messages and guide cell-cell interaction.

Vladimir Rudashevsky commented in the conference that diffusion is a process of mutual exchange. That statement can certainly be fruitfully reflected upon by our biologist colleagues, who would probably agree and have many examples at the cellular and genetic level. Back at the level of human society, Marchetti made the remark that people diffuse in and out of scientific and other sectors based on the rates of difficulty and success in finding new things. As the effort to develop a theoretical basis for diffusion studies continues, individual researchers must look beyond their own disciplinary borders to areas of communication theory, biology, and other fields.

There was much discussion about the importance of diversity. Diversity again recalls biological issues, this time those current in the vast undertaking proposed to map the human genome. Most biologists think that much of the genome may contain little information, or is *junk* as the mappers sometimes say. There may be only particular segments, or sequences of genes, that are vitally important. There may be some metaphors here with economic and social processes that are worth exploring. Dosi and others noted the importance in economics of the heterogeneity of preferences, expectations, and competencies, and that the heterogeneity is required for evolution. What seems to be true at the genetic level needs to be understood and appreciated better at the economic and industrial levels.

Now this seemingly inefficient evolutionary model introduces some awkwardness for traditional economic theory. It runs counter to some of the preferred assumptions in economics for the last three or four generations about perfect information, rationality, and existence of equilibria. Another aspect that came through quite clearly in several talks is that the role of prices from a diffusion point of view is unclear. Prices convey certain kinds of information and send certain signals, but their roles and causal function from a diffusion perspective would seem to be less important, especially in the early phases of diffusion, than mainstream economic theory would say. What emerges is the importance of specific history, captured in the phrase that Brian Arthur (1988) has popularized and that was mentioned by several speakers, including Yuri Kaniovsky, "path dependence."

Castiglione was an early supporter of path-dependency. Let me share another quote, which relates to some of the marketing questions discussed in the conference. "Custom often makes the same things pleasing and displeasing to us. Whence it comes about that customs, dress, ceremonies, and fashions that were once prized become despised and contrariwise the despised become prized." History matters. Where you are will affect where you will be the next day. Where you are now is determined by where you were before. You do not and cannot reshuffle all the cards everyday.

Continuing on these general issues of methods and theory, certainly one of the debates is about the literal use and exercise of curve-fitting. Many examples of curve-fitting are seen in the diffusion literature. There are simple examples taken from biology, the growth-to-limits behavior of bacteria. There are the Fisher-Pry competition models. There are the multiple substitution models. There are logistic functions, Gompertz functions, and modified exponential functions. There are also questions raised about the applicability of each kind of curve-fitting to each case. Moreover, does the curve-fitting in practice usually apply to cases that are so simple as to be trivial? There is no resolution of that debate, other than to say, use a model that makes sense and enjoy the fact that some simple relationships are deep and important.

1.4 Predictability

The sight of good fits of data and the scent of a theory raise hope of predictability. There was disagreement in the conference about the extent to which understanding of diffusion enhances ability to predict. A majority was on the positive side, saying there is quite a lot of predictability, but there were several strong caveats. Haustein reported one example where prediction based on a diffusion theory failed. Was it a failure of diffusion theory in general or of a particular model or application? Robert Ayres presented a possible counter-example in the area of motor fuels, where the product under study did not follow a simple S-shaped path. John Tilton shared the problems of using diffusion for prediction in multiple substitution, with a case of six competing kinds of beverage containers. How strong are current tools in the face of real world complexity?

At the same time, there were promising papers about methods in predictability. Vacca and Valerio Franchina presented a method which is a virtually completely automatic procedure using triplets – three data points, including early data points out of the chaotic, turbulent, initial phase of technology innovation – to estimate growth processes, and showed several striking examples, excellent fits. They say it appears to be a more objective process than others that have been employed. Alain Debecker also presented some interesting rules of thumb about the accuracy of prediction, certainly handy for practitioners to use. The prerequisite for most analyses though is that the process already has to be visible, signalling to the human eye. There has to be a reasonably high level of signal. The crystal ball cannot be completely empty or cloudy. There has to be a reasonable number of data points in it. Underlying any look into the crystal ball are diverse views about the extent to which natural and social systems are chaotic, tending toward certain forms of (self-) organization, or more strongly deterministic.

1.5 Time and Space

The question of predictability naturally raises the question of time. Following is a quote from another distinguished physicist, Robert Herman. Back in the late 1940s and early 1950s, Herman and Ralph Alpher were responsible for predicting the background radiation of the universe, the black body radiation, which led to the *Big Bang Theory.* This was certainly a discovery that would reinforce the arrow of time. Herman is said to have commented subsequently, "Time is the disgusting coordinate of the universe." The conference participants heard a lot about time, mostly as a strict clock ringing the hours of diffusion.

Scientists who build models of the global climate talk about processes, like the changes in the ice caps, as having slow physics and other processes, like formation of thunder clouds and storms, as having fast physics. Social processes may also be said to have slow and fast physics. While transportation infrastructures spread over intervals of fifty years or more, as shown by Nebojša Nakićenović, clothing fashions diffuse worldwide in a few months. In fact, it was suggested by Grübler that diffusion processes have a hierarchical structure, perhaps a fractal structure. Alternatively, there is a continuum of parameters on short and long time scales. To what extent is there measurable structure within the temporal dimension of technological diffusion?

Perhaps the most widely discussed feature of temporal diffusion is the bunching of innovations. Like a *baby boom* there appear to be periods of increased birth activity, when clusters of technologies are initiated together for certain reasons that we may not understand. Under what circumstances are there multiple innovations or clusters? Similarly, is there a focusing of diffusion phenomena at the end of certain time periods? On the one hand, there were proposals that there are bouquets of innovations, and on the other hand, there are forces that act as lenses or cones that appear to concentrate diffusion processes. So, there are most interesting phenomena in time. Of course, the question was raised if technological life-cycles are becoming shorter. Is there an acceleration taking place such that each successive innovation spreads more rapidly? Is there an acceleration taking place within the hypothesized 50-year *long-waves* or pulses of economic growth such that innovations taking off later within such a time frame diffuse more rapidly?

Time is an important issue not only for research. Time is important because it has implications for equity. It is not only that everybody gets things, but who gets something first, that matters a great deal. The time issues are also important for education, as was raised by Maryellen Kelley and some others. There may be mismatching of technology diffusion processes with the educational system. To what extent does this mismatch stem from ignorance that might be overcome?

Returning to the issue of whether the average life-cycle is becoming shorter, I, for one, am unconvinced by the evidence shared at the conference. It is possible there is such a shortening. However, the situation may be confused by an increase in the number of diffusion processes. To give an illustration, the typical supermarket in America now has an average of 18,000 items for sale, whereas 40 years ago it had 2,000 items. It may well be that the life-span of an average product has not changed much. However, because there are so many products, there is the illusion of acceleration. Perhaps changes in quantity and complexity of processes are being mistaken for an acceleration. This is a question that should be researched.

One more question in the time area relates to the notion of appropriability. This is the ability to prevent others from taking or making use of technology without authority or right and thereby relates to the capability of innovators to internalize some of the economic benefits of technical progress. Appropriability was raised several times by economists. Is appropriability just a diplomatic word used by economists for the control of diffusion over time? It seems to be a word that landlords would like and renters might not.

To turn from time to space, there was lively discussion about spatial diffusion. Helga Nowotny asked whether, when diffusion researchers talk about space, is the meaning of space metaphorical? The meaning appears to go beyond traditional geographical coordinates. Anderson talked about technology dissolving the role of contiguity. Several speakers, including Lawrence Brown, Hägerstrand, Kamann, and Sonis pointed out the need to examine spatial and temporal diffusion together. It was said that more complete coordinate systems are needed. This view recalls the question of networks. Clearly, spatial issues need to be revisited.

Yet, the traditional notion of space can still be important. There is the question, for example, of diffusion within the single Europe that is foreseen (Charles Edquist and William Peirce). So, even as new kinds of spatial or space/time relationships are important, space by itself may still be meaningful. Anderson offered a reminder about the parts of the world that are not included in rapid diffusion. The relationship of diffusion and development is obviously critical. Sonis also emphasized special spatial niches, wombs, one might say, that are needed, areas where technological innovations take hold. The conclusion is that it would be a loss to abandon completely the study of space as simply metaphorical or illusory.

1.6 Innovation and Niche Definition

There was forceful debate about what is an innovation and what is a niche. Anderson warned about perils in theories based on ill-classified phenomena. There was also discussion about the differences between fundamental and incremental innovation. For example, there was debate about, what is Computer-Integrated Manufacturing (CIM), what is a Flexible Manufacturing System (FMS), what is *just-in-time inventory.* Are these innovations? If not, what are the boundaries of innovations? Certainly, such concepts are not as clear-cut as are innovations in genetics or the invention of the automobile. Modelski, with his analysis of democracy as an innovation, raised more generally the question of forms of government and forms of social and institutional innovation.

The argument about niches had several facets. The work by Dominique Foray on metal casting raised the issues of the appropriate definition of the diffusing object and the relationship of the occupant to the niche. There is also the important question of how to modify the niche. The fascinating case of the car and the horse was presented. At first, the niche for the car is replacement of the horse. Then all the horses are gone, and the niche for the car becomes a new expanding niche which the car itself largely creates. In short, innovations and niches are themselves interacting and dynamic.

1.7 Selection Dynamics

Analysis of the population of innovations naturally leads to the issue of selection. Selection was addressed in a variety of ways. There is the canonical statement from the Bible, "Many are called, few are chosen." Mahajan said that the rule of thumb in marketing is that 70 percent to 90 percent of all new products fail. Marchetti said that from his studies looking earlier in the process, one percent of innovations succeed. Then the question is, what drives us to experiment under such bad odds? It seems crazy.

Silverberg pointed out that many innovations are inferior and more expensive than their competitors at the outset. Bruce Guile noted that perhaps the way out is to recognize the triumph of action over analysis. Maybe some combination of ignorance coupled with a general attitude of optimism begins to explain the yearly parade of bankruptcies. Dosi and Silverberg argued that diversity of expectations, including those that may lead to failure, are required to explain diffusion processes. Also striking was Theodore Modis' statement that studies done about the history of diffusion are usually the history of winners. It is a bloody history, and Modis suggested more effort in counting the casualties, the deaths, the lunatics, and losers. It is important to understand more about what has happened at the end of life-cycles and in aborted processes.

Dosi made the perceptive comment that product markets select technologies, but financial markets select firms. The two processes are not identical. There were several papers relevant in this area. A paper by Charlie Karlsson that examined why enterprises adopt technologies emphasized information channels and frequency of exposure. In a way, his view is similar to a marketing perspective. Edquist looked at empirical differences in diffusion within the countries of the Organisation for Economic Cooperation and Development (OECD) and emphasized social and cultural factors. Kelley and Brooks offered a study in the manufacturing technology area trying to understand a particular sequence of adopters among firms. Of course, there is the question of the role of entrepreneurs. It would seem that there are many open questions in the area of selection.

1.8 Role of Marketing

By now marketing has been mentioned several times. Marketing is an attempt to change or modify selection. Marketing people may be considered true aspiring bio-engineers of human society. But, one may also raise the question whether marketing matters. We saw several comparisons between the Western economies and the Soviet Union with some strikingly similar diffusion data. One has to ask the question, given such different marketing and distribution strategies and channels, how is it that some results are so similar? Perhaps marketing is the pageantry, the flags flying in the procession of the king. To use a different metaphor, perhaps it is the navigation system for a largely pre-set trajectory. Marketing may only rarely change the niche to be filled.

It is also important to ask how marketing changes over technology lifecycles. Thomas Lee raised this issue well. One idea is that it is necessary to market in pulses in order to fill successive niches, with information that needs to be distributed through time in certain ways. There were several comments, some flippant and others serious, about the importance of word-of-mouth. This appears to remain the dominant way for people to communicate decisively. Technologies may make the shelves of our marketplaces look modern, but diffusion processes may show how close today's humans are to our chattering, oral ancestors.

Cross-cultural comparisons would help show what is deeply similar and what is superficially different, given similar outcomes of diffusion processes. It should be possible to do revealing comparative studies of the effects of marketing strategies for the same products or technologies in different countries. In such studies, it would be useful to explicitly employ hypotheses from the diffusion literature. For example, one could explore reasons that early penetration may be associated with high ultimate levels of market saturation, while cases of later penetration are associated with lower ultimate levels of saturation, as found by Nakićenović in the analysis of diffusion of transportation technologies.

1.9 Social Aspects

Quite a few participants raised the question of social factors. Everyone recognizes that the diffusion of an innovation or several innovations can have many social effects. Several people noted that there are many causal chains leading to an overall effect. To give an example, automobiles, which were mentioned a number of times, replaced horses, diminished the number of stables, and reduced the number of flies and the amount of solid waste. They thus may have lessened somewhat the spread of communicable diseases. They helped the growth of suburbs. They reduced railroad traffic and transformed shipping. They changed the nature of the hotel business. They diminished the employment of domestic servants. They changed marketing areas. They caused international difficulty over oil resources. They affected rural life. They made drive-in movies and new kinds of vacations possible.

There were suggestions that more of the social dimensions of diffusion should be explored. At the same time, Silverberg and others made the point that social change represents the combined contributions of many inventions. While the automobile made possible the suburbs, the suburbs may also have required the telephone. What is visible is almost always the result of an accumulation of influences, in many cases of smaller innovations. These smaller innovations are a significant part of the process. The force of any particular invention or innovation might be quite weak. The phrase that Silverberg used was the collective nature of technological progress. In fact, several participants referred to the notion of development trajectories and the concept of *techno-economic paradigms* consisting of clusters of interrelated technical, organizational, and social innovations that has been associated with Christopher Freeman and Carlota Perez (1988).

One issue that might be evaluated more in the future is a social, even a moral and an aesthetic, one. Diffusion itself (or competition) has countervailing effects, and the balance changes through time. On the one hand, diffusion is a force for homogenization. On the other hand, it multiplies difference. We have both the increase in the number of inventions and innovations, and at the same time the possibilities for greater or wider standardization. In the discussion sometimes people were talking about innovation and diffusion as something quite subversive that would overthrow and would change the society. This was evident in discussions and papers of Soviet colleagues about innovation in the context of *perestroika.* At other times, there was a sense that technology and diffusion are a force of standardization, a conservative force in a certain way. These offsetting tendencies might be explored more. A starting proposition may be that fluctuation generates diversity, but propagation leads to homogenization.

1.10 Globalization

Another social, and also political and economic, issue that was discussed was that of globalization. This was raised by Hägerstrand, Lee, Marchetti, and Modelski. Two aspects of globalization were talked about. There is the global diffusion of technology, and there is the more specific process of technology transfer between nations. Research presented indicated isolation of technology, technological protectionism, fails in the long run. Technology simply does diffuse globally. The question is then how much one can abet or retard it. In the East-West context especially, are there more positive ways to handle the diffusion of technology?

At an abstract level, the issue may be phrased, is there a widely acceptable way for nations to capture income globally that is attributable to research and development? It appears that nations behave, or would like to behave, much like the successful firms described in the Silverberg model. It is therefore not surprising that there is growing international debate about equitable and optimal national levels of investments in R&D. It is the appropriability question in another guise or at a higher level of the system. The critical point is whether the diffusion of innovation and its benefits remain ordered in a civilized way, without wars and other violent conflicts referred to by Modelski and others.

In globalization, with regard to technology transfer, the issue is *catch-up.* To what extent can the introduction and diffusion of innovations be accelerated, especially to developing countries like China? Sergei Glaziev's paper had a good, provocative phrase, "the advantages of the backward." There is a need to understand better what the potential advantages of the backward are. Some examples of acceleration of diffusion processes were shown; these appeared to follow a kind of learning or experience curve in which late adopters were spared some of the time-consuming experimentation of the early adopters. To what extent does this hold for sophisticated, as well as trivial, cases?

Ultimately, it must be asked, "After *catch-up* – What?" It is unrealistic to think that everybody, at some moment in future history, whether firms or countries, will be at a relatively even point in the adoption of technologies. It is even less realistic to expect that such parity could be sustained. Diffusion phenomena in a way seem to be elastic, with some leaders always pulling away, but then a pack of followers periodically coming nearer, only for a leader, sometimes a new one, to pull away. Although not frequently, the ordering of diffusion processes does change among nations, and the desire to lead, to take the lead, is the essence of competition in politics, as well as business.

To return to the status of the less developed countries, one of the troubling features of the conference was that, while there is a large library of evidence on diffusion processes, quantitatively documented cases from developing countries are scarce. The past behavior of countries like the United States, Sweden, the United Kingdom, and increasingly the Soviet Union has been studied extensively. In contrast, there are few data and analyses from India, China, Argentina, and other such countries. On the empirical side it is urgent to do more work on less developed countries. It would help crucially to answer questions about whether and how *catch-up* occurs.

1.11 Applications

The last area for comment is applications of diffusion for policy, especially in firms and the national economy. The prospects for applications were addressed by several people, including Brian Sullivan and Guile. Judgements here are closely related to views on the predictability question. Sullivan points out that application of diffusion methods is itself diffusion and adoption. It might be called the diffusion of diffusion. Presumably there should be a competitive edge from employing the kind of ideas and analyses shared at the conference, at least until they reach saturation. There was debate about whether there is a vicious circle in use of diffusion research, in that if the information is widely publicly available, it may no longer be valuable. This debate resembles other debates about the value of information in markets.

At the same time, there are broader and probably useful guidelines about behavior related to diffusion that appear to emerge from the presentations and papers. For example, there is the evidence from Grübler and Nakićenović that late adoption is associated with fast diffusion, but a relatively low saturation level. There is also evidence of *seasons of saturation*, when many diffusion processes concurrently reach their culmination and create both economic stress and windows of opportunity for initiation of new growth. Rudashevsky explored how such macro patterns of diffusion link to issues of restructuring or *perestroika*. Thus, at a conceptual level, there may be useful notions for policy and applications apart from specific predictions.

At the micro-level, there are, however, potentially specific applications of diffusion analysis. Among these is the remarkably precise application from Vacca and Franchina in the AIDS area. The question is if it would be socially robust to accept or use such a prediction? How much should it be relied upon? Should the health minister of Italy risk the entire government health care system on this prediction? Would even supporters of diffusion theory prefer to place, say, half of their resources on their own forecast and then diversify a little? Even if one has a high level of confidence in the diffusion-based prediction, strategically and tactically, how is it best to deal with this faith?

Both at the macro and micro level, the system almost always seems to keep away from homogeneity. Dosi made the playful comment that for growth it is important to have many economic agents grossly uninformed about the future. This appears to be a general prerequisite for evolution. It may also assuage any fears that experts may have that everyone will adopt or act upon their ideas. Linstone emphasized the benefits of multiple perspectives, of exploration of several paths. Shunsuke Mori provided a valuable caution about the extent to which one enterprise, or presumably one nation, can effectively manage the whole process from invention through innovation and diffusion. There may always be gaps between intention and performance in application, whether due to inherent capability or external surprise. The environment may be so turbulent that one is unable to pursue a strategy over a long enough period for it to matter.

Guile stressed the delicacy of timing for successful applications of diffusion analyses. He talked about possible mismatches, the need for matching time scales, and the need for the fertility of the receptor. To illustrate, here is a quotation about the fax machine, which has proliferated rapidly worldwide since about 1987. "One possible extension of electrical invention ... is facsimile transmission. Newspapers have been thus sent from New York to San Francisco ... Other uses of the same mechanism are for sending news pictures, identifications of criminals, x-ray photographs, weather maps, signed documents, chemical formulae, graphs and messages in other alphabets and symbols." This was an exactly correct statement by an analyst, a technology assessor, a distinguished sociologist ... in the year 1933 (Ogburn and Gilfillan, 1933). Of course, the simple fax machine concept was 54 years too early from the point of view of diffusion. A related cluster of innovations, including high quality telecommunications, was required to potentiate the fax. Certainly, timing is critical in commercial application.

Finally, in the applications area, one must raise again the question that has been posed in the work of Henry Ergas (1987) and also raised by Edquist. At the government level, what specifically constitute *diffusion-oriented policies*? Peirce wondered whether governments have at their disposal appropriate policies that can influence diffusion processes at all. *Diffusion-oriented policies* is a most tempting term and needs to be explored and clarified.

To conclude, let me recall two of the best phrases used in the conference. One participant (Kamann) referred to diffusion processes as "rat-race dynamics." Another (Dosi) commented on the necessity for "crazy companies." Tilton asked the question what has changed in 20 years of diffusion studies in economics and other fields. My answer is that we have sorted a lot of order from the apparent chaos of social behavior, but have also recognized better the necessity of disorder, and we are trapped as ever in the race. But, as shown in *Figure 1.1*, our paradigm is gaining.

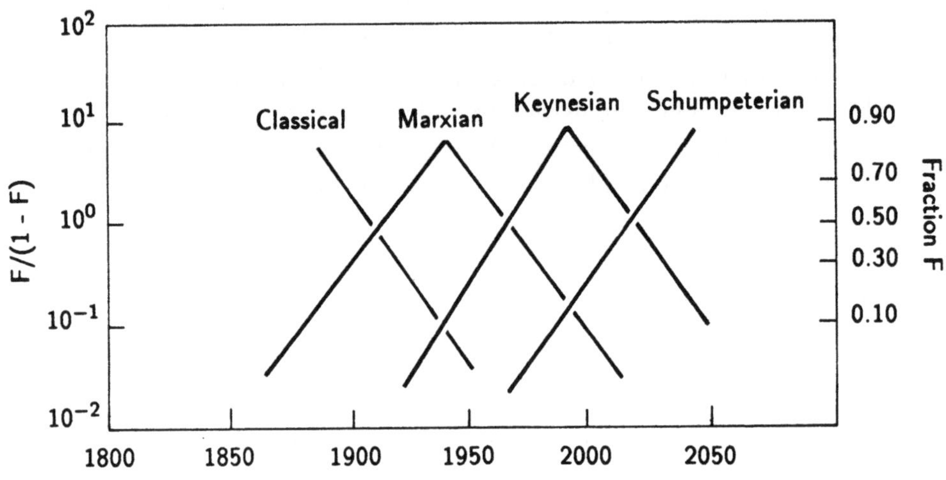

Figure 1.1. Diffusion of techno-economic paradigms.

Acknowledgments

The author gratefully acknowledges suggestions from Arnulf Grübler, Robert Herman, Peter de Jánosi, Nebojša Nakićenović, and Jerry Weisbach and the assistance of Margret Holland.

References

Arthur, W.B., 1988, Urban systems and historical path dependence, in J.H. Ausubel and R. Herman, eds., *Cities and Their Vital Systems: Infrastructure Past, Present, and Future*, National Academy Press, Washington, DC, USA.

Castiglione, B., 1528, *The Book of the Courtier*, translated from the Italian by Charles S. Singleton, Anchor Books edition: 1959, New York, NY, USA.

Ergas, H., 1987, Does technology policy matter? in B.R. Guile and H. Brooks, eds., *Technology and Global Industry*, National Academy Press, Washington, DC, USA.

Freeman, C. and Perez, C., 1988, Structural crises of adjustment: Business cycles and investment behaviour, in G. Dosi *et al.*, eds., *Technical Change and Economic Theory*, Pinter Publishers, London and New York.

Ogburn, W.F. with Gilfillan, S.C., 1933, The Influence of Invention and Discovery, Report of the President's Research Committee on Social Trends, in *Recent Social Trends in the United States*, McGraw-Hill, London and New York.

Montroll, E.W., 1978, *Social Dynamics and the Quantifying of Social Forces*, Proceedings of the US National Academy of Sciences, Vol. 75, No. 10.

Chapter 2

Democratization from a Long-term Perspective

George Modelski and Gardner Perry III

2.1 Introduction

Democracy is a mechanism of collective choice and a form of social organization that can be considered a superior substitute for other such mechanisms or forms of organization. As such, democracy may be expected to grow, or diffuse, over time, amongst the world's population, and the question posed in the present study is: does that growth follow a regular pattern, according to the Fisher-Pry substitution model of technological change?

Our inquiry finds that, prior to 1800, democratic development was experimental in character, but it has been growing fairly rapidly since the middle of the 19th century, generally fitting the model of diffusion quite closely. At the present time, about forty percent of the world's people live in democracies; by extrapolation the model suggests that the democratic community might reach the 90 percent level toward 2100.

2.2 Democracy and Democratization

Democracy literally means rule of the people. More analytically speaking it might be seen both as a technology of collective choice, and also as a type of social behavior or community organization. The technical, or instrumental, aspects of democracy are best recognized in the election process, a form of macro-decision by which a community selects some individuals to positions of public leadership. In this sense, practical democracy consists of a number of rule-governed devices – media, parties, voting, and majority rule – and of policies executed via working mechanisms of representation. The substantive, or expressive, aspects of democracy, on the other hand, relate to its characteristics of equality and freedom, and they reflect the optimum conditions in which democracy operates effectively: a community of equals (symbolized by *one man, one vote*) in which, in a state of autonomy, all live as free individuals. The first recorded discussion of democracy aptly combines these two aspects in the phrase "equality under law".[1]

Democratization is the process of building, or creating, democracy. That process, also, moves along two distinguishable, if related, paths: wherever democratic techniques of macro-decision are discovered and spread, a process of diffusion of these innovations occurs. Alternatively, we see it also as the process by which democratic communities grow, via a form of clustering (or concentration) into larger communities of democracy; for the evolution of new types of community, too, is a form of innovation. The two processes are obviously interdependent: the diffusion of democratic procedures will produce no more than *formal* democracy unless rooted in, and nourished by, conditions of greater equality and freedom, where procedural rights are effectively exercised and are seen to work; a society of equality and freedom cannot last without observing democratic procedures.

Viewed from another angle, democratization may be either intensive or extensive. By intensive (or vertical) democratization we understand the change in the quality of the democratic experience in a given community; that is, we ask, how good is this particular democracy? We may wish to judge it, e.g., by the extent of voting rights, the working of representative institutions, or the presence of democracy at the several levels or fields of social organization (such as national, local, political, economic, etc.). We might also want to judge it by some general standard of *perfect democracy* (analogous to *perfect competition* in economics). That means that nation-states, and communities, might be seen as less, or more democratic, and that the quality of democracy is subject to change over time, in both positive and negative directions. Conceptions as to what is democratic might also change, and the point at which a given community is to be called democratic might therefore be a variable one.

Extensive (or horizontal) democratization, on the other hand, measures the quantitative extension of democratic communities, and their global spatial reach. It asks: how much democracy is there in the world? The present chapter addresses this question.

The process of democratization occurs in an *environment* (or niche) of finite *carrying capacity.* We conceive of it as unfolding within the limits of the world system, in the context of its evolution, and as part of it. That is, democratic procedures are diffused inside a *market* that extends to, and is limited by, the social organization of the human race on this planet, and democratic communities grow as part of world system evolution. We assume that democracy is not limited in its reach to certain regions or cultural areas or special circumstances. Hence the limit of the process of democratization, its potential, is the world population; however, we also need to bear in mind that this population has been a steadily expanding one, by more than one order of magnitude in the period under discussion.

As a set of techniques, and principles of community organization, democracy is in some respect a superior substitute for other technologies and forms of community organization. In its classic Greek form, democracy was the basic alternative to monarchy, with aristocracy an intermediate solution. In the modern era republics and (increasingly absolute) monarchies emerged as the two principal forms assumed by the rising nation-state. Gradually, republics (and constitutional monarchies) evolved into democracies as substitutes for absolute rule, or for narrowly based, arbitrary, despotic, or dictatorial forms of organization, and they have been shown to have wide potential appeal, on such grounds as flexibility, accommodation of variety, self-legitimating properties, and capacity to civilize conflict. On such a view democratization has been a process of gradual substitution of democratic for nondemocratic forms of organization; or else as a persistent problem of choice between alternative and competing forms available to communities for dealing with their problems.

Democratization is, moreover, a time-bound process. We do not expect it to be completed, all at once, worldwide, but rather to spread in stages, gradually, and slowly at first. We would expect it to emerge following a series of experiments, some of which are bound to falter, and only a few to succeed. We know that the first trials with direct democracy occurred in relatively small urban communities in ancient Greece, prominently in Athens, *ca.* 500–300 BC. They did not last, but their failure was a noble one and is remembered to this day. Another series of attempts at popular rule materialized in Italy, after the year 1000, among self-governing cities of which Venice was the most successful; but there, too, most experiments deteriorated into tyrannies, and even Venice ultimately turned into an oligarchy.

The seeds of modern democracy began to sprout after 1500 as alternatives to the unfettered monarchy, which was then prevailing as the answer to the question of the character of the modern nation-state: i.e., was the state to have unlimited or limited powers?[2] The critical events were the wars of religion in Western Europe after 1561: they produced both the absolute monarchies of Spain and France, and the republican United Provinces of the Netherlands. In limited monarchies, this republican *strain* went under the banner of *commonwealth* (as in John Locke's *Two treatises of government*). Helped though it was by the Venetian model, now transposed to the larger scale of nation-states, the republican way advanced slowly at first. Eventually, it acquired a strong liberal flavor and came to be strongly tied (as in the British case) to representative institutions.

Democratization took a leap forward in the 19th century in the form of representative democracy. But it could not have done so without suitable techniques of representation that had been previously devised in the republican and liberal states. That is why the achievements of the 16th–18th centuries are an essential part of the master trend of democratization.

The propensity to view democracy from a long-term perspective can hardly be claimed as altogether new. Having observed the American experience with Jacksonian democracy, Alexis de Tocqueville (1835, p. 9) announced more than a century and a half ago that "a great democratic revolution ... is taking place in our midst" and added that some saw it as "irresistible" because it is "the most continuous, the most ancient, and the most permanent tendency known to history".[3] But this revolution, if we are to credit de Tocqueville's term, is still in progress, and its precise shape remains to be determined.

2.2.1 Democracy as innovation

From a long-term perspective, democratization can therefore be regarded as a process of structural change, or innovation or, better still, as a cluster of innovations. An innovation is an idea or practice that is new to an individual or a community. Democracy certainly was such an idea when its techniques originally evolved in a simple but persuasive form; it was again innovative in its modern version, when it had to be adapted to a larger scale of social organization and (by continuous evolution) it has come to form a repertoire of some complex new practices. While in the "old democracies" this inventive aspect is apt to be forgotten, it is much in evidence in those communities where democratic practices are still unfamiliar.

The question arises: is it not likely that democratization, as a process of innovation diffusion, is in fact also a learning process?

We know that innovations spread in a diffusion process involving regular patterns in social systems over time. According to Everett Rogers (1962, p. 67) "the process by which innovations are adopted by individuals is essentially a limited example of how any type of learning takes place". More generally, we might assume, with Cesare Marchetti (1980, p. 267), that "society is a learning system", that learning is "random search with filters" that brings about lasting changes in behavior, and that such "random searches" are characterized by logistics (a curve representing a function involving an exponential and shaped like the letter S). Logistic functions have been used to study the growth of human and other populations.

"A general finding of past investigations," Rogers argues, is that the pattern characterizing the way innovations are adopted (that is, "adopter distributions") follows "a bell-shaped curve over time and approaches normality. ... this type of distribution is essentially S-shaped when plotted on a cumulative basis".

Three reasons have been adduced why adopter distributions might be expected to approach normality: early sociologists, including Stuart Chapin, who studied the city manager form of government in the United States, observed that the adoption of an idea followed a bell-shaped pattern. The learning of individuals follows a normal curve, and the gain in learning per trial is proportional to the product of the amount already learned and the amount remaining to be learned before the limit of learning is reached. As the individual learning curve is extended to the case of a social system, experience with the innovation is gained as each successive member adopts it. Last, the interaction effect is the process through which adopters of an innovation influence those members who have not yet adopted; the group pressure for adoption rises as the number of adopters increases (Rogers, 1962, pp. 152–155).

This time pattern of diffusion in turn suggests a categorization of adopters. The simplest might be one that distinguishes between innovators, early adopters, late adopters, and laggards. We might regard innovators as the first 10 percent of adopters, and the next forty percent as early adopters.

At this point (50 percent, the population mean, or the flex point) adoption ceases to increase at an increasing rate, brings in the *late adopters*, and ultimately levels off for the *laggards* (the last ten percent).

In this chapter we shall apply one basic form of the diffusion model, the Fisher-Pry (Fisher and Pry, 1971) model of technological substitution, on the assumption that democratization may be considered as a competitive substitute for other methods of social organization. Basic human needs for organization and collective choice need to be satisfied, and democracy provides that better substitute. The Fisher-Pry model expresses that process, and its most convenient form is the equation (Fisher and Pry, 1971, p. 77):

$$F/(1-F) = \exp[2\alpha(t-t_0)], \tag{2.1}$$

where F represents the fraction of substitution (in our case, the *fraction democratic*), and 1 stands for the *size of the market* (in our case, the world population). Then 2α is the slope of the curve, and t (in years) is the time constant (called by Fisher and Pry the take-over time), defined as the number of years required to go from 10 percent to 90 percent saturation level. The mid-point of that range (50%) is the flexpoint t_0. Then a plot on semilog paper of $F/(1-F)$ as a function of time allows one to fit a straight line through the data and make appropriate extrapolations.

In the remaining portion of this chapter we shall address the following questions: Can democratization be shown to be a major pattern of the world system? Might it be no more than a series of unrelated national developments governed by local conditions? If not, then how do we document it and what is the evidence for it? What has been the shape of (horizontal) democratization over time?

This is a large field; for our present brief, we propose the following course:

- To operationalize the question by asking what proportion of the world population has lived in democracies at various times since about 1500?
- To further ask does this information fit the substitution (or learning) model and, if so, what might be the implications?

For if the cumulative growth of the share of world population living in democratic communities (that is, in the global democratic community) can be described by a logistic-type (S-shaped) distribution (that in a semilogarithmic plane appears as a straight line), then we have grounds for arguing that democratization is a social learning process (of innovation-diffusion/ community growth).

2.3 The Data

We propose to test our hypothesis against two sets of data now available on the population of democratic communities since about 1500. Our data on these communities are in two parts:

(1) For the period 1800–1986 our basic source is the POLITY II dataset (Gurr *et al.*, 1989); the background to that collection is explained in (Gurr *et al.*, 1990).

The POLITY II survey covers "all independent members of the international system", those that have attained independence by 1975 and whose population exceeded one million by the mid-1980s.[4] It gives, for each such polity, an annual score of "institutional democracy", on a scale ranging from zero to ten. The score is constructed from three codings, those for (a) competitiveness of political participation – via the party system; (b) openness and competitiveness of political recruitment, that is quality of the electoral system; and (c) constraints on the chief executive (checks and balances). There are no coded data on human rights or political liberties.

The "institutional democracy" score is constructed by adding these three indicators, so that party competitiveness rates a maximum weight of three, executive recruitment via elections also a maximum of three, and restraints on the chief executive, four. It measures the degree of intensity of democracy for each polity.

Selected for inclusion in this survey since 1800 are all polities that had, in any year since 1800, an "institutional democracy" score of *six and above*. There was one in 1800 (United States, with a score of seven); 14 in 1900, and 46 in 1986 (of which 29 scored 10). The most populous national polities and their POLITY II scores in 1986 are shown in *Table 2.1*.

(2) For the period between 1450 and 1800 we apply the same criteria (parties, elections, checks, and balances) and the same scoring weights and ask: what are the polities that can reasonably be described, on these criteria, as "democratic experiments" – that is, those exploring a variety of paths of trial and error, at first in the republican, and later also in the liberal traditions, all generally tending in the democratic direction.

This yields a tentative list of eight independent polities as follows[5]: Venice (1450–1640); Swiss Confederation (1499–1531)[6]; Poland (1505–1605) (Poland-Lithuania after 1569); the Dutch Republic (1579–1787); Britain (1688–1800); Sweden (1718–1771); United States (1776–1800);

Table 2.1. Major national polities 1986.*

Institutional Democracies	Score	Others	Score
India	(7)	China	(1)
United States	(10)	U.S.S.R.	(1)
Brazil	(7)	Indonesia	(1)
Japan	(10)	Nigeria	(0)
Germany, F.R.	(10)	Bangladesh	(1)
Italy	(10)	Pakistan	(0)
United Kingdom	(10)	Mexico	(3)
France	(10)	Vietnam	(1)
Thailand	(8)	Philippines	no data
		Egypt	(3)
		Turkey	(4)
		Iran	(2)

Parentheses show POLITY II institutional democracy scores for 1986.
*Populations (49 million or greater) exceeding 1 percent of the world total in 1986, arranged in descending order of size. Polities shown here account for 75 percent of the world population.

France (1789–1799). Each rates a score of at least *four* on the POLITY II scale; a somewhat lower threshold is used to take into account the experimental character of these early cases. Only two (Britain and the United States) remain in 1800, which is a time of crisis amidst global war, and only the United States carries over to the POLITY II survey, where it rates a seven. Britain keeps a score of four until 1837.

The data on democratic communities are then combined with population figures, for the relevant communities, and for the world, based on McEvedy and Jones (1978) and other sources cited in Perry (1987). The populations are those of the metropolitan territories, and do not include dependencies.

2.4 The Analysis

Table 2.2 summarizes in Part (a) the data for the period 1450–1800, basically the era of the innovators. The period up to about 1750 is one of experimentation, with a world share of population (*fraction democratic*) ranging between one and two percent. The same data are presented graphically in *Figure 2.1*, Experimental Democracy; the trend line fits the data with a correlation (R^2) of 0.7.[7]

Table 2.2. World population fractions (*fraction democratic*).

	No. of polities	*F* (× *100*)	*F/(1–F)*
(a) Experimental Democracy			
1450	1	0.05	0.0005
1500	2	0.24	0.002
1550	2	0.78	0.008
1600	3	1.96	0.020
1650	1	0.37	0.004
1700	2	1.11	0.011
1750	3	1.56	0.016
1790	3	4.74	0.050
1800	2	1.87	0.019
(b) Institutional Democracy			
1800	1	0.58	0.006
1810	1	0.74	0.008
1820	1	0.93	0.009
1830	3	1.37	0.014
1840	3	3.88	0.040
1850	6	7.52	0.081
1860	7	5.43	0.057
1870	10	6.75	0.072
1880	13	11.56	0.131
1890	12	11.91	0.135
1900	14	12.32	0.141
1910	15	13.07	0.150
1920	29	20.63	0.260
1930	26	18.33	0.224
1940	15	11.35	0.128
1950	29	34.45	0.526
1960	41	41.32	0.704
1970	37	36.43	0.573
1980	43	38.07	0.615
1986	46	39.61	0.656

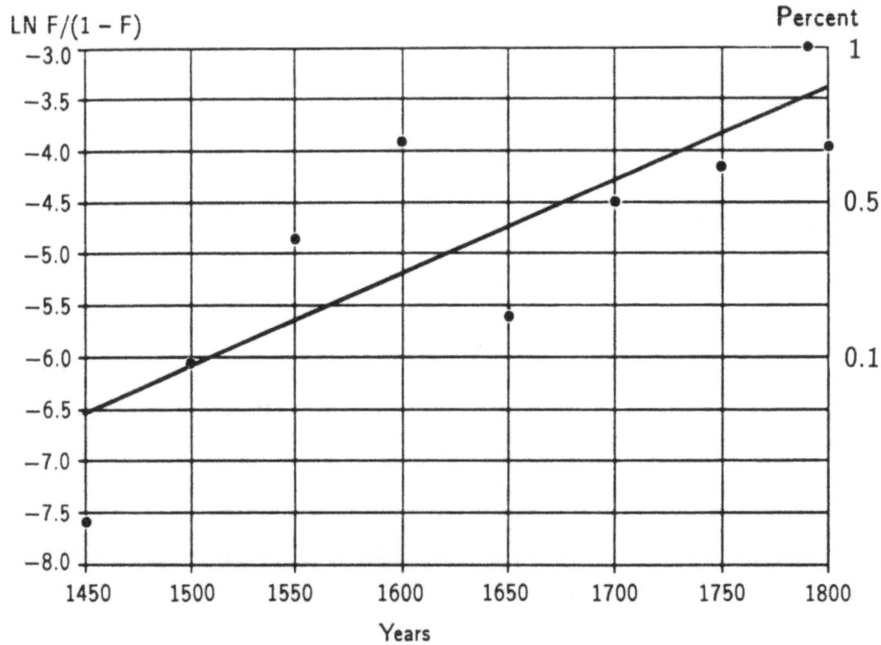

Figure 2.1. Experimental Democracy (1450–1800).

Table 2.2, Part (b), summarizes the data for the period 1800–1986, the era of the *early adopters*. The same data are presented graphically in *Figure 2.2*, Institutional Democracy; the R^2 of the trend line is 0.915. But starting the analysis in 1837 (*Figure 2.3*) yields a trend line with a somewhat lower growth rate.[8]

The American and the French Revolutions launch, after 1776, the stage of early adoption, even though the American Revolution is, in the first place, seen mainly as a republican achievement, and the French Revolution soon loses its democratic character. But the democratic trend takes off strongly in the mid-19th century, moves past the 10 percent range, and in the late 20th century approaches the flexpoint. This theoretical half-way mark, when 50 percent of the *market* could be expected to be saturated with democratic practices, is, for *Figure 2.3*, the year 2003; the actual *fraction democratic* was approaching the 40 percent level in the late 1980s.

The overall directionality of the process is unmistakable. But the process is not unilinear; both peaks and setbacks are strewn along the way. The

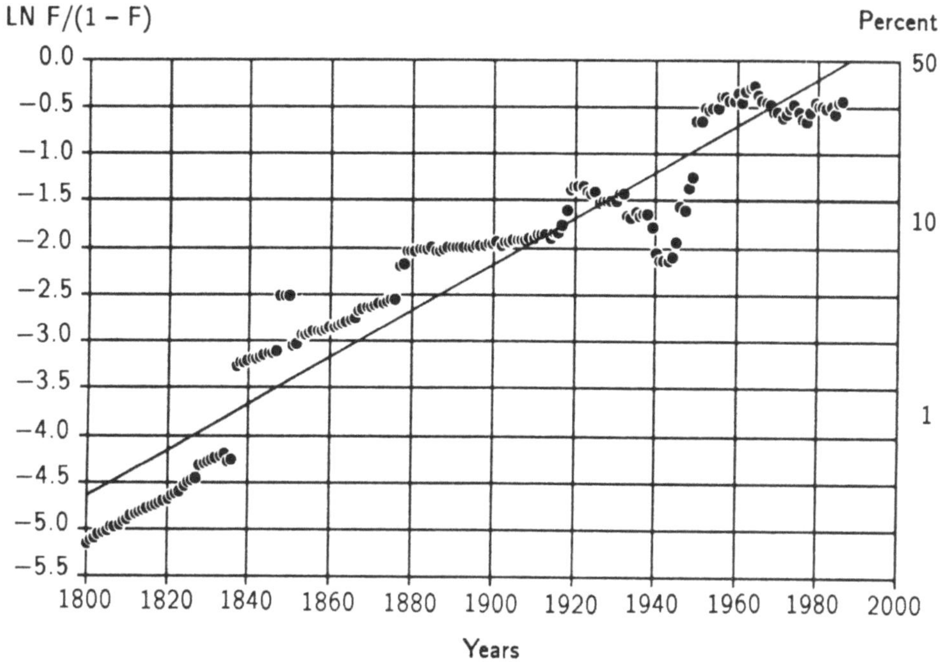

Figure 2.2. Institutional Democracy 1800–2000 (1800–1986 data).

early record, the (a) series, shows two peaks (in 1600 and in 1790) each followed by declines, when the proportion fell as compared with the preceding decades. In the more detailed and more recent series (b), we observe four abrupt increases; a major spurt beginning with the years 1837 (when Britain enters), and going on to 1877 (when France enters), 1918–1921 (post-World War I), and 1945–1950 (post-World War II). A perceptible decline marks the years after 1930 and World War II. In other words, democratization shows fluctuations in a generally upward movement; the fluctuations are due to *lumpiness* (when large countries become – or cease to be – democratic); the process is also adversely affected by global wars.

2.5 Discussion

Our test has been conducted with two sets of data, but with the same classification and scoring. Should other data become available, using somewhat different criteria of democracy, such as those incorporated in the annual surveys conducted by Freedom House after 1972 (see Gastil, 1989a and 1989b)

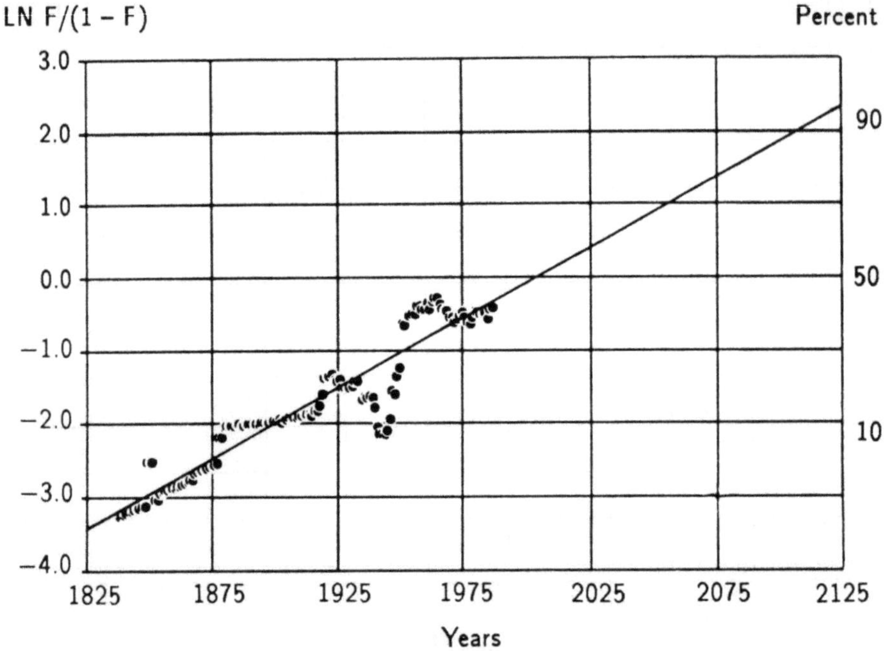

Figure 2.3. Institutional Democracy 1825–2125 (1837–1986 data).

and that also give additional weight to civil or political liberties, then the test must obviously be replicated. It appears though that these two particular surveys are not that dissimilar in their rankings. For the year 1986, the difference between the POLITY II and the Freedom House surveys, expressed in *fractions democratic*, as shown in *Table 2.3*, is only 2.4 points.

Our analysis shows that the cumulative growth in the fraction of the world's population that comprises its democratic communities conforms to the Fisher-Pry model. Is this no more than a descriptive finding? How might we explain this unexpected regularity?

For such an explanation (as already indicated), we reach to theories dealing with innovation diffusion and learning, phenomena that in societal dimensions and in long perspective might best be viewed as collective evolutionary processes. Democratization is a pervasive process of structural change at the global level that effects a host of other developments and serves, indeed, as an envelope curve for most of what we commonly view as technological and economic innovation. It is also a long-range process,

Table 2.3. Two surveys compared.

Survey	Fraction Democratic (percent) 1986	1988
POLITY II (Institutional Democracies)	39.6	
Freedom House (Free Countries)	37.2	39

The differences between the two surveys are chiefly due to the following discrepancies in ratings. Thailand, Malaysia, Sudan, Zimbabwe, South Africa, El Salvador, and Guatemala rate six or higher on the POLITY II scale, but are coded only *partly free* by Freedom House; Honduras, Bolivia, and the Dominican Republic, coded *free* by Freedom House, do not make a score of six in POLITY II for 1986. Included as well in the Freedom House listings are a score of very small countries that do not meet the POLITY II population criteria.

Accounting for the change between 1986 and 1988 in the Freedom House fraction is mostly the movement of the Philippines and Korea (South) from *partly free* to *free* status.

with a time constant on the order of 200 years. If the trend line for "institutional democracy 1800–1986" is extrapolated into the future, the world will be passing the 90 percent mark in 2075 (*Figure 2.2*); if we take as our guide the analysis that treats the current pulse of democratization as having begun in the mid-19th century, and we extrapolate our data starting with 1837, then the 90 percent level is attained in 2117 (as shown in *Figure 2.3*). Either way the status of democratization as a major evolutionary process is unmistakable.

The process has, in the first place, the aspect of the diffusion of a technology of collective choice, and can be easily understood as such. The wide dissemination of improved artifacts of life (or *better mousetraps*) is, after all, one of the most obvious facets of civilization. It can be studied, for example, by asking who are the experimenters and the innovators, how are they followed by early adopters, and how do they come to be engulfed in the great majority. It can also be studied by establishing the mechanisms and conditions facilitating or obstructing diffusion: proximity, similarity, or opportunity for interaction.

The process might also be viewed as one of community formation: one by which cooperation gradually evolves into more stable and institutional forms, first by trial and error experimentation, and then by a species of clustering (or nucleation), which via coalescing, branches out into structures of global cooperation. This second aspect, about which we know less than about diffusion proper, makes it clear that the process must be a slow one because communities change only at a deliberate pace and grow but gradually.

What lends stability to this process, and credibility to our extrapolations? Might it not be said that the Fisher-Pry model assumes an unchangeable environment, whereas changes are bound to affect future structural change in unfathomable ways?

Democratization is one of a family of global, collective, evolutionary processes.[9] Other members of that family are the Kondratieff wave, and the long cycle of global politics. The Kondratieff, with a time constant of more than 50 years, governs global economic innovation and exchange. The long cycle, with a period of just over 100 years, and centered on the roles of global leadership and challenge, shapes the structure of global politics (Modelski, 1987 and 1990). These are coupled processes that are mutually reinforcing, jointly ensuring a dynamically stable development.

The role of global politics in all this is essential. The mechanisms of diffusion and clustering are set in motion by the long cycle. In that process the role of global leadership has served as a source of innovation and, via demonstration effects, as a center of innovation diffusion, as well as the nucleus of the emerging global democratic community. From the success of nation-states performing that role (the Dutch Republic, Britain, and the United States) has sprung the process of reinforcement that is essential to learning (the key to which is the proposition that *reinforced behavior becomes more frequent*). The world powers have been the preferred models of imitation, and, successively, the centers of gravity for community organization.[10] When these centers, at times, ceased to *hold* (as in the 1930s or about 1800, and in the 1670s or the 1580s), the prospects of democratic organization dimmed. The long cycle of global politics, itself a learning process, intermeshes with, supports, and in turn derives strength from the evolution of the global democratic community (Modelski, 1987 and 1990; Huntley, 1980).

Viewing democratization in long perspective has enabled us to see it as the slow and gradual substitution of republican, liberal, and democratic regimes for monarchical, absolutist, and dictatorial (or autocratic) forms of government and society. The question might be asked: do we expect this process to reach some form of equilibrium between democratic and dictatorial (or autocratic) forms, or should we look forward to the steady expansion of democracy through its entire global *market*, that is through the bulk of the world's population?

Our view inclines toward the latter interpretation, and significantly on grounds having to do with the present phase of global politics that calls for worldwide "socialization", or "civilizing" of conflict. If global nuclear war is indeed unthinkable, then alternatives to it need to be put in place. The obvious forms of such socialization, or civilizing, are the forms that have been nurtured in the emerging democratic community. Those which democracy has been substituting for, forms of absolute and dictatorial rule, have been too closely linked to the origins of the global wars of the past and do not offer a sound basis for development, or for dealing with other global problems including environmental ones. It is from within the democratic community that substitutes for global war as a mechanism of collective choice are likely to emerge, and it is from within this mechanism that new forms will materialize that will be the substance of competitive substitution in the future.

Acknowledgments

We wish to thank Ted Gurr and the Center for Comparative Politics at the University of Colorado for making data available from the POLITY II dataset, and Raymond Gastil and James Huntley for their comments.

Notes

[1] In Herodotus' *The Histories* (Book III, 83) (written *ca.* 450–430 BC), where it is compared and contrasted with monarchy and aristocracy. Writing a century later (in *Politics* III, 7), Aristotle added new distinctions to these terms but, like Plato, he also put a negative spin on the term democracy.

[2] Early modern writers on politics were familiar with the concept of democracy from classical sources; prominent examples include Niccolò Machiavelli (*Discourses*, 1531), Gasparo Contarini (1599) (whose book on the Venetian constitution originally published in 1543 quickly became a classic of republican and constitutional thought), Andrzej Frycz Modrzewski (1551), Hugo Grotius (1600), Thomas Hobbes (1651), John Locke (1689), Jean-Jacques Rousseau (1762), James Madison (1787), and Immanuel Kant (1795). But they all saw it as direct democracy and doubted its practicality, except perhaps for small states (Rousseau), and might have been influenced by the negative spin put on it by Plato and Aristotle; in a republic, on the other hand, they saw at work the principle of representation, together with the idea of *commonwealth* (a whole body of people united by common consent to form a political community, a *res publica*). For Madison a republic was "a government in which the scheme of representation takes place" and it was to "modern Europe" that the "great principle of representation" was owed (*The Federalist*, Nos. X, XIV). Not until the publication of *Democracy in America* in 1835 did the concept of democracy regain wide acceptance, and this time, in a representative form.

[3] Woodrow Wilson (1918, p. 35) wrote that "democracy seems about universally to prevail. Ever since the rise of popular education in the last century ... the spread of democratic institutions ... promise(s) to reduce politics to a single form ... by reducing all forms of government to Democracy". When James Bryce (1921, pp. 24, 42) posed the same question: "whether the trend toward democracy ... is a natural trend due to a general law of social progress" his answer was less sanguine: "although democracy has spread ... we are not yet entitled to hold ... that it is the natural and therefore ... the inevitable form of government".

[4] This means that a number of small communities (such as Barbados, Belize, Kiribati, Malta, Solomon Islands, etc.) have thus been omitted, many of which are democratic.

[5] *Cf.* Cole's (Cole, 1987, p. 88) table of "Liberal Regimes in the World System since 1600" to which two earlier cases have been added: Venice (factions, elective chief executive, checks on chief executive by Grand Council) and Poland-Lithuania (factions, elective chief executive, checks on chief executive by Diet; described in Reddaway *et al.*, 1950, p. 440, as "Gentry Democracy"). Portugal *ca.* 1500 might be a borderline case (proto-parties, limits on chief executive), as well as Florence (1494–1512). This listing is entirely tentative and is meant to highlight representative trends at the national level of organization rather than draw up an exhaustive and definitive inventory of free and self-governing communities.

[6] The Swiss Confederation, emerging out of the Everlasting League of 1291, became virtually independent from the Empire in 1499. By 1513 it consisted of 13 cantons, the government of some of which, including Schwyz, Uri, and Unterwalden, was a form of direct democracy (by an assembly of all male citizens of full age). By 1531 (war of Catholic cantons against Zürich) religious conflict divided it and "common action became impossible" (Langer, 1972), yet this did not exhaust the persuasive power of the Swiss example. Such models were a source of inspiration, i.e., for political writers in the Dutch republic *ca.* 1600.

[7] The equation for the trend line in *Figure 2.1* is $y = 0.009x - 19.6$.

[8] The equation for the trend line in *Figure 2.2*, Institutional Democracy 1800–1986 is $y = 0.025x - 49.6$. The time constant Δt (10–90 percent) is 176 years; the flexpoint is 1987, and 90 percent is reached in 2075. Calculating a seven-year moving average raises R^2 only slightly, to 0.919. For the years 1837–1986 only (*Figure 2.3*), the regression line is $y = 0.019x - 38.54$, and the time constant is 228 years. The 10 percent level was reached in 1889; the flexpoint is 2003, and 90 percent saturation is reached in 2117; R^2 is 0.893.

[9] Writing in 1795, Immanuel Kant proposed the hypothesis of a self-organizing social process tending to bring about the condition of perpetual peace via, e.g., the creation of republican regimes.

[10] For a study of the post-1945 growth of the "Atlantic-Pacific" system as a basis for a community of democracies, see Huntley (1980).

References

Bryce, J., 1921, *Modern Democracies*, Macmillan, London, UK.

Cole, T., 1987, *United States Leadership and the Liberal Community of States*, Ph.D. Dissertation, University of Washington, WA, USA.

Contarini, G., 1599, *The Commonwealth and Government of Venice*, Edmund Mattes, London, UK.

de Tocqueville, A., 1835, in J.P. Mayer, ed., 1969, *Democracy in America*, Anchor Books, New York, NY, USA.

Fisher, J.C. and Pry, R.H., 1971, A simple substitution model of technological change, *Technological Forecasting & Social Change* **3**:75–88.

Gastil, R., 1989a, *Freedom in the World: Political Rights and Civil Liberties 1988–1989*, Freedom House, New York, NY, USA.

Gastil, R., 1989b, *The Future of Democratic Progress and the Prospect for Peace*, Paper presented to the Pacific Northwest Colloquium on International Security, Seattle, WA, USA.

Gurr, T.R., Jaggers, K., and Moore, W.H., 1989, *POLITY II Codebook*, Center for Comparative Politics, University of Colorado, CO, USA.

Gurr, T.R., Jaggers, K., and Moore, W.H., 1990, The transformation of the western state: The growth of autocracy, democracy, and state power since 1800, *Studies in Comparative International Development* (forthcoming).

Huntley, J.R., 1980, *Uniting the Democracies*, New York University Press, New York, NY, USA.

Langer, W., 1972, *The Encyclopedia of World History*, Houghton Mifflin, Boston, MA, USA.

Marchetti, C., 1980, Society as a learning system: Discovery, invention, and innovation cycles revisited, *Technological Forecasting & Social Change* **18**:267–282.

McEvedy, C. and Jones, R., 1978, *Atlas of World Population History*, Penguin, New York, NY, USA.

Modelski, G., 1987, *Long Cycles in World Politics*, University of Washington Press, Seattle, WA, USA.

Modelski, G., 1989, Is America's decline inevitable? *The Bridge* **19**(2):11–18, Summer.

Modelski, G., 1990, Is world politics evolutionary learning? *International Organization* **44**(1):1–24, Winter.

Perry, G. III, 1987, *A Time Series Study of Developing Liberalism in the World*, unpublished manuscript.

Reddaway, W.F., Penson, J.H., Halecki, O., and Dyboski, R., 1950, *The Cambridge History of Poland*, Cambridge University Press, Cambridge, UK.

Rogers, E.M., 1962, *Diffusion of Innovations*, The Free Press, New York, NY, USA.

Wilson, W., 1918, *The State: Elements of Historical and Practical Politics*, D.C. Heath, Boston, MA, USA.

Chapter 3

Innovation and Diffusion in a Single Europe: Institutional Structure and Industrial Prospects for the European Communities

William S. Peirce

3.1 Introduction

The recent flurry of motion toward the goal of a *Single Europe* (at least for the 12 members of the European Communities) has been marked by considerable optimism about the effects of the elimination of internal barriers on the economy of the European Communities (EC). Indeed, the high expectations of economic gains are inseparable from the progress toward integration because the expected economic gains are a principal benefit that compensates for the costs of conveying some control over national ways of life to the strange quasi-government of the EC. Because the integration process will cause some losses in the national individuality that people value, progress toward integration would be unlikely to occur if people did not expect economic gains.

The initial economic effects of integration do offer prospects for real gains (Cecchini, 1988; ECC, 1988). These include such direct benefits to consumers as the reduced prices that can be expected when the slightly differentiated products of national monopolists are subjected to more intense competition from rivals in other nations. In addition, there is the hope that the larger market will permit longer production runs and economies of scale. Rationalization of production in a smaller number of plants specially designed to capture economies of scale might be expected, but that evokes consideration of the pain of closing the inefficient plants. Moreover, the US and Japanese firms that are building plants in the EC because of the attractions of the single market or the fear of possible European protectionism, may speed up the diffusion of technology. Of course, border crossing formalities themselves consume resources, and these losses can be greatly reduced.

More to the point of this chapter, however, is the hypothesis of Mancur Olson (Olson, 1982) that jurisdictional integration reduces the power of some of the existing distributional coalitions or rent-seeking groups, which try to use the powers of government to redistribute income or wealth in their own favor. One of Olson's examples was the original Common Market of Six, which did seem to have had a more dramatic influence on economic growth than could be explained with the tools of static economic analysis. This experience was consistent with the idea that opening the borders destroyed some of the power of the distributional coalitions that had worked closely with the separate national governments to obtain benefits at the expense of poorly organized groups, including consumers and taxpayers (McCormick and Tollison, 1981; Olson, 1965). Not only can consumers benefit directly when a redistribution is eliminated but, more important, if the payoff for trying to obtain redistributive favors is reduced, more effort will be devoted to productive activities.

That argument raises the more general question of whether the gains from integration consist of a one-time improvement in consumer welfare (perhaps realized over a decade or two), or whether the benefit also includes an increase in the long-term rate of economic growth. In the Olson argument, that depends, in part, on the speed with which the interest groups can reorganize to suit the new political arrangements and thereby resume their redistributive successes. This is a point to which we will return in a later section.

In the long run, of course, growth in consumption per capita is limited by the growth in productivity, which leads to the question of whether jurisdictional integration will lead to more rapid adoption of innovations. It turns out that this is related to rent-seeking behavior, but that takes us too far ahead of our story. The immediate question is whether any aspect of integration, including the diffusion policies that may be adopted by the EC, will lead to more rapid productivity increases in the long run. It is not clear that this is a reasonable expectation, given our knowledge of the diffusion process, the political process, and the instruments that are available to government to influence diffusion.

One crucial aspect of this is the effect of jurisdictional integration on competition. Most discussions have focussed on the competition among firms and have, correctly, stressed that expanding the size of the market permits the market to include enough firms to yield the advantages of rivalry without eliminating the possibility for firms to enjoy economies of scale.

The other form of competition, however, is competition among governments, analyzed in a different context by Backhaus and Wagner (1987), which prevents the development of extremely inefficient laws. As the jurisdiction increases in size, it faces less stringent constraints on its behavior because those who bear the costs of inefficient laws find it harder to escape. If the Netherlands or Austria passes an inefficient law or operates with an institutional structure that is conducive to inefficiency, the damage is limited by the fact that the small country must export to survive and laws that make that impossible are quickly amended in formal or informal ways.

As long as relatively open borders are nearby, individuals will contrive to buy goods in the cheapest market and move capital to the least repressive environment. As the national boundaries within the EC are increasingly opened to the movement of goods, labor, capital, and firms, the economic powers of national governments will therefore be increasingly constrained. Perhaps this process will lead to a libertarian Economic Community, but in view of the traditions in most of the member countries it seems more plausible that the locus of regulatory activity will shift to the EC level. Since the constraints on the repressiveness and inefficiency of government and on its redistributive activities are relaxed as the jurisdiction becomes larger, the net effect could be an increase in inefficient regulatory and redistributive activity.

Much of the discussion of innovation and productivity in the European Communities refers to the United States and Japan as large, innovative economies. One might, however, turn to the USSR, instead, and see a large economy that has had difficulty in introducing rapid rates of innovation and an increase in productivity. The point, of course, is that the specific institutional factors, in addition to the size of the market, may have something to do with the initial adoption and the rate of diffusion of innovations.

The European Communities have a very peculiar political structure. While that structure can be explained historically, it sets limits on what can be done efficiently by that quasi-government. In particular, the structure of the EC abets the reformation of the distributional coalitions. That fact combines with the attenuation of international competition by integration to produce a situation in which most of the instruments by which governments have tried to speed up innovation and diffusion become ineffective or counterproductive. The following section develops this topic further.

3.2 The Instruments of Diffusion Policy

Ideally a list of instruments of diffusion policy would be based on a well developed theory of diffusion. This list has less ambitious origins in observations about the actions of governments that have influenced diffusion or been adopted in an attempt to do so. However, if one examines a comprehensive framework for factors influencing diffusion rates, such as that given by Rosegger (1986), it would appear that the direct interventions of government play a relatively limited role. That is also an implication of studies that show similar diffusion rates in dissimilar countries and times.

It would obviously be possible for the EC to refrain from an active diffusion policy; it could allow all innovation to be determined by the "animal spirits" of the entrepreneurs combined with whatever policies the individual member countries choose to adopt. As the EC pursues integration more actively, however, the range of policy choices for member nations will be restricted to a subset of those observed by Vickery and Blair (1987). Moreover, a passive approach by the EC to diffusion is not consistent with the expectations that have been raised or with the traditions of most of the countries involved.

3.2.1 The background for diffusion

Diffusion is a complex process, and it may be that some of the general measures that governments can take to improve the environment for technological change and investment are as important as direct approaches. For example, in some circumstances an effort to improve the technical capacity of the population by a major educational investment could increase the rate of diffusion of technological innovations. In the case of Western Europe, however, the common complaint has been that the results as measured by industrial innovation are not commensurate with the investment in high quality scientific and technical education. Moreover, under current plans, the educational systems will remain the responsibility of the national governments, so the EC has little positive influence on this variable.

Indeed, the influence of integration on education could become negative for two reasons. First, the tax competition, discussed in the next paragraph, places all expenditures under increasingly heavy pressure. Second, countries will not capture the full returns from investments in education because some of the benefits spill over to other countries as the educated people move. Any countries that have been concerned with a "brain drain" in the past will find that the increasingly open borders within the EC will exacerbate the problem.

Roughly similar comments apply to the other set of background measures, the general measures designed to increase the profitability of investments or of research and development in particular. The traditional techniques include such manipulations of the corporate tax law as tax credits for investments and rapid depreciation (US Congress, 1985). These, of course, cannot be EC measures because the EC has no corporate profits tax. Moreover, the national governments are beginning to yield to competitive pressure to reduce corporate tax rates, which is part of the phenomenon that has been called the "Delawarization of Europe" (Aretz, 1988). One result of a reduction in tax rates is to reduce the value of tax breaks for specific activities. Of course, governments can, and do, subsidize investments directly, without using the tax mechanism. This, too, is likely not to become a general activity at the EC level because of the large costs and the political problems involved. At the national and local level, where such activities have been common, governments are now encountering opposition from the EC because such subsidies interfere with the completion of the internal market.

3.2.2 Targeting public spending toward diffusion

In addition to improving the general environment for diffusion, governments can take more specific measures to encourage technological change and the diffusion of innovations. The most direct method is to subsidize research and development expenditures, but this does not necessarily lead to innovation, to say nothing of diffusion. Nevertheless, governments have difficulties in focusing resources very far along the spectrum that begins with pure research and ends where the innovation has become a routine investment. If the diffusion stage is defined as what happens after the first commercial adoption, then any direct subsidy of the diffusion process raises the question of why a subsidy is necessary if the project is worth doing. Innovation, even after the first adoption, is risky, but much of this risk reflects the real difficulties and ignorance surrounding an innovation, so it is not clear that efficiency is served by allowing the costs to be borne by anyone other than the decision maker. Payment of a share of the costs by the public sector can be justified only if there are external benefits from adoption or if the public sector decision makers are better informed than are the people in the enterprise. Neither condition seems likely in most cases.

The other difficulty with direct aid at the diffusion stage is that it is likely to be very expensive. The EC does not have a large enough budget to do much of this. Moreover, any national efforts at direct subsidy would be called into question in the interests of maintaining a "fair" internal market. Quigley's analysis (Quigley, 1988) of decisions by the Commission (ECC, 1971–1985) suggests that, as a rough rule, subsidies for research and development are acceptable to the Commission, while subsidies for operations are not. It would require an unprecedented degree of bureaucratic subtlety to subsidize only the innovative aspects of new investments and to defend the decision rules before the Commission.

The experience with the ESPRIT program is illustrative of some other problems that can be expected with EC programs of this sort (Sharp, 1987; Sharp and Shearman, 1987). ESPRIT subsidizes precommercial research in information technology, so one could argue that the results are not directly applicable to diffusion. Nevertheless, there are certain structural features that would be relevant for any direct subsidy program. In particular, at this stage in the process of European integration the political problem of geographic distribution of any subsidies must be considered explicitly in the legislation or implicitly in the administration. This has been a continuing theme of personnel policy in the Commission, the Common Agricultural Policy, and, of course, regional policy where, as late as 1987, 45% of the population of Germany lived in assisted areas (EC, 1987).

The technique for handling this in ESPRIT was to require that each project involve firms or other organizations from at least two EC countries. This helps to solve the political problems of geographic distribution and especially of integration, but it makes the program inaccessible to small firms, as Sharp and Shearman (1987) noted. A study by Mytelka and Delapierre (1987) found that only 34% of the ESPRIT projects for 1983–1985 did not involve the 12 major electronics firms that helped devise the program (ICL, GEC, Plessey, AEG, Nixdorf, Siemens, Thomson, Bull, CGE, Olivetti, STET, and Philips). Many of the other participants were also very large firms from other industries. Moreover, to the extent that ESPRIT really does change the patterns of cooperation on research and development, one might raise the question of whether the Communities will gain any economic benefits. Studies by Hagedoorn and Schakenraad (1988) and Valls-Pasola and Valles (1988) have indicated that European firms are more likely to cooperate with firms from the USA than with firms from other European countries, and presumably this reflects the best business judgment about long-term costs and benefits. Of course, it is to be expected that when government uses an instrument of policy to pursue a non-economic goal, the performance of purely economic measures of success will suffer.

In addition to making direct grants, a policy which is better suited for research than for diffusion, governments can speed the diffusion process for some products by specifying them in the requirements for products purchased by government (Katz and Phillips, 1982; Levin, 1982; Mowery and Rosenberg, 1982). One of the classic examples of this is the assistance given to the computer and electronics industries in the USA, not only by research contracts, but also by specifications for military and space systems that pushed firms beyond what was routinely available (Flamm, 1987 and 1988). However, the EC itself does not purchase enough to make much difference. The traditional sector in which to hide activities related to research and development is the military, both because it has a large enough budget to dwarf most research activities and because opponents traditionally have found it difficult to argue against military strength. The member countries have enough government procurement to influence diffusion, but are now under EC pressure not to use procurement as a device for favoring domestic producers. The case that has attracted a great deal of attention recently is the relationship between national telephone companies and the producers of telecommunications equipment.

3.2.3 The government as a regulator

The powers of a government as a purchaser are limited by its budget, while the powers of a government as a regulator are limited only by imagination.

There is no doubt that governments can have substantial effects on the diffusion process through regulation. In the most direct cases, it can require or forbid particular products or processes. The interesting cases are more complex than, for example, the requirement that seatbelts be fitted in automobiles that results in the diffusion of seatbelts. If the requirement is that automobile emissions be reduced to a particular level, the regulation does not specify a technique, but it serves as an inducement to adopt particular methods of meeting the regulation. Such regulations are usually adopted in the name of health, safety, or the environment, so one would not expect them to improve standard economic measures of productivity. As the EC takes on more of these functions from the national governments, however, some have hoped for indirect improvements in productivity.

The reasoning here is straightforward. If each country has a set of regulations that covers the same items, but does so in ways that differ, eliminating the country regulations and replacing them with one set of regulations for all twelve countries would contribute to efficiency in a variety of ways. Competition, as viewed by any consumer, would increase because any item manufactured in any part of the EC could be sold in any other part. Manufacturers could expect longer production runs for the standard product that met specifications throughout the EC. Furthermore, the design problem would be simpler when new models were introduced if only one set of regulations had to be met. Pelkmans (1987) analyzed past and present practice in the EC regarding technical harmonization.

It would seem, however, that providing the consumer with more choice and the manufacturer with longer production runs may not be fully consistent. One must make some specific assumptions about the rationalization process, the elasticity of demand, and the value to the consumer of different types of product differentiation to show that benefits are fairly general in the short to intermediate run. In the long run, of course, the larger market should permit economies of scale for the surviving producers; and consumers should benefit from lower prices if the surviving producers compete with one another or with imports.

Before proceeding to consider the implications for diffusion, however, it is necessary to examine one of the crucial assumptions. Is it likely that moving regulation from the level of the country to the level of the EC will keep the total amount of regulation unchanged? Tumlir (1983) noted that the amalgamation of national agricultural policies into the Common Agricultural Policy produced a more inefficient transfer scheme covering more crops than any of the separate national policies. Similarly, the process of negotiating health, safety, and environmental regulations may result in more detailed coverage of a greater number of products and processes than most countries had regulated independently. In any event, one can be certain that some formerly unregulated items become regulated somewhere. This may have implications for innovation and diffusion.

In short, in the area of technical regulations and standards, the possibility exists that the European Communities can prepare the way for more rapid diffusion, but whether that happens depends on how things are done. Thus it is necessary to examine the process by which the EC is attempting to achieve technical harmonization. The early efforts put the Council of Ministers in the untenable position of trying to reach unanimous agreement on detailed questions on the design of lawn mowers and tire pressure gauges. In this era, Pelkmans (1987) has suggested, new discrepancies in technical standards were accumulating among the member countries of the EC faster than the old ones were being eliminated.

The "new approach to technical harmonization and standards" was spelled out in a Council Resolution of 7 May 1985 (EC, 1985). Under the new approach the Council will set the broad requirements for health, safety, environmental protection, and consumer protection. The detailed technical interpretation of these general requirements will be published as standards by separate organizations. The designated organizations are CEN (the European Committee for Standardization) and CENELEC (for electrical equipment). Manufacturers or importers meeting the CEN or CENELEC standards will be deemed to have met the basic requirements legislated by the Council and will thus be free to sell throughout the EC. Those who do not meet the standards are invited to prove to a national government that they do meet the requirements.

The existence of a way around the standards will probably prevent this procedure from being used to build a "Fortress Europe." Major US and Japanese firms can adapt their designs for European markets or prove the case for their own methods to meet the requirements. The more important question is whether the standards will be used in the future to inhibit entry into the market by small European firms with innovative ideas. From a public choice perspective that would appear to be inevitable. The process of standard setting is so obscure to the public, and the particular process established by the Council is so remote from electoral control, that the eventual capture of the standard-setting organization by the established producers would seem to be as close to a sure thing as is any institutional forecast.

The net result would be to inhibit diffusion relative to that in a world with no controls, but the more important comparison is to the realities of the European Communities without a centralized harmonization process. Davidson (1983) argued that the internal technical barriers to trade would gradually have eroded anyway as a result of the strong position taken by the European Court of Justice in the *Cassis de Dijon* and similar cases. Such an approach to harmonization would have been slow, of course, but as borders become more open within the EC, it becomes easier for consumers, labor, capital, and firms to escape from the most repressive national regulatory regimes.

In one type of case the EC provides what should be a superior mechanism to increase the rate of diffusion. This is in the choice among rival standards in situations where system interdependency requires a unique choice. Choice of the particular type of high definition TV system to be adopted or the particular standards for the telecommunications system are examples that come to mind. Yet, the evidence from an earlier era suggests that conscious choice, not difficulties with the decision process, were the source of such inconsistencies as did occur in systems such as railroads. The European Communities provide a permanent forum within which such choices of system can be made, but it remains to be seen whether the decisions will in fact be made more quickly than by the old style of first-mover advantage and *ad hoc* negotiation.

3.2.4 Summary: The effect of the EC on diffusion

This discussion has exhausted the list of the most important policies available to speed up diffusion. The EC could, of course, provide information about innovations, but information is rarely the constraint to diffusion of industrial innovations among large firms and private advertisers seem able to tell the public about consumer innovations. If special efforts must be made to reach many small firms, national governments seem better equipped for that task. The more important direct action may be the dismantling of some existing barriers to diffusion by either the regulatory or the legal route noted above.

The generally negative tone of this discussion in part reflects the fact that diffusion is difficult for any government to accelerate. To the extent that adoption can be analyzed in the same way as routine investment, governments generally have crude macro techniques mainly incorporated into the tax code or resulting from monetary policy that may make a difference. These are not available to the EC and the existence of the EC increasingly constrains their use at the national level. To the extent that any adoption decision can be viewed as innovation, it is not clear that government has much influence, except for the barriers that it raises or removes. If something about the increasing integration of Europe is expected to increase innovation and diffusion rates in the Communities, the mechanism must be indirect and subtle.

3.3 Innovation and Rent-Seeking in the European Communities

As noted earlier, Mancur Olson (1982) used the original Common Market of Six as one of his examples of how the distributional coalitions that had gained increasing power in the individual nations were disrupted by the jurisdictional integration. Similar bursts of energy could probably be detected in the other members after they joined the EC. During such eras of innovation and growth, one would expect measures of diffusion of technological innovation to show improvement. Actual measurement of diffusion rates is difficult enough for single, identifiable products and processes and, of course, becomes essentially impossible in terms of aggregates for an economy. One would probably have to settle for such aggregate measures as the growth rate of GNP and the aggregate productivity measures associated with it.

Traditional justifications for customs unions are not phrased in terms of the Olson hypothesis, however (Gowland, 1983). The most satisfying theoretical argument is the argument for free trade combined with some political constraint that limits the free trade to a selected group of countries. In the case of European integration, obviously, the political arguments are very strong, but one must presume that most people expect some economic gains as well, at least compared with the condition of non-integration, although perhaps not when compared with the (unattainable) free trade. Perhaps that explains the stress on scale in discussions of the EC, because the source of economies of scale is specialization of functions within the firm that is closely parallel to the classical specialization of functions among nations (Gold, 1981).

If scale is the source of the expected gains, then rapid technical harmonization and reduction of other internal barriers become goals of the highest priority. In some modern versions of the theory of trade, moreover, scale is particularly important in keeping firms at the forefront of technology because the costs of research and development can be recovered from a larger output before the firm's advantage is eroded away by other innovations.

The two views – one emphasizing size of firm and size of market while the other emphasizes distributional coalitions – have different policy implications. If one believes that size brings innovation and progress, one uses whatever tools are available to break down the barriers in the internal market and one tolerates or even encourages mergers and cooperative arrangements among firms. In contrast, if one believes that Europe has been inefficient because of a legacy of guild-type restrictions, special privileges, and subsidies and transfers to interest groups, then one takes whatever steps are available to discourage the counterattack of the rent-seeking groups that lost some power when the market was widened.

The relationship between size of firm and innovation has generated too much literature to be considered here, but the rapidity with which the interest groups have formed around the European Communities has received less attention. *Table 3.1* presents the basic data regarding the number of interest groups by year of founding. The cumulative percentage would make a respectable diffusion curve. Of the 546 officially recognized interest groups, 462 have a known date of founding. More than half of these were already in existence before 1965. If Mancur Olson's model is correct, the European Communities may have already enjoyed the golden age of growth.

Table 3.1. Interest groups of the European Communities (number by year of founding).

Years	Number founded	Percent of known	Cumulative
Through 1949	19	4	4
1950–1954	40	9	13
1955–1959	119	26	38
1960–1964	93	20	58
1965–1969	53	11	70
1970–1974	54	12	82
1975–1979	55	12	94
1980–1984	29	6	100
Date not reported	84		
Total	546		

Source: Compiled from European Communities (1986).

The three reasons for the rapid formation of groups to lobby the EC are that representation of interest groups is required by the basic treaties establishing the EC, that the task of organizing at the EC level was easy, and that the structure of the EC is conducive to strengthening the influence of such groups. The interest groups are formally represented in the Economic and Social Committee and thus have an official status and a slight amount of formal power. Organization was easy for those EC interest groups that are federations of existing national groups. Most important, however, is the structure of decision making in the European Communities (Peirce, 1991). The absence of clearly defined political authority and the necessity of approaching consensus for major explicit decisions means that many decisions must be made within the bureaucracy or delegated to private groups. In either event, the group that has its representative actively engaged in the whole process has much to gain.

3.4 Conclusions

Firms are responding to the increasing economic integration of Europe by mergers and rationalization of production within Europe. Firms that had no capacity within the EC are building or acquiring plants within its boundaries. In the long run, however, economic progress of the EC depends on the diffusion of innovations within its borders. The most common policies used by governments to encourage innovation and diffusion, notably public procurement and incentives built into the tax law, are not available to the European Communities. Moreover, the reduction of internal barriers to the movement of goods, labor, capital, and firms will increasingly inhibit the use by member countries of such policies. The structure and procedures of the EC are conducive to the dominance of the interests of established producer groups. Although this has been traditional in many European countries, it becomes a more serious problem for the economy as the jurisdiction increases in size because of the attenuation of competition among jurisdictions. The established producer groups are well-positioned to inhibit innovation by new entrants. Thus the EC will rely heavily for its innovations on the same group of firms that the individual countries have relied upon in the past. This leads to the question of whether the larger market will provide these firms with sufficient stimulation to increase the rate of diffusion.

Acknowledgments

I wish to thank my colleagues at CWRU for making my sabbatical possible and the faculty of the University of Limburg, Maastricht, The Netherlands, and especially Professor Juergen Backhaus, for providing a productive environment for research.

References

Aretz, E., 1988, Seminar paper, University of Limburg, Maastricht, The Netherlands, October.

Backhaus, J. and Wagner, R.E., 1987, The cameralists: A public choice perspective, *Public Choice* **53**:3–20.

Cecchini, P., *et al.*, 1988, *The European Challenge, 1992: The Benefits of a Single Market*, Gower, London, UK.

Davidson, S., 1983, Legal aspects of the common market in goods, in J. Lodge, ed., *Institutions and Policies of the European Community*, Frances Pinter, London, UK.

European Communities, Official Journal, Series C, 4 June 1985, No. 136:1–9 (Resolution 85/C136/01).

European Communities, Commission, 1986, *Directory of European Community Trade and Professional Associations*, 3rd ed., Brussels, Belgium.

European Communities, Commission, 1987, *XXIst General Report on the Activities of the European Communities*, Brussels, Belgium.

European Communities, Commission, 1988, *Research on the "Cost of Non-Europe"*, Office for Official Publications of the European Communities, Luxembourg.

European Communities, Commission, 1971–1985, *Annual Reports on Competition Policy*, Luxembourg.

Flamm, K., 1987, *Targeting the Computer: Government Support and International Competition*, Brookings Institution, Washington, DC, USA.

Flamm, K., 1988, *Creating the Computer: Government, Industry, and High Technology*, Brookings Institution, Washington, DC, USA.

Gold, B., 1981, Changing perspectives on size, scale, and returns: An interpretive survey, *Journal of Economic Literature* **XIX**(1):5–33, March.

Gowland, D., 1983, *International Economics*, Croom Helm, London, UK.

Hagedoorn, J. and Schakenraad, J., 1988, *Strategic Partnering and Technological Co-operation*, Paper presented at the 15th Annual Conference of the European Association for Research in Industrial Economics, Rotterdam, The Netherlands, August 31–September 2, 1988.

Katz, B.G. and Phillips, A., 1982, The computer industry, in R.R. Nelson, ed., *Government and Technical Progress: A Cross-Industry Analysis*, Pergamon Press Inc., New York, NY, USA.

Levin, R.C., 1982, The semi-conductor industry, in R.R. Nelson, ed., *Government and Technical Progress: A Cross-Industry Analysis*, Pergamon Press Inc., New York, NY, USA

McCormick, R.E. and Tollison, R.D., 1981, *Politicians, Legislation, and the Economy*, Martinus Nijhoff, Dordrecht, The Netherlands.

Mowery, D.C. and Rosenberg, N., 1982, The commercial aircraft industry, in R.R. Nelson, ed., *Government and Technical Progress: A Cross-Industry Analysis*, Pergamon Press Inc., New York, NY, USA.

Mytelka, L.K. and Delapierre, M., 1987, The alliance strategies of European firms in the information technology industry and the role of ESPRIT, *Journal of Common Market Studies* **XXVI**(2):231–253.

Olson, M., 1965, *The Logic of Collective Action*, Harvard University Press, Cambridge, MA, USA.

Olson, M., 1982, *The Rise and Decline of Nations: Economic Growth, Stagflation, and Social Rigidities*, Yale University Press, New Haven, CN, USA.

Peirce, W.S., 1991, Decision making in the EC Council of Ministers, in R. Vaubel and T.D. Willett, eds., *The Political Economy of International Organizations*, Westview Press, Denver, CO, USA (forthcoming).

Pelkmans, J., 1987, The new approach to technical harmonization and standardization, *Journal of Common Market Studies* **XXV**(3):249–269, March.

Quigley, C., 1988, The notion of a state aid in the EEC, *European Law Review* **13**(4):242–256, August.

Rosegger, G., 1986, *The Economics of Production and Innovation*, 2nd ed., Pergamon Press, Oxford, UK.

Sharp, M., 1987, Europe: Collaboration in the high technology sectors, *Oxford Review of Economic Policy* **3**(1), Spring.

Sharp, M. and Shearman, C., 1987, *European Technological Collaboration*, Routledge and Kegan Paul, Chatham House Papers, London, UK.

Tumlir, J., 1983, Strong and weak elements in the concept of European integration, in F. Machlup *et al.*, eds., *Reflections on a Troubled World Economy: Essays in Honor of Herbert Giersch*, Macmillan for the Trade Policy Research Centre, London, UK.

US Congress, Congressional Budget Office, 1985, Federal Support for High Technology Industries, *CBO Study*, Washington, DC, USA, June.

Valls-Pasola, J. and Valles, R.S., 1988, *Cooperation between Firms on an International Scale: Towards an Evaluation of the Empirical Evidence*, Paper presented at the 15th Annual Conference of the European Association for Research in Industrial Economics, Rotterdam, The Netherlands, August 31–September 2, 1988.

Vickery, G. and Blair, L., 1987, Diffusing new technologies: Microelectronics, *STI Review* **2**:41–80, September.

Chapter 4

Multiple Perspectives on Technological Diffusion: Insights and Lessons

Harold A. Linstone

Since World War II there has been a remarkable proliferation in the use of mathematical modeling under the labels of systems analysis, decision analysis, operations analysis, management science, and policy analysis. Technology forecasting, impact, and transfer have also reflected the trend. However, serious limitations are inherent in these reductionist approaches and they have created a serious gap between the model and the real-world decision maker. This discussion focuses on the problem and a means to overcome it.

4.1 The Analyst's Perspective

Science and technology represent the most successful "religion" of modern times. From Galileo to the Apollo lunar landing, from Darwin to recombinant DNA, the paradigms of science and technology have yielded dazzling triumphs. This world view is typified by the following characteristics:

- We define *problems* by abstraction from the world around us, with the implicit assumption that such problems can be *solved.*

However, in the living world a solution nearly always creates new problems; we shift problems rather than solve them. Public health measures have cut the death rate drastically; however, they have also led to a global population explosion and, indirectly, to starvation.

- We seek the *best* or optimal solution.

  Cost-benefit analysis and linear programming are typical of this search. But complex living systems strive to maximize their options rather than confine them by selecting the best one. They seek to minimize the cost of failure rather than the likelihood of failure. They recognize that we learn more from our failures than from our successes. Ecological systems sacrifice efficiency for resilience; they trade avoidance of failure for the ability to survive failure, the *fail-safe* strategy of the engineer for a *safe-fail* strategy.

- Reductionism is the norm; a system is defined in terms of a very limited number of variables and the relationships are often linearized.

  Complexity begets nonlinearity. But the theorems are in the abstract linear domain. The analyst is caught between two unpalatable choices: solving linear, irrelevant problems and struggling unsuccessfully with nonlinear relevant ones.

- Reliance is placed on data and models, and combinations thereof, as the only legitimate modes of inquiry and as a basis for theories.

  The analysts' emphasis on certain types of models easily leads to a kind of "groupthink." For example, as system dynamics has proliferated and the number of modelers has multiplied, conferences, papers, and books create a community. Shared interest and mutual reinforcement increasingly focus attention on baroque model improvements and compulsive extensions. Econometrics is another case in point. In its most extreme form modeling becomes an end rather than a means (*the Pygmalion complex* – the modeler falls in love with his model and believes it to be the reality). A look beyond the realm of traditional science and engineering opens our eyes: there are other important modes of inquiry; indeed, the lawyer and the government executive make effective use of them.

- Quantitative analyses drive out qualitative analyses.

  We confuse dollars spent with effectiveness, because money is easier to count. We produce masses of numerical trend extrapolations but shy away from probing the underlying assumptions. We find comfort in the six-figure precision of output data although it masks the real uncertainties. The faith in econometrics reflects this tendency.

- The conviction persists that the analyst is an objective observer and that truth is observer-invariant.

  In the complex real world virtually everything interacts with everything, and this includes the observer. Without the observer there are no descriptions; the observer's faculty of describing enters, by necessity, into his description.

- The individual may be considered as type, but not as a unique person.

  Complexity has been defined as the ability to hold conflicting world views at the same time and to benefit therefrom, to see the world globally *and* in terms of unique individuals. Abstraction and generalization are not a substitute for specific case studies.

- Time moves linearly at a universally accepted pace, with no consideration of differential time perceptions, planning horizons, and discount rates.

  Recent experiments demonstrate how human beings apply a psychological discount rate to their own past and thus distort the integration of their own experience, that is, their subjective probability. Recent events are overstressed in comparison to more remote ones. Similarly, we look at the future as if through the wrong end of a telescope: distant crises and opportunities appear smaller than they actually are, so they are ignored. Such discounting of the future drastically affects the choice among alternatives in technology transfer. It particularly downgrades local training and product improvement (R&D) with its long-term payoff in favor of imported turn-key operations.

The characteristics discussed here suffice to explain the traditional perspective of the engineer and scientist. *Figure 4.1* schematically shows the application of this *technical* (or **T**) perspective to well-structured engineering systems and in *Figure 4.2* we see its extension to the study of ill-structured or *messy* systems, typically involving people – as in technology diffusion. Agricultural development or modernization constitutes such an ill-structured system. Modern farming techniques are worthless if they are not adopted by the farmers. Cash or commodity crops do not produce cash if false assumptions are made about the market. Ethnic concerns, such as Bhumiputras versus

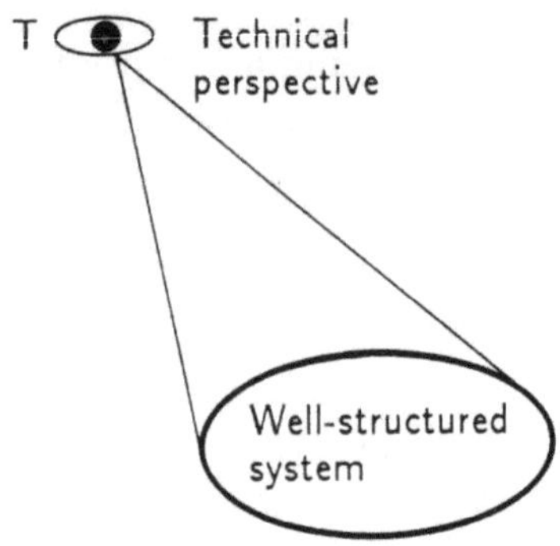

Figure 4.1. Viewing a technological system.

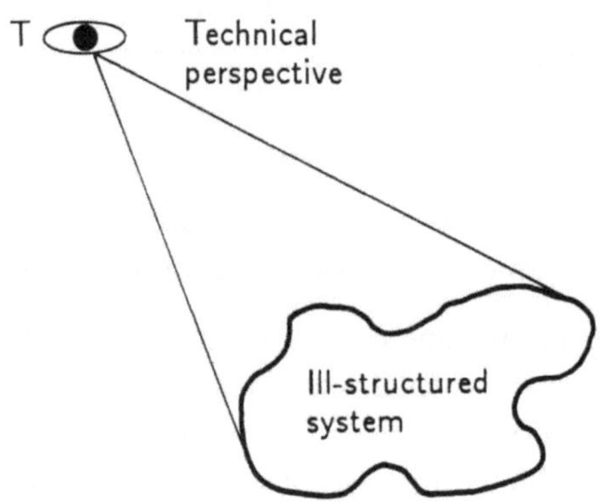

Figure 4.2. Extension of the same perspective to a system involving people and technology.

Chinese and Indians in Malaysia, lead to special operational constraints on technology diffusion that are easily missed by outsiders.

If the **T** perspective is not adequate, how do we proceed?

4.2 The Multiple Perspective Concept

The following two examples suggest an answer to the question raised above.

- The engineering technique of system dynamics (Forrester, 1961) provides important insights about an enterprise, particularly its material and money flows. Machiavelli also provides valuable insights about organizations and how to control them (*cf.* Jay, 1968). Both are looking at organizations, but from very different angles. Each perspective presents insights not obtainable with the other.

- A regional government leader must make a decision on allocating major development funds under his control. He has an analysis from his technical staff concluding that a certain distribution is the best of several alternatives. But he does not make his decision solely on the basis of this report. He talks to various bureaus to determine whether there is strong support or opposition. At an unrelated meeting with leaders of other regions, he talks to a close friend from a distant region whom he has known for thirty years and whose judgment he values. He also has his own intuition and experience to draw upon. Then he decides. He has integrated in his mind – without a weighting formula – several different, probably conflicting, perspectives: technical, organizational, and personal.

We also find it desirable, indeed essential, to call on several perspectives in addressing systems which are ill-structured, which deal with people as well as artifacts (Allison, 1971; Linstone *et al.*, 1981; Linstone, 1984 and 1985). We emphasize that we are *augmenting, not replacing*, the **T** perspective. Specifically, we draw on three types:

T: the technical perspective (see Section 4.1),
O: the organizational or societal perspective, and
P: the personal or individual perspective.

The different perspectives force us to distinguish *how* we are looking from *what* we are looking at. They do not represent different mathematical models, but rather different sets of paradigms. *Table 4.1* summarizes the features of these distinct world views and *Figure 4.3* is a schematic that may be compared with *Figures 4.1* and *4.2*. As the figure suggests, there are usually several **O** and **P** perspectives appropriate in any one situation.

The following points are stressed:

- Any complex problem may be viewed from any perspective. For example, an organizational decision may be seen from a **T** perspective, as decision analysis does; technology may be viewed from a **P** perspective, as does *The Existential Pleasures of Engineering* (Florman, 1976).
- Our perspectives differ in their underlying paradigms, inexorably moving us beyond those associated with science and engineering. Experimental design and validation of hypotheses are *intraparadigmatic*: they operate

Table 4.1. Characteristics of the three perspectives.

	Technical (T)	Organizational (O)	Personal (P)
Goal	Problem solving, product	action, stability, process	Power, influence, prestige
Mode of inquiry	Modeling, data, analysis	Consensual and adversary	Intuition, learning, experience
Ethical basis	Rationality	Justice, fairness equity	Morality
Planning horizon	Far	Intermediate	Short, with exceptions
Other characteristics	Cause and effect	Agenda (problem of the moment)	Challenge and response
	problem simplified, idealized	problem delegated and factored	hierarchy of individual needs
	need for validation, replicability	political sensitivity, loyalties	need for certainty
	claim of objectivity	reasonableness	need for beliefs
	optimization (seek best solution)	satisficing (first acceptable solution)	cope only with few alternatives
	quantification	incremental change	fear of change
	trade-offs	standard operating procedures	leaders and followers
	use of averages, probabilities	compromise and bargaining	creativity and vision by the few
	uncertainties noted, (on one hand ...)	avoid uncertainties	filter out inconsistent images
Communication	Technical report, briefing	Language differs for insiders, public	Personality important

Source: Linstone (1984).

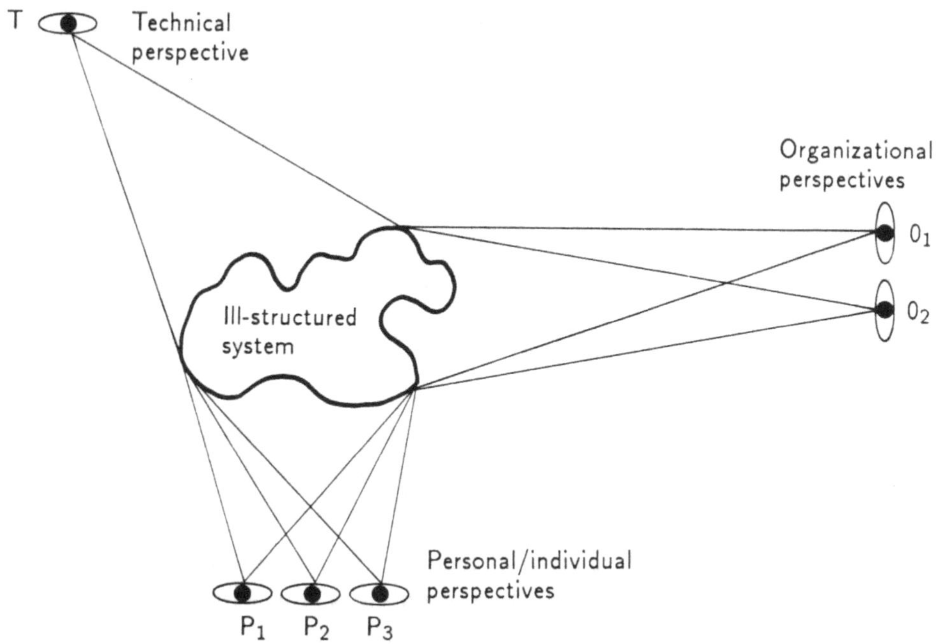

Figure 4.3. **T**, **O** and **P** perspectives on a system involving people and technology.

within the framework of a perspective. They cannot prove that a model gives the most "correct" representation of reality; they cannot give assurance that the variables chosen are sufficiently inclusive or appropriate.

- We cannot prove that a set of perspectives is the "right" set any more than an executive can prove he listened to the right input before making his decision. We cannot derive the "proper" weighting in integrating perspectives any more than a jury can in integrating the testimony of different witnesses.
- Two perspectives may reinforce each other or cancel each other out; they frequently "cross-cue" each other. As Churchman (1977) observes:

  > The mature individual is [one] who can hold conflicting world views together at the same time, and act and live, and that his or her life is enriched by that capability – not weakened by it.

- Taken together, the multiple perspectives constitute a meta-inquiring system (Churchman, 1971). As such it is pragmatic and includes application of all other inquiring systems, for example, data and model based or dialectic, as needed.

- In real life situations, technology diffusion management consists of three activities: (a) finding paths, (b) decisions, and (c) implementation. The **T** perspective focuses most strongly on (a) and least on (c), hence the "gap" between analysis and action. But implementation depends first and foremost on the use of human resources and that means **O** and **P** become crucial as we move from (a) to (c).

In sum, the justification for the use of multiple perspectives is basically twofold:

(1) *Each perspective yields insights not obtainable with the others, and*
(2) *The* **O** *and* **P** *perspectives are essential in bridging the gap between analysis and action in tasks such as technology diffusion.*

The multiple perspective concept has been applied successfully to a wide spectrum of problems in the private and public sector, such as military system decisions, education planning, policy analysis, technology assessments of national hydropower development and new agricultural crops, as well as health care planning (Linstone, 1984).

4.3 Illustrations

(a) Strategic Planning and Decision Making: The American Experience

In his survey of strategic management in 25 major US corporations Halal (1980 and 1984) found that

> skillful executives do not rely primarily upon the outcome of formal planning .... The decision maker continually gathers opinions, pieces of data, new ideas, etc., through exchanges with persons that are trusted and respected [Halal, 1980, pp. 57–58].

Peters and Waterman (1982) in their analysis of 43 particularly well-run US corporations similarly stressed that success is correlated to the ability to go beyond the "rational" model, in other words, the **T** perspective. A more recent survey by *Business Week* (1984) similarly attributed the failure of the majority of strategic plans to number-crunching professional planners:

> The quantitative, formula-matrix approaches to strategic planning developed ... in the 1960s are out of favor ... [Mead's former chief strategic

> officer says:] "The old process was just too mechanized. The real world is just too complicated for that." [The vice-president for corporate planning at Westinghouse adds:] "The notion that an effective strategy can be constructed by someone in an ivory tower is totally bankrupt".

The results of these three surveys all point directly to the inadequacy of the **T** perspective.

Intuition, a facet of the **P** perspective, is well appreciated by top executives:

R.P. Jensen, chairman of General Cable Corp.:

> On each decision, the mathematical analysis only got me to the point where my intuition had to take over [Rowan, 1979].

R. Siu, management consultant:

> Effective CEO's ... are aware that rationality and the scientific method provide critical inputs to only one of three questions overarching key decisions. These are: (a) Does it add up? (b) Does it sound OK? and (c) Does it feel right? Logic and science contribute primarily to the first question, less to the second, and even less to the third [Siu, 1978].

The personal perspective has historically played a key role in US enterprises, being instrumental in entrepreneurship and leadership. Recent (1987) business writings place renewed attention on the latter.

> What's required now ... are not merely managers, but *leaders* .... The new paragon is an executive who can envision a future for his organization and inspire colleagues to join him in building that future .... Corporate America has always maintained at least a nodding interest in the subject of leadership, but recently the exigencies of global competition, deregulation, and accelerating technological change have whipped that interest into an anxious search for new answers to old questions: Can leadership be taught? How do you spot potential leaders? And what, precisely, sets leaders apart from everyday managers? [Main, 1987]

The strong difference in Japanese and American approaches to strategic planning can be traced to cultural traits. The Japanese have tended to submerge the personal to the societal view, the American the societal to the personal view. In Japan **O** and **T** have become tightly bonded, while **P** is minimized. In America **P** plays a much stronger role. Japan's Ministry of International Trade and Industry (MITI) undertakes to do long-range planning for entire industry sectors, while there is no equivalent concern in the USA with collective long-range industrial policy and strategy. On the other hand, the relative strength of the **P** perspective in the USA is reflected in its individual creativity output or basic research dominance: 135 US Nobel Prizes in science to 4 for Japan.

(b) Risk Evaluation and Management: American and Soviet Experiences

Technology management is always concerned with risks. We consider here only two types, physical risk and political risk.

At the Three Mile Island nuclear plant the errors included inadequate training of the utility company operators and supervisors, toleration of poor control room practices, and failure of the construction engineers (Babcock & Wilcox) to inform their nuclear reactor customers of persistent failures of pilot operated relief valves. The President's Commission concluded

> ... the fundamental problems are people-related problems and not equipment problems ... wherever we looked, we found problems with the human beings who operate the plant, with the management that runs the key organization, and with the agency that is charged with assuring the safety of nuclear power plants. [*Report of the President's Commission on the Accident at Three Mile Island*, 1979].

At Chernobyl, a mishap on April 26, 1986 occurred in the context of a turbine test. Faulty actions caused a loss of the water that continuously cools the uranium fuel rods in the reactor's core. This led to a partial core meltdown. As of August 1986, 31 fatalities were noted and over 200 were hospitalized with radiation sickness. A total of 135,000 people had to be evacuated and the long-term effect is estimated by one source at 4,000 additional cancer deaths in Europe, by another at 5,000 to 24,000 in the Soviet Union alone over the next 70 years (*Washington Post*, August 30, 1986). While Westerners have pointed to technical flaws in the reactor design (*New York Times*, 1986), the Soviet report to the International Atomic Energy Agency focused on a series of human errors, mistakes that violated safety regulations and, in some cases, common sense. Andronik Petrosyants, head of the State Committee for the Use of Atomic Energy, said:

> For almost 12 hours the reactor was functioning with the emergency cooling system switched off .... It is quite possible that the (previous) smooth operations brought on complacency and that this led to irresponsibility, negligence, lack of discipline, and caused grave consequences [*Washington Post*, August 22, 1986].

Valeri Legasov, first deputy director of the principal Soviet atomic research institute, added:

> If at least one violation of the six would be removed, the accident would not have happened. The engineers psychologically did not believe that such a sequence of improper actions would be committed. Such a sequence of human actions was so unlikely that the engineer did not include (it) in the project. Is that human or technical? [*Washington Post*, August 22, 1986]

Table 4.2 shows how the three perspectives illuminate different views of risk. The **T** perspective undertakes probabilistic calculations and draws up fault trees; the **O** perspective deals with standard operating procedures and threats to organizational integrity; the **P** perspective perceives personal fears and images of horror. Not surprisingly, there is a dramatic difference between actuarial, societal, and personally perceived risk rankings. According to a recent survey, the typical individual views the risk of auto accidents as equal to those of nuclear power, while the actual annual mortality rate of the former is over 500 times that of the latter. Strokes kill 85% more people than do accidents; yet people estimate that accidents take 25 times as many lives as strokes.

Table 4.2. Perspectives on Risk.

T	O	P
Probabilistic	Threat to organization survival	Time for consequences to materialize
Actuarial/ statistical	Threat to product line or image	Personal experience or history
Quantification vital	Political sensitivity	Age of individual
Margin of safety design	Ease of diffusing blame	Peer esteem
One definition of risk for all	Definition varies with organization	Definition varies with personal fears
Statistical inference	Standard operating procedures	Ethical values
Fault trees	No single decision maker	Risk is danger to some, opportunity to others

In political risk forecasting for international enterprises, Ascher and Overholt (1983) move

> beyond a tradition of studying forecasting primarily as a series of discrete mathematical methods .... An exclusive emphasis on formal methods, particularly complex quantitative methods, will often prove self-defeating .... We affirm the importance of studying forecasting in the context of the actual behavior of people and institutions rather than in a formalistic manner.

A clear distinction is drawn between the policymaker's "rational information needs" and his "political needs". The former refer to the meaning and content of the information, the degree of certainty, and the policy recommendations embedded in, or implied by, the information. The latter focus on the ability (Ascher and Overholt, 1983):

- To be a convincing advocate of preferred policies, hence to have access to appropriate information.
- Whenever possible to be correct, that is, to choose policies that produce positive results.
- When wrong, not to be disastrously so – thus, to make conservative decisions that avoid major risks even at the expense of foregoing certain opportunities.
- When wrong, to avoid adverse political repercussions for the policymaker.
- To maintain his decision-making discretion at all times.

It is evident how important a role is allotted to organizational loyalties (**O**). The authors are also convinced that long-range strategic thinking is qualitatively different from short-range tactical thinking, frequently to the extent of requiring different personalities (**P**).

(c) Energy Forecasting

A careful analysis of accuracy in academic, government, and industry forecasts of population, energy, and economic trends (Ascher, 1978) yields a clear and consistent pattern: the core assumptions underlying the forecast are the major determinant of forecast accuracy. They prove to be far more crucial than the sophistication of the model used. A back-of-the-envelope model with good core assumptions is preferable to a sophisticated computer model with obsolete core assumptions. In other words, the methodology cannot "save" the forecast when the core assumptions are poor. An example of a poor core assumption is that used in the early ultimate petroleum reserve forecasts: no significant change in the technology of recovery.

A recent study (Sapp, 1987) examined the energy demand forecasting process at Bonneville Power Administration in the Pacific Northwest region of the United States. Such analysis is a vital input used by utility companies for planning resource acquisitions and for financing decisions. The econometric models inevitably favored by the economists who constitute the forecasting group are very complex. Many core assumptions underpinning the models were accepted without question. For example, they assumed that there would be no major economic, social, and political discontinuities or structural changes over a 25 year period. Therefore, the long-term projections were inevitably biased by short-term trends. A five percent average annual rate of growth of regional electricity demand was accepted by all utilities in the region as realistic until the late 1970s. Major power supply shortages were anticipated in the mid-1980s on the basis of an assumed continuation of the regional economic boom experienced in the 1960s and early 1970s as well as a continuation in old customer behavior patterns. Ascher (1978) calls this tendency "assumption drag". Uncertainties and possible surprise events were submerged in a sea of quantitative model output and "best estimates."

The decision to build the five nuclear power plants (the Washington Public Power Supply System) was based on such forecasts. The result was a planning disaster: major changes in the forecasts due to altered assumptions could simply not be accommodated in the construction program without enormous financial losses. Analysts are not accustomed from their background to maximize adaptability to unanticipated changes, that is, to disaster-avoidance planning. Even the leadership of the Bonneville Power Administration is a salient factor. One head administrator saw the impending power shortage as a technical problem; his successor saw it as a political problem; the third saw it as a business problem. Different foci lead to different solutions. The **O** and **P** perspectives are important for this reason: who is doing the forecasting is as important as what is being forecast.

The central nature of the **O** perspective for the decision process in energy facility siting has been illuminated in detail for four specific cases of liquified natural gas (LNG) projects – in the United States, Scotland, the Federal Republic of Germany (FRG), and the Netherlands (Kunreuther and Linnerooth *et al.*, 1983). The comparison shows very different institutionalized styles of risk handling that reflect cultural distinctions. The United Kingdom process was characterized by trust in experts and informal inspections as well as deference to top-down leadership. This consultative, consensual style contrasted with the American adversary, statutory bottom-up leadership style.

In all cases two insights stand out: the siting decision process was political and it was sequential. The final outcome depended strongly on the actors' styles, their interactions, and the way the agenda was set.

The dialectic approach characteristic of the **O** perspective is also reflected in the history of energy resource forecasts in the United States (Wildavsky and Tenenbaum, 1981). The deep division between industrial interests and conservationists on oil and gas resources was already apparent in the early 1900s. In 1908 the US Geological Survey (USGS) forecast total US oil resources between 10 and 24.5 billion barrels and indicated that supplies of oil would be depleted between 1935 and 1943. Each side seized on these estimates to confirm its policy stand. Many forecasts have been made since then and, except for the World War I and II periods, each faction habitually accuses the other of manipulating the forecasts for its own purposes. *Table 4.3* suggests the different **O** views on resource forecasts. It becomes clear that the forecasts are the servants of policies already determined or preferred rather than being prerequisites for policy formulation. The **T** perspective quests for more accurate forecasts in this area are thus only of marginal relevance.

Table 4.3. O perspectives on oil reserve forecasts.

	When prices are high	When prices are low
Industrialists favor	High forecasts "Major new supplies can be found if prices are high"	Low forecasts "Higher prices are needed to bring on more supplies"
Consumers favor	Low forecasts "Oil is no longer the solution"	High forecasts "No need to raise prices"
Conservationists favor	Low forecasts "High prices encourage overproduction"	Low forecasts "Low prices encourage overconsumption"

Source: Wildavsky and Tenenbaum (1981).

(d) Trade Deficits

Technology diffusion is linked to trade balances and the problem of trade deficits is of intense concern to policymakers. Udwadia and Agmon argue that perceptions of the trade deficit and its effect on decisions that lead to national trade policies can only be understood through the incorporation of economic, political, and moralistic arguments, where the compound explanation goes beyond that which can be obtained from any one of these separate fields of study [Udwadia and Agmon, 1988].

We can readily identify the economic argument with the **T** perspective, the political one with **O**, and the moralistic one with **P**. The arguments arising from each perspective are summarized in *Table 4.4*. The perspectives interact with each other and must therefore all be taken into account in policy planning.

Table 4.4. Perspectives on trade deficits.

Economic (**T**)	Political (**O**)	Moralistic (**P**)
Rational choices	Trade surplus = national strength	Western culture precepts: – "Only enjoy what you can afford"
If we maximize current consumption (max. net pres. value, high discount rates, min. current R&D) result is trade deficit	Trade deficit = weakness Power has precedence over profit	– "Don't live beyond your means" – Practice "fair play" – "Save for a rainy day"
If we maximize current production (low discount rate, high consumption later) result is trade surplus	Use of political intervention to maintain power – industrial policy – trade barriers	Should a positive or negative discount rate be used? Am I more important than my grandchildren, or v.v.? Is it moral to burden later generations?
Simply a free choice: produce now and consume later, or v.v.		Senior's bumper sticker: "I am spending my children's inheritance"
Deficit no problem	**Deficit undesirable**	**Deficit undesirable**

Source: Udwadia and Agmon (1988).

We have sampled the menu of application areas. As the technology diffusion process inevitably involves systems that are not purely technological in nature, the multiple perspective concept is being called upon with increasing frequency in this domain. In the next section a few basic guidelines for interested users are summarized. Following that discussion, we will turn to applications involving technology diffusion (Section 4.5).

4.4 Guidelines for Users

The Multiple Perspective Concept is not simply another methodology to add to the analyst's tool kit. There is no 6-step procedure, no formula to weight perspectives. *Table 4.1* makes it clear that the **O** and **P** perspectives use inquiring modes and paradigms not natural to the **T**-trained engineer or economist – although quite familiar to managers, lawyers, politicians, bureaucrats, and even journalists. In this section we propose some guidelines to assist in applying the concept. Those who are already effectively bridging the gap between analysis and action obviously do not have need for them. It is hoped that the guidelines will help the many others who are struggling to link theory with implementation in the real world.

(a) Balancing T, O and P

Strive consciously for a balance between **T**, **O**, and **P** perspectives. Either use an individual who exhibits a good balance of **T**, **O**, and **P** (an uncommon breed) or create a team with diverse backgrounds. We do not refer here to a philatelic mix, say, an engineer, an economist, and a computer systems analyst. They are all **T**-driven and concerned with model detail and precision. They are likely to spend 90% of the available time on the **T** perspective with which they are comfortable and 10% on **O** and **P** with which they are not. A better mix is an engineer, a lawyer, and a businessman.

(b) Choice of O and P Perspectives

Do not confuse *what* you are looking at with *how* you are looking at it. There are as many **O** perspectives as there are affected or affecting organizations and interest groups. Within a company or agency each department has its unique **O** perspective. You cannot include all; the choice is necessarily judgmental. Do not be surprised if perspectives are in conflict – this is, after all, the real world. The same cautions apply to the selection of **P** perspectives. Experience will make it apparent that the hierarchy or organization chart is not always a good guide; some key individuals in the decision process do not appear on the charts. An in depth understanding of the organization will illuminate its myths, standard operating procedures, and actual decision process. The **P** perspectives pose the most difficult challenge; they lie at the deepest and least accessible level. Look particularly for individuals who are likely to act outside of an institutional role and would affect outcomes.

(c) Use of Interviews for O and P

The **T** perspective is developed using traditional data and model based analysis. We have stressed that more of the same will not yield the **O** and **P** perspectives. Rather, they depend strongly on personal immersion, on digging below the surface, on really understanding what makes the actors "tick." Interviews are of great value in gaining **O** and **P** perspectives. But they require talent: the interviewer must be a good listener and sensitive to nuances and nonverbal communications. What is *not* said may be as important as what is said. Volunteered asides may be as significant as answers to questions. The effective interviewer recognizes that structured questionnaires or Delphi are no substitutes for such exchanges.

Language and cultural differences must be understood for the interviewing process. Our recent experience in China [see Section 4.5(c)] showed that the Chinese well understood what was being probed with **O**-type questions. Clearly the Chinese culture is bureaucratic and hierarchic, so that **O** games and strategies are known to all. Power relationships are enshrined in all kinds of slogans, such as "two down, one up" and "the pyramid of power." **P** perspectives presented more of a problem. We found that it was very important to keep pushing for concrete examples and anecdotes to flesh out and interpret the often too general and spare answers.

Since interviews play such a central role in the multiple perspective approach, the quality of information generated by key interviewees is of major concern. All translation becomes interpretation, and this requires a sophisticated knowledge of the local culture. Since simple word-for-word translations are not possible and the Chinese language contains many untranslatable metaphors, similes, and allusions, very well trained and sophisticated interpreters are essential. Often an interviewee's apparently peripheral response may bring forth valuable insights not anticipated by the questioner.

For those who wish to apply this "new systems analysis" to their own sociotechnical systems, the obvious answer to the interview problems is to learn to develop the **O** and **P** perspectives themselves through interview techniques uniquely suited to their dialog and interpersonal communications.

But for cross-cultural systems work and joint ventures, there is no easy solution. Detailed guidelines for interviewing are found in Linstone (1984).

(d) Integration of Perspectives

The question is often asked: Should the perspectives be integrated into one picture before submission to the decision maker or should the set of different perspectives be presented? In answer to this question, it is useful to call on the analogy of the trial courtroom. The jury hears the testimonies of the various witnesses (perspectives) *and* summations by the prosecutor and defense attorney. The jury can accept one or the other integrated version or use the original perspectives in arriving at its decision. The executive has similar options. We recommend displaying the different perspectives and possibly our own "prototype" integration. We must keep in mind that our cross-cuing and weighting of perspectives cannot simulate that of the decision maker. There is no way we can predict this mental process; indeed he or she often cannot articulate this crucial decision process even *a posteriori*. As President Kennedy wrote:

> The essence of ultimate decision remains impenetrable to the observer – often, indeed, to the decider himself .... There will always be the dark and tangled stretches in the decision-making process – mysterious even to those who may be most intimately involved [Kennedy, 1963].

However, the presentation of the several perspectives encourages cross-cuing among them. For example, a manager's vision of the company's future (his **P** perspective) may become the organization's **O** perspective if he has the flair to engage others in sharing that vision. Such interplay also leads to consideration of important facets that are not captured by any one perspective. The willingness of a corporation or government to balance projects having only long-term payoff with those providing near-term payoff requires a conjunction of quite distinct perspectives.

(e) Communication

The technical report or briefing is ideal for communicating the **T** perspective. The **O** perspective often involves a private insiders' language in combination with a hortatory one for the public. However, as any successful artist, dramatist, and media producer knows, the personal level of the **P** perspective is the most effective of all. Even in the industrial world,

> we are more influenced by stories (vignettes that are whole and make sense in themselves) than by data (which are, by definition, utterly abstract) [Peters and Waterman, 1982].

It is hardly surprising to find that the **T**-type analyst is not the most persuasive scenario writer. Recognition that, to a degree, the medium is indeed the message, is the first step to the skillful communication of perspectives.

(f) Implementation

The inherent process orientation of the **O** perspective virtually assures avoidance of a trap commonly encountered with the **T** perspective: walking away from problems of implementation, problems that focus on the role of human beings, both collectively and individually.

4.5 Cross-Cultural Applications: Asia

In the process of technology diffusion, be it domestic or transnational, vertical or horizontal, we must obviously deal with multiple perspectives: technical, organizational/societal, and individual. The preceding discussion points to the desirability of applying our concept to develop insights that help to bridge the wide gaps inevitably created by the different perspectives of the diverse actors in the process (societies, bureaucracies, and individuals with unique cultural, educational, and economic backgrounds).

As noted earlier, Kunreuther and Linnerooth *et al.* (1983, pp. 235–240) have examined the liquified natural gas facility siting process in four Western countries and found significant culturally-induced differences that can be determined using distinct perspectives. Ascher and Overholt (1983) have found the organizational perspective to be "the central metaphor" in strategic forecasting for Mexico and South Korea. The institutional infrastructure is crucial for any culture, and the differences between cultures are strongly reflected in the system change process.

We shall now consider application of the multiple perspective concept in the framework of three settings: Nepal, India, and China. The first deals with technology indirectly (environmental problems), the second with a consequence of technological transfer, the third with technological modernization.

(a) Nepal

Thompson and Warburton (1985) were commissioned by the United Nations Environmental Programme to construct a systems framework for the apparently severe environmental problems in that developing region. They found it exceedingly difficult to gather factually precise information, but

> the information that is gathered does quite accurately reflect the various social forces that are at work in the Himalayas. In this sense, it is *institutionally* accurate and, as we shall argue later, perhaps institutional accuracy is more valuable (and more accessible) than factual accuracy ... it is the institutional forces that muddle [the scientist's] attempts to analyze and solve what, at first glance, appear to be technical problems. In many ways, it seems to us, the institutions *are* the facts ... [Thompson and Warburton, 1985, pp. 6, 11].

In the case of the Himalayas (Nepal) there is remarkable uncertainty in the physical facts (values of variables such as per capita fuel wood consumption) as well as in the causation (interactions among variables, i.e., the system structure). On the other hand, there is reasonable certainty about the institutions. Each has its own perception of the world, its own **O** perspective. They exist and function; indeed they are what we have to work with, like it or not.

> The process of institutional development is inherently unplannable. But it does, at certain places and at certain times, offer points of leverage – localized opportunities for facilitating and integrating the development of institutions in the desired direction [Thompson and Warburton, 1985, p. 13].

The authors remind us that the classic development approach has been to sound the alarm and then tell the country what it will have to do.

> It has not worked. It has not because it ignored (as if it were a mere detail of implementation) the deep political, economic and cultural structure that is, in fact, what determines the country's attention and lack of attention. What is needed is a more sensitive approach; an approach that places the "mere details" – the institutions that constitute the deep structure – at the very center of the stage and relegates to the wings the alarm bellringers and their immaculate prescriptions .... Grand designs are appropriate only when there is a shared understanding of "the problem" and complete knowledge of the causes of the problem [Thompson and Warburton, 1985, pp. 10, 17].

In 1950 the Nepalese government increased its centralization of power. In the past thirty years there has been an enormous increase in the total amount of power in the system. Now the government is devolving more control back to the local levels while retaining much central power.

The situation facing a proposed project may be described as in *Table 4.5*. Many projects fall into category 3, few into category 4.

Table 4.5. Planning alternatives.

Institutional support: match or mismatch	Probable result
Neither top nor bottom level support	Dead loss
Bottom level support only (grass roots enthusiasm but no funds)	Strangled at birth
Top level support only (top-down planning)	Difficulties in implementation
Top and bottom level support (in China labelled "top-down, bottom-up")	No problem

As systems analysts the authors consciously part company with their **T**-focused professional peers by recognizing the centrality of the **O** perspective for regional planning in developing countries.

> Though what we have done is applied systems analysis, it may not look like it. There is, we concede, a fair-sized break between the traditional single problem/single solution approach and the one we have developed here. There are many ways to characterize this break but perhaps the best is in terms of the shift it makes from product thinking to process thinking. The systems frame is no longer a model of the problem but simply an evaluative mechanism. When the problem is to know what the problem is, we need more than one perspective. The approach by way of plural institutions and divergent perceptions meets this need. It gives us problems and solutions that are multiple but not infinite; certainties that are contradictory but not chaotic [Thompson and Warburton, 1985, p. 33].

Thompson and Warburton conclude that:

- Anything that increases the security and local control of the peasant farmer will also increase the total quantity of power within the wider system.
- Top-down development is in the nature of a project (an intervention); "bottom-up" development is in the nature of a process; the project comes down from the top, but must be modified from the bottom.
- The meshing of top-down and bottom-up requires constructive intervention at the "right" points of leverage; it may be described as tinkering in contrast to a grand design approach; finding the "right" points is not

easy, indeed, it is an art for they may be few and far apart, invisible to the unsensitized viewer.

(b) India

A recent multiple perspective examination of the Bhopal disaster is also illuminating. On December 2–3, 1984, a catastrophic leak of methyl isocyanate (MIC) occurred at the Union Carbide plant in Bhopal. Each perspective applied draws forth elements which contribute to an understanding and development of recommendations (Bowonder, 1987; Bowonder and Linstone, 1987). Some examples:

T: The technical perspective shows that causal factors include:

(1) Hard errors such as bad structural design (vent gas scrubber, water sprays), defective pressure gauge, and poor instrumentation.
(2) Human errors such as failure to recognize entry of water into the MIC tank during line and valve cleaning, and failure to communicate a major pressure rise in the MIC tank.
(3) System errors such as lack of total system audits and lack of follow-up to these audits.

As in the case of Three Mile Island and Chernobyl [Section 4.3(b)], we are dealing here with a system involving (a) man + machine, as well as (b) an event characterized by the combination of very low probability + severe consequence. Such combinations are termed by Perrow (1984) "normal accidents" and the engineers' standard **T**-type analysis, e.g., fault trees and expected value computations, are of very dubious validity.

O$_1$: The corporate perspective informs us about:

(1) Proprietary aspects, such as the inadequate dissemination of information on the toxicity and clinical treatment of exposure to MIC.
(2) "Stonewalling" as the standard initial corporate reaction to a disaster.
(3) The need for training in "crisis management" capability.
(4) Blind technology transfer.

> (Technology transfer) takes place from one societal/cultural setting to another. In the case of the United States and India, there are vast differences between these settings and they affect the success of the transfer .... Therefore application to an Indian facility of safety rules and procedures developed for American use is naive. The **T**-focused technical audits of the

> Union Carbide staff failed to consider vital aspects of Indian culture, such as attitudes toward preventive maintenance and precise adherence to rules of operation ... [Bowonder and Linstone, 1987].

$\mathbf{O}_2$: The governmental perspective shows:

(1) Poor enforcement of worker safety and environmental rules – 15 factory inspectors to monitor more than 8,000 plants, two mechanical engineers with little knowledge of chemical hazards assigned as inspectors in Bhopal.
(2) Ignoring of the Bhopal development plan requiring plants manufacturing pesticides to be relocated to an industrial zone 15 miles away – the existing plant received an MIC license just two months after the issuance of the development plan.

P: Personal perspectives indicate

(1) Filtering out of input conflicting with ingrained views, a plant manager stated: "we do not know of any fatalities either in our plant or in other carbide plants due to MIC."
(2) The importance of an effectively trained leader – a neighboring plant suffered no losses because the manager, a former brigadier in the Indian army, efficiently evacuated 1,400 workers upon detection of the leak.

Integration and cross-cuing of the perspectives exhibit their interactiveness. For example, the safety audit (**T**) and the financial priorities of the company (**O**) affect the correction of problems. The combination of cultural differences (**O**) and standard operating procedures (**O**) foil adequate training. Most important, they lead to a basic reconsideration of the action levels in the treatment of risk:

Level 1. Seek a means to reduce the likelihood of catastrophic consequences with the existing system. Examples:

T: Try to make the system *safe-fail* by decoupling of subsystems so that an accident can be bounded or limited to one subsystem.
O: Revise instruction manuals to allow for cultural differences, resulting in equivalence in practice rather than merely literal equivalence in language.
P: Give investigative reporters and "whistleblowers" more protection in exposing poor practices, thus anticipating potential catastrophes.

Level 2. Redesign the system to reduce dangers. Give added weight to:

T: Decoupling of subsystems in the design stage rather than after installation.

O/P: Partial customization of the system to the local culture for increased safety.

Level 3. Probe conceptually different system solutions that avoid catastrophic consequences altogether.

T: Alter the production process or use materials in such a way that the same need is met in a technologically new way which excludes the possibility of catastrophic accidents.

Level 4. Ask the final question: can the consequence of a catastrophic accident be tolerated or not?

O: This is clearly a societal, more than a technical, question. With rapidly moving technology, the possible kinds of low likelihood/severe consequence accidents will grow. Example: Today AIDS makes the possibility of a future bioengineering error creating a deadly virus which sweeps over the world with lightning speed seem far less remote than it did twenty years ago. Also, the demands for management capability are becoming more severe. History has shown that the human and ecological resilience to disaster is enormous – but there are limits!

(c) China

Cultural Background

Let us begin by sampling some characteristics indicative of cross-cultural distinctions that have a major impact on technology transfer.

- Primacy of the group

  In China the *danwei*, one's group or unit, is most important for each person, counting more than one's name; without a *danwei* a person's existence is barely recognized. This concept captures a vital difference between Chinese and Western cultures: the relative importance of the group and the individual. The traditional view implies that individual lives virtually do not exist separately from the life of the unit and privacy is not recognized in the same way as in the West (Bonavia, 1980; Butterfield, 1982). Individuals do not like to express personal preferences, but prize conformity. A set of guidelines for doing business in the PRC advises:

> The foreign businessman should not focus on the individual Chinese person, but rather on the group of individuals who are working for a particular goal. If a Chinese individual is singled out as possessing unique qualities, this could very well embarrass the person .... In discussions with Chinese people, the foreigner should avoid "self-centered" conversation in which the "I" is excessively used. The Chinese view with contempt the individual who strives to display personal attributes, as Chinese are much more oriented to the group [Harris and Moran, 1987, p. 406].

The value of the primacy of the group lies in the ability to undertake all kinds of collective activities and projects. In a country of one billion, the resulting harmony and cohesion are of tremendous significance.

- Language

For thousands of years the Chinese language, composed of more than 8,000 characters, has evolved to express the most complex ideas and has united the people. But,

> ... despite its beauty and subtlety, it has increasingly become an obstacle to modernization. The language does not absorb new ideas readily and new characters are rarely invented. As a geneticist at the Chinese Academy of Sciences described the situation, "We are being held prisoners by our language. There is an information explosion in the world that we cannot cope with in our language as it is now" [Los Angeles Times, 1982].

- Analysis

In the past six years the "opening" of China has led to a growing number of professional contacts between Chinese and Americans. It is interesting to note the comments of several American analysts:

On decision analysis:

> Our impression is that the three fundamental activities underlying decision analysis (generating alternatives, accounting for uncertainties, and eliciting decision makers' preferences), at least as espoused in Western countries, are not being accepted nor are they likely to soon be accepted by the Chinese bureaucracy .... A distinguishing feature of decision analysis is the explicit consideration of uncertainty. What we found in China, however, was a decision-making environment that was almost completely devoid of a formal concern for uncertainty. Within a planned society, deviation from the established goals is anathema. This seemed to make it particularly uncomfortable for our Chinese colleagues to accept uncertainty. Our impression was that any analysis would proceed under

> the assumption of complete certainty about almost all important aspects: costs, technological availability, construction timing, weather, effects of treatment on the measures of pollution, social benefits, governmental policies, and so forth .... In a formal, ritualistic culture that values the expected (certainty), assigning probabilities to events must seem strange indeed .... Our experience in China was therefore bound to be frustrating .... Perhaps it was presumptuous of us to expect the Chinese – with a bureaucracy and decision-making culture highly evolved over a period of more than 3,000 years – to accept with more than polite acknowledgment a way of looking at problems that is so obviously Western: reductionist, empirical, quantitative, democratic in its approach [Pollock and Chen, 1986, pp. 34–37].

On operations research:

> MS/OR [management science/operations research] problem solving may suffer from the traditional Chinese tendency to categorical formalism reflected by the fondness for such constructs as "the Four Modernizations," "the Ten Major Relationships," and "the Three Fundamental Principles." Indeed the one weakness of my otherwise exceptional students was their tendency to force problems into a "type" like those types studied in class. They displayed great ingenuity in solving formal problems, but when faced with less formal problem descriptions, it rarely occurred to them to think "outside" of the problem, to change it, or to wonder whether there was a problem at all. Even more tellingly, they seemed to lack skepticism about the tools, methods, and applicability of MS/OR [Bartholdi, 1986, pp. 29–30].

Wei Bei Agricultural Development

A recent joint China-US project funded by the National Science Foundation (Linstone *et al.*, 1987) considers the issue of the agricultural development of the semi-arid Wei Bei region in Shanxi Province. We sketch here a summary of the clues provided by the use of multiple perspectives.

Figure 4.4 shows the nature of cross-cuing among some of the major elements or factors contributing to the improvement of the Wei Bei economy. There is a strong likelihood that "everything interacts with everything," although the time to do so may vary greatly from one impact to another.

Wei Bei regional development has three primary goals:

(1) Maintenance of the grain crop at a satisfactory level.
(2) Modernization of agriculture, including high value cash crops, animal husbandry, and use of uncultivated lands for forestry and grazing.

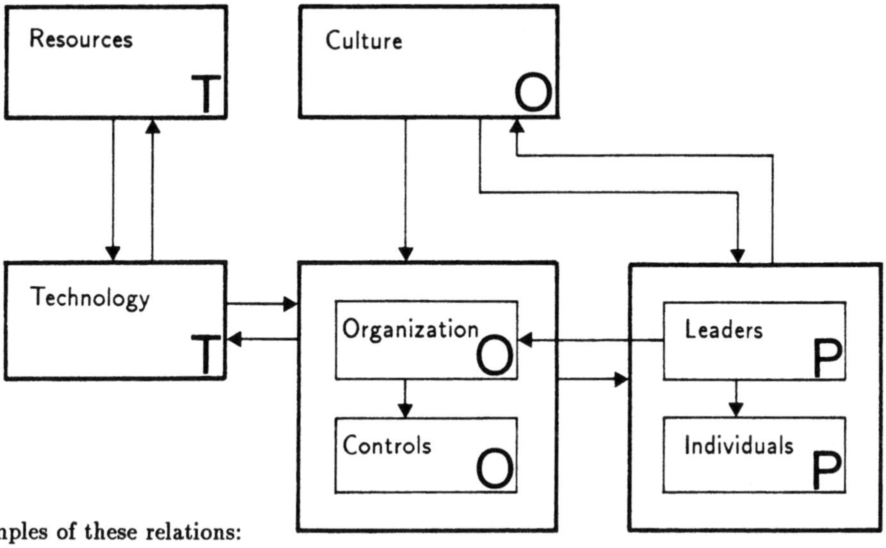

Examples of these relations:

Resources → Technology
inadequate farm land → need for fertilizers, water management
minimal financial resources → inadequate quality of research and training

Technology → Resources
pressure resulting from resource shortages increases research support → new resources (materials, energy systems) (Goeller and Weinberg, 1976)

Culture → Leaders
centrality of *danwei* → use of power pyramids

Leaders → Culture
Mao Tsetung → irrevocable change in culture effected

Culture → Organization
conformism tradition → strong hierarchy and centralization

Organization → Control
centralization → strict controls on hiring and movement

Culture → Individuals
importance of blood relations → family connections used

Organization → Individuals
sudden policy shifts → peasants' desire to keep individual land plots for security

Organization → Technology
horizontal isolation → poor diffusion of research among institutions

Technology → Organization and Control
information technology → unprecedented centralization *and* decentralization possible

Leaders → Organization
little modern management experience → redundancy at all levels of hierarchy

Figure 4.4. The nature of cross-cuing.

(3) Diversification into profitable non-agricultural enterprises at the county, prefecture, and township levels.

Goal (1) corresponds to what the corporate planning literature terms a "disaster-avoidance" strategy (as contrasted to an optimization strategy). It is a desirable direction when operating in an uncertain environment, i.e., a *safe-fail* strategy (see Section 4.1).

We shall concentrate our attention on goals (2) and (3). We recall the point made by Thompson and Warburton: "the problem is to know what the problem is." This is certainly apt when Wei Bei agricultural development is seen as a "three-dimensional" (**T**+**O**+**P**) task rather than a "one-dimensional" **T** problem. It becomes apparent at once that there is not a shared understanding of the problem. The actors tend to blame each other – agency people insist government is good, but the farmers do not listen; the farmers feel the government is filled with incompetents and scientists propose impractical schemes; the technicians see the problem as the closed minds of the people at the rural level. This is the "responsibility merry-go-round." We now comment on several of the factors brought to the surface by the multiple perspective approach.

- Reorganization for modernization (**O**)

  A very important and basic insight provided by the perspectives involves the nature of the reorganization itself. A policy change at the top was necessary, but is hardly sufficient to make it happen. Modernization involves non-traditional kinds of crops and activities, such as fruit orchards, dairy products, forestry, and tobacco processing. The profitable local enterprises or non-agricultural sideline businesses also demand non-traditional functions of the Wei Bei labor force. It is one thing to shift responsibility to lower levels of government as part of the decentralization policy. As long as the activities to be conducted are familiar ones, it is reasonable to expect such a shift to proceed smoothly. An example of this is the introduction of the responsibility system. It permitted individual peasants and groups of households considerable freedom of decision in agricultural production. But it is another thing to expect the shift to proceed smoothly when the activities involve non-traditional tasks that are unfamiliar to either local government or peasants, management or labor.

  It is easy to be lulled by the clear evidence of success of the policy in raising food production in the past eight years. Indeed, the shift has had uneven success. Many villages, especially those near towns and cities, have prospered remarkably under the decentralization policy. They planted familiar food crops, sold to markets within easy reach, and developed sideline activities (e.g., brick yards, cement plants) that they well understood. Even so, there were wasteful surges of overproduction (e.g., watermelons, rabbits). At the same time the vast hinterlands saw many communities remain impoverished.

With non-traditional activities the needed skill levels in production, marketing, and management as well as support services (financial, transport, equipment servicing, technology, and administration) are simply not available at the local level. The tasks cannot be organized into a working system, let alone, an efficient one. The development of non-traditional activities at the local level is quite analogous to the more general challenge of technology transfer from advanced countries to China. This has been the subject of a recent statistical study by the Science and Technology Policy Research Division in Beijing (Technology Import Project Group, 1986). Sampling 220 projects, the study concluded that the leading factors affecting the "digestion and absorption" of a technology in China involved not only the obvious ones of available funds and technical hardware, but weakness of the management system, lack of supporting policies, and lack of qualified technical personnel.

Our own **O** and **P** insights indicate:

- There are clashes of interest among stakeholders – between levels of government, between technical agencies, between government and technical agencies, between government or institute and peasant producer. This clouds the fulfillment of the projected goals. There is confusion as to which level and agency of government is to provide a resource such as capital, managerial oversight, entrepreneurial expertise, and technical advice for a given project.
- Experimental or demonstration plots run by agricultural institutes, "sparking" funding for small innovative projects, or "lead" families are not the answer in tackling non-traditional, unfamiliar agricultural or sideline enterprises. There is a need for all kinds of support services not readily accessible to the peasants.
- The need for information is particularly strong when dealing with non-traditional, unfamiliar production activities. Much essential information is unavailable or inaccessible due to compartmentalization, a lack of coordination, and poor communications. The result

is duplication of effort, waste of limited skilled manpower, inability to make decisions, and failure to build on on each other's findings.

In Wei Bei agricultural modernization and non-traditional sideline businesses in towns and villages require outside support on a sustained basis until these activities have proven their worth to the local peasants and local expertise is at hand.

It was beyond the scope of this study to make specific recommendations to overcome these difficulties. But there are some clues on the direction to be taken. We point to a reorganization analogy in the United States. Hierarchical business organizations were highly effective from 1850 to 1950. After World War II, complexity grew and hierarchical organizations were increasingly transformed into matrix organizations. This means that the project tasks determined the team composition, in contrast to the traditional procedure of assigning a project to an existing functional department. The project organization exists only for the duration of the project; it does not have the permanence of the departments. The most suitable expertise is selected from various departments and placed under the direction of the project leader.

A modernization project in Wei Bei should be viewed as a series of tasks that must be accomplished. There are managerial/entrepreneurial, administrative, and production tasks. There are questions of jurisdiction and authority, and each task creates resource requirements: information, skills, material, and services. The tasks and work organization vary from project to project. The following questions must be addressed and answered in pragmatic, not ideological, terms:

- Which entity (government agency, town, or village) has overall responsibility for the project?
- For administrative tasks, does the entity have appropriate jurisdiction? Does it have the skills and information?
- With regard to managerial/entrepreneurial tasks, does the manager/entrepreneur have the authority to act? Bounds of authority should be clear and appropriate. Conflicts between authority and jurisdiction can also be avoided by negotiation. Does the manager have the skills and information? The managerial function should not be assigned to government personnel whose training is political rather than managerial.
- For production tasks, does the entity have the information, skills, material, and service resources? If not, where are the resources

located that can fill the need for the project? They may be scattered throughout the province, in agencies and institutes, or they may only exist outside the province. There is no need to place the necessary resources, for example, services, at each town or village that has a project requiring them. If, say, a prefectural-level service has the resource, it can assign one of its persons to be a project team member to supply the expertise or coordinate its availability to the project.

The interviews and workshop elicited many instances of mismatch between needed and accessible resources. Examples: dairy projects are hampered by the lack of county-supplied veterinarians; farmers have problems obtaining county pest-control services and market information. Another complaint was that the county or prefecture does not have the expertise to market its products in other provinces. A matrix organization provides the flexibility to maximize effectiveness in the use of resources which is essential in moving beyond the current level of agricultural development. We stress that a matrix organization does not imply the abolition of existing functional departments. (In the United States the introduction of matrix organizations in corporations did not mean the end of their functional departments.) In the case of Wei Bei, we are suggesting a concept somewhat similar to the matrix during the transition phase. As capability develops at the lower levels, there will come a time when the services will be either internalized in the production enterprise or they will become enterprises themselves and contract their support directly to the production enterprise, a common feature in advanced service economies.

It is interesting to recall that the problem of reorganization in the face of technological modernization is one that confronted Chinese and Japanese society already in the nineteenth century when Western technology was introduced (Fried and Molnar, 1978, p. 92). The primary managerial responsibility was officially assigned to the provincial civil servant category in China (particularly the Mandarins). In Japan it was given by the Meiji reformers to the *daimyo*, or feudal lords (who later merged with merchant groups to form the *zaibatsu* oligarchy). It failed in China largely because the civil servants were not experienced managers or entrepreneurs; it succeeded in Japan because the *daimyo* were.

- Managers and entrepreneurs (**P**)

  China is the richest country in the world with respect to one resource: human brains. So there must be more potential leadership material in China than anywhere else. Switzerland and Israel have a tradition of drawing on the military for leaders in the civilian sector. In these countries the military establishment with its high standards constitutes a superb training ground for leadership. Senior officers retire early and often become top managers in nonmilitary enterprises. Does China with its immensely greater manpower resources have any such reservoir of leadership capability?

  Many Chinese professors and bright students now go to Western universities for advanced management training. This is an important long-term strategy. But it is not nearly enough. For one thing, such academic upgrading misses the practical, hands-on learning that business and industry experience provide.

  Let us pause to consider the American experience. After one of the best known academic authorities on organization theory and management, Richard M. Cyert, co-author of the classic text *A Behavioral Theory of the Firm* (Cyert and March, 1963), became a university president, he wrote:

  > As a professor of organization theory and management, I used to wonder about the practical value of these academic fields. For the last eight years I've had some first-hand experience finding out .... And I've concluded that the study of management makes a useful, but only limited, contribution to the practicing manager [*Wall Street Journal*, 1980].

  Similarly, the foreign educational institutions are not likely to be an adequate provider of managerial/entrepreneurial talent for China. An effort should be made to place more Chinese as apprentices or management trainees in Western enterprises. Alternatively, it may be very fruitful to invite retired Western managers and entrepreneurs to spend time, preferably the duration of a project, with the Chinese enterprises in order to give them the benefit of their experience. Finally, there is the possibility of having successful foreign agricultural companies set up local enterprises. Here local managerial capability can be developed more effectively (see below).

- Technical training, outside expertise, and the brain drain (**T/O/P**)

  The need for major improvements in technical education and training is self-evident. Two impeding factors are evident: the short-term focus of the farmer and the poor horizontal communications. We now pause to consider a fundamental problem, one that is implied by the entire *raison d'être* of the multiple perspective concept. In moving beyond familiar crops and simple sideline activities, we also obviously move into more complex technology transfer at the local level. An implication of all that has been said about the need for multiple perspectives is that training in universities and institutes, be it in China or Western countries, is not going to encompass an adequate basis of preparation for undertaking such activities in Wei Bei (or anywhere else). The total is more than the sum of its parts. A prestigious Harvard MBA (Master of Business Administration) degree does not constitute an adequate preparation for operating a business enterprise.

The same limitation applies in Wei Bei. The availability of academically trained technical and management personnel does not insure a successful project. Actual system operation provides an essential "total-system," multiple perspective learning experience. What can be done? Turn-key operations as a mode of technology transfer are a familiar concept. A foreign company may set up a plant in China that duplicates its own enterprise. Their disadvantage is that they often do not provide a proper learning environment for the local people and may not be suitable in terms of **O/P** in the local cultural setting. In dealing with rural areas there are other interesting possibilities to effect development of a balanced local capability.

China has drawn in foreign private sector expertise in its industrial development; it is therefore not beyond reason to consider drawing in integrated (**T+O+P**) foreign private sector expertise in agricultural development. There have been some cases involving private sector companies from developed countries effectively performing agricultural technology transfer in Third World areas where agriculture is pursued on small family farms (one to five hectares). In this approach the companies, processors and marketers of foodstuffs and cash crops, develop an integrated operation with the strong involvement of local farmers. They reach agreement with these small producers guaranteeing a fair market price, providing credit, technology, inputs such as fertilizers, herbicides, and seeds, assistance in soil preparation, harvesting, and storage, and servicing in moving the product to processing and market. Successful operations cited by Freeman and Karen (1982): sugarcane and tobacco enterprises in the Dominican Republic, a banana enterprise in the Philippines.

- Information technology and technical support (**T**)

The technical information available is severely limited and often inaccurate or unreliable. Even if it is correct and meaningful, it does not circulate easily. Information retrieval systems leave much to be desired. A particularly significant problem impeding development of technical know-how and provision of services in rural areas is the poor horizontal communication capability. It is not clear to us whether information technology can overcome the cultural/societal patterns in Wei Bei, but the potential for movement of information with today's state of the art is enormous. In particular, such concepts as (a) the use of satellite communication links and videotape for rural education as well as training, and (b) desk top publishing for rapid, low-cost spread of information need to be explored. But China is faced with at least four serious handicaps:

- There is a lack of funds for information systems and telecommunications infrastructure development.
- There is an acute electricity shortage. Per capita consumption of electric energy remains at a very low level. (In 1984 China ranked 112th in the world in per capita electricity generation.)
- There is the Chinese language itself. It vastly complicates the use of today's state of the art information technology.
- There is culturally a disinclination to share information. (This may be tied to the dominance of the **O** perspective in the culture: it is a common characteristic of organizational thinking.) Consequently, computers and available links are underutilized.

One result is the incredible redundancy in research. Often there are impermeable information barriers. According to the Chinese *Liberation Daily* (1984):

> research units were carrying out duplicate research in 27 out of 53 projects begun between 1973 and 1974 and in 28 out of 63 projects begun between 1978 and 1979.

It is important to remind ourselves that information technology greatly facilitates both centralization and decentralization. Indeed, it will have a more pervasive impact on the entire society than any other technology between now and 2000. The first challenge is to avoid a widening of the gap between China and the developed information-age world before 2000; the second challenge is to close the gap after 2000.

More generally, we have noted that in China **O** tends to dominate **T** and **P**. Thus, more emphasis on **T** and **P** is obviously indicated. But the partial decentralization policy in progress today also necessitates a major reorientation of **O** on the part of the Chinese. Traditional concepts such as the hierarchy and power pyramid, the cadre/peasant and city/countryside divisions need to be reassessed. The new policy implies higher organizational sophistication and complexity. So there is a challenge to all three perspective types.

(d) Addendum

It may be of interest that perhaps the most striking example of the relevance of the **P** perspective in technology diffusion is represented by Dr. Hyung-Sup Choi of the Republic of Korea. This dynamic US-trained engineer played a central role in the creation of the Korean Institute of Science and Technology and served as Minister of Science and Technology for seven years. His policy of concentrating on a few strategic industry sectors, assimilating and adapting technologies, and proceeding to higher sophistication levels has had much to do with Korea's remarkable success. Close linkages between research institutes and industry are emphasized. The theme is creation, not merely imitation (Choi, 1986a and 1986b). The point to be made here is that Choi himself is a critical factor in understanding the effectiveness of the diffusion process. As in the case of Theodore Judah and the US transcontinental railroad and Wernher von Braun in rocketry, one person can make a major difference in technology diffusion.

Indeed, a culturally-based distinction between Korea and Japan is evidenced by Sharif's finding that the relative strengths of **O** and **P** differ: **P** is stronger in Korea, **O** in Japan (Sharif, 1989). Both foster a strong **T** perspective, with the respective **P+T** and **O+T** combinations creating two distinct paths to success.

4.6 Information Technology: A Unique Diffusion Factor

Information technology is the most dynamic, pervasive, and influential technology of our times. It may be said that we are moving from an industrial society to an information society. In the terms of Perez (1983), information technology is the overarching style of the present Kondratieff cycle as oil, steel, and railroads were the dominant technologies of the three preceding cycles, respectively. As such, this technology will alter our labor, business, governmental, and other socio-institutional structures. *The degree and speed of diffusion of this technology will affect the diffusion of all other technologies.* Countries, government agencies, and corporations that adapt to it will be able to facilitate the technology diffusion process, while those that resist will be relatively impeded. Illustrative of the problem/opportunity are:

- The support or resistance to the use of personal computers by individuals in the Soviet Union.
- The readiness or hesitation to share information in China – horizontal communications are traditionally much poorer than vertical communications [see Section 4.5(c)].

The explosion of information technology accentuates the growing mismatch between the rates of technological and socio-institutional change. Even the most advanced countries are confronting 21st century technology with 19th century institutions. There is growing concern that this widening gap may lead to a slowdown of the diffusion of information technology and consequently also of other technologies.

This technology can move a country or a firm toward power concentration and uniformization, or toward power distribution and diversity (Linstone, 1989; Perez, 1983). Experience underscores the dangers of the former and the advantages of the latter for technology diffusion. By the beginning of the modern era (around 1500), Ming China and the Ottoman Empire had become centers of power with scientific and technological strength, while Europe was decentralized and technologically backward. However, the Eastern empires

> suffered from the consequences of having a centralized authority which insisted upon a uniformity of belief and practice, not only in official state religion but also in such areas as commercial activities and weapons development [Kennedy, 1989].

It was decentralized Europe that benefited from a stimulating competitive, entrepreneurial environment and overtook the existing empires. A startling example of technology diffusion is the Gutenberg printing press: only 23 years passed from the time of the single Mainz press of 1457 to the establishment of presses in 110 towns. Pocket books, "how to" books, and translations of the Greek classics soon became available at low cost. Democratized knowledge and mass education played a decisive role in the subsequent diffusion of other technologies.

4.7 Implications for Policy Planning

As we noted at the outset, the multiple perspective approach is by no means a new concept to the successful decision maker. Effective leaders and managers use it intuitively. But it is not natural to most technologists or government officials. Scientists and engineers are at home with the **T** perspective and ignore or downplay **O** and **P**. This is the reason they overemphasize analysis and underemphasize implementation, overemphasize product and underemphasize process. And process is the essence of technology diffusion. They are even prone to misinterpret the multiple perspective approach itself as another **T**-type methodology for the analyst's tool kit. One objective of our study has therefore been to rectify the analysts' and technicians' misconceptions and to show that, beyond an awareness of multiple perspectives, conscious use of them yields very practical benefits. Many bureaucrats have the reverse problem: they focus on **O** and avoid **T**.

In any cross-cultural situation we face another problem. The **O** and **P** perspectives bring to the surface subtle cultural distinctions. These affect technology diffusion in important ways. They are certainly masked in a one-dimensional **T**-type analysis and may even elude a Westerner's effort at **O** or **P** (Westerners produce Westernized perspectives). Wherever possible, those fully familiar with the local culture should be involved in developing the **O** and **P** perspectives. For example, our experience in using multiple perspectives in the case of agricultural modernization in the Wei Bei region of China [Section 4.5(c)] should have relevance to planning for Malaysia's priority industrial sectors, i.e., food, rubber, palm oil, timber, metal, and plastics. But Malaysia's unique mixture of races – Malay (Bhumiputra), Chinese, and Indian – has an important influence on its culture. And the culture (values, attitudes, etc.) plays a very significant role in technology diffusion (Fallows, 1988).

Our work corroborates the system view of Thompson and Warburton (1985): (a) we should not expect a clear problem definition, and (b) the institutions rather than the data may constitute the "facts." A deep understanding of the institutional actors is a prerequisite for dealing with implementation, specifically, with determining points of leverage. This means that "systems analysis" must move beyond its traditional bounds if it is to be useful in decisions on complex, ill-structured sociotechnical systems. Its core is no longer a model of the problem but an evaluation process.

We must cope with wide gaps – between theoretician and practitioner, between technology donor and recipient, between national technology planner

and decision maker. Our first aim in introducing multiple perspectives is to raise the awareness of the need for a better balance among* T, O, *and* P. *The second is to promote a reduction of these gaps through the effective use of this approach in technology diffusion policy planning.

References

Allison, G., 1971, *Essence of Decision: Explaining the Cuban Missile Crisis*, Little Brown and Co., Boston, MA, USA.

Ascher, W., 1978, *Forecasting: An Appraisal for Policy-Makers and Planners*, The Johns Hopkins University Press, Baltimore, MD, USA.

Ascher, W. and Overholt, W., 1983, *Strategic Planning and Forecasting: Political Risk and Economic Opportunity*, John Wiley & Sons Inc., New York, NY, USA.

Bartholdi III, J.J., 1986, Operations research in China, *Interfaces* **16**(2):29–30.

Bonavia, D., 1980, *The Chinese*, Lippincott & Crowell Publishers, New York, NY, USA.

Bowonder, B., 1987, The Bhopal accident, *Technological Forecasting & Social Change* **32**(2):169–182.

Bowonder, B. and Linstone, H.A., 1987, Notes on the Bhopal accident: Risk analysis and multiple perspectives, *Technological Forecasting & Social Change* **32**(2):183–202.

Business Week, 1984, *The New Breed of Strategic Planner*, pp. 63–64, September 17.

Butterfield, F., 1982, *China: Alive in the Bitter Sea*, Times Books, New York, NY, USA.

Choi, H.S., 1986a, *Technology Development in Developing Countries*, Asian Productivity Organization, Tokyo, Japan.

Choi, H.S., 1986b, Science and technology policies for industrialization of developing countries, *Technological Forecasting & Social Change* **29**:225–240.

Churchman, C.W., 1971, *The Design of Inquiring Systems*, Basic Books, New York, NY, USA.

Churchman, C.W., 1977, A philosophy for planning, in H.A. Linstone and W.H.C. Simmonds, eds., *Futures Research: New Directions*, Addison-Wesley, Reading, MA, USA.

Cyert, R. and March, J., 1963, *A Behavioral Theory of the Firm*, Prentice-Hall, Englewood Cliffs, NJ, USA.

Fallows, J., 1988, Comments in *The Atlantic*, pp. 10, 11, 22–26.

Florman, S.C., 1976, *The Existential Pleasures of Engineering*, St. Martins Press, New York, NY, USA.

Forrester, J.W., 1961, *Industrial Dynamics*, MIT Press, Cambridge, MA, USA.

Freeman, O.L. and Karen, R., 1982, The farmer and the money economy: The role of the private sector in the agricultural development of LDC's, *Technological Forecasting & Social Change* **22**:183–200.

Fried, J. and Molnar, P., 1978, *Technological and Social Change: A Transdisciplinary Model*, Petrocelli Books, New York, NY, USA.

Goeller, H.E. and Weinberg, A.M., 1976, The age of substitutability or what do we do when the mercury runs out, *Science* **191**:683–689.

Halal, W., 1980, *Strategic Planning in Major US Corporations*, Study prepared for General Motors Corporation at George Washington University, Washington, DC.

Halal, W., 1984, Strategic management, the state of the art and beyond, *Technological Forecasting & Social Change* **25**(3):239–261.

Harris, P.R. and Moran, R.T., 1987, *Managing Cultural Differences*, Gulf Publishing Company, Houston, TX, USA.

Jay, A., 1968, *Management and Machiavelli*, Holt, Rinehart, and Winston, New York, NY, USA.

Kennedy, J.F., 1963, Preface to *Decision in the White House* by T. Sorensen, Columbia University Press, New York, NY, USA (quoted in G. Allison *op. cit.*, p. vi).

Kennedy, P., 1989, *The Rise and Fall of the Great Powers*, Vintage Books, New York, NY, USA.

Kunreuther, H., Linnerooth, J., *et al.*, 1983, *Risk Analysis and Decision Processes: The Siting of Liquified Energy Gas Facilities in Four Countries*, Springer-Verlag, Berlin, Heidelberg, New York.

Liberation Daily, 1984, quoted in W.A. Fisher, Scientific and technical planning in the People's Republic of China, *Technological Forecasting & Social Change* **25**:189–207.

Linstone, H.A., 1984, *Multiple Perspectives for Decision Making*, North Holland Publishing Company, Amsterdam, New York, Oxford.

Linstone, H.A., 1985, Multiple perspectives: Overcoming the weaknesses of MS/OR, *Interfaces* **15**(4):81.

Linstone, H.A., 1989, Mediacracy, mediocracy, or new democracy, *Technological Forecasting & Social Change* **36**(1–2):153–169.

Linstone, H.A., Fried, J., Wang Yinglin, and Shu Hui, 1987, *Multiple Perspectives in Cross-Cultural Systems Analysis: The China Case*, Systems Science Ph.D. Program, Portland State University, Portland, OR, USA.

Linstone H.A., Meltsner, A.J., Adelson, M., Mysior, A., Umbdenstock, L., Clary, B., Wagner, D., Shuman, J., 1981, The multiple perspective concept, *Technological Forecasting & Social Change* **20**(4):275–325.

Los Angeles Times, April 4, 1982.

Main, J., 1987, Wanted: Leaders who can make a difference, *Fortune*, p. 92, September 28.

New York Times, August 26, 1986.

Perez, C., 1983, Structural change and assimilation of new technologies in the economic and social systems, *Futures* **15**(5):357–375.

Perrow, C.E., 1984, *Normal Accidents*, Basic Books, New York, NY, USA.

Peters, T.J. and Waterman, Jr., R.H., 1982, *In Search of Excellence*, Harper and Row, New York, NY, USA.

Pollock, S.M. and Chen, K., 1986, Strive to conquer the black stink: Decision analysis in The People's Republic of China, *Interfaces* **16**(2):31–42.

Report of the President's Commission on the Accident at Three Mile Island (The Kemeny Commission), 1979, Pergamon Press Inc., New York, NY, USA.

Rowan, R., 1979, Those business hunches are more than blind faith, *Fortune*, p. 112, April 23.

Sapp, J., 1987, *Electricity Demand Forecasting in a Changing Regional Context: The Application of the Multiple Perspective Concept to the Prediction Process*, Ph.D. Dissertation, Systems Science Ph.D. Program, Portland State University, Portland, OR, USA.

Sharif, M.N., 1989, Technological leapfrogging: Implications for developing countries, *Technological Forecasting & Social Change* **36**(1–2):201–209.

Siu, R.G.H., 1978, Management and the art of Chinese baseball, *Sloan Management Review* **19**(3):85.

Technology Import Project Group, 1986, *Analysis of Unfavorable Factors in Digesting-Absorbing Imported Technologies in China*, Report of Science and Technology Research Division, National Research Center for Science and Technology for Development, Beijing, China, July 28.

Thompson, M. and Warburton, M., 1985, Decision Making Under Contradictory Certainties: How to Save the Himalayas When You Can't Find Out What's Wrong With Them, *Journal of Applied Systems Analysis* **12**:3–33.

Udwadia, F.E. and Agmon, T., 1988, Trade Deficits: A Look Beyond the Economic View, *Technological Forecasting & Social Change* **33**:109–118.

Wall Street Journal, April 7, 1980.

Washington Post, August 22, 1986.

Washington Post, August 30, 1986.

Wildavsky, A. and Tenenbaum, E., 1981, *The Politics of Mistrust*, Sage Publications Inc., Beverly Hills, CA, USA.

Chapter 5

Technogenesis: Origins and Diffusion in a Turbulent Environment

Dirk-Jan F. Kamann and Peter Nijkamp

5.1 Introduction

In recent years a wide variety of different terms has come into being: technological innovation, spatial dynamics, economic restructuring, industrial rejuvenation, urban incubation, and many others. All such expressions mirror the drastic changes that took place in the "roaring eighties".

The offspring of this sudden interest in the triangle of technology, economy, and space in a sociocultural setting is to be found in the reorientation of general economics which has realized that contextual factors (technological progress, urban and regional incubation) are of critical importance to the trajectory of the economy as a whole, which in turn is shaping new conditions for technological and spatial dynamics.

Spatial development patterns have in the past decade exhibited dramatic changes leading to more diversification; even within the same country growth and decline seem to accompany each other. Therefore, it is extremely important to analyze the *conditions* under which such structural changes may occur. This is especially relevant, because spatial dynamics is marked by nonlinear evolutionary patterns, which may incorporate bifurcations and even chaotic behavior (Nijkamp *et al.*, 1988).

Particularly in recent years, various researchers have claimed that recent spatial trends exhibit a clean break with the past (*cf.* Bluestone and Harrison, 1982). A major analytical question is whether such structural changes are due to endogenous spatial developments or to exogenous spatial key forces. This problem runs parallel to the current scientific debate on the existence of long waves, and in particular on the Schumpeterian viewpoint regarding the endogeneity of Kondratieff cycles with their major phases of takeoff, rapid growth, maturation, saturation, and decline.

In such debates the *supply* side of the economy (which is intertwined with technical progress and spatial dynamics) is of critical importance. The recent literature in this area has made plausible the idea that industrial innovations (either basic or process innovations) may be regarded as major driving forces for structural changes in the space-economy. In this context, the so-called *depression-trigger* hypothesis is an important analytical departure, as this hypothesis takes for granted that a down-swing phase of the economy will induce the invention and implementation of radically new (often clustered) technologies. The demand side of the market can, in this framework, be incorporated by means of the so-called *demand-pull* hypothesis (see Ewers and Wettman, 1980). The depression-trigger hypothesis, which is essentially based on a challenge-response type of economic behavior, is extremely relevant in a spatial (urban or regional) setting, as it states that a stimulus for economic restructuring requires basic innovations in the productive sector. Such innovations need not only the production of new commodities, but also the provision of new locations for innovative entrepreneurs. This also implies that the implementation of new urban or regional infrastructures is a *sine qua non* for spatial economic dynamics. Thus the combination of productive capital, public overhead capital, R&D capital, and the emergence of new markets are critical conditions for creating radical technological changes.

In light of the foregoing remarks two major questions seem to emerge. First, technological innovation is not a *deus ex machina* which serves to save a malfunctioning economy without active efforts by all actors involved. Thus this leads to the question which are the driving forces that stimulate *technogenesis* (economic mechanisms, urban and regional seedbed conditions, etc.)?

Second, even when technological innovations have materialized (or when they are potentially available), this does not mean that all firms or regions are able to "reap the fruits" of a new technogenesis. Apparently, there are many bottlenecks to be overcome. Thus this leads to the question, which *transfer mechanisms* (e.g., networks) are favorable for ensuring a smooth diffusion and adoption of new technologies?

These two questions are central in this chapter. The chapter is organized as follows. Section 5.2 is devoted to a review of findings regarding the interwoven relationship between spatial incubation and technological innovation. In Section 5.3 the relationship between industrial dynamics and the impact of technological progress is further taken up, among others by referring to diffusion mechanisms. Next, Section 5.4 deals with an often neglected but extremely important interaction and diffusion pattern, *viz.*, networks. The various concepts used in literature will be reviewed and synthesized into an approach that seems, from preliminary tests, to be a useful analytical instrument.

5.2 Technogenesis: A Spatial Perspective

Regions and cities are the "workfloor" of technological developments. In general, regions and cities are competitive geographical units which will try to obtain an economic advantage by either generating technologically advanced products (or processes) or attracting a maximum share of the pool of available technologies. Thus the question of the spatial *selection environment* which induces technogenesis is at stake here.

It is noteworthy that the technological performance of a region (or city) is dependent upon two factors: (1) its sectoral structure in terms of industrial composition, firm size, industry-technology life cycle, R&D investments, industrial networks, etc., and (2) its incubation potential in terms of agglomeration economies, information networks, accessibility, labor market, cultural amenities, production milieu, and similar.

It is evident that the sectoral structure and the incubation potential are not entirely independent factors, as innovative entrepreneurs are operating in an open economic *and* spatial system. According to Ewers and Wettman (1980) the task environment of a firm and its internal behavior are mutually interlinked phenomena.

In general, the innovative performance of a firm can be characterized by answering various questions, such as:

- Is the firm creating a new technology or only adopting an existing technology?

- Is the firm a member of an innovative sector (e.g., high-tech) or of a traditional (e.g., old-line) sector?
- Is the firm located in a creative environment (e.g., a science park) or in a conventional industrial climate?

In Malecki and Nijkamp (1988) it has been argued that the blend of entrepreneurial spirit, technologically sensitive sectoral structures, and creative environments is of critical importance for a successful technological transformation process. Since new technology is an important weapon in a competitive market, firms will consider a favorable geographical location as being an important dimension of their entrepreneurial strategy. Consequently, the locational aspects of technogenesis have become an important aspect of current technology research. In this context, six types of studies may be distinguished, *viz.*, technological production factors, technological performance, industrial sectors, regional seedbeds, longitudinal evolution of firms, and regional (technology) policy. These categories will now briefly be discussed.

5.2.1 Technological production factors

In these types of analysis the attention is focused on the inputs which are necessary to generate technological progress. Particular emphasis is usually placed upon R&D. From a geographical perspective this implies that spatial variations in R&D investments may explain to a large extent the technological differences among regions in a national economy. Since in general new technologies increase the productivity of firms, it is evident that investments in such technologies may be regarded as efficiency enhancing production factors.

5.2.2 Technological performance

Studies on technological performance focus attention on the output side of production. Various criteria for measuring the technological performance of a firm may be distinguished, such as numbers of product or process innovations, numbers of patents, etc. It is evident that the technological performance of a firm will depend on both the composition of the sector it belongs to and the spatial incubation potential of the place it is located. In most studies on technological performance a causal link with the regional selection environment is usually assumed (Oakey *et al.*, 1980).

5.2.3 Industrial sectors

Analysis of technological evolution from a sectoral angle is usually based on the question, which *sectors* are concentrated in which areas? A good example of such studies from this point of view can be found in the many attempts at identifying the geographical pattern of high-tech industries. The regional selection environment – or in a broader setting the regional incubation potential – is then confronted with the spatial concentration pattern of given sectors, among others by using the location quotient method. Special attention is often given to the question whether city centers, urban rings, intermediate areas or semi-peripheral areas show a certain concentration of innovative industrial sectors.

5.2.4 Regional seedbeds

Analysis of regional seedbeds starts with a comprehensive inventory of geographic factors which in a given area favor technological progress. Examples are the functioning of the regional labor market, the availability of university research facilities, a favorable quality of life, etc. On the basis of a *regional* angle the technological performance of an area is then analyzed and judged. In general, the hypothesis is then tested whether a region with a promising incubation potential is technologically more advanced than others (for a critical review see also Davelaar and Nijkamp, 1987).

5.2.5 Longitudinal evolution of firms

In recent years there has been increasing awareness that the time horizon of technological transformations is fairly long, so that static or cross-sectional studies have by definition many shortcomings. This has led to a new research endeavor, called company life-history analysis. The aim of this type of longitudinal analysis is to trace the most critical decisions of a firm (e.g., investment decisions, new markets, product or process innovations) over a long period of time (20 to 40 years) and to confront these with locational developments and general economic circumstances. Instead of individual firms, it can also be applied to networks of firms. This type of analysis does more justice to the evolutionary process of industrial developments and places technological evolution in a comprehensive and broader setting.

5.2.6 Regional (technology) policy

In recent years, there has been considerable debate about the impact of public policy on technological innovation. This debate has mainly centered on two major questions, namely, the conditions for a "technological takeoff", and the degree of diffusion and pervasiveness of new technologies. The performance of technological innovation was measured not only in terms of new firms or new jobs (the incubator profile), but also in terms of the modernization of existing plants, social equity, regional development, and resource impacts.

Nowadays it is a widely held belief that technological innovation is not "manna from heaven" but can be generated by well-focused public policies at national, regional or local levels (in close cooperation with private initiatives). And hence the question has emerged: which type of policy is needed for which type of technology? The microelectronics technology, telecommunications technology, biotechnology, agrotechnology or off-shore technology all may have different seedbed conditions, and hence there is no uniform technology policy which can generate all kinds of new technologies. Technology policy is essentially more a tailor-made endeavor to favor the creation of specific innovative activities in specific sectors and at specific locations. This implies that technology policy cannot be separated from other fields of public policy such as socioeconomic policy and physical planning, nor can it be implemented only from one (national) level; in this context, much emphasis is being placed upon the self-organizing potential of a region. Until recently technology policy was mainly pursued in isolation from socioeconomic policy and physical planning. In recent years, however, an increasing awareness has grown that technology, economics, and space have an interwoven triangular interrelationship with close interactions between all three components. It is noteworthy that in several countries there has been a shift towards a more coherent approach to technology policy, economic policy, and physical planning, and this shift is also reflected in regional-economic technology research.

This concise overview of various studies in the area of spatial dimensions of technology shows that there is a wide variety of different research endeavors seeking to link economic performance, technological progress, and regional selection environments. Both analytically and empirically many advances have been made, although the long-term perspective of spatial technogenesis is still underdeveloped. In the next section, the issue of the spatial diffusion of new technologies, seen from an industrial dynamics viewpoint, will be discussed.

5.3 Diffusion of Technologies in an Industrial System

The spatial picture of technological evolution has been given due attention in the past years. Space not only acts as a dimension upon which techno-economic processes are projected (e.g., in terms of locational dynamics) but also provides the medium through which economic dynamics is generated and transferred (witness the importance of both incubation and diffusion theories). The spatial economic structure appears to provide the vehicle for the generation and adoption of innovations in an industrial system (e.g., on the basis of producer services). In this context, the increasing importance of (spatial and socioeconomic) networks is noteworthy.

It should be noted that differences in the rate of economic growth between regions in advanced economic systems are explained not only from the sector structure and the selection environment of those regions, but also from the diffusion pattern of new technologies. Such diffusion patterns are partly influenced by indigenous competitive forces in a spatial economy, partly by information transfer channels, and partly by institutional structures (e.g., financial agencies, subcontracting systems, university attitudes). A major role is played by organizational structures of industrial and commercial companies and by the organization of labor within companies. In this context, the emergence of neo-Fordist modes (e.g., based on flexible organizational structures, economies of scope, customization of products, and rapid adjustment to market demand) has to be seen. The pathway of new techno-economic structures is marked by much variation, ranging from conventional *hardware* investments to company migration (or branch plants) and from internal upgrading of skills to purchase (or transfer) of skills from elsewhere (or even takeovers).

The impacts of new technologies may be twofold, either as direct spinoffs in a local economy (e.g., via inter-firm interactions) or as transfers in a broader spatial system (e.g., via new information technologies). Thus a favorable seedbed function of a region is a necessary, but by no means a sufficient, condition: information channels ensuring an appropriate transfer of technological achievements are equally important (Townroe, 1988).

Diffusion analysis has in recent years become an important field of research in industrial economics. This analysis not only focuses on the distribution and adoption of new technologies, but also on the business services related to technological transformations. According to Cappellin (1990) there are at least three reasons to regard business services explicitly in the context of technological innovation:

- The transfer of new producer services to industrial firms may be regarded as a major organizational (or process) innovation at the firm level.
- The creation of such services (either as new products of existing firms or in the form of the birth of new service firms) in a certain area may be regarded as a product innovation.
- Service firms may adopt such organizational (or process) innovations in order to increase their productivity or to improve the quality of services offered.

In most diffusion studies the S-shaped curve forms a central component. Both the adoption time and the adoption rate can be pictured in this curve. The precise shape of the S-curve can then be explained from firm size, market structure, profitability of innovations, etc. An important role can be played in this context by (barriers to) information transfer in a spatial system (see Giaoutzi and Nijkamp, 1988). It is evident that the adoption of innovations via a spatial transfer mechanism brings also the demand side of innovations back into the picture. In this context various social-spatial communication linkages/patterns are often distinguished, such as hierarchical and contagious diffusion patterns.

Diffusion models can then be used to mirror the techno-economic landscape of a spatial system, although normally such models are hampered by a major shortcoming, i.e., the presupposed stability of parameters. This is a major flaw in diffusion analysis, as in this case the Schumpeterian swarming effects of new basic technologies and the feedback effects from adopted innovations on spatial structures cannot be adequately taken into account (see also Alderman, 1990).

Another field where operational research would still need further development is the *channels along which diffusion takes place*. Information transfer systems, technology and R&D centers, industrial institutional agencies, consultancy and banking institutions, and many other (formal or informal) channels serve to enhance the flexible transmission and adoption of (information on) new technological innovations. However, which channels are economically most efficient and politically most effective is still an open research question in diffusion research. Therefore, it is worth exploring these points by dealing more explicitly with *networks* of relations – potential channels of diffusion – between *actors*.

5.4 Networks

Networks are becoming an increasingly popular issue in the scientific literature. There is a growing awareness that firms should not be seen as individual organisms that live their own lives independently from other actors in their economic, social, and cultural environment. Waves of merger activities and takeovers have resulted in noticeable effects (and sometimes disruption), not only in financial circles, but also in local employment situations and welfare. Technological change, e.g., in telecommunications – information technology – has resulted in locational changes of parts of multinational organizations and changing activity patterns. Government efforts to stimulate innovative industries, and the growing interest for autonomous growth potential, territorial industrial complexes, and local initiatives have induced a great deal of interest in the networks of relations between firms, including the diffusion of innovations.

After a brief review of existing theoretical paradigms, we will explain the behavior of the actor in his contextual situation. We do so, in order to deal with two main issues already raised in Section 5.1, i.e., the impact of (the configuration of) actors in a network upon technogenesis, and the impact of (the configuration of) these actors upon adoption and diffusion inside the network.

Two main streams of theorizing play a major role in the network literature: *field theory* and *systems theory*. For instance, Perroux (1955) describes how actors meet on a plan in abstract economic space, where they try to dominate other actors and the relevant plan in economic space. Some actors may be active on various plans – they are *multidimensional*. The position of an actor in the arena on the plan – in the field – is determined by external forces and internal forces. External forces are demand, public policies, etc., internal forces are the – nonrational – drives of actors.

Systems theory may be used to describe attempts to externalize internal problems of actors (resulting in negative external effects) and to internalize external problems (by means of, e.g., internal labor markets, information systems, capital supply). It also assists in describing behavior to resist threats and challenges. A process comparable with "negentropic" behavior:

the capability to attain stability in a time-dependent steady state. In our context, we would use stability in terms of market power, aggregate power, and network power. The major drawback of this approach is the assumed equilibrium that ignores Schumpeterian disequilibria. The continuous shifts in coalitions between actors in a network would call for a more dynamic approach, like chaos theory, game theory combined with Markov-type decision models.

After this brief overview, we will deal with the adoptive behavior of the actor and the contextual role.

5.4.1 Cause for change: The competitive challenges

The major reason for actors to reorganize their activities, to change their routines, originates from the *competitive forces* to which they are subjected. Actors that fail to respond to competitive challenges, threats or opportunities, will find themselves on the loser's side. Actors that opt for the wrong strategy may face the same fate, or may be the subject of takeover raids. Although a proper response to competitive challenges is of vital importance, three major categories of causes for weak performance may occur:

(1) Actors are *not aware* of the challenge or alarmingly poor performance.
(2) Actors are aware of the challenge, but *do not know the proper strategy* to follow.
(3) Actors are aware of the challenge, know the proper response but are *unable* to implement the proper strategy.

The first two cases are *information gap* problems. The reason for this gap can be manifold. Ignorance, poor information retrieval systems, value perception gaps, false routines, inertia, no gatekeepers, isolated location, poor management, and so on are examples (*cf.* Kamann, 1985). Or, the configuration of actors, the network, one participates in may be a combination of poor quality. This may be caused by geographical backwardness or isolation, but also by selecting the wrong companions. However, the actor also may be a victim of deliberate exclusion. Strategic information may be withheld by other actors in an attempt to gain a more advantageous position. This aspect also causes the third bottleneck: an actor is unable to seize opportunities or fight threats because of external circumstances beyond his control. Therefore, when we want to study the innovative behavior and performance of individual actors, we have to do so in the proper *context* of external relations.

5.4.2 Three dimensions of space

In the preceding sections we have dealt with two spatial dimensions: geographical space and economic space. We will now add a third dimension: sociocultural space. Actors have to work in a given sociocultural setting. This local sociocultural environment may favor certain technologies, *ways of producing goods and services* (Kamann, 1988), like Tayloristic production techniques, and by doing so may hamper other technologies. It favors a particular *régime d'accumulation*, a particular *régime de régulation* (Aglietta, 1986). The finding that technologies and their organizational and ideological or cultural dimensions are closely related elements, also means that changes in any element requires suitable or adequate adaptations in the other elements. In network and organizational terms: a network with a *traditional* – Tayloristic – organization and matching culture, will be a slow adopter of technologies that are alien to that organization and culture, and vice versa.

Part of the influence of sociocultural space is the process of socialization and culturalization that individuals experience. This moulds the internalized value system and behavior, the *personality* (Kamann, 1988) of the actors, given their biological and genetic inheritance.

All three spaces show a differentiation. Geographical space is polarized with large metropolitan areas and rural areas. Economic space is polarized for many types of activities: Toyota, Nissan, GM, Ford as giants in the segment – the plan – of automobile industries versus the peripheral sweat-shops in dual production situations. Sociocultural space also is polarized: "western" societies, with their emphasize on individual action versus societies with more "traditional" or even feudal value systems. As a result, actors in some societies are better able to cope with western competitive systems or decentralized decision-making procedures and production systems than others, in other societies. Even inside western societies, the large cities show distinct differences from the rural areas; both in sociocultural environment and technology. It results in different types of personalities and characteristics; different managerial types. These elements, which play a role when starting new *high-risk* enterprises or when drastically changing routines, show a distinct geographical differentiation.

To the differentiation in the three dimensions of space we have to add their *dynamics*. The environment is not static, it changes. Some actors may even have the perception to live in a chaotic environment. In light of the diffusion theories described above, we may say that these dynamics do not occur at all places at the same time; some actors are facing turbulence earlier than others. It also means that some actors have adapted to the new situation at the same time other actors are in a state of shock, which makes them vulnerable to attacks from the first category. Furthermore, the nature of the environment shows a significant differentiation in geographical space, apart from the diffusion aspects. Whereas actors in one area may have a very stimulating environment, other actors, in other areas may find their environment very hostile. They have to use much of their energy to solve the problems arising out of that environment.

The spatial differentiation of the three dimensions is one cause for differences in performance between actors. A second cause can be found when studying the nature of external relations with other actors. We could term this the *network performance* of actors. The question is, *why do actors participate in networks at all*?

5.4.3 Motives for participation and types of partnership

The answer to the question why actors participate in networks is a rather easy one: since they have suppliers and buyers, they *automatically* are participants in a network. Networking *not* only consists of cooperative projects, partnerships and relations based on trust. Network relations also include those with actors one considers to be "the enemy". Apart from this automatic – *passive* – network participation, we will describe the motives for *active* network participation.

Keeping or increasing the market share is one of the most important strategies of actors. Therefore, given the information gap we mentioned, we will find that a significant part of network participation of actors is directed towards collecting *information* about customers, suppliers, the competitors and his products, prices, R&D programs and policies, new markets and changes in market shares of existing markets. Hence, the information *contents* of relations are important.

In his attempts to increase his share of the market or simply to meet the challenge posed by competitors or newcomers, an actor is likely to run into obstacles or may well find some bottlenecks when implementing a proper and adequate strategy. These bottlenecks can occur in research, development, finance, production, distributing, marketing, organization, and so on. Actors have become aware that technological developments these days very rarely take place in a single firm. It is an "interplay between different organizations where independent activities are taking place simultaneously in different parts of the network" (Laage-Hellman, 1987, p. 31). This implies that firms realize that in a number of cases it is better to co-develop product and process innovations with suppliers, buyers or even competitors. "Different units have different resources and skills which are complementary in nature" (Laage-Hellman, 1987, p. 37). The value of the network in this case is *the combination of "resources" and "skills" which as such is unobtainable for each of the actors involved.* This positive value remains positive, even when the individual actor has to increase his dependency on other actors. Active network participation and distribution of activities enable increased specialization of each actor, while increasing the need for interaction because of the required coordination. "The establishment of development relationships may in other words be a pre-condition for increasing specialization of the in-house development process" (Axelsson, 1987, p. 131). Actors start partnerships depending on the bottleneck they experience and the type of specialization they prefer.

When network performance is defined as the way an actor is able to use external relations to achieve his goals and to solve bottlenecks, the following two questions follow: Who are the other actors? What is the nature of external relations?

As a *collective* operator, one network may compete with other networks in the same market. However, the more - different - actors that participate, the more important coordination becomes. Efficient networks show fast information diffusion - when such is required - and coordinated innovative efforts. As we will see later on, we may find that control and coordination often coincide.

5.4.4 The other actors

The question "who are the other actors" relates to matters such as effectiveness of relations. Does an actor have suppliers that provide him with the latest products, machines, know-how, and information? Is he in contact with producers of new substitutes? Are the suppliers and buyers *reliable and trustworthy*? What about the competitors? To answer these questions, we first have to make a brief detour and describe the class of *potential* external actors an actor can relate to. This class of entities external to the individual actor and determining his behavior and performance is known as the *selection environment*, *milieu innovateur* or "atmosphere". Because of its dynamic character, terms such as turbulent fields or even chaos are used. We distinguish four sub-categories of *actors* in the selection environment (Kamann, 1988):

- Friends and relatives.
- Actors related to *locational* factors.
- *Institutional* – usually *non-market* – actors.
- Actors that are specific to the *product market combination* of an actor and his *corporate linkages*.

Family and relatives play a role in the socialization process, act as a peer group and sometimes advise in business matters. Here, we find a cultural differentiation between countries and areas. In some areas, social and business contacts are mixed, in others, they are not.

Locational factors make a location attractive or unattractive. Actors, related to these factors, for instance, represent infrastructural services, water works, energy suppliers or railroad companies. Examples of actors of the third category are trade union representatives, public administrators, and lobbies. Two organizational elements included here are the organization of firms and the organization of the built-up environment. The former is better known as *industrial agglomeration effects* in economic space, while the latter is known as *urbanization effects* in geographical space. Economic and geographical space *used to* coincide for agglomeration effects. Because of improved transport and communication techniques and modes, we find that agglomeration effects in economic space no longer necessarily coincide with geographical space. The widespread phenomenon of the branch-plant that links to a corporate network of communication, information and transport is a good example from the 1960s. Presently, we find that *just-in-time* deliveries allow for a hundred miles – for the Dutch case – between suppliers and their main buyer, even for *vital* parts and components. In spite of this, we still find examples of agglomeration effects in geographical space (Storper and Scott, 1990). However, it is *not* a necessary condition any more for survival of especially the *niche* operating firm, particularly when it is linked up to a proper network. As to urbanization effects, apart from the usual public services, they also include business services and as such are a locational factor for, e.g., office activities.

The fourth category of actors in the selection environment we mentioned contains units of the corporate network and competitors, suppliers, customers, producers of potential substitutes or possible entrants.

We can project the differentiation in economic and sociocultural space on geographical space, and add these to the existing differentiation in *locational* features of its selection environment. Together, they show the differences in opportunities for existing firms, the formation of new firms and the innovative behavior of actors.

5.4.5 The nature of relations

Having introduced the potential set of actors one can relate to, we face the second question: what is the nature of these relations?

Markets and Hierarchies

Following Williamson's (1975, 1979, and 1987) transaction cost theory, we could use a continuum between "markets" and "hierarchies" or intra-organizational arrangements. The choice between the two extremes – the *make or buy* question – is based on an optimalization process of the organizational form and a minimalization of costs. Three aspects play a role:

(1) *Asset specificity*. Durable goods and R&D tie up large investments. This reduces the flexibility in changing policies. The most common types of asset specificity are:

- *Site specificity*, resulting in minimizing transport costs.
- *Physical specificity*, related to machines or products that can be used for only one particular product or item.
- *Human asset specificity*, related to the built-up routines of employees, researchers, shop-floor workers and directors.

Investments in machines and employees required to satisfy the demand of a large buyer may result in a situation of *bilateral monopoly*; the supplier and buyer are mutually dependent. The supplier has no other alternative demand for his products, nor can he shift to other products without a great loss, while the buyer has no alternative supplier, nor can he make the product himself without heavy investments in machines and human capital and at a cost of lost loss of sales because of transitional problems. Asset specificity is no exceptional case, although buyers try to prevent such by means of "dual sourcing". Because of recent technological developments, asset specificity has become a rule rather than an exception in numerous intermediate markets.

(2) *Uncertainty about the future*. Like other human beings, actors try to reduce risks, allowing for maximum freedom to break up relations when they become burdensome. The price to be paid is determined by information, negotiating, and quality assurance costs. The choice between

long-term relationships and short-term relationships is what we would call the ***relation specificity***. Investments in mutual trust, knowing each other, understanding each other, and especially knowing whom to address for information or to fix things.

(3) *Frequency*. Transaction chains show a repetitive pattern. This reduces the costs of breaking up a relation, although this is difficult to quantify. This problem of quantifying costs of a relation coincides with the problem of quantifying the costs and benefits when externalizing certain activities. Quantifying the immediate production-related costs may be quite possible. Quantifying the costs of increased uncertainty in information, quality, and reliability is another thing.

The choice between markets and hierarchy, in-house production, is rather easy for the two extreme cases: the one-off job versus eternal production. However, these extremes are not always the case in real life.

The transaction cost theory assumes that hierarchy does not exist in markets, and markets do not exist in organizations. Real-life experience shows that the opposite is true: a market transaction between completely equal partners – atomistic actors – is an exception rather than a rule. Further, it ignores that, as Johanson and Mattson (1984) correctly state, today the goals of many actors cannot be achieved *but* through other actors. In other words, actors depend on their network partners for the realization of their goals. This implies a minimal *mutual trust* in relations. It means that actors have to recognize the goals of their partners.

In large corporations, an additional problem occurs when the decision to externalize or internalize has to be taken. Externalizing an activity means that the internal unit that used to perform that activity will lose power. Therefore, that unit is likely to oppose such a decision and will lobby to prevent such. Or, in the opposite case, the unit will give favorable information about its abilities to perform activities that previously were done externally and is tempted to give low-cost figures.

Given the fact that the theory largely ignores spatial differentiation – in any space – and knowledge, skills, machines, products, and sites are considered to be homogeneous, one is tempted to declare transaction cost theory to be a neoclassical equilibrium theory (Johanson and Mattson, 1984). Still, the basic question of whether to make or buy is valid; when less unrealistic assumptions are included, the theory could become rather valuable.

Between Markets and Hierarchies: Networks

We just discussed the polarity *markets* and intra-organizational *hierarchies*. We now focus on the part in between: network relations. We express the degree of dependency, *hierarchy*, induced by these remaining relations by a scale. On one side we find completely symmetric exchange relations between equal partners. On the other side of the continuous scale we find the asymmetric relations of complete dependency/dominance in *market relations*.

The ideal case of a symmetric *transaction* relation is the exchange of a product for money between two atomistic actors. An example of a symmetric relation *in partnership* is the exchange of production of certain products by Dutch oligopolistic dairy companies. Neither partner would reach economies of scale for either of the two products. When exchanging the production of the two products, each firm produces one product for both partners involved; thereby achieving economies of scale.

An example of the other extreme case can be found in the "dual production organization"; for instance, the case of the Japanese *kanban* system of a many-tiered hierarchy of sub-contractors (Storper and Scott, 1990).

Hence, we find a continuum of partnerships:

- *Exchange* of production between two or more competitors enable all actors involved to obtain scale effects.
- The *joint venture*, a form useful for:
  - Development of products and penetration of markets.
  - Expansion in markets.
  - Consolidation of markets by horizontal integration.
  - Retreat from markets.
- *Long-term contracts* between suppliers and one of their important buyers. They should guarantee timely and reliable deliveries, and enable suppliers to invest in innovations required by their buyer.
- *Joint production planning* between suppliers and buyers to enable zero-stock-inventories and just-in-time deliveries.
- *Joint planning* of suppliers and their industrial buyers where the latter give active support to R&D and capital.
- *Dual production organization*; the sweat shops, sub-contractors, and other supliers; here a *one-sided* monopoly exists. For the buyer, it is easy to shift to other suppliers; for the supplier, this is impossible.

Licence agreements can be part of any of the partnerships mentioned, including straightforward market transactions without any partnership involved.

Type of Dependency

As we said before, various activities between network participants are linked together in transaction chains that tend to repeat established configurations. This may induce the following dependencies in relations:

- *Technical dependency* occurs when products and services fit technically together and result in inter-industry standards. Therefore, within networks, coordination mainly focuses on rigid transaction chains, while actors are allowed more freedom in less rigid links (Kamann and Strijker, 1990).
- *Knowledge dependency* means that the supplier has to know the requirements of users of their products, while users have to know what they actually can do with their input materials, machines, hardware and software.
- *Continuity dependence* occurs when a supplier sells a large share of his output to a single buyer. The actual percentage is a function of the power of the actors involved, the profit margins and the profitability of other activities of the firm. The reverse occurs when producers depend on a single supplier for a particular product or service they require. Firms will try to prevent this situation by applying *dual sourcing*. However, in a large number of cases, increased specialization has led to single sourcing.
- *Social dependency* is the result of normal social group behavior, where participants are likely to cooperate with other participants before establishing contacts with actors outside the network. Axelsson (1987, p. 159) uses the term *soft distance* to indicate the sociocultural distance between actors or entire networks: attitudes, values, norms, culture. The term *hard* distance indicates the distance in kilometers.
- *Logistical and administrative dependency* means that suppliers and buyers have to use the same system to be able to communicate. High switching costs prevent small suppliers to switch easily to other buyers with different systems.
- *Innovative dependency* exists in three varieties: (i) user dominant, (ii) supplier dominant, (iii) research dominant. Increased cooperation between suppliers, users, and research institutes or consultants makes this division less valid today. The new question is who took the initiative –

the *impetus* – for new developments, rather than who actually carried out the development work, produced or used the product or process.

- *Financial dependency* may have rather important effects, ranging from profits and strategic information leaking away to enforced purchase of licences, goodwill, and products. *Complete* takeovers tend to lead to centralization of production activities in a particular location and centralization of overhead activities in another location. The actor involved runs the risk of converting his plant into a standardized production branch plant. Even when he remains relatively independent as a business unit, corporate planning, investment decisions, and tax management are transferred to the parent company.

The various types of dependency, usually in a user dominant relationship, may result in the *dual production organization.* Based on the dual segmentation theory this theory assumes a primary and a secondary sector. Large and technically sophisticated corporations are part of the primary sector or *core* of the economy. They operate in stable and safe segments of markets applying modern, capital intensive Fordist mass-production techniques. In the secondary sector or peripheral sector, relatively small firms operate with flexible technologies, catering for fluctuating and risky markets. Companies in the core sector *externalize uncertainty and labor costs* to the peripheral firms. In secondary firms, wages are low, prospects are poor, working conditions are bad. Alan Scott (1985, p. 17), for instance, found that many secondary firms in Orange County employ Mexican and Asian workers since these "cannot perceive or are unable to demand ... recognition of rights". Although it increases local linkages compared to the situation with branch plants, it dominates local firms and segments the labor market. It is a new variant of the *cumulative causation* theory.

Most of the relations we discussed so far were of a *dyadic* nature: between two actors. We will now broaden our scope to more relations: the network.

5.4.6 Plans in space

Returning to Perroux's description, we see that actors meet on an abstract plan in economic space. For the time being, we will assume, that an actor coincides with the entire firm: a holistic concept of the firm.

Some actors only deal with actors in their own production column: their upstream and downstream partners. We would label these actors *one-dimensional* actors. They operate on a single plan in space [*Figure 5.1*(*a*)].

A second category deals with actors that operate on two unrelated plans. For instance Philips microelectronics, and formerly Philips-Duphar pharmaceutics. These actors can be named *multidimensional unrelated actors* [*Figure 5.1(b)*]. Because of the present increased emphasis on *core* activities, many actors of the latter type are separated into two independent actors. Finally, a third category operates on more plans in space *at the same time*. For instance the transport firm, specialized in transporting cut flowers, operates on the transportation plan, facing large carriers, car manufacturers, logistical services and so on. It also operates on the horticultural plan, facing large wholesalers, auctions, state agencies, and so on. The transport firm in fact operates in an *intersection* of the two plans. These actors can be named *multidimensional related actors* [*Figure 5.1(c)*].

Role of Dimensions

The effect of being one-dimensional, multidimensional related or multidimensional unrelated is twofold:

- The performance of one-dimensional actors is determined by the nature and configuration of the network on a single plan. The performance of multidimensional actors on a particular plan is also determined by the nature of the networks and the position of the particular actors in those networks on other plans. This implies that an actor may be hampered to pursue certain strategies on one of the plans, networks, it operates on since it has to put all its resources in its struggle for survival on another plan. Or the reverse side could occur in that an actor may improve its position on one plan because of *cross-subsidizing* its efforts with the profits earned on another plan.
- Diffusion of innovations may be facilitated from one plan to another plan through *bridge* actors that operate on both plans. This may give this actor a competitive edge. It also means that actors not only have to be aware of new developments – innovations, potential substitutes – on their own plan, but also on all related plans. This is, with the limited time available, a severe handicap for the average small firm.

5.4.7 Type of network

Actors meet other actors on plans, sometimes passively, sometimes actively. The question is, when are we allowed to say they are part of a *network* and when is it just *a dyadic relation*? A rather general definition of network is

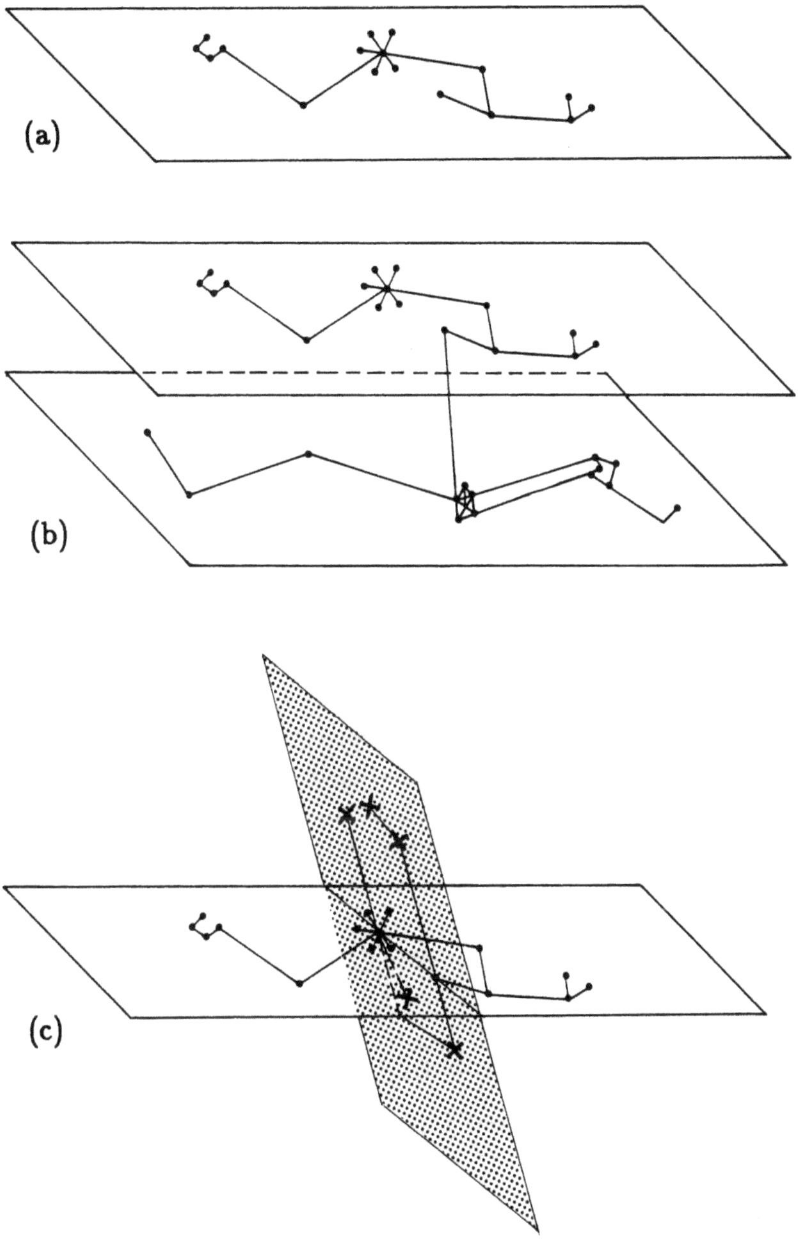

Figure 5.1. One-dimensional and multidimensional actors in space.

the "totality of all units connected by a certain type of relationship" (Jay, 1964, p. 138) or Cook and Emerson's (1978, p. 725) statement that between actors, "*sets of two or more connected exchange relations*" are established, shaping the network. The obvious question here is, how many relations have to exist before we can use the term *network*. From the literature we derive five categories:

(1) The *dyadic relation*, when based on information "tend to grow out of interpersonal associations between organizational representatives" (Whetten, 1987, p. 240), both agency theory and transaction cost theory focus on a dyadic relation.
(2) The cluster of relations, forming a *clique* or *coalition* inside the organization.
(3) *Organization sets*, constituted around a focal actor, "it is not a real network, because ... the relations between the interacting organizations are ignored" (Whetten, 1987, p. 240).
(4) The *action set* is a coalition or clique between organizations; "purposive networks ... working together to accomplish a specific purpose" (Whetten, 1987, p. 241), compare Galbraith's (1963) theory of countervailing power.
(5) *Networks*, "consist of all interactions between organizations in a population, regardless of how the population is organized into dyads, organization sets or action sets" (Whetten, 1987, p. 242). This broad definition can easily become an "every actor has to do with every other actor". However, "a network ... is a construct created by an investigator" (Aldrich and Whetten, 1981, p. 387), which means in practice that networks are limited in size because the researcher usually has to limit his scope.

5.4.8 Conspiracy, opportunism, and trust

The difference between the various types of clusters that may be found in networks sometimes is a matter of the *number of relations*. In some types, we find an element of *conspiracy*. However, as with most conspiracies, not every participant will receive an equal share of the cake. Given the inequality among the conspirators – inequality in market force, financial strength, know-how, and especially, information – the winner may try to take it all. This may induce a new series of attempts to conspire more successfully by those who find themselves less privileged than they had hoped.

The reason for using the term conspiracy is that in our perception this is a *combination of mutual trust and opportunistic behavior*. It is reflected in the cooperation paradox: actors realize that they need each other to pursue their objectives, but at the same time do not want to become dependent on others. They are worried that the partner gets more out of it or runs off with all the benefits and secrets. This paradox causes cooperative agreements, e.g., joint ventures, to show an unstable pattern of behavior; exist until the conditions that favored cooperation change. This cooperation paradox not only exists in relation to production economics; all kinds of social systems show such unstable cooperation patterns.

In conclusion, we say that actors show an opportunistic behavior in some relations, establish mutual trust in others and finally mix the two elements in conspiracies with a third set of actors to which they relate.

5.4.9 The selection process: From potential reservoir to actual segment

Given the *potential* number of actors with which contacts could be made, an actor selects a certain network segment. This term coincides with the term *microposition* (Johanson and Mattson, 1984), and stands for the *actual configuration of relations between actors*. In such a segment, there are some actors that one particular actor very rarely meets, while he may have intensive contacts with others. When an actor has no direct contacts with certain actors, but assumes that other persons, who are included in his list of contacts, meet these actors, then this is termed *indirect contacts*.

> I know someone, and they know someone, but I don't know who they know. The power of the network is that the participants all know it exists. We all know that we know lots of people in the Valley ... the rate of rumor-passing in Silicon Valley is simply phenomenal. Reputations, successes, people leaving a firm, new products. The mill grinds out these rumors at a prodigious rate [Rogers and Larssen, 1987, p. 80].

We find two types of *weak* contacts in literature:

- Infrequent contacts or weak relations that are "kept in storage" for moments they become of importance: public agents in charge of, e.g., building permission, grants or subsidies. They are of no importance for the regular operation of the firm. Relations with frequent and intensive contacts are termed *intensive*.

- Sole contacts with actors that belong to another network or cluster or plan. Breaking up the relation ends all contacts between two clusters. These contacts may provide important information about new technologies and increase diffusion from one cluster to the actor's; the *bridges* between clusters. When they are sole operators, not closely related to any specific cluster or network, they are termed *liaisons* – e.g., knowledge brokers.

5.4.10 The role of personality

The choice of the actual network segment – whether contacts will be direct or indirect, strong or weak – is a *strategic choice*. Since this choice is very rarely made with full information about the potential network, the resulting incompleteness is aggravated by a geographical limitation of the actor's scope and his sub-sector fixation.

This stage of the selection process is largely determined by the personality of the actor, his business routines, and goals. A multiple actor organization has many actors in different functions, roles, and tasks. Each actor has his own personality, and groups of actors demonstrate subcultures and coalition behavior.

5.4.11 Leaving the holistic concept of the firm

In the last paragraph, we actually left the *holistic concept of the firm*. Actors in each functional department of a firm will have their own interest and type of information required and their own sets of actors that are important to them. For instance, actors in the sales department focus on buyers while production workers focus on machine suppliers (Hamfelt and Lindberg, 1987). In the management literature, this view is supported by Porter (1985). It makes our understanding of networking more complete.

5.4.12 Porter's "value chain" and "value system"

Porter (1985) introduced his "value system" and "value chain" to analyze competitive advantage. The value chain disaggregates a firm into its strategically relevant activities (Porter, 1985, p. 33). These activities are *primary activities*, such as inbound logistics, operations, outbound logistics, marketing and sales, and service; and *support activities*, like firm infrastructure, human resource management, technology development, and procurement.

The primary activities form the columns, the support activities the rows in a matrix.

Firms now may start "coalitions", that "involve coordinating or sharing value chains with coalition partners" (Porter, 1985, p. 34). In other words, a firm may try to increase its competitive advantage by starting cooperative partnerships for certain segments of its value chain: the cells of the matrix to which we referred. The actors involved are also required to understand the jargon of each segment, the segmental "culture" and, to use the words of Crevoisier and Maillat (1989), the *"savoir-faire"*. It implies that:

- Firms are a *vector* of network segments in which they operate, representing functions and activities.
- This vector contains *different* segments.
- The firm as unit is to be a *multidimensional* actor.
- Some of the network relations may show conflicting interests.
- When the size of a firm is an indicator for functional segmentation and diversification, it also means that large firms may run into coordination problems between the various networking interests involved.
- Large firms are likely to pick up innovations from related network segments that are valuable for its competitive position.

The required coordination is a matter of *balancing control*; a delicate balance between centralized authority and decentralized decision-making power. The optimal balance varies between products, markets, and especially technologies used. It determines the success of the entire corporate organization. In fact, a very similar argument goes for networks. Therefore, methods of *running* a network bear resemblance with the strategic management and control of a corporate firm. However, this assumes an unequal distribution of power over the participating actors. The issue to study therefore is whether the so-called *symbiotic* networks, with flat networks, are as successful as *controlled* hierarchical networks. Or, perhaps they are not as flat as has been assumed.

5.4.13 Network orientation and geographical orientation

We concluded that activities in the organization differ in their mix and types of internal and external contacts and with different actors involved. From research in *contact systems* and urban orientation, we know that *some* of the external actors are exclusively located in metropolitan areas (Goddard, 1973). This implies that a *single actor* firm with restricted time available will find it hard to have proper network contacts. A large multiple actor organization can specialize and distribute the various contacts. Although this requires an *internal* network that also consumes time, special *gatekeepers* act as the interface with the environment and provide relevant external information, translated into the jargon of the internal network. As a result of this, activities that draw their information from the internal network are less dependent on the location in the source area of their relevant information, the focus (Kamann, 1988), as long as their gatekeepers are there. To extend the parallel between corporations and networks to this point: we may say that proper networking may give the same results to the participants, and may make actors relatively "footloose".

Using relocation costs and opportunity costs of missed information, an organization may optimize its location. However, persons have a mix of activities and therefore may have different priorities and preferences for certain localities. In an organization with numerous actors the usual social processes will take place, where coalitions or cliques and power rather than individual ratio will decide where the total group will locate. These social processes with their culturally determined characteristics should be incorporated in an attempt to describe and predict organizational behavior in this context.

5.4.14 Measuring control: The distribution of power

When visualizing and measuring relations between actors, we will use the terms *manifest* and *latent*. Manifest characteristics of relations are observable entities. They consist of:

(1) *Material* relations: flows of goods, services, capital, printed information, human embodied information, and skills.
(2) *Immaterial* exchange relations: information through personal contacts.

The first category can be visualized by applying input/output analysis or by means of the expanded *filière* approach (Kamann, 1988). The second category can be visualized through contact systems and interlocking systems.

The latent characteristics of relations are not directly observable but may be derived or inferred from observable manifest characteristics. Latent characteristics mainly deal with the "strategic value of a relation related to the issue of dominance and performance. Manifest relations are the materialized dimension of latent relations" (Kamann and Nijkamp, 1990, p. 15). Features of latent relations are classified under three headings:

(1) The already mentioned dependencies.

(2) Instability; at the micro level found in the growth or decline of the individual firm; at the meso level found in takeovers, mergers, market-concentration, and collusive and conglomerate behavior; at the macro level found in the general reduction of life cycles, the international shift in economic power in geographical space, and the duality between mass production and flexible, customized production.
(3) Paradigm fixation; such aspects as inertia (Kamann, 1986) of entire networks may occur (compare asset specificity). The social aspect of this implies that areas with an activity dominating the social network "repulse" actors with activities perceived as "alien" to those in power.

5.4.15 Techniques to visualize latent relations

A common method of visualizing networks uses *directed graphs and matrices* (Aldrich and Whetten, 1981, p. 397; this also includes a discussion on methodological problems involved). Terms such as "density", "reachability", and "hierarchy" or "centrality" are used and operationalized. Its static orientation, however, makes this type of analysis less suitable for our purpose.

Following the work of Simon (1962) and others, the *linking pin* theory became a popular object of study. "Having ties to more than one action-set or subsystem, linking pin organizations are the nodes through which a network is loosely joined" (Aldrich and Whetten, 1981, p. 390). The three main functions of linking pin organizations are:

- Communicating between organizations.
- Providing general services interlinking third parties.
- When dominant, serving as an example.

A popular technique to visualize the structure of loosely connected subsystems is "*block modeling*" based on a transaction matrix. This technique seems to be useful for identifying pockets of intense interaction between members, but does not examine relationships between the clusters (Whetten, 1987, p. 242). This makes it less suitable for our purposes.

We propose a *triple technique* as a methodology to visualize dominance in networks:

(1) Draw up an expanded *filière*, containing all material links that the researcher judges as significant; in order to increase a proper judgment, the researcher should be familiar with the sector and actors involved.
(2) Visualize immaterial links between actors; compare this picture with (1).

(3) Use the technique of shocks: *threats and challenges* (Kamann and Strijker, 1990). Using a *dynamic* analysis of the network, the enforcement of power and possible shifts in power can be traced. The outcome of (3) feeds back to (1) and (2) to check whether all significant actors and their links are included. It checks the initial arbitrary judgment of the researcher.

To this triple technique, we suggest to add a fourth layer, covering the *cultural dimension*: the symbols, myths, stories, heroes, customs, script, shared values, and uses of the network.

5.4.16 How to observe power, dynamics, and paradigm fixation

In order to operationalize the various types of dependencies, dynamics, and paradigm fixation, we first of all have to deal with terms such as "control", "power" or "dependency". In a static situation, power supposedly leads to an unequal distribution of profit margins over the actors involved. Furthermore, one could use market share or the market concentration ratio, and the aggregate concentration ratio. Other potential useful indicators are the share of input or output that an actor, a single supplier, or a buyer accounts for and, related to this, the situation of dual sourcing of input versus a single buyer of output. However, in our opinion, power of control should be measured in a dynamic context. Kamann and Strijker (1990) describe it as *the ability to*:

- Prevent a partner from terminating the relation, the no exit power.
- Prevent a partner from duplicating relations with other actors, the power of exclusiveness.
- Exclude potential newcomers from the network or even from the market involved, the no entry power.
- Start and induce innovations in products, processes, materials used and organizational setup, set standards, dictate research agendas, in other words, dictate the technological trajectory or even the entire paradigm.
- Prevent innovations, standards, and research projects.
- Internalize external threats to the network without giving up central control.
- Dictate network responses to outside dangers and opportunities.
- "Control" the interlocking system, to have one's representatives replace representatives of other actors.

- Increase shareholding and/or board functions.
- Dictate the social paradigm.

An analysis that tries to trace centrality of control – power – has to be dynamic; attention should be paid to the various types of dependencies and paradigm fixations or shifts as mentioned earlier.

5.4.17 Conclusions about networks

Actors do not live in an atomistic world, they depend on other actors. Because of cooperation between actors, specialization is possible and projects can be pursued that otherwise would not be possible for each of the actors involved. This network behavior creates a *synergetic surplus*.

The resulting degree of dependence is determined by the nature of the mutual relations between actors. It affects the strategic value of flows of information, goods, and services an actor is able to obtain. It involves the freedom to act in the network and to select products, processes and markets, suppliers and buyers, partners or competitors. Those actors who manage to *dominate* network relations or even entire networks will try to consume the synergetic surplus of the network, at the cost of an equal distribution over all participants.

Whether this is organized in the style of the spatial product life cycle – head office, pilot plant, standardized branch plants – or with more autonomous business units is important for actors that are part of a corporate network. For "independent" actors, it is important whether they are *equal* partners or participants of a dual production organization, to mention the two extreme cases of a scale. Takeovers of actors in a region by actors that operate within the same region will result in job losses because of rationalization of production and overheads, but may increase the region's competitiveness with actors from other regions. External control however, from minority shareholding to complete takeovers may, in its worst form, result in the closure of actors in the network and because of this, loss of jobs and capital.

Therefore, we conclude that network domination not only affects the locational choice of actors, but also the intrinsic value of activities in an area. It therefore determines the present incomes generated in an area *and* the possible incomes in the future.

5.5 Conclusions

The spatial differentiation in the selection environment is an important aspect of the origin of new ideas and innovations. It also plays an important role in the opportunities actors have to be informed about new developments of relevance to them. Whether they can actually *use* new developments is to a large extent determined by their network *freedom* and participation. Network cooperation between actors creates a synergetic surplus. Those actors who succeed to dominate other actors or even entire network segments will consume that synergetic surplus at the cost of others.

Dominant actors will therefore, through these network relations, dominate those areas where they are situated. Distribution of power over the participants of a network and the ability to monopolize strategic information in a network are important for the diffusion of innovations and for the related distribution of incomes generated.

Unfortunately, so far only a few analytical models have been devised that seek to cover the above-mentioned conflicting patterns of spatial development.

References

Aglietta, M., 1986, *A Theory of Capitalist Regulation*, Left Books, London, UK.

Alderman, N. and Thwaites A., 1990, Methodological issues in the development of predictive models of innovation diffusion, in R. Cappellin and P. Nijkamp, eds., *The Spatial Context of Technological Development at the Local Level*, Gower, Aldershot, UK.

Aldrich, H. and Whetten, D.A., 1981, Organization-sets, action-sets and networks: Making the most of simplicity, in P.N. Nystrom and W.H. Starbuck, eds., *Handbook of Organizational Design*, Oxford University Press, Oxford, UK.

Axelsson, B., 1987, Supplier management and technological development, in H. Håkansson, ed., *Industrial Technological Development: A Network Approach*, Croom Helm, London, UK.

Bluestone, B. and Harrison, B., 1982, *The Industrialization of America*, Basic Books, New York, NY, USA.

Cappellin, R., 1990, *The diffusion of producer services in the urban system*, Paper presented at the European Advanced Institute of the Regional Science Association Arco, Italy, July.

Cook, K.S. and Emerson, R.M., 1978, Power, equity and commitment in exchange networks, *American Sociological Review* **43**:721–739, October.

Crevoisier, O. and Maillat, D., 1989, Milieux, organization industrielle et système productif territorial: vers une typology, *Groupe de Recherche Europeen sur les Milieux Innovateurs* (GREMI), Barcelona, Spain, March.

Davelaar, E.J. and Nijkamp, P., 1987, The urban incubation hypothesis: Old wine in new bottles? *Research Memorandum* 1986-39, Free University, Amsterdam, The Netherlands.

Ewers, H.J. and Wettman, R.W., 1980, Innovation-oriented regional policy, *Regional Studies* **14**:161–179.

Galbraith, J.K., 1963, *American Capitalism, The Concept of Countervailing, Power*, Penguin, New York, NY, USA.

Giaoutzi, M. and Nijkamp, P., 1988, *Informatics and Regional Development*, Gower, Aldershot, UK.

Goddard, J., 1973, Office employment, urban development and regional policy, *Office Location and Regional Development*, An Foras Forbartha, Dublin, Ireland.

Hamfelt, C. and Lindberg, A.K., 1987, Technological development and the individual's contact network, in H. Håkansson, ed., *Industrial Technological Development: A Network Approach*, Croom Helm, London, UK.

Jay, E.J., 1964, The concepts of "field" and "network" in anthropological research, *Man* **64**:137–139.

Johanson, J. and Mattson, L.-G., 1984, *Marketing Investments and Market Investments in Industrial Networks*, Paper presented at the International Research Seminar on Industrial Marketing, Stockholm School of Economics, Stockholm, Sweden.

Kamann, D.J.F., 1985, Technologie en ruimte: steeds nieuwe paradigma's, in M.T. Brouwer and H.W. ter Hart, eds., *Ondernemen in Nederland*, Kluwer, Deventer, The Netherlands.

Kamann, D.J.F., 1986, Industrial organization, innovation and employment, in P. Nijkamp, ed., *Technological Change, Employment and Spatial Dynamics*, Springer-Verlag, Berlin, Heidelberg, New York.

Kamann, D.J.F., 1988, *Spatial Differentiation in the Social Impact of Technology*, Avebury (Gower), Aldershot, UK.

Kamann, D.J.F and Nijkamp, P., 1990, Technogenesis: Incubation and diffusion, in R. Cappellin and P. Nijkamp, eds., *The Spatial Context of Technological Development at the Local Level*, Gower, Aldershot, UK.

Kamann, D.J.F. and Strijker, D., 1990, In search of the spider in the web, *European Review of Agricultural Economics* (forthcoming).

Laage-Hellman, J., 1987, Process innovation through technical cooperation, in H. Håkansson, ed., *Industrial Technological Development: A Network Approach*, Croom Helm, London, UK.

Malecki, E.J. and Nijkamp, P., 1988, Technology and regional development: Some thoughts on policy, *Environment & Planning C* 6(4):383–399.

Mowery, D. and Rosenberg, N., 1979, The influence of market demand upon innovation: A critical review of some recent empirical studies, *Research Policy* **8**:102–153.

Nijkamp, P., Poot, J., and Rouwendal, J., 1988, R&D policy in space and time: A nonlinear evolutionary growth model, *Research Memorandum* 1988-30, Department of Economics, Free University, Amsterdam, The Netherlands.

Oakey, R.P., Thwaites, A.T., and Nash, P.A., 1980, The regional distribution of innovative manufacturing establishments in Britain, *Regional Studies* **14**:235–253.

Perroux, F., 1955, Note sur la notion pôle de croissance, *Economie Appliqué*, Série D, **8**.

Porter, M.E., 1985, *Competitive Advantage*, Collier-MacMillan, New York, NY, USA.

Rogers, E.M. and Larssen, J.K., 1987, *Silicon Valley Fever*, Basic Books, New York, NY, USA.

Scott, A.J., 1985, *High Technology Industry and Territorial Development: The Rise of the Orange County Complex, 1955–1984*, Working Paper Series, Nr. 85, Department of Geography, University of California, Los Angeles, CA, USA.

Simon, H.A., 1962, The architecture of complexity, *Proceedings of the American Philosophical Society* **106**:467–482.

Storper, M. and A.J. Scott, 1990, The geographical foundations and social regulation of flexible production complexes, in J. Wolch and M. Dear, eds., *The Power of Geography*, Urwin Hyman (forthcoming).

Townroe, P., 1988, *Regional Development Potentials and Innovation Capacities*, Paper presented at the Conference on Innovation and Regional Development, Berlin, mimeo.

Whetten, D.A., 1987, Interorganizational relations, in J.W. Lorsch, ed., *Handbook of Organizational Behavior*, Prentice-Hall, Englewood Cliffs, NJ, USA.

Williamson, O.E., 1975, *Markets and Hierarchies: Analysis and Antitrust Implications*, The Free Press, New York, NY, USA.

Williamson, O.E., 1979, Transaction-cost economics: The governance of contractual relations, *The Journal of Law and Economics* **2**:233–261.

Williamson, O.E., 1987, Transaction-cost economics: The comparative contracting perspective, *Journal of Economic Behaviour and Organization* **8**:617–625.

Chapter 6

New Product Diffusion Models in Marketing: A Review and Directions for Research*

Vijay Mahajan, Eitan Muller, and Frank M. Bass

6.1 Introduction

The diffusion of an innovation traditionally has been defined as the process by which that innovation is "communicated through certain channels over time among the members of a social system" (Rogers, 1983, p. 5). As such, the diffusion process consists of four key elements: innovation, communication channels, time, and the social system.

As a theory of communications, diffusion theory's main focus is on communication channels, which are the means by which information about an innovation is transmitted to or within the social system. These means consist of both the mass media and interpersonal communications. Members

*Reprinted with permission from the *Journal of Marketing*, Vol. 54 (January 1990), 1–26. Published by the American Marketing Association, Chicago, IL 60606, USA.

of a social system have different propensities for relying on mass media or interpersonal channels when seeking information about an innovation. Interpersonal communications, including nonverbal observations, are important influences in determining the speed and shape of the diffusion process in a social system.

Since its introduction to marketing in the 1960s (Arndt, 1967; Bass, 1969; Frank *et al.*, 1964; King, 1963; Robertson, 1967; Silk, 1966), innovation diffusion theory has sparked considerable research among consumer behavior, marketing management, and management and marketing science scholars. Researchers in consumer behavior have been concerned with evaluating the applicability of hypotheses developed in the general diffusion area to consumer research (Gatignon and Robertson, 1985). The marketing management literature has focused on the implications of these hypotheses for targeting new product prospects and for developing marketing strategies aimed at potential adopters (see, e.g., Engel *et al.*, 1986, Chapter 20; Kotler and Zaltman, 1976; McKenna, 1985, Chapter 4). Researchers in management and marketing science have contributed to the development of diffusion theory by suggesting analytical models for describing and forecasting the diffusion of an innovation in a social system. More recently, this literature also has been concerned with developing normative guidelines for how an innovation should be diffused in a social system.

We focus on the contributions of management and marketing science literature to the cumulative understanding of the dynamics of innovation diffusion. The main impetus underlying these contributions is a new product growth model suggested by Bass (1969). The Bass model and its revised forms have been used for forecasting innovation diffusion in retail service, industrial technology, agricultural, educational, pharmaceutical, and consumer durable goods markets (Akinola, 1986; Bass, 1969; Dodds, 1973; Kalish and Lilien, 1986a; Lancaster and Wright, 1983; Lawton and Lawton, 1979; Nevers, 1972; Tigert and Farivar, 1981). Representative companies that have used the model include Eastman Kodak, RCA, IBM, Sears and At&T (Bass, 1986).

Since publication of the Bass model, research on the modeling of the diffusion of innovations in marketing has resulted in an extensive literature. Contributions of this literature through the 1970s were reviewed by Mahajan and Muller (1979). However, in the ensuing decade a plethora of studies has contributed to our understanding of the structural, estimation, and conceptual assumptions underlying diffusion models. Though some of these recent developments have been documented by Mahajan and Peterson (1985) and

Table 6.1. Emergence of diffusion modeling literature in marketing.

Research areas	Time period 1960s	1970s	1980s
Basic diffusion models	Formulation of relationship between imitators and innovators over time. Saturation effect		Unbundling of adopters. Definition of innovators/imitators. Development of diffusion models from individual-level adoption decisions
Parameter estimation considerations	Estimation when data are available: Ordinary least squares estimation procedure		Estimation when no prior data are available: Algebraic estimation. Product/market attribute-based analogical estimation. Estimation when data are available: Time-invariant parameter estimation (maximum likelihood, nonlinear least squares). Time-varying parameter estimation (Bayesian estimation, feedback filters)
Flexible diffusion models			Systematic (or random) variation in diffusion model parameters over time. Flexible diffusion patterns in terms of timing and magnitude of peak of adoption curve
Refinements and extensions		Dynamic diffusion models: market saturation changes over time. Multi-innovation diffusion models: other innovations influence diffusion of an innovation. Space/time diffusion models: diffusion of an innovation occurs simultaneously in space and time. Multistage diffusion models: adopters pass through a series of stages in the innovation-decision process	Multigeneration models: timing and adoption of different generations of an innovation. Multistage diffusion models: effect of negative word of mouth in the innovation decision process. Diffusion models with marketing mix variables: effect of price (linkage with experience curves), advertising, personal selling, distribution, and timing of new product introduction on diffusion patterns. Product/market attribute-based diffusion models: effect of social system characteristics and perceived product attributes on diffusion patterns. Controlled diffusion models: effect of supply restrictions on diffusion patterns. Multiadoption diffusion models: incorporation of repeat sales and replacement sales in diffusion patterns. Competitive diffusion models: effect of competitive actions in terms of pricing, advertising, and number of brands on diffusion patterns
Use of diffusion models	Forecasting	Forecasting: problems in use of diffusion models for forecasting	Forecasting: problems in the use of diffusion models for forecasting. Descriptive: testing of hypotheses related to diffusion of innovations across countries, effect of product/market attributes on diffusion patterns, relation between innovation diffusion and market structure factors such as the experience curve phenomenon and proliferation of number of brands. Normative: derivation of optimal pricing, advertising, and timing strategies in monopoly and oligopoly markets

Mahajan and Wind (1986a), we now extend these efforts by presenting a critical evaluation of the cumulative developments since the Bass (1969) and Mahajan and Muller (1979) articles. *Table 6.1* is a summary of these developments over the last two decades across five subareas: basic diffusion models, parameter estimation considerations, flexible diffusion models, refinements and extensions, and use of diffusion models.

6.2 The Basic First-Purchase Diffusion Models

Mahajan and Muller (1979) have stated that the objective of a diffusion model is to present the level of spread of an innovation among a given set of prospective adopters over time. The purpose of the diffusion model is to depict the successive increases in the number of adopters and predict the continued development of a diffusion process already in progress. In the product innovation context, diffusion models focus on the development of a life cycle curve and serve the purpose of forecasting first-purchase sales of innovations. That is, in the first-purchase diffusion models one assumes that, in the product planning horizon being considered, there are no repeat buyers and purchase volume per buyer is one unit. The number of adopters defines the unit sales for the product. Diffusion models, by definition, are concerned with representing the growth of a product category.

The best-known first-purchase diffusion models of new product diffusion in marketing are those of Bass (1969), Fourt and Woodlock (1960), and Mansfield (1961). These early models attempted to describe the penetration and saturation aspects of the diffusion process. After briefly reviewing the original formulations of these models, we review the recent developments that further evaluate their basic structure.[1]

6.2.1 The Bass model

The main impetus underlying diffusion research in marketing is the Bass model. Subsuming the models proposed by Fourt and Woodlock (1960) and Mansfield (1961), the Bass model assumes that potential adopters of an innovation are influenced by two means of communication – mass media and word of mouth. In its development, it further assumes that the adopters of an innovation comprise two groups. One group is influenced only by the mass-media communication (external influence) and the other group is influenced only by the word-of-mouth communication (internal influence). Bass termed the first group "Innovators" and the second group "Imitators."

Unlike the Bass model, the model proposed by Fourt and Woodlock (1960) assumes that the diffusion process is driven primarily by the mass-media communication or the external influence. Similarly, the model proposed by Mansfield (1961) assumes this process is driven by word of mouth.

Figure 6.1 is a plot of the conceptual and analytical structure underlying the Bass model. As noted in *Figure 6.1(a)*, the Bass model conceptually assumes that "Innovators" or buyers who adopt exclusively because of the mass-media communication or the external influence are present at any stage of the diffusion process. *Figure 6.1(b)* shows the analytical structure underlying the Bass model. As depicted, the noncumulative adopter distribution peaks at time T^*, which is the point of inflection of the S-shaped cumulative adoption curve. Furthermore, the adopter distribution assumes that an initial pm (a constant) level of adopters buy the product at the beginning of the diffusion process. Once initiated, the adoption process is symmetric with respect to time around the peak time T^* up to $2T^*$. That is, the shape of the adoption curve from time T^* to $2T^*$ is the mirror image of the shape of the adoption curve from the beginning of the diffusion process up to time T^* (Mahajan *et al.*, 1990).

The Bass model derives from a hazard function (the probability that an adoption will occur at time t given that it has not yet occurred). Thus $f(t)/[1 - F(t)] = p + qF(t)$ is the basic premise underlying the Bass model. The density function of time to adoption is given by $f(t)$ and the cumulative fraction of adopters at time t is given by $F(t)$. This basic premise states that the conditional probability of adoption at time t (the fraction of the population that will adopt at time t) is increasing in the fraction of the population that has already adopted. Therefore, the basic premise states that part of the adoption influence depends on imitation or "learning" and part of it does not. The parameter q reflects that influence and parameter p reflects an influence that is independent of previous adoption. If q is zero, $f(t)$ will follow the negative exponential distribution. If m is the potential number of ultimate adopters, the number of adopters at time t will be $mf(t) = n(t)$ and the cumulative number of adopters at time t will be $mF(t) = N(t)$. The basic premise of the Bass model can be manipulated, along with the definitions just provided, to yield:

$$n(t) = \frac{dN(t)}{dt} = p[m - N(t)] + \frac{q}{m}N(t)[m - N(t)]. \tag{6.1}$$

The first term, $p[m - N(t)]$, in equation (6.1) represents adoptions due to buyers who are not influenced in the timing of their adoption by the number

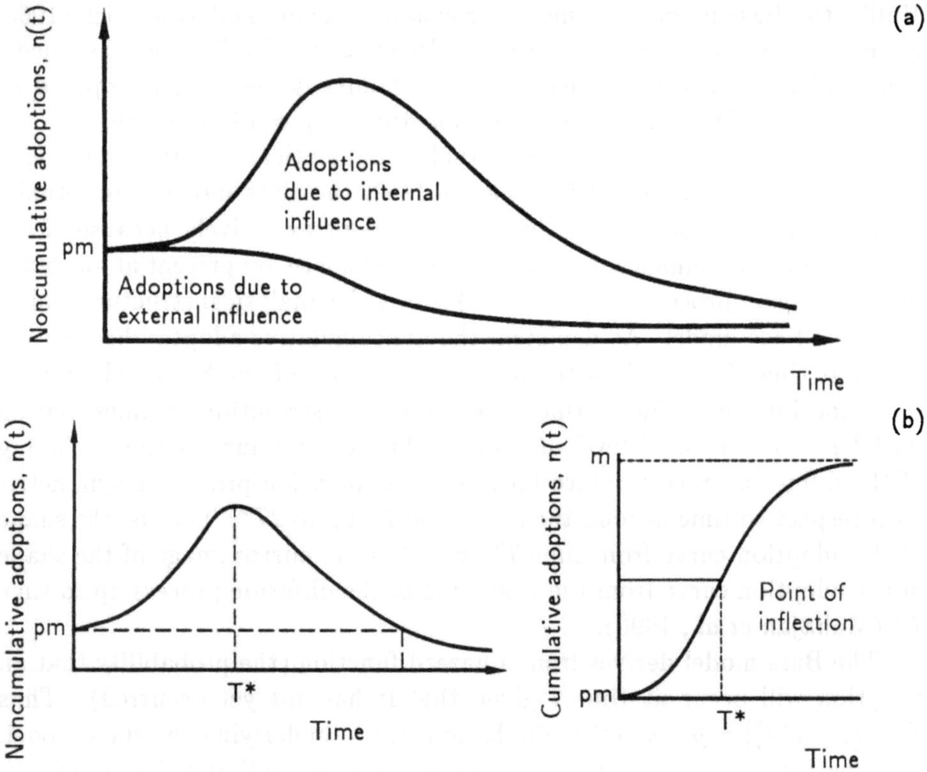

Figure 6.1. The Bass new product diffusion model. (a) Adoptions due to external and internal influences in the Bass model. (b) Analytical structure of the Bass model.

of people who have already bought the product. Bass (1969) referred to p as the "coefficient of innovation." The second term in equation (6.1), $q/mN(t)[m-N(t)]$, represents adoptions due to buyers who are influenced by the number of previous buyers. Bass (1969) referred to q as the "coefficient of imitation." Note in equation (6.1) that at time $t = 0, n(0) = pm$.

Equation (6.1) is a first-order differential equation. It can be integrated to yield the S-shaped cumulative adopter distributor, $N(t)$. Once $N(t)$ is known, further differentiation yields expressions for the noncumulative number of adopters, $n(t)$, and the time (T^*) and magnitude $[n(t^*)$ and $N(t^*)]$ of the peak of the adoption curve.[2]

Given the basic structure of the Bass diffusion model, three questions can be raised:

- How does the Bass model compare with the classical normal distribution model proposed by Rogers (1983)?
- Is the Bass model complete in capturing the communication structure between the two assumed distinct groups of innovators and imitators?
- How can the Bass model, which captures diffusion at the aggregate level, be linked to the adoption decisions at the individual level?

Recent developments that address these three questions are discussed next.

6.2.2 Unbundling of adopters

Rogers (1983, p. 244) has articulated that because of the interpersonal interaction, the adoption curve should have a normal distribution. In fact, using two basic statistical parameters of the normal distribution – mean and standard deviation – Rogers has proposed an adopter categorization scheme dividing adopters into five categories of Innovators, Early Adopters, Early Majority, Late Majority, and Laggards.

To establish the linkage between the Bass model and the classical normal distribution model, Mahajan *et al.* (1990) compared the two approaches. In their comparison, they highlight two points. First, they argue that adopters termed "Innovators" in the Bass model should not be called innovators because they are not necessarily the first adopters of an innovation, as defined by Rogers. Following Lekvall and Wahlbin (1973), they suggest that because the Bass model captures the spread of an innovation due to the mass media and interpersonal communication channels, the Bass model coefficients p and q should be referred to as the coefficient of external influence and the coefficient of internal influence, respectively. (We will use these labels in the rest of this article.) They also provide an explicit expression to estimate the total number of adoptions due to external influence at any time in the diffusion process:

$$N_1(t) = m\frac{p}{q}\ln\left[\frac{1+\frac{q}{p}}{1+\frac{q}{p}\,e^{-(p+q)t}}\right],$$

where $N_1(t)$ represents adoptions due to external influence. Hence, adoptions due to internal influence are, $N_2(t) = N(t) - N_1(t)$.

Second, Mahajan *et al.* (1990) suggest that because one standard deviation away from the mean of the normal distribution represents its points of inflection (the analytical logic underlying the categorization scheme proposed by Rogers), the same analytical logic can be used to develop adopters categories for the Bass model. This scheme also yields five adopter categories with the number of buyers (pm) who initiate the Bass model being defined as innovators. Examining the diffusion of personal computers, Mahajan *et al.* (1990) show how the adopter categories based on the Bass model can be used to study differences among their profiles.

6.2.3 Innovators versus Imitators

Irrespective of the term "Innovators" used to label buyers who adopt because of external influence in the Bass model, a question can be raised as to whether the Bass model really captures the communication structure between two assumed groups of adopters called "Innovators" and "Imitators." Emphasizing this argument, Tanny and Derzko (1988) suggest that the communication structure assumed in the Bass model is not complete. They propose an extension of the Bass model wherein (1) potential adopters are divided into two distinct groups of Potential Innovators (say m_1) and Potential Imitators (say m_2), (2) both Potential Innovators and Potential Imitators are influenced by the mass-media communication, and (3) *only* Potential Imitators are influenced by word of mouth due to Innovators and Imitators. To appreciate the linkage between the Bass model and its extension proposed by Tanny and Derzko (1988), consider the following rate equations they proposed:

$$\text{Innovators}: \frac{dN_1(t)}{dt} = p_1[m_1 - N_1(t)] \tag{6.2}$$

$$\text{Imitators}: \frac{dN_2(t)}{dt} = p_2[m_2 - N_2(t)] + q_2[N_1(t) + N_2(t)] [m_2 - N_2(t)] \tag{6.3}$$

If we assume that $p_1 = p_2 = p$ (i.e., the coefficient of external influence is the same for both the groups), the total adoptions can be represented by summing the two rate equations [and noting that $m_1 + m_2 = m$ and $N(t) = N_1(t) + N_2(t)$].

$$\begin{aligned} \frac{dN(t)}{dt} &= p[m_1 + m_2 - N_1(t) - N_2(t)] + q_2N(t)[m_2 - N_2(t)] \\ &= p[m - N(t)] + q_2N(t)[m_2 - N_2(t)] \end{aligned} \tag{6.4}$$

Note that equation (6.4) is identical to the Bass model, equation (6.1), except for the fact that equation (6.4) considers the word-of-mouth influence on the potential adopters that are Potential Imitators rather than on all of the potential adopters as is done in the Bass model. In their empirical work on some of the consumer durable products analyzed by Bass (1969), Tanny and Derzko (1988) did not find satisfactory results for their proposed extension (the model either reduced to the Bass model or it failed to provide estimates for the additional model coefficients). These empirical results are not surprising because as the diffusion process progresses, the population of potential adopters mostly comprises Potential Imitators, justifying the parsimonious model suggested by Bass.

6.2.4 Diffusion models from individual adoption decisions

A key aspect of the Bass model is that it addresses the market in the aggregate. The typical variable measured is the number of adopters who purchase the product by a certain time t. The emphasis is on the total market response rather than an individual customer. This approach is convenient in practical terms but it raises the following issue: Can the diffusion model be built by aggregating demand from consumers who behave in a neoclassical microeconomic way? That is, assume that potential adopters are smart and are not just carriers of information. They therefore maximize some objective function such as expected utility or benefit from the product, taking into account the uncertainty associated with their understanding of its attributes, its price, pressure from other adopters to adopt it, and their own budget. Because the decision to adopt the innovation is individual-specific, all potential adopters do not have the same probability of adopting the product in a given time period. Is it possible to develop the adoption curve at the aggregate market level, given the heterogeneity among potential adopters in terms of their probability of adopting the product at any time t? Development of a model that answers this question can potentially assist in ascertaining the effect of marketing mix and other variables on demand for the product via their effect on individual consumers.

In recent years, attempts have been made by Hiebert (1974), Stoneman (1981), Feder and O'Mara (1982), Jensen (1982), Oren and Schwartz (1988), Chatterjee and Eliashberg (1989), and Lattin and Roberts (1989) to develop diffusion models by specifying adoption decisions at the individual level. In these models one assumes that, at any time t, a potential adopter's utility for an innovation is based on his uncertain perception of the innovation's

Table 6.2. Characteristics of diffusion models based on individual adoption decisions.

	Hiebert (1974)	Stoneman (1981)	Feder and O'Mara (1982)	Jensen (1982)	Oren and Schwartz (1988)	Chatterjee and Eliashberg (1989)	Lattin and Roberts (1989)
Nature of innovation studied	High yielding seed varieties (agricultural)	New technology (industrial)	New technology (agricultural)	Exogenously developed innovation (industrial)	Any new product that is potential substitute for current product	Durable goods	Durable goods
Perceptions							
Type of uncertain attribute/benefit	Net income	Return	Profit	Success rate	Performance	Benefit	
Perceptual uncertainty model (distributional assumption)	No specific model: uncertainty due to imperfect information about yield response to inputs (e.g., fertilizer)	Normal distribution	Normal distribution	Discrete: binary (innovation is profitable or unprofitable; uncertainty captured via subjective probability of innovation being profitable)	Beta distribution	Normal distribution	Normal distribution
Preference structure							
Attribute(s) incorporated in utility function	Net income	Returns from new and old technologies, adjustment costs	Profit	Expected return	Success rate	Performance price	Benefit
Assumption about attitude toward risk	No specific assumption: different attitudes considered	Risk averse	Risk neutral	Risk neutral	Risk averse	Risk averse	Risk averse

Table 6.2. Continued.

	Hiebert (1974)	Stoneman (1981)	Feder and O'Mara (1982)	Jensen (1982)	Oren and Schwartz (1988)	Chatterjee and Eliashberg (1989)	Lattin and Roberts (1989)
Adoption decision rule	Maximize expected utility (partial adoption of innovation possible)	Maximize utility to determine proportion of output produced on new technology	Expected profit from new technology exceeds profit from current technology	Expected return from adoption is greater than expected value of continuing waiting for additional information	Expected utility for new product exceeds expected utility for current product	Expected utility for new product exceeds expected utility for status quo	Expected utility for new product exceeds expected utility for status quo
Learning							
Learning model	Not explicit: learning reduces uncertainty	Bayesian	Bayesian	Bayesian	Bayesian	Bayesian	Bayesian
Source of information		Previous experience with new technology	Internal (previous adopters)	External	Internal (previous adopters)	Both internal and external	Internal (previous adopters)
Aggregation							
Heterogeneity criterion on which aggregation is done across potential adopters	No aggregation	Not applicable: considers only intrafirm diffusion	Mean of initial perception about profitability	Initial subjective probability of innovation being profitable	Risk aversion parameter (note: model assumes constant flow of consumers so that aggregation yields market share rather than cumulative penetration)	Initial perceptions (both expectation and degree of uncertainty); perceived reliability of information; risk aversion parameter; price/performance tradeoff	Difference in mean of perceptions about benefit from status quo

performance, value, or benefits. The potential adopter's uncertain perceptions about the innovation, however, change over time as he learns more about the innovation from external sources (e.g., advertising) or internal sources (e.g., word of mouth). Therefore, because of this learning, whenever his utility for the innovation becomes greater than the status quo (he is better off with the innovation), he adopts the innovation. Aggregation across the various potential adopters yields the cumulative adoption curve.

Table 6.2 contrasts the various individual-level diffusion models on several dimensions. Of all the models compared in *Table 6.2*, only three provide explicit functions for aggregate diffusion models. Depending on the assumptions made about the distribution of parameters that measure heterogeneity across individuals, the model by Chatterjee and Eliashberg (1989) yields several basic diffusion models. If risk aversion across potential adopters is assumed to follow a negative exponential distribution, the model by Oren and Schwartz (1988) reduces to the Mansfield model. If the perceived differences in the potential benefits of the product across potential adopters are assumed to follow a uniform distribution, Lattin and Roberts (1989) suggest the following model.

$$N(t) = a + bN(t-1) - \frac{d}{c + N(t-1)} \tag{6.5}$$

where a, b, c, and d are constants. Using the data on several consumer durables, they indicate that their model provides a better fit to the data than the Bass model. Their model contains four parameters, however (vs. three in the Bass model) and, unlike the Bass model, it does not provide $N(t)$ as an explicit function of time, which limits its long-term forecasting ability.

6.3 Parameter Estimation Considerations

The use of the Bass model for forecasting the diffusion of an innovation requires the estimation of three parameters: the coefficient of external influence (p), the coefficient of internal influence (q), and the market potential (m). Though the estimate for the market potential of a new product can be derived from the diffusion time-series data, recent applications of diffusion models have obtained better forecasting results by using exogenous sources of information (such as market surveys, secondary sources, management judgments, or other analytical models) for estimating m (see, for example, Heeler and Hustad, 1980; Mesak and Mikhail, 1988; Oliver, 1987; Souder and Quaddus, 1982; Teotia and Raju, 1986).

In the 1980s, several estimation procedures were proposed to estimate the Bass model parameters (see *Table 6.1*). Meta-analyzing the results of 15 such diffusion studies, Sultan *et al.* (1990) report average values of 0.03 and 0.38 for the coefficients of external influence and internal influence, respectively. Their analyses further suggest that the values of these coefficients are influenced by the type of estimation procedure used to estimate them. For a practitioner, the main question is which of the several estimation procedures should be used and why. The answer to this question depends partially on the amount of data available to estimate these parameters. We review estimation procedures that are designed to develop estimates both in the absence of and in the presence of time-series diffusion data.[3]

6.3.1 No prior data available

If no data are available, parameter estimates can be obtained by using either management judgments or the diffusion history of analogous products.

One procedure that exclusively uses management judgments to estimate the diffusion parameters is an algebraic estimation procedure suggested by Mahajan and Sharma (1986). The implementation of this procedure requires information from managers on three items: (1) the market size (m), (2) the time of the peak of the noncumulative adoption curve, and (3) the adoption level at the peak time (n^*). That is, the key information required by the estimation procedure is the peak of the noncumulative adoption curve. Knowing this information, one can estimate the coefficients of external influence and internal influence. Though the algebraic estimation procedure has been implemented in actual applications by some firms (e.g., Institute for the Future), Bass (1986) has questioned its desirability, suggesting that one of the key outputs of the diffusion model is the prediction of the timing and magnitude of the peak. Therefore, if one can guess these items, there is no need to estimate model parameters.

An alternative algebraic estimation procedure has also been suggested by Lawrence and Lawton (1981). This procedure also involves obtaining information from management on three items: (1) the potential market size (m), (2) the number of adoptions in the first time period, and (3) an estimate of the sum of the coefficients of external influence and internal influence, that is, the $p + q$ value. Though managers may be able to guess the adoption level for the first time period, how does one guess the $p + q$ value? A record of the parameter values of earlier new products may provide a basis, by analogy, for guessing $p + q$. From an analysis of the diffusion patterns of several products, Lawrence and Lawton (1981), for example, recommend using a value of 0.66 for industrial product innovations and a value of 0.50 for consumer product innovations (for an application of this procedure to consumer durables, see DeKluyver, 1982). Such a recommendation may be too general, however, and does not consider idiosyncratic characteristics of a particular diffusion situation. Thomas (1985) therefore has recommended that, for a new product under consideration, the parameters can be estimated by taking a weighted sum of the parameters of analogous products where weights are determined by establishing the similarity/dissimilarity relationships between the new product and the various analogous products on five bases of comparison: environmental situation, market structure, buyer behavior, marketing mix strategy, and characteristics of innovation itself. In fact, to consider idiosyncratic characteristics of a new product in a particular social system, recent analogical approaches estimate its coefficients of external influence and internal influence from regression models that express a historical empirical relationship between these coefficients and product or market attributes of several current products. Once this relationship has been established, the values for the coefficients of a new product can be estimated by knowing its characteristics. Four such approaches for the Bass model have been suggested by Srivastava *et al.* (1985), Gatignon *et al.* (1989), Sultan *et al.* (1990), and Montgomery and Srinivasan (1989).

In studying parameter estimates of the Bass model, Lawrence and Lawton (1981) have found that $p + q$ ranged from 0.3 to 0.7 over several innovations. They note that first year sales, S_1, may be expressed as $m[1 - e^{-(p+q)})/(1 + (q/p)e^{-(p+q)}]$ and hence q/p can be expressed as $[m(1-e^{-(p+q)}) - S_1]/S_1 e^{-(p+q)}$. It is possible to use judgment in guessing m and S_1. In strategic terms, probably the most critical forecast deriving from the Bass model is the time of peak of adoptions, T^*. This value is given by $[1/(p+q)]Ln(q/p)$. Because $p + q$ varies over a relatively narrow range and has a mode around 0.5, for consumer products, guesses of $p+q$, m, and S_1 may provide good estimates of T^*. Lawrence and Lawton (1981, p. 535) report good results with this method.

6.3.2 Availability of data

Since the Bass model contains three parameters (p, q, and m), adoption data for a minimum of three time periods are required to estimate these parameters. Recent empirical studies, however, have documented that estimates of these parameters, and hence the adoption forecasts, are sensitive to the number of datapoints used to estimate them (see, e.g., Hyman, 1988; Tigert and Farivar, 1981). In fact, these studies suggest that stable and robust parameter estimates for the Bass model are obtained only if the data under consideration include the peak of the noncumulative adoption curve (Heeler and Hustad, 1980; Srinivasan and Mason, 1986). Because of these concerns, attempts have been made in recent years to develop estimation procedures that update parameter estimates as additional data become available after the initiation of the diffusion process. These procedures include Bayesian estimation procedures and adaptive filtering approaches that provide time-varying parameter estimates.

Time-invariant estimation procedures. One of the first procedures suggested to estimate the diffusion parameters is the ordinary least squares (OLS) procedure proposed by Bass. The OLS procedure involves estimation of the parameters by taking the discrete or regression analog of the differential equation formulation of the Bass model [i.e., equation (6.1)]. In fact, rearrangement of equation (6.1) yields:

$$\begin{aligned} N(t+1) - N(t) &= pm + (q-p)N(t) - \frac{q}{m}N^2(t) \\ n(t+1) &= \alpha_1 + \alpha_2 N(t) + \alpha_3 N^2(t) \end{aligned} \tag{6.6}$$

where $\alpha_1 = pm$, $\alpha_2 = q - p$, and $\alpha_3 = -q/m$. That is, regression analysis is used to estimate α_1, α_2, and α_3 in equation (6.6). Once α's are known, p, q, and m can be estimated. If one has reason to believe that all datapoints in the diffusion time series should not have an equal weighting in the least squares procedure, discounted least squares may be used for estimating α's (for an application of discounted least squares to the discrete analog of the Bass model, see Young and Ord, 1985).

The OLS procedure, however, has three shortcomings (Schmittlein and Mahajan, 1982):

- Because of the likely multicollinearity between independent variables in equation (6.6), that is, $N(t)$ and $N^2(t)$, the procedure may yield parameter estimates that are unstable or have wrong signs.
- The procedure does not directly provide standard errors for the estimated parameters p, q, and m (and, hence, statistical significance of these estimates cannot be assessed).

- There is a time-interval bias since discrete time-series data are used for estimating a continuous model (i.e., the solution of the differential equation specification of the Bass model).

To overcome these shortcomings, Schmittlein and Mahajan (1982) have suggested a maximum likelihood estimation procedure to estimate the parameters directly from the solution of the differential equation specification of the Bass model. This procedure also has limitations, however. For example, Srinivasan and Mason (1986) point out that because the maximum likelihood procedure considers only sampling errors and ignores all other errors such as the effects of excluded marketing variables that influence the diffusion process, it underestimates the standard errors of the estimated parameters, resulting in possible wrong inferences about the statistical significance of the parameters. To overcome this shortcoming, they suggest a formulation by means of which estimates of p, q, and m can be obtained by using any appropriate nonlinear regression package (a similar formulation has also been suggested by Jain and Rao, 1989). This formulation also uses the solution to the differential equation specification of the Bass model for parameter estimation.

From the preceding descriptions, it is clear that both the maximum likelihood and the nonlinear estimation procedures offer better choices over the OLS procedure. An empirical comparison of these estimation procedures (along with the algebraic estimation procedure suggested by Mahajan and Sharma, 1986) by Mahajan *et al.* (1986) suggests an overall superiority of the nonlinear estimation procedure, but the maximum likelihood procedure performs equally well when survey-type diffusion data are used to estimate the parameters because of the dominance of sampling errors (Mahajan *et al.*, 1986; Srinivasan and Mason, 1986).

Parameter estimation for diffusion models is primarily of historical interest; by the time sufficient observations have developed for reliable estimation, it is too late to use the estimates for forecasting purposes. The estimates can be used for model testing and for comparison across products. Considered in such a context, the methods often yield estimates that do not differ greatly.

Time-varying estimation procedures. These procedures are designed to update parameter estimates as new data become available.[4] The updating of parameters is achieved either with the Bayes procedure or the feedback filters.

Such procedures have been applied in various diffusion settings by Sultan *et al.* (1990), Lenk and Rao (1989), and Bretschneider and Mahajan (1980).

All of these procedures have two elements in common: (1) they require an initial estimate of the diffusion model parameters before the diffusion data become available and (2) they specify an updating formula to upgrade the initial estimates as additional diffusion data become available.

In the Bayesian estimation procedure advocated by Sultan *et al.* (1990), statistical results of their meta-analysis study are used to develop initial estimates for the coefficients of external influence and internal influence for a new product. For each of these two coefficients, the procedure updates the initial estimates by taking a weighted sum of its two values, the initial estimate and the estimate developed from the actual data (by using any procedure such as the nonlinear estimation procedure of Srinivasan and Mason, 1986). The weights in the updating formula are expressed as a function of the variation in the parameter estimates from the actual data so that as these time-varying estimates stabilize, the weight for the initial estimate based on the meta-analysis goes to zero. A Bayesian estimation procedure has also been reported by Lenk and Rao (1989). Their procedure explicitly considers the between-product and within-product variations in establishing initial estimates for the new product.

An alternative approach to updating the diffusion model parameters for the Bass model has been demonstrated by Bretschneider and Mahajan (1980). It estimates the time-varying values of the Bass model parameters by updating the regression coefficients in the discrete analog of the Bass model, equation (6.6). The updating formula is based on a feedback filter suggested by Carbone and Longini (1977). This feedback filter estimates an adjustment to the current values of parameters at time t based on the error between the actual and the predicted values of the noncumulative number of adopters at time t. Though the procedure provides time-varying estimates for the diffusion model coefficients, it has the same shortcomings as the ordinary least squares procedure.[5]

6.4 Flexible Diffusion Models

The basic structure of a diffusion model can be characterized in terms of two mathematical properties, point of inflection and symmetry. The point of inflection on a diffusion curve occurs when the maximum rate of diffusion is reached. If the diffusion pattern after the point of inflection is the mirror image of the diffusion pattern before the point of inflection, the diffusion curve is characterized as being symmetric. For example, as depicted in

Figure 6.1(b), the adopter distribution for the Bass model peaks at time T^*, which is the point of inflection of the S-shaped cumulative adoption curve, and is symmetric with respect to time around the peak time T^* up to time $2T^*$. Furthermore, the Bass model assumes that the maximum penetration rate cannot occur after the product has captured 50% of the market potential. In practice as well as in theory, the maximum rate of diffusion of an innovation should be able to occur at any time during the diffusion process. Additionally, diffusion patterns can be expected to be nonsymmetric as well as symmetric.

Easingwood *et al.* (1983) have suggested that the flexibility in the diffusion models can be achieved by recognizing an important underlying assumption. In most of the diffusion models, the impact of the word of mouth on potential adopters is assumed to remain constant throughout the diffusion span. This assumption is tenuous because, for most innovations, the word of mouth is likely to increase, decrease, or remain constant over time (Hernes, 1976). Easingwood *et al.* (1983) suggest that the time-varying nature of the word-of-mouth effect can be incorporated in the Bass model by specifying the coefficient of internal influence as systematically varying over time as a function of penetration level. That is,

$$w(t) = q\left[\frac{N(t)}{m}\right]^{\alpha} \tag{6.7}$$

where α is a constant and $w(t)$ is the time-varying coefficient of external influence. Substitution of equation (6.7) into the Bass model, equation (6.1), yields the non-uniform influence (NUI) model suggested by those authors (in terms of the cumulative fraction of adopters):

$$\frac{dF(t)}{dt} = [p + qF^{\delta}(t)][1 - F(t)] \tag{6.8}$$

where $F(t) = N(t)/m$ and $\alpha + 1 = \delta$. When $p = 0$ (i.e., coefficient of external influence is zero), equation (6.8) or (6.9) yields a flexible extension of the Mansfield model termed nonsymmetric responding logistic (NSRL) by Easingwood *et al.* (1981). An interesting alternative interpretation of the NSRL model in terms of experience curve and price elasticity has been provided by Sharp (1984).

In addition to the NUI and NSRL models, *Table 6.3* reports characteristics of nine other diffusion models. The following observations are warranted from this table:

- In addition to the NUI and NSRL models, only two models offer complete flexibility in capturing diffusion patterns (i.e., point of inflection can occur from 0% to 100% penetration and the diffusion patterns can be symmetric or nonsymmetric). These are the models proposed by von Bertalanffy (1957) (an identical model has been proposed by Nelder, 1962 and Bewley and Fiebig, 1988).
- Like the NUI and NSRL models proposed by Easingwood *et al.* (1981 and 1983), the model by von Bertalanffy expresses the coefficient of internal influence as systematically changing over time as a function of penetration level, that is,

  $$w(t) = \frac{q(1 - F^{\phi})}{(1 - F)} \tag{6.9}$$

  where ϕ is a constant. Unlike the NUI and NSRL models, the differential equation used to specify the diffusion process by the Von Bertalanffy model has a closed-form solution enabling one to represent cumulative adoption as an explicit function of time. This model, however, assumes that the word-of-mouth effect decreases over time. The NUI and NSRL models can accommodate the word-of-mouth effect that increases, decreases, or remains constant over time.
- In comparison with the models suggested by Easingwood *et al.* (1981 and 1983) and von Bertalanffy (1957), the FLOG (flexible logistic growth) model suggested by Bewley and Fiebig (1988) expresses the systematic variation in the coefficient of internal influence as a function of time, that is,

  $$w(t) = q[(1 + kt)^{1/k}]^{\mu - k} \tag{6.10}$$

  where k and μ are constants. The FLOG model offers a closed form solution and, like the NUI and NSRL models, can accommodate the time-varying word-of-mouth effect.

Though some evidence suggests that, in comparison with the basic diffusion models (such as the Bass model), the flexible models provide a better fit to diffusion data (see Easingwood, 1987 and 1988; Lattin and Roberts, 1989; McGowan, 1986; Rao, 1985), this advantage is obtained by incorporating additional parameters. Hence these models are more difficult to use in the absence of diffusion time-series data (using the historical data on existing products, however, Easingwood (1989) has demonstrated how the NUI model can be used to develop analogical parameter estimates for a new product).

Table 6.3. Flexible diffusion models.

Model	Model equation (dF/dt =)	Model solution (F =)	Point of inflection (F*)	Symmetry[a]	Coefficient of internal influence	Illustrated reported applications
1. Bass (1969)[b]	$(p+qF)(1-F)$	$\frac{1-e^{-(p+q)t}}{1+\frac{q}{p}e^{-(p+q)t}}$	0.0–0.5	NS	Constant	Consumer durable goods; retail service, agricultural, education, and industrial innovations; electronics, photographic products, industrial processes
2. Gompertz curve[c] (see Hendry, 1972; Dixon, 1980)	$qF\ln(\frac{1}{F})$	$e^{-e(c+qt)}$	0.37	NS	Constant	Consumer durable goods, agricultural innovations
3. Mansfield (1961)	$qF(1-F)$	$\frac{1}{1+e^{-(c+qt)}}$	0.5	S	Constant	Industrial, high technology, and administrative innovations
4. Floyd (1962)	$qF(1-F)^2$	*	0.33	NS	Decreasing to zero	Industrial innovations
5. Sharif and Kabir (1976)[d]	$\frac{qF(1-F)^2}{1-F(1-\sigma)}$	*	0.33–0.5	S or NS	Constant or decreasing to zero	Industrial innovations
6. Jeuland (1981)[e]	$(p+qF)(1-F)^{1+\gamma}$	*	0.0–0.5	S or NS	Constant or decreasing to zero	Consumer durable goods
7. Nonuniform influence (NUI) (Easingwood *et al.*, 1983)	$(p+qF^{\delta})(1-F)$	*	0.0–1.0	S or NS	Increasing, decreasing, or constant	Consumer durable, retail service, and education innovations
Nonsymmetric responding logistic (NSRL, p = 0 in NUI) (Easingwood *et al.*, 1981)	$qF^{\delta}(1-F)$	*	.0–1.0	S or NS	Increasing, decreasing, or constant	Medical innovations

Table 6.3. Continued.

Model	Model equation ($dF/dt =$)	Model solution ($F =$)	Point of inflection (F^*)	Symmetry[a]	Coefficient of internal influence	Illustrated reported applications
8. Nelder[f,g] (1962; see McGowan, 1986)	$qF(1-F^{\phi})$	$\frac{1}{[1+\phi e^{-(c+qt)}]^{1/\phi}}$	0.0–1.0	S or NS	Decreasing to a constant	Agricultural innovations
Von Bertalanffy[h] (1957; see Richards, 1959)	$\frac{q}{1-\theta}F^{\theta}(1-F^{1-\theta})$	$[1-e^{-(c+qt)}]^{1/1-\theta}$	0.0–1.0	S or NS	Decreasing to a constant	
9. Stanford Research[i] Institute (e.g., Teotia and Raju, 1986)	$\frac{q}{t}F(1-F)$	$\frac{1}{1+(\frac{T^*}{t})^q}$	0.0–0.5	NS	Decreasing to zero	Energy-efficient innovations
10. Flexible logistic[j] growth (FLOG); (Bewley and Fiebig, 1988)	$q[(1+kt)^{1/k}]^{\mu-k}$	$\frac{1}{1+e^{-[c+qt(\mu,k)]}}$	0.0–1.0	S or NS	Increasing, decreasing, or constant	Telecommunication innovations

[a] S = symmetric, NS = nonsymmetric.
[b] The model is symmetric around the peak time T^* up to $2T^*$.
[c] c is a constant.
[d] $0 \leq \sigma \leq 1$.
[e] $\gamma \geq 0$.
[f] The model suggested by Nelder (1962) is identical to the model originally suggested by Von Bertalanffy (1957). The equivalence between the two can be shown by substituting $\phi = \theta - 1$ in the Van Bertalanffy model.
[g] c is a constant; model reduces to Mansfield model for $\Phi = 1$ and the Gompertz curve as Φ approaches zero.
[h] c is a constant; $\theta \geq 0$; model reduces to Mansfield model when $\theta = 2$ and the Gompertz curve as θ approaches 1.
[i] The model is not invariant to the choice of time scale. A linear transformation of time t is required to make it time-scale independent. T^* is time of 50% penetration. See Bewley and Fiebig (1988).
[j] μ and k are constants and $t(\mu,k) = \{[(1+kt)^{(1/k)}]^{\mu} - 1\}/\mu, \mu \neq 0, k \neq 0, t(\mu,k) = (1/k)\log(1-kt), \mu = 0, k \neq 0, t(\mu,k) = (e^{\mu t}-1)/\mu, \mu \neq 0, k = 0, t(\mu,k) = t, \mu = 0, k = 0$.

6.5 Refinements and Extensions of the Bass Diffusion Model

Several assumptions underlie the Bass model. Most are simplifying assumptions that provide a parsimonious analytical representation of the diffusion process. However, recognition of these assumptions is important to properly understand and interpret the dynamics of innovation diffusion captured by the Bass model. *Table 6.1* lists several of these assumptions that have been of concern to diffusion modelers in the 1970s and 1980s. Nine of these assumptions warrant attention.

Market potential of the new product remains constant over time. The Bass model assumes that the market potential (m) of a new product is determined at the time of introduction and remains unchanged over its entire life (Kalish, 1985; Mahajan and Peterson, 1978; Sharif and Ramanathan, 1981). Theoretically, there is no rationale for a static potential adopter population. Instead, a potential adopter population continuously in flux is to be expected.

Extensions of the Bass model that address this assumption have attempted to relax by specifying the market potential as a function of relevant exogenous and endogenous variables – controllable as well as uncontrollable – that affect the market potential. Examining the diffusion of a durable, Kalish (1985), for example, specified the dynamics of the market potential as a function of price of the product and reduction of uncertainty associated with the product with its increased adoption. Assuming full product awareness in the population, he specified

$$m(t) = m_0 \exp \left[- dP(t) \left(\frac{a+1}{a + \frac{N(t)}{m_0}} \right) \right] \tag{6.11}$$

where a and d are constants, m_0 is the size of the market potential at the time of production, $P(t)$ is the product price, and the term $[(a+1)/(a+N(t)/m_0)]$ represents the effect of market penetration in increasing the size of market potential due to the word-of-mouth effect. Other applications have represented the market potential as a function of growth in the number of households (Mahajan and Peterson, 1978), population growth (Sharif and Ramanathan, 1981), product profitability (Lackman, 1978), price (Chow, 1967; Jain and Rao, 1989; Kamakura and Balasubramanian, 1988), growth in the number of retailers making the product available to potential customers

(Jones and Ritz, 1987), and income distribution, price, and product uncertainty (Horsky, 1990).

Diffusion of an innovation is independent of all other innovations. The Bass model assumes that the adoption of an innovation does not complement, substitute for, detract from, or enhance the adoption of any other innovation (and vice versa) (Peterson and Mahajan, 1978). In reality, however, an innovation is not introduced into a vacuum nor does it exist in isolation. Other innovations are present in the marketplace and may have an influence (positive or negative) on its diffusion. Consideration of simultaneous diffusion of multiple innovations is especially critical if the diffusion of one innovation is contingent upon the diffusion of another innovation (e.g., compact disc software and compact disc hardware) or if the diffusion of one innovation complements the diffusion of another innovation (e.g., washers and dryers).

Following the contingent diffusion model suggested by Peterson and Mahajan (1978), Bayus (1987), for example, conducted an empirical study examining the diffusion dependence between compact disc software and compact disc hardware. In the contingent diffusion model, the market potential of the dependent product is contingent upon the diffusion of the primary product. That is, in the Bass model representation of its growth, equation (6.1), its market potential is specified as $[N_1(t) - N_2(t)]$ where $N_1(t)$ is the cumulative number of adopters of the primary product (e.g., compact disc hardware) and $N_2(t)$ is the cumulative number of adopters of the contingent product (e.g., compact disc software).

Nature of an innovation does not change over time. Manufacturers of high technology products usually achieve diffusion in the marketplace by offering successive generations of an innovation. Each generation is positioned to be better than its predecessors on relevant product attributes. Assessment of market penetration therefore is critical for successive generations of a high technology product. In addition to creating its own demand, each generation of the product cannibalizes the diffusion of its predecessors. This important application of diffusion models for assessing technological substitution has been demonstrated by Norton and Bass (1987) for the growth of two basic types of integrated circuits, memory and logic circuits. If τ_2 represents the introduction of the second generation, Norton and Bass suggest that the word-of-mouth effect within each generation and substitution effects across successive generations can be represented by the following extension of the Bass model:

$$S_1(t) = m_1F_1(t) - m_1F_1(t)F_2(t - \tau_2) \tag{6.12}$$

$$S_2(t) = m_2F_2(t - \tau_2) + m_1F_1(t)F_2(t - \tau_2) \tag{6.13}$$

where equation (6.12) represents the diffusion equation for the first generation product and equation (6.13) represents the second generation product, S_1 and S_2 are their shipments at time t, and $F_1(t)$ and $F_2(t)$ are fractions of adoptions for each generation and are given by the Bass model [solution of equation (6.1)]. In equations (6.12) and (6.13), the term $m_1F_1(t)F_2(t - \tau_2)$ represents the cannibalization or substitution effect.

The geographic boundaries of the social system do not change over the diffusion process. Despite the fact that the diffusion of an innovation occurs simultaneously in space and time, research on these two dimensions of diffusion seldom has been integrated in a marketing context. For example, the new product rollout is clearly a popular option used by many firms to diffuse their products from market to market over time (in both national and international markets). Such a new-product launch strategy enables a firm to capitalize on word-of-mouth communication, referred to as the "neighborhood effect" (Brown, 1981; Gore and Lavaraj, 1987), across markets. Simultaneous assessment of market penetration within a market and across markets therefore is necessary.

One application that addresses diffusion from a joint space and time perspective has been reported by Mahajan and Peterson (1979). In examining the adoption of tractors in 25 states in the central agricultural production region of the United States for the period 1920–1964, they extend the Bass model by assuming that (1) the innovation is introduced initially in one market and (2) the relative number of total adoptions is greater in markets that are closest to the market of innovation origination (i.e., the neighborhood effect diminishes with increased distance from the market of innovation origination, decreasing the size of market potential across markets).

The diffusion process is binary. The Bass model assumes that potential adopters of an innovation either adopt or do not adopt the innovation. As a consequence of this assumption, the Bass model does not take into account stages in the adoption process (e.g., awareness, knowledge, etc.). Some of the attempts to extend the two-stage models to incorporate the multistage (or polynomial) nature of the diffusion process include models by Midgley (1976), Dodson and Muller (1978), Sharif and Ramanathan (1982), Mahajan *et al.* (1984), and Kalish (1985). Most of these extensions tend to characterize stages in which positive, negative, or neutral information is communicated about the product. The implementation of these models is rather cumbersome as they require detailed information about the customer flow across the various stages. In empirical applications, the developers of these models therefore either collapse the various stages (Kalish, 1985 assumes full product awareness), attempt to derive the population in various stages by decomposing the time-series diffusion data (Midgley, 1976; Sharif and Ramanathan, 1982) with too many parameters to be estimated with the limited available data (Silver, 1984), or trace the innovation diffusion with the panel data (Mahajan *et al.*, 1984a; Mahajan *et al.*, 1984b).

Diffusion of an innovation is not influenced by marketing strategies. Since the pioneering work of Robinson and Lakhani (1975) that incorporated the impact of price in the Bass model, several efforts have been made to study systematically the impact of marketing mix variables such as price, advertising, promotion and personal selling, and distribution on product growth (efforts related to price and advertising are reviewed extensively by Kalish and Sen, 1986). As the Bass model contains three parameters (coefficients of external influence and internal influence, and the market potential), the impact of marketing mix variables has been incorporated into the Bass model by representing these parameters as a function of relevant variables. Attempts have been made to represent the market potential as a function of price (e.g., Kalish, 1983 and 1985) and the distribution growth (Jones and Ritz, 1987). Other attempts to incorporate marketing mix variables have been concerned with representing the coefficients of external influence and internal influence as a function of diffusion-influencing variables. Though analytically very elegant, most of these modeling efforts lack empirical validation (Mahajan and Wind, 1986b). However, they can be useful in establishing working hypotheses to examine the likely impact of marketing mix variables on innovation diffusion. As these hypotheses are presented in the next section, we briefly comment here on studies that have provided some empirical support for their extensions.

Two empirical studies by Horsky and Simon (1983) and Simon and Sebastian (1987), respectively, have examined the impact of advertising on innovation diffusion. Studying the diffusion of a new banking service, Horsky and Simon argue that because advertising provides information to innovators, the coefficient of external influence in the Bass model should be represented as a function of advertising expenditures (with diminishing returns). Their empirical results provided a good fit to their diffusion data, supporting their argument. Studying the diffusion of new telephones in the Federal Republic of Germany (FRG), Simon and Sebastian suggest that, though advertising may influence innovators (and hence the coefficient of external influence) in the early stage of the product life cycle, it is more likely to influence the coefficient of imitation in the intermediate life cycle stage of a new product because the objective of the advertising content in the intermediate stage is to influence potential customers through evaluation by customers and social pressure. Furthermore, the advertising effect is cumulative over time. They report a good fit to their diffusion data, supporting their arguments about incorporation of the cumulative effect of advertising into the coefficient of imitation.

The question of the inclusion of price in the Bass model intrigued diffusion analysts in the 1970s and 1980s. Examining the diffusion of a (unmentioned) durable, Kalish (1985) suggested that price affects the market potential of a product [see equation (6.11)]. However, recent empirical studies by Kamakura and Balasubramanian (1988) and Jain and Rao (1989), employing data on several consumer durable products, show that price affects the rate of diffusion (via the coefficients of external influence and internal influence) rather than the market potential.

Product and market characteristics do not influence diffusion patterns. The Bass model does not consider explicitly the impact of product and market characteristics on diffusion patterns. Empirical studies reported in the innovation diffusion literature, however, have found that product and market characteristics have a substantial impact on innovation diffusion patterns (Rogers, 1983; Tornatzky and Klein, 1982). Three empirical studies (Gatignon *et al.*, 1989; Kalish and Lilien, 1986a; Srivastava *et al.*, 1985) have attempted to incorporate product and market characteristics into the Bass model by expressing the coefficients of external influence and/or internal influence as a function of these characteristics. Whereas Srivastava *et al.* (1985) and Kalish and Lilien (1986a) examine the impact of product characteristics on diffusion patterns, Gatignon *et al.* (1989) study the impact of market characteristics on the diffusion of a product across markets. Only Kalish and Lilien (1986a), however, explicitly consider the changing consumer perceptions of the product characteristics as the product is accepted over time. They define the coefficient of imitation as changing over time due to changes in the product characteristics.

They are no supply restrictions. The Bass model is a demand model. If the demand for a product cannot be met because of supply restrictions, such as the unavailability of the product due to limitations on production capacity or difficulties in setting up distribution systems, the excess unmet demand is likely to generate a waiting line of potential adopters (Simon and Sebastian, 1987). In such a situation, the adopter distribution is same as the supply distribution and applying the Bass model to these adoption data is inappropriate. Therefore the Bass model must be extended to integrate the demand-side dynamics with the supply-side restrictions.

A model that captures innovation diffusion dynamics in the presence of supply restrictions has been suggested by Jain *et al.* (1989). Their model conceptualizes diffusion as a three-stage process: potential adopters → waiting adopters → adopters. They have demonstrated the application of their model for the diffusion of new telephones in Israel.

There is only one adoption by an adopting unit. The objective of a diffusion model is to represent the level or spread of an innovation among a given set of prospective adopters. For a great many product innovations, the increase in the number of adopters may consist of first-time buyers as well as repeat buyers. The Bass model, however, captures only the first-time buyers.

In recent years, five empirical studies have been reported that capture the repeat/replacement dynamics of innovation diffusion. Two of these studies, by Lilien *et al.* (1981) and Mahajan *et al.* (1983), include repeat purchase in the Bass model to examine diffusion of ethical drugs. The other two studies, by Olson and Choi (1985) and Kamakura and Balasubramanian (1987), include product replacements in the Bass model to assess long-term sales for consumer durable products. Norton and Bass (1987) assume that adopters continue to buy and that the average repeat buying rate over the population of adopters is constant.

6.6 Uses of Diffusion Models

Innovation diffusion models traditionally have been used in the context of sales forecasting. However, as pointed out by Mahajan and Wind (1986b) and Kalish and Lilien (1986a), sales forecasting is only one of the objectives of diffusion models. In addition to forecasting, perhaps the most useful applications of diffusion models are for *descriptive* and *normative* purposes. Because diffusion models are an analytical approach to describing the spread of a diffusion phenomenon, they can be used in an explanatory mode to test

Table 6.4. Illustrative descriptive applications of diffusion models.

Study by	Hypothesis tested	Diffusion model used	Remarks
Bass (1980)	As a result of learning and accumulated experience, declining patterns of costs and prices result for technological innovations	Bass	Reports results for six durable products. The hypothesis is generally confirmed for most of these products. Similar results are provided by DeKluyver (1982).
Olshavsky (1980)	Product life cycles of consumer durable goods are shortening because of rapidly accelerating technological developments	Mansfield	Study uses data from 25 consumer durable products. Hypothesis is tested by examining relationship between coefficient of imitation and time of introduction of an innovation. Findings confirm hypothesis.
Kobrin (1985)	The pattern of oil production nationalization across countries is a "social interaction" phenomenon	Bass	Study examines patterns of number of countries per year that nationalized oil production from 1960 to 1979. Supplementing quantitative results with detailed qualitative analyses, study confirms hypothesis.
Srivastava *et al.* (1985)	Potential adopters' perceptions of innovation attributes explain the diffusion patterns of a product	Bass	Study examines diffusion of 14 investment alternatives. To explain diffusion patterns across investment alternatives, coefficient of imitation is expressed as a function of perceived product attributes. Two attributes of perceived information cost and perceived likelihood of loss of principal/negative return explain those differences. Findings confirm hypothesis.
Mahajan *et al.* (1988)	Adoption of the M-form organizational structure by the US firms resulted from an imitation behavior	Bass	Study examines adoption of M-form organization structure among 127 US firms from 1950 to 1974. Findings question validity of the hypothesis.

Table 6.4. Continued.

Study by	Hypothesis tested	Diffusion model used	Remarks
Modis and Debecker (1988)	There is a relationship between the number of new computer models and the number of new computer manufacturers.	Mansfield	Study uses data on number of new models introduced and number of new manufacturers that emerged in computer market from beginning of 1958 to end of 1984. Study is also done for personal computers. By examining relationship between growth patterns of number of new computer models and number of new computer manufacturers, the authors conclude that, on average, a new computer manufacturer emerges for every five new models that appear on the market. For the personal computers market, this relationship is around one for every six.
Rao and Yamada (1988)	Potential adopters' perceptions of innovation attributes explain the diffusion pattern of a product	Lilien *et al.* (1981)	Study examines diffusion of 21 ethical drugs using repeat-purchase diffusion model suggested by Lilien, Rao, and Kalish. Model coefficients are expressed as a function of six perceived attributes of ethical drugs. Findings confirm hypothesis.
Takada and Jain (1988)	Cultural differences among countries will lead to different diffusion patterns	Bass	Study examines diffusion of eight consumer durable products in Japan, Korea, and the USA. By testing differences between coefficients of innovation and imitation across the three countries, the authors conclude that among the three countries analyzed, a product is adopted in Korea at a much faster rate than in either the USA or in Japan. No significant differences are found between the diffusion patterns in Japan and the USA.
Gatignon *et al.* (1989)	Three dimensions explain the differences in the diffusion patterns across countries: level of cosmopolitanism of a country, mobility, and the role of women in the society	Bass	The study examines the diffusion of six consumer durable products in 14 European countries. Coefficients of imitation and innovation are expressed as a function of variables measuring the three hypothesized dimensions, and their impact on the two coefficients is determined simultaneously across products and across countries. Findings confirm hypothesis.

specific diffusion-based hypotheses. Further, because diffusion models are designed to capture the product life cycle of a new product, they can be used for normative purposes as the basis of how a product should be marketed.

6.6.1 Descriptive uses

Table 6.4 is a listing of nine illustrative studies in which the diffusion modeling framework has been used to test hypotheses. Srivastava *et al.* (1985) and Rao and Yamada (1988) use diffusion models to test hypotheses related to the impact of perceived product attributes on diffusion patterns. Kobrin (1985), Takada and Jain (1988), and Gatignon *et al.* (1989) use diffusion models to test hypotheses related to innovation diffusion across countries. Bass (1980), Olshavsky (1980), and Modis and Debecker (1988) use diffusion models to test hypotheses related to the life-cycle dynamics of a new product. Finally, Mahajan *et al.* (1988) evaluate the hypothesis that any S-shaped curve may not be a result of the imitation process. The preceding studies clearly demonstrate how the diffusion models can be used to evaluate hypotheses related to the dynamics of innovation diffusion.

6.6.2 Normative uses

Though diffusion models are concerned with representing the growth of a product category, that growth can be influenced by the individual or by collective actions of competitors that have long-term effects on the growth or decline of the market. Alternatively, even if there is only one firm in the industry, it must consider the life cycle dynamics over time to determine optimal marketing mix strategy for maximizing its profitability. That is, it must find out what trajectory (pattern or strategy) of the relevant marketing mix variables it should follow to maximize its discounted profits over the planning period given the constraint that the life cycle of the product follows a certain growth pattern. It therefore solves the following dynamic optimization problem:

$$\text{Maximize } \pi = \text{Total discounted profits over the planning period} \tag{6.14}$$

$$\text{Subject to : A given life cycle growth pattern} \tag{6.15}$$

The dynamic optimization formulation outlined in expressions (6.14) and (6.15) is the general framework that has been used by several authors in the 1980s to develop optimal marketing mix strategies, especially for price and advertising. Most of these studies use the Bass model, and its extensions incorporating marketing mix variables, in expression (6.15) to represent the life cycle dynamics over time. They usually consider a single marketing mix variable, such as price, to isolate its effects on product growth. Before we comment on these studies, a further elaboration on expressions (6.14) and (6.15) is warranted.

Note that the determination of trajectory of the marketing mix variable(s) that maximizes expression (6.14) depends on the specification of the growth model used to specify the life cycle growth pattern in expression (6.15). Therefore, though most of the studies use the Bass model to capture the word-of-mouth effect in expression (6.15), different optimal strategies can be obtained depending on how the relevant marketing mix variables are incorporated in the Bass model. To highlight this point, we consider here derived optimal pricing strategies for new durable goods.

When launching a new product, a firm usually can choose between two distinct pricing strategies, market skimming and market penetration. A market-skimming strategy uses a high price initially to "skim" the market when the market is still developing. The market penetration strategy, in contrast, uses a low price initially to capture a large market share.

Introduction of the impact of price in the Bass model framework generally has resulted in two types of normative pricing strategies. One derived pricing strategy posits that price will increase at introduction, peak, and decrease later (Dolan and Jeuland, 1981; Jeuland and Dolan, 1982; Kalish, 1983; Robinson and Lakhani, 1975). Kalish and Sen's (1986, p. 94) intuitive explanation for this pricing strategy is that if early adopters have a strong positive effect on late adopters, a low introductory ("subsidized") price should encourage them to adopt the product. Consequently, once a product is established, price can be raised because the contribution to sales due to additional adopters decreases over time. Studies deriving this pricing strategy generally assume that price does not affect the population of potential adopters and produces a *multiplicative* effect on the rate of diffusion. That is, from equation (6.1)

$$\frac{dN(t)}{dt} = \left[p + \frac{q}{m}N(t)\right](m - N(t)g(P) \tag{6.16}$$

where $g(P)$ is the price response function for the dynamic price P at time t. Equation (6.16) assumes that price affects the rate of diffusion.

The second derived pricing strategy posits that price is more likely to decrease over time, supporting the market-skimming strategy (Kalish, 1983).

Table 6.5. Optimal marketing mix strategies for innovation diffusion.

Issue and marketing mix variable	Major assumptions/comments	Major normative results	Illustrative references
Industry setting: Monopoly *Price*			
How should a monopolist price a new product over its life cycle?	1. Price interacts with diffusion (rate of adoption) 2. Demand saturation effect causes decline in price over time and diffusion effect causes price to increase over time; experience effect (learning by doing) causes a decline in price over time	For a long planning horizon, if imitation effect is dominating, price first increases and then decreases	Robinson and Lakhani (1975), Dolan and Jeuland (1981), Kalish (1983), Clarke *et al.* (1982)
	Price has a multiplicative effect on diffusion and interacts with experience curve	Price declines over time	Kalish (1983), Bass and Bultez (1982)
	Price affects market potential	Price declines over time	Kalish (1983)
How should a monopolist price over time a new product that can be copied?	Monopolist produces a new product that can be copied. Market potential is affected by price	If product is not protected against copying, price is initially high and then decreases as copying increases	Nascimento and Vanhonacker (1988)
How should a monopolist price a repeat-purchased product over its life cycle?	Price affects market potential	1. Price increases monotonically over time 2. With experience effect, price may first increase (strong imitation effect) then decrease (strong experience effect)	Feichtinger (1982), Jorgensen (1983 and 1986), Kalish (1983), Jeuland and Dolan (1982)
How should a monopolist price over time a new product or service whose consumption value increases with the expansion of the "network" of adopters referred to as a network externality (e.g., electronic mail)?	Price and cumulative adoption affect market potential	Price increases over time	Dhebar and Oren (1985)

Table 6.5. Continued.

Issue and marketing mix variable	Major assumptions/comments	Major normative results	Illustrative references
Advertising			
How should a monopolist advertise a new product over time?	Advertising affects coefficients of innovation and/or imitation; three types of response functions can be used to represent this effect: linear, concave (diminishing returns), S-shaped (increasing and then diminishing returns)	1. Linear response function implies a blitz followed by a constant maintenance level 2. If advertising affects only innovators, concave response function implies a policy whereby advertising decreases over time, gradually approaching the maintenance level; if advertising affects imitators, concave response function implies a policy whereby advertising increases over time 3. S-shaped response function implies a high intensity blitz level followed by a pulsing policy	Horsky and Simon (1983), Dockner and Jorgensen (1988a), Mahajan and Muller (1986)
Timing			
When should a monopolist introduce a product if both positive and negative word of mouth affect diffusion process? How should product be advertised over time?	Members in a social system pass through three stages of innovation decision process: unaware, potential customers, and adopters; both potential customers and adopters circulate positive as well as negative word of mouth	Optimal timing calls for advertising before product is introduced and withdrawal of product after end of advertising period	Mahajan *et al.* (1984a)
When should a monopolist introduce a second generation product? Should firm shelve it or introduce it as soon as it is available?	No pricing or advertising effect	In most cases, optimal timing decision is "now or never"; if optimal introduction time exists, it is early in life cycle of first product	Wilson and Norton (1989)

Table 6.5. Continued.

Issue and marketing mix variable	Major assumptions/comments	Major normative results	Illustrative references
Industry setting: Oligopoly			
Price			
How will firms in an oligopoly price their products over their life cycle?	Price has a multiplicative effect on diffusion; demand saturation causes decline in price; diffusion effect causes price increase; experience effect causes a decline in price	Monopoly results extend to oligopoly case: if imitation effect is strong, price increases initially, and if planning horizon is long, it decreases toward end of planning horizon	Thompson and Teng (1984), Clarke and Dolan (1984), Dockner and Jorgensen (1988b)
How does an industry set a price of a new product class over time?	Market price is a function of quantities set by oligopolists	Same as above	Rao and Bass (1965)
Advertising			
How would firms in an oligopoly advertise their products over time?	Advertising affects innovators or imitators: linear or concave advertising response	1. In many cases, advertising starts with a high level that decreases to a constant maintenance policy 2. Emphasis on final market shares causes an increase in advertising toward end of planning horizon 3. In some cases, advertising may increase; moreover, in some cases advertising for one competitor may increase while that of the second competitor may decrease, or both may increase	Deal (1979), Teng and Thompson (1983), Thompson and Teng (1984), Erickson (1985)
Timing			
Does order of entry affect long-term market share? How would anticipation of entry affect investment decision of a monopolist?	Multiplicative price effect on diffusion	1. Order of entry has no long-term effects on final market shares 2. Monopolist who does not foresee entry overcapitalizes in contrast to foresighted monopolist who anticipates entry	Fershtman *et al.* (1990)
How would anticipation of entry affect pricing decision of a monopolist?	Multiplicative price effect on diffusion	Foresighted monopolist who anticipates entry reduces price in contrast to a surprised monopolist who does not foresee entry	Eliashberg and Jeuland (1986)

In deriving this optimal strategy, some researchers have assumed that price affects the market potential. That is:

$$\frac{dN(t)}{dt} = \left[p + \frac{q}{m}N(t)\right][m(P) - N(t)]. \tag{6.17}$$

The preceding analyses illustrate that we must be cautious about the normative policies derived from the diffusion-based dynamic optimization framework because the derived policies could be simply an artifact of the underlying assumptions made for analytical convenience. Despite this observation, the diffusion modeling framework has provided an excellent opportunity to develop a "theory" of life cycle analysis for empirical validation.

Table 6.5 is a summary of some of the major results from various studies for optimal strategies for three variables: pricing, advertising, and product introduction time. We summarize these results for two industry settings, monopoly and oligopoly. The major results reported for each study reflect the issue raised in the study.

6.7 Conclusions and Discussion

From our review of the emerging literature on innovation diffusion modeling in marketing, we can highlight research issues that must be addressed to make these models theoretically more sound and practically more effective and realistic. We discuss such research possibilities related to the five subareas of recent developments.

6.7.1 Basic diffusion models

Though several assumptions underlying the Bass model have been of concern in the 1980s (Mahajan and Wind, 1986a), we believe five issues warrant further investigation.

Adoptions due to internal influence. One of the key features of the Bass model is that it explicitly considers the influence of internal (word of mouth) as well as external sources of communication on innovation diffusion. As depicted in *Figure 6.1(a)*, the Bass model assumes that adopters whose purchase decisions are influenced by external sources of information are present at any stage of the diffusion process. Such adopters, however, should not be labeled "innovators" because innovators, by definition, are characterized as the first adopters of an innovation (Mahajan *et al.*, 1990). The question now are: What are the characteristics of adopters who, despite a large product penetration in the marketplace, are predominantly influenced by external sources? How do they differ from innovators and other adopter categories on those characteristics? Because, within a certain time period in the diffusion process, the Bass model implies the presence of adopters due to both internal influence and external influence, how do those two groups differ from each other?

In a recent empirical study, Feick and Price (1987) suggest that in any social system there are individuals who assimilate and disseminate information on products (and therefore influence others) and tend to rely on external sources of information. They label these individuals "market mavens." On the basis of their empirical results, however, they conclude that "the concepts of the market maven and the innovative consumer are distinct" (1987, p. 90). Their findings raise research questions about the linkage in the Bass model between market mavens and adopters who buy as a result of external influence.

Multiple adoptions. The Bass model has been developed to represent the conversion of potential adopters to adopters. It explicitly assumes that each potential adopter buys only one unit of the product. However, certain innovations are bought in multiple units by potential adopters (e.g., multiple units of scanners by a supermarket and multiple units of personal computers by a firm). For these innovations, the sales data must be linked with the number of adopters by using a function that explicitly takes into consideration the multiple-unit-adoption behavior of the potential adopters (see Norton and Bass, 1987).

Effect of consumer expectations. For certain innovations (e.g., computers), consumer expectations about an innovation's future characteristics (e.g., price) influence purchase intentions (see, e.g., Holak *et al.*, 1987; Winer, 1985). For such innovations, in addition to influencing the nature of the adoption curve, consumer expectations can also influence the optimal marketing mix strategy used by a firm. For example, incorporating consumer expectations related to price in the Bass model, Narasimhan (1989) suggests that the optimal pricing strategy for a monopolist cycles over time and within each cycle the price increases at introduction, peaks, and decreases later. Given the importance of consumer expectations in understanding diffusion dynamics, we expect future research to incorporate them into the Bass model.

Exploration of recent developments in hazard models. The different diffusion models can be viewed as making different assumptions about the "hazard rate" for non-adopters as a function of time (the hazard rate being the likelihood that an individual who has remained a non-adopter through time t will become an adopter in the next instant of time). The Bass model specifies this rate as a linear function of previous adopters. Since the publication of the Bass model, however, much work developing and applying hazard models has appeared in the statistics, biometrics, and econometrics literatures (e.g., Cox and Oakes, 1984; Kalbfleisch and Prentice, 1980; for possible marketing applications of hazard models, see Helsen and Schmittlein, 1989; for interpretation of diffusion models as hazard models, see Lavaraj and Gore, 1990). The key development in hazard models over the last decade has been in the area of understanding covariate effects on the hazard rate (and consequently on duration times). This development is particularly important because attempts to incorporate marketing mix variables (and other covariate effects) in diffusion models have to date have been very limited in scope and *ad hoc* in their choice of model specifications for those effects. Exploration of recent developments in the hazard modeling framework may provide a unifying theme for understanding of covariate/marketing mix effects in diffusion models.

Understanding of diffusion processes at the micro (individual) level. Diffusion models based on individual-level adoption decisions offer an opportunity to study the actual pattern of social communication and its impact on product perceptions, preferences, and ultimate adoption. The empirical evidence provided by Chatterjee and Eliashberg (1989) on the development of aggregate diffusion models from individual-level adoption decisions, though limited, is encouraging. Further empirical work on such models may assist in developing the aggregate diffusion models prior to launch.

6.7.2 Parameter estimation considerations

In comparison with the other subareas we review, parameter estimation considerations for the Bass model probably received the most attention in the 1980s. These developments are timely and encouraging, but further empirical work on the validation of meta-analysis procedures (Montgomery and Srinivasan, 1989; Sultan *et al.*, 1990), Bayesian estimation procedures (Lenk and Rao, 1989; Sultan *et al.*, 1990), and procedures that capitalize on the information provided by managers and potential adopters (e.g., Randles,

1983; Souder and Quaddus, 1982) is important. An emerging body of literature in the forecasting area suggests that combining parameter estimates from different estimation procedures can yield better forecasting results (see Mahajan and Wind, 1988). Empirical studies that explore the feasibility of such findings for diffusion models are desirable (Lawrence and Geurts, 1984).

6.7.3 Flexible diffusion models

Flexible diffusion models have the advantage of capturing penetration patterns that are symmetric as well as nonsymmetric with no restrictions on the point of inflection. However, among all the models reviewed in *Table 6.3*, only the models by von Bertalanffy (1957) (or Nelder, 1962) and Bewley and Fiebig (1988) offer closed-form solutions to the differential equations used to specify the diffusion dynamics (i.e., express the number of adopters as an explicit function of time, which is desirable for long-term forecasting). Furthermore, these models have a flexibility advantage by requiring estimation of additional numbers of parameters. However, two important questions remain: How much additional long-term forecasting accuracy is provided by the flexible models, in comparison with the basic diffusion models such as the Bass model, when controlled for the number of parameters? Given the parameter estimation considerations discussed here, how can parameters in these models be calibrated prior to launch for long-term forecasting? Further empirical work related to these questions is desirable.

6.7.4 Refinements and extensions

We briefly discuss below a number of possibilities for further refinement and extension of the Bass model.

- A decade ago, Mahajan and Muller (1979) concluded that it was not clear how marketing mix variables should be incorporated into the Bass model. The few empirical studies reported in 1980s still do not provide conclusive guidelines on this question. Despite the arguments made in the favor of including price in the market potential, empirical studies on consumer durable products by Kamakura and Balasubramanian (1988) and Jain and Rao (1989) suggest that price affects the rate of diffusion (by influencing the coefficients of external influence and internal influence). Similarly, in relation to the inclusion of advertising in the Bass model, the two reported empirical studies suggest different alternatives. Horsky and Simon (1983) recommend that it be included in the coefficient of external influence whereas Simon and Sebastian (1987) report better results by including it in the coefficient of internal influence. Interestingly, though both of these studies examine the effect of advertising on the diffusion of a service (a banking service by Horsky and Simon and a telephone service by Simon and Sebastian), they are conducted in two different markets (USA and the FRG) and under different market conditions (there was a supply problem with the availability of telephones in the FRG). Whether these differences had an impact on the reported results is an empirical question. Given the importance of including marketing mix variables in understanding diffusion dynamics, we expect more empirical work including other marketing mix variables such as distribution.

- Several of the empirical studies reported in *Table 6.5* have incorporated product attributes in the Bass model. A natural extension of these studies is to develop procedures to determine optimal product design to obtain the desirable penetration rate.
- For high technology products the time interval between successive generations of technologies has been decreasing. Norton and Bass (1987) have shown how diffusion of successive generations interacts within the context of the Bass model. Forecasting possibilities stemming from this work appear to be promising. Extensions involving pricing of generations of technology would be desirable and feasible.
- When should a firm introduce a second generation product? Though the analytical results of Wilson and Norton (1989) suggest the answer is "now or never", they exclude the impact of other variables such as price. Further theoretical and empirical work addressing this question would be welcome.
- For high-technology products, the product offering of a firm generally includes both hardware and software, such as Nintendo hardware (keypad) and Nintendo software (video games) for children. Because of the contingency inherent in the relationship, it is important to develop diffusion models that examine the diffusion of the entire bundle of product offerings. In addition to forecasting, normative questions may relate to its optimal pricing and distribution. For example, how should a monopolist (e.g., Nintendo) manufacture and distribute its hardware and software? Should it keep a monopoly on both of them? Should it keep a monopoly on hardware and create an oligopoly for software to increase demand for the hardware?

- How do the number of competitors and the rivalry among them influence the growth of a product category? Does the growth affect the entry/exit patterns of competitors? Answers to these questions are within the domain of the diffusion modeling framework and provide a linkage with the strategic planning literature. Theoretical and empirical work on these questions will further enhance the utility of diffusion models.
- Supply restrictions influence diffusion patterns. For certain type of products (e.g., prescription drugs), it may be desirable to retard the diffusion process by controlling their supply and distribution. Further empirical and theoretical work, on this linkage would enable managers to control the life cycle of a product by managing the supply.
- Market interventions (e.g., patent violations) represent externalities that can influence the growth pattern of a new product. Though the use of intervention analysis is well established in the time-series analysis literature, no attempt seems to have been made to conduct intervention analysis with the diffusion models (Mahajan *et al.*, 1985). Theoretical and empirical work in this area could assist in assessing the impact (e.g., assessing patent violation damages in a legal case) of market interventions on the product life cycle.
- Though integration of the time and spatial dimensions has been of interest to geographers, their integration is equally important in marketing to evaluate alternative product distribution strategies across markets. Such extensions of the Bass model could assist in evaluating the impact on the growth of a new product of how and where the product is made available.
- The diffusion literature has emphasized consistently the importance of negative word of mouth on the growth of a new product (Mahajan *et al.*, 1984a). The multistage extensions of the Bass model offer an avenue for considering its impact on the growth pattern. These extensions lack empirical validation, however. Data collection and estimation procedures should be developed to make these extensions practicable.
- Not all new products are accepted by consumers at the time of their introduction. Some products are much slower than others in being accepted by potential adopters. That is, they differ in terms of how long it takes them to "take off." The "takeoff" phenomenon is not considered explicitly by the Bass model. The Bass model assumes the presence of a certain number of consumers before "takeoff" (i.e., pm). Extensions of the Bass model that explicitly consider this phenomenon will be useful in explaining and predicting the take-off behavior of a new product.

6.7.5 Use of diffusion models

One of the critical uses of diffusion models has been for forecasting the first-purchase sales volume curve. In recent years, questions have been raised about the forecasting accuracy of diffusion models (Bernhardt and Mackenzie, 1972; Heeler and Hustad, 1980). We sympathize with such concerns and believe that further empirical work is needed to identify conditions under which diffusion models work or do not work. For example, recent work by Jain *et al.* (1989) suggests that the use of the Bass model is inappropriate in international settings where the supply of the product is restricted. Furthermore, as the diffusion models capture the dynamics of innovation diffusion for the first-time buyers, it is not clear that the same diffusion dynamics are applicable to replacement sales. Therefore the use of diffusion models for such adoption data may be inappropriate (see, e.g., Bayus, 1988; Bayus *et al.*, 1989). Finally, diffusion models are imitation models. Any S-shaped curve, however, may not be a result of the imitation process, and alternative time-series models may be more appropriate for such data (Mahajan *et al.*, 1988). Even in the presence of the imitation effect, it may be necessary to examine various diffusion models systematically to identify the one that best describes the data (Rust and Schmittlein, 1985). There is also a growing body of literature on "chaos theory" suggesting that for certain parameter values, diffusion models generate persistent chaotic behavior within predictable boundaries (Gordon and Greenspan, 1988). Understanding of such phenomena may be essential to decipher the impact of changes that affect the diffusion dynamics.

The use of diffusion models to test diffusion-based hypotheses is very encouraging. The empirical studies documented in *Table 6.4* clearly attest to their potential for such applications. We expect additional empirical work employing a diffusion modeling framework to test hypotheses related to life cycle dynamics (for example: How does the number of competitors change over the life cycle of a product? How does the number of brands available in a market influence the growth of a product? How does the rivalry among competitors in an industry affect the life cycle of a product?).

The use of diffusion models to derive normative results for the dynamics of innovation diffusion received considerable attention in the 1980s. However, as summarized in *Table 6.5*, these results are simply working hypotheses. Furthermore, the nature of these results is contingent on the assumptions made in their analytical derivation. For most of these studies, the analytical elegance surpasses the empirical validation of the derived results. Empirical evidence is needed to find out if and when the firms use the derived normative strategies.

Finally, it is important to acknowledge that several firms have used diffusion models for forecasting the demand of a new product. By sharing their experiences, industry users can contribute to the further validation of diffusion models.

Acknowledgments

The author would like to thank Roger Kerin, Dipak Jain, David Schmittlein, Rabikar Chatterjee, Subatra Sen, Mike Hanssens, and Jon Eliashberg for their helpful comments.

Notes

[1] Related to the Mansfield model is the model suggested by Fisher and Pry (1971) and the Gompertz curve. For applications of the Gompertz curve and its comparison with the Mansfield model, see Hendry (1972), Dixon (1980), and Ziemer (1988). Several other growth models also have been proposed in the marketing, economics, and technological substitution literatures to depict the growth phenomenon (e.g., the Weibull distribution). As some of these models either do not explicitly consider the diffusion effect in their formulation or combine other models, they are not included in our review. For applications of such models to new product growth situations, see DeKluyver (1982), Sharif and Islam (1980), Meade (1984), Lee and Lu (1987), and Skiadas (1985 and 1986).

[2] These expressions are given by:

$$N(t) = m\left[\frac{1-e^{-(p+q)t}}{1+\frac{q}{p}e^{-(p+q)t}}\right], \quad n(t) = m\left[\frac{p(p+q)^2 e^{-(p+q)t}}{(p+q\; e^{-(p+q)t})^2}\right]$$

$$n(T^*) = \frac{1}{4q}(p+q)^2, \qquad N(T^*) = m\left[\frac{1}{2} - \frac{p}{2q}\right]$$

$$T^* = -\frac{1}{(p+q)}\ln\left(\frac{p}{q}\right)$$

[3] A brief analytical description of these procedures is given in an appendix in the unabridged version of this paper which can be obtained from the authors.

[4] The idea that coefficients of a market response model should change over time is not new in marketing. In fact, several theoretical approaches that assist in developing market response models when model coefficients have a time-varying behavior have been applied and documented in the marketing literature (see, e.g., Mahajan *et al.*, 1980; Wildt and Winer, 1978). Two such approaches also have been examined in the context of diffusion models: the systematic parameter variation methods and the random coefficient methods. The systematic parameter variations assume *a priori* the time path of the model coefficients. These methods have generated a new set of diffusion models termed "flexible diffusion models" (Mahajan and Peterson, 1985) that are reviewed here.

In the random coefficient methods, the random parameters are assumed to constitute a sample from common multivariate distribution with an estimated mean and variance-covariance structure. Following Karmeshu and Pathria (1980a and 1980b), Eliashberg *et al.* (1987) explored the applicability of these methods to the Bass diffusion model. They consider the coefficients of innovation (p) and imitation (q) in the Bass model as stochastic and hence time-varying by assuming that $\tilde{p} = p + \epsilon_p(t)$ and $\tilde{q} = q + \epsilon_q(t)$, where p and q denote constant means and ϵ_p and ϵ_q denote normally distributed error terms surrounding those means such that their means are zero and the variances are constant. Their empirical results suggest that their stochastic formulation of the Bass model does as well as the deterministic version of the Bass model. There are other types of stochastic diffusion models, but they are not included in our review. Reviews of such models are given by Bartholomew (1982), Eliashberg and Chatterjee (1986), and Boker (1987).

[5] For an application of this approach to the diffusion of robotics in the State of New York, see Bretschneider and Bozeman (1986). Other feedback filters can also be used to estimate time-varying diffusion parameters. For example, the use of the Kalman filter to estimate the time-varying coefficients for the Mansfield model has been reported by Meade (1985).

References

Akinola, A.A., 1986, An application of the Bass model in the analysis of diffusion of coco-spraying chemicals among Nigerian cocoa farmers, *Journal of Agricultural Economics* **37**(3):395–404.

Arndt, J., 1967, Role of product-related conversations in the diffusion of a new product, *Journal of Marketing Research* **4**:291–295, August.

Bartholomew, D.J., 1982, *Stochastic Models for Social Processes*, 3rd edition, John Wiley & Sons Inc., New York, NY, USA.

Bass, F.M., 1969, A new product growth model for consumer durables, *Management Science* **15**:215–227, January.

Bass, F.M., 1980, The relationship between diffusion rates, experience curves, and demand elasticities for consumer durable technological innovations, *Journal of Business* **53**:51–67, July, part 2.

Bass, F.M., 1986, The adoption of a marketing model: Comments and observations, in V. Mahajan and Y. Wind, eds., *Innovation Diffusion of New Product Acceptance*, Ballinger Publishing Co., Cambridge, MA, USA.

Bass, F.M. and Bultez, A.V., 1982, A note on optimal strategic pricing of technological innovations, *Marketing Science* **1**:371–378, Fall.

Bayus, B.F., 1987, Forecasting sales of new contingent products: An application to the compact disc market, *Journal of Product Innovation Management* **4**:243–255, December.

Bayus, B.F., 1988, Accelerating the durable replacement cycle with marketing mix variables, *Journal of Product Innovation Management* **5**:216–226, September.

Bayus, B.F., Hong, S., and Labe, R.P., Jr., 1989, Developing and using forecasting models for consumer durables, *Journal of Product Innovation Management* **6**:5–19, March.

Bernhardt, I., and Mackenzie, K.M., 1972, Some problems in using diffusion models for new products, *Management Science* **19**:187–200, October.

Bewley, R. and Fiebig, D.G., 1988, A flexible logistic growth model with publications in telecommunications, *International Journal of Forecasting* **4**(2):177–192.

Boker, F., 1987, A stochastic first-purchase diffusion model: A counting approach, *Journal of Marketing Research* **24**:64–73, February.

Bretschneider, S.I. and Bozeman, B., 1986, Adaptive diffusion models for the growth of robotics in New York state industry, *Technological Forecasting & Social Change* **30**:111–121, September.

Bretschneider, S.I. and Mahajan, V., 1980, Adaptive technological substitutions model, *Technological Forecasting & Social Change* **18**:129–139, October.

Brown, L.A., 1981, *Innovation Diffusion: A New Perspective*, Methuen and Co. Ltd., New York, NY, USA.

Carbone, R. and Longini, R.L., 1977, A feedback model for automated real estate assessment, *Management Science* **24**:241–248, November.

Chatterjee, R. and Eliashberg, J., 1989, *The Innovation Diffusion Process in a Heterogeneous Population: A Micromodeling Approach*, Working Paper, Marketing Department, Krannert Graduate School of Management, Purdue University, Lafayette, IN, USA.

Chow, G.C., 1967, Technological change and the demand for computers, *American Economics Review* **57**:1117–1130, December.

Clarke, D.G. and Dolan, R.J., 1984, A simulation model for the evaluation of pricing strategies in a dynamic environment, *Journal of Business* **57**:S179–S200, January.

Clarke, F.H., Darrough, M.N., and Heineke, J.M., 1982, Optimal pricing policy in the presence of experience effects, *Journal of Business* **55**(4):517–530.

Cox, D.R. and Oakes, D., 1984, *Analysis of Survival Data*, Chapman and Hall, London, UK.

Deal, K.R., 1979, Optimizing advertising expenditures in a dynamic duopoly, *Operations Research* **27**:682–692, July–August.

DeKluyver, C.A., 1982, A comparative analysis of the Bass and Weibull new product growth models for consumer durables, *New Zealand Journal of Operations Research* **10**:99–130, July.

Dhebar, A. and Oren, S.S., 1985, Optimal dynamic pricing for expanding networks, *Marketing Science* **4**:336–351, Fall.

Dixon, R., 1980, Hybrid corn revisited, *Econometrica* **48**:1451–1461, September.

Dockner, E. and Jorgensen, S., 1988a, Optimal advertising policies for diffusion models of new product innovations in monopolistic situations, *Management Science* **34**:119–130, January.

Dockner, E. and Jorgensen, S., 1988b, Optimal pricing strategies for new products in dynamic oligopolies, *Marketing Science* **7**:315–334, Fall.

Dodds, W., 1973, An application of the Bass model in long-term new product forecasting, *Journal of Marketing Research* **10**:308–311, August.

Dodson, J.A. and Muller, E., 1978, Models of new product diffusion through advertising and word of mouth, *Management Science* **24**:1568–1578, November.

Dolan, R.J. and Jeuland, A.P., 1981, Experience curves and dynamic demand models: Implications for optimal pricing strategies, *Journal of Marketing* **45**:52–62, Winter.

Easingwood, C.J. 1987, Early product life cycle patterns forms for infrequently purchased major products, *International Journal of Research in Marketing* **4**(1):3–9.

Easingwood, C.J. 1988, Product life cycle patterns for new industrial products, *R&D Management* **18**(1):23–32.

Easingwood, C.J. 1989, An analogical approach to the long term forecasting of major new product sales, *International Journal of Forecasting* **5**:69–82.

Easingwood, C.J., Mahajan, V., and Muller, E., 1981, A nonsymmetric responding logistic model for technological substitution, *Technological Forecasting & Social Change* **20**:199–213, October.

Easingwood, C.J., Mahajan, V., and Muller, E., 1983, A nonuniform influence innovation diffusion model of new product acceptance, *Marketing Science* **2**:273–296, Summer.

Eliashberg, J. and Chatterjee, R., 1986, Stochastic issues in innovation diffusion models, in V. Mahajan and Y. Wind, eds., *Innovation Diffusion Models of New Product Acceptance*, Ballinger Publishing Co., Cambridge, MA, USA.

Eliashberg, J. and Jeuland, A.P., 1986, The impact of competitive entry in a developing market upon dynamic pricing strategies, *Marketing Science* **5**:20–36, Winter.

Eliashberg, J., Tapiero, C.S., and Wind, Y., 1987, *Innovation Diffusion Models with Stochastic Parameters: Forecasting and Planning Implications*, Working Paper No. 87-003, The Wharton School, University of Pennsylvania, Philadelphia, PA, USA.

Engel, J.E., Blackwell, R.D., and Miniard, P.W., 1986, *Consumer Behavior*, Dryden Press, Hinsdale, IL, USA.

Erickson, G.M., 1985, A model of advertising competition, *Journal of Marketing Research* **22**:297–304, August.

Feder, G. and O'Mara, G.T., 1982, On information and innovation diffusion: A Bayesian approach, *American Journal of Agricultural Economics* **64**:145–147, February.

Feichtinger, G., 1982, Optimal pricing in diffusion model with concave price-dependent market potential, *Operations Research Letters* **1**:236–240.

Feick, L.F. and Price, L.L., 1987, The market maven: A diffuser of marketplace information, *Journal of Marketing* **51**:83–97, January.

Fershtman, C., Mahajan, V., and Muller, E., 1990, Market share pioneering advantage: A theoretical approach, *Management Science* (forthcoming).

Fisher, J.C. and Pry, R.H., 1971, A simple substitution model for technological change, *Technological Forecasting & Social Change* **2**:75–88, May.

Floyd, A., 1962, Trend forecasting: A methodology for figure of merit, in J. Bright, ed., *Technological Forecasting for Industry and Government: Methods and Applications*, Prentice-Hall, Englewood Cliffs, NJ, USA.

Fourt, L.A. and Woodlock, J.W., 1960, Early prediction of market success for grocery products, *Journal of Marketing* **25**:31–38, October.

Frank, R.E., Massy, W.F., and Morrison, D.G., 1964, The determinants of innovative behavior with respect to a branded, frequently purchased food product, in L.G. Smith, ed., *Proceedings of the American Marketing Association*, American Marketing Association, Chicago, IL, USA.

Gatignon, H. and Robertson, T.S., 1985, A propositional inventory for new diffusion research, *Journal of Consumer Research* **11**:849–867, March.

Gatignon, H., Eliashberg, J., and Robertson, T.S., 1989, Modeling multinational diffusion patterns: An efficient methodology, *Marketing Science* (forthcoming).

Gordon, T.J. and Greenspan, D., 1988, Chaos and fractals: New tools for technological and social forecasting, *Technological Forecasting & Social Change* **34**:1–26, August.

Gore, A.P. and Lavaraj, V.A., 1987, Innovation diffusion in a heterogeneous population, *Technological Forecasting & Social Change* **32**:163–167, September.

Heeler, R.M. and Hustad, T.P., 1980, Problems in predicting new product growth for consumer durables, *Management Science* **26**:1007–1020, October.

Helsen, K. and Schmittlein, D.C., 1989, *Analyzing Duration Times in Marketing Research*, Working Paper, Marketing Department, The Wharton School, University of Pennsylvania, Philadelphia, PA, USA.

Hendry, I., 1972, The three parameter approach to long-range forecasting, *Long Range Planning* **51**:40–45, March.

Hernes, G., 1976, Diffusion and growth: The non-homogeneous case, *Scandinavian Journal of Economics* **78**:427–436.

Hiebert, L.D., 1974, Risk, learning and the adoption of fertilizer responsive seed varieties, *American Journal of Agricultural Economics* **56**:764–768, November.

Holak, S., Lehmann, D.R., and Sultan, F., 1987, The role of expectations in the adoption of innovative consumer durables: Some preliminary evidence, *Journal of Retailing* **63**:243–259, Fall.

Horsky, D., 1990, The effects of income, price and information on the diffusion of new consumer durables, *Marketing Science* (forthcoming).

Horsky, D. and Simon, L.S., 1983, Advertising and the diffusion of new products, *Management Science* **1**:1–18, Winter.

Hyman, M.R., 1988, The timeliness problem in the application of Bass-type new product growth models to durable sales forecasting, *Journal of Business Research* **16**(1):31–47.

Jain, D.C. and Rao, R.C., 1989, Effect of price on the demand for durables: Modeling, estimation and findings, *Journal of Business and Economic Statistics* (forthcoming).

Jain, D.C., Mahajan, V., and Muller, E., 1989, *Innovation Diffusion in the Presence of Supply Restrictions*, Working Paper, Cox School of Business, Southern Methodist University, Dallas, TX, USA.

Jensen, R., 1982, Adoption and diffusion of an innovation of uncertain profitability, *Journal of Economic Theory* **27**:182–193.

Jeuland, A.P., 1981, *Parsimonious Models of Diffusion of Innovation: Part A, Derivations and Comparisons*, Working Paper, Graduate School of Business, University of Chicago, IL, USA.

Jeuland, A.P. and Dolan, R.J., 1982, An aspect of new product planning: Dynamic pricing, in A.A. Zoltners, ed., *TIMS Studies in the Management Sciences*, Vol. 18: *Marketing Planning Models*, Elsevier Science Publishing Co., New York, NY, USA.

Jones, J.M. and Ritz, C.J., 1987, *Incorporating Distribution into New Product Diffusion Models*, Working Paper, Marketing Department, University of North Carolina, Chapel Hill, NC, USA.

Jorgensen, S., 1983, Optimal control of a diffusion model of new product acceptance with price-dependent total market potential, *Optimal Control Applications and Methods* **4**:269–276.

Jorgensen, S., 1986, Optimal dynamic pricing in an oligopolistic market: A survey, *Lecture Notes in Economics and Mathematical Systems* **265**:179–237.

Kalbfleisch, J.D. and Prentice, R.L., 1980, *The Statistical Analysis of Failure Time Data*, John Wiley & Sons Inc., New York, NY, USA.

Kalish, S., 1983, Monopolist pricing with dynamic demand and production cost, *Marketing Science* **2**:135–160, Spring.

Kalish, S., 1985, A new product adoption model with pricing, advertising and uncertainty, *Management Science* **31**:1569–1585, December.

Kalish, S. and Lilien, G.L., 1986a, A market entry timing model for new technologies, *Management Science* **32**:194–205, February.

Kalish, S. and Lilien, G.L., 1986b, Applications of innovation diffusion models in marketing, in V. Mahajan and Y. Wind, eds., *Innovation Diffusion Models of New Product Acceptance*, Ballinger Publishing Co., Cambridge, MA, USA.

Kalish, S. and Sen, S.K., 1986, Diffusion models and the marketing mix for single products, in V. Mahajan and Y. Wind, eds., *Innovation Diffusion Models of New Product Acceptance*, Ballinger Publishing Co., Cambridge, MA, USA.

Kamakura, W.A. and Balasubramanian, S.K., 1987, Long-term forecasting with innovation diffusion models: The impact of replacement purchase, *Journal of Forecasting* **6**(1):1–19.

Kamakura, W.A. and Balasubramanian, S.K., 1988, Long-term view of the diffusion of durables, *International Journal of Research in Marketing* **5**:1–13.

Karmeshu and Pathria, R.K., 1980a, Stochastic evolution of a nonlinear model of diffusion of information, *Journal of Mathematical Sociology* **7**:59–71.

Karmeshu and Pathria, R.K., 1980b, Diffusion of information in a random environment, *Journal of Mathematical Sociology* **7**:215–227.

King, C.W., Jr., 1963, Fashion adoption: A rebuttal to the *trickle down* theory, in S.A. Greyser, ed., *Proceedings of the American Marketing Association*, American Marketing Association, Chicago, IL, USA.

Kobrin, S.J., 1985, Diffusion as an explanation of oil nationalization or the domino effect rides again, *Journal of Conflict Resolution* **29**:3–32, March.

Kotler, P. and Zaltman, G., 1976, Targeting prospects for a new product, *Journal of Advertising Research* **16**:7–20, February.

Lackman, C.L., 1978, Gompertz curve forecasting: A new product application, *Journal of Marketing Research Society*, **20**:45–47, January.

Lancaster, G.A. and Wright, G., 1983, Forecasting the future of video using a diffusion model, *European Journal of Marketing* **17**(2):70–79.

Lattin, J.M. and Roberts, J.H., 1989, *Modeling the Role of Risk-Adjusted Utility in the Diffusion of Innovations*, Working Paper 1019, Graduate School of Business, Stanford University, CA, USA.

Lavaraj, U.A. and Gore, A.P., 1990, On interpreting probability distributions fitted to times of first adoption, *Technological Forecasting & Social Change* (forthcoming).

Lawrence, K.D. and Geurts, M., 1984, Converging conflicting forecasting parameters in forecasting durable new product sales, *European Journal of Operational Research* **16**:42–47.

Lawrence, K.D. and Lawton, W.H., 1981, Applications of diffusion models: Some empirical results, in Y. Wind *et al.*, eds., *New Product Forecasting*, Lexington Books, Lexington, MA, USA.

Lawton, S.B. and Lawton, W.H., 1979, An autocatalytic model for the diffusion of educational innovations, *Educational Administration Quarterly* **15**:19–53.

Lee, J.C.O. and Lu, K.W., 1987, On a family of data-based transformed models useful in forecasting technological substitution, *Technological Forecasting & Social Change* **31**:61–78, March.

Lekvall, P. and Wahlbin, C., 1973, A study of some assumptions underlying innovation diffusion functions, *Swedish Journal of Economics* **75**:362–377.

Lenk, P.J. and Rao, A.G., 1989, New models from old: Forecasting product adoption by hierarchical Bayes procedures, *Marketing Science* (forthcoming).

Lilien, G.L., Rao, A.G., and Kalish, S., 1981, Bayesian estimation and control of detailing effort in a repeat purchase diffusion environment, *Management Science* **27**:493–506, May.

Mahajan, V. and Muller, E., 1979, Innovation diffusion and new product growth models in marketing, *Journal of Marketing* **43**:55–68, Fall.

Mahajan, V. and Muller, E., 1986, Advertising pulsing policies for generating awareness for new products, *Marketing Science* **5** :89–106, Spring.

Mahajan, V. and Peterson, R.A., 1978, Innovation diffusion in a dynamic potential adopter population, *Management Science* **24**:1589–1597, November.

Mahajan, V. and Peterson, R.A., 1979, Integrating time and space in technological substitution models, *Technological Forecasting & Social Change*, **14**:231–241, August.

Mahajan, V. and Peterson, R.A., 1982, Erratum to: Innovation diffusion in a dynamic potential adopter population, *Management Science* **28**:1087, September.

Mahajan, V. and Peterson, R.A., 1985, *Models for Innovation Diffusion*, Sage Publications Inc., Beverly Hills, CA, USA.

Mahajan, V. and Sharma, S., 1986, Simple algebraic estimation procedure for innovation diffusion models of new product acceptance, *Technological Forecasting & Social Change* **30**:331–346, December.

Mahajan, V. and Wind, Y., 1986a, *Innovation Diffusion Models of New Product Acceptance*, Ballinger Publishing Co., Cambridge, M, USA.

Mahajan, V. and Wind, Y., 1986b, Innovation diffusion models of new product acceptance: A reexamination, in V. Mahajan and Y. Wind, eds., *Innovation Diffusion Models of New Product Acceptance*, Ballinger Publishing Co., Cambridge, MA, USA.

Mahajan, V. and Wind, Y., 1988, New product forecasting models: Directions for research and implementation, *International Journal of Forecasting* 4(3):341–358.

Mahajan, V., Bretschneider, S.I., and Bradford, J.W., 1980, Feedback approaches to modeling structural shifts in market response, *Journal of Marketing* **44**:71–80, Winter.

Mahajan, V., Mason, C.H., and Srinivasan, V., 1986, An evaluation of estimation procedures for new product diffusion models, in V. Mahajan and Y. Wind, eds., *Innovation Diffusion Models of New Product Acceptance*, Ballinger Publishing Co., Cambridge, MA, USA.

Mahajan, V., Muller, E., and Kerin, R., 1984a, Introduction strategy for new products with positive and negative word of mouth, *Management Science* **30**:1389–1404, December.

Mahajan, V., Muller, E., and Sharma, S., 1984b, An empirical comparison of awareness forecasting models of new product acceptance, *Marketing Science* **3**:179–197, Summer.

Mahajan, V., Muller, E., and Srivastava, R.K., 1990, Using innovation diffusion models to develop adopter categories, *Journal of Marketing Research* (forthcoming).

Mahajan, V., Sharma, S., and Bettis, R.A., 1988, The adoption of M-form organizational structure: A test of imitation hypothesis, *Management Science* **34**:1188–1201, October.

Mahajan, V., Sharma, S. and Wind, Y., 1985, Assessing the impact of patent infringement on new product sales, *Technological Forecasting & Social Change* **28**:13–27, August.

Mahajan, V., Yoram, W., and Sharma, S., 1983, An approach to repeat purchase diffusion models, in P.E. Murphy *et al.*, eds., *AMA Proceedings*, Series 49, American Marketing Association, Chicago, IL, USA.

Mansfield, E., 1961, Technical change and the rate of imitation, *Econometrica* **29**:741–766, October.

McGowan, I., 1986, The use of growth curves in forecasting market development, *Journal of Forecasting* **5**(1):69–71.

McKenna, R., 1985, *The Regis Touch*, Addison-Wesley, Reading, MA, USA.

Meade, N., 1984, The use of growth curves in forecasting market development: A review and appraisal, *Journal of Forecasting* **3**(4):429–451.

Meade, N., 1985, Forecasting using growth curves: An adaptive approach, *Journal of the Operational Research Society* **36**:1103–1115.

Mesak, H.I. and Mikhail, W.M., 1988, Prelaunch sales forecasting of a new industrial product, *Omega* **16**(1):41–51.

Midgley, D.F., 1976, A simple mathematical theory of innovative behavior, *Journal of Consumer Research* **3**:31–41, June.

Modis, T. and Debecker, A., 1988, Innovation in the computer industry, *Technological Forecasting & Social Change* **33**:267–278, May.

Montgomery, D.B. and Srinivasan, V., 1989, *An Improved Method for Meta Analysis: With Application to New Product Diffusion Models*, Working Paper, Graduate School of Business, Stanford University, CA, USA.

Narasimhan, C., 1989, Incorporation consumer price expectations in diffusion models, *Marketing Science* (forthcoming).

Nascimento, F. and Vanhonacker, W.R., 1988, Optimal strategic pricing of reproducible consumer products, *Management Science* **34**:921–937, August.

Nelder, J.A., 1962, An alternative form of a generalized logistic equation, *Biometrics* **18**:614–616, December.

Nevers, J.V., 1972, Extensions of a new product growth model, *Sloan Management Review* **13**:78–79, Winter.

Norton, J.A. and Bass, F.M., 1987, A diffusion theory model of adoption and substitution for successive generations of high technology products, *Management Science* **33**:1069–1086, September.

Oliver, R.M., 1987, A Bayesian model to predict saturation and logistic growth, *Journal of Operations Research Society* **38**(1):49–56.

Olshavsky, R.W., 1980, Time and the rate of adoption of innovations, *Journal of Consumer Research* **6**:425–428, March.

Olson, J.A. and Choi, S., 1985, A product diffusion model incorporating repeat purchases, *Technological Forecasting & Social Change* **27**:385–397.

Oren, S.S. and Schwartz, R.G., 1988, Diffusion of new products in risk-sensitive markets, *Journal of Forecasting* **7**:273–287, October–December.

Peterson, R.A. and Mahajan, V., 1978, Multi-product growth models, in J. Sheth, ed., *Research in Marketing*, JAI Press Inc., Greenwich, CT, USA.

Randles, F., 1983, On the diffusion of computer terminals in an established engineering environment, *Management Science* **29**:465–476, April.

Rao, S.K., 1985, An empirical comparison of sales forecasting models, *Journal of Product Innovation Management* **2**:232–242. December.

Rao, R.C. and Bass, F.M., 1985, Competition, strategy, and price dynamics: A theoretical and empirical investigation, *Journal of Marketing Research* **22**:283–296, August.

Rao, A.G. and Yamada, M., 1988, Forecasting with a repeat purchase diffusion model, *Management Science* **34**:734–752, June.

Richards, F.J., 1959, A flexible growth function for empirical use, *Journal of Experimental Botany* **10**:290–300.

Robertson, T.S., 1967, Determinants of innovative behavior, in R. Moyer, ed., *Proceedings of the American Marketing Association*, American Marketing Association, Chicago, IL, USA.

Robinson, B. and Lakhani, C., 1975, Dynamic price models for new product planning, *Management Science* **10**:1113–1122, June.

Rogers, E.M., 1983, *Diffusion of Innovations*, 3rd edition, The Free Press, New York, NY, USA.

Rust, R.T. and Schmittlein, D.C., 1985, A Bayesian cross-validated likelihood method for comparing alternative specifications of quantitative models, *Marketing Science* **4**:20–40, Winter.

Schmittlein, D.C. and Mahajan, V., 1982, Maximum likelihood estimation for an innovation diffusion model of new product acceptance, *Marketing Science* **1**:57–78, Winter.

Sharif, M.N. and Islam, M.N., 1980, The Weibull distribution as a general model for forecasting technological change, *Technological Forecasting & Social Change* **18**:247–256.

Sharif, M.N. and Kabir, C., 1976, A generalized model for forecasting technological substitution, *Technological Forecasting & Social Change* **8**:353–364.

Sharif, M.N. and Ramanathan, K., 1981, Binomial innovation diffusion models with dynamic potential adopter population, *Technological Forecasting & Social Change* **20**:63–87.

Sharif, M.N. and Ramanathan, K., 1982, Polynomial innovation diffusion models, *Technological Forecasting & Social Change* **21**:301–323.

Sharp, J.A., 1984, An interpretation of the non-symmetric responding logistic model in terms of price and experience effects, *Journal of Forecasting* **3**(4):453–456.

Silk, A.J., 1966, Overlap among self-designated opinion leaders: A study of selected dental products and services, *Journal of Marketing Research* **3**:255–259, August.

Silver, S.D., 1984, A simple mathematical theory of innovative behavior: A comment, *Journal of Consumer Research* **10**:441–444, March.

Simon, H. and Sebastian, K.-H., 1987, Diffusion and advertising: The German telephone company, *Management Science* **33**:451–466, April.

Skiadas, C.H., 1985, Two generalized rational models for forecasting innovation diffusion, *Technological Forecasting & Social Change* **27**:39–62.

Skiadas, C.H., 1986, Innovation diffusion models expressing asymmetry and/or positively or negatively influencing forces, *Technological Forecasting & Social Change* **30**:313–330.

Souder, W.E. and Quaddus, M.A., 1982, A decision-modeling approach to forecasting the diffusion of Longwall mining technologies, *Technological Forecasting & Social Change* **21**:1–14.

Srinivasan, V. and Mason, C.H., 1986, Nonlinear least squares estimation of new product diffusion models, *Marketing Science* **5**:169–178, Spring.

Srivastava, R.K., Mahajan, V., Ramaswami, S.N., and Cherian, J., 1985, A multiattribute diffusion model for forecasting the adoption of investment alternatives for consumers, *Technological Forecasting & Social Change* **28**:325–333, December.

Stoneman, P., 1981, Intra-firm diffusion, Bayesian learning and profitability, *The Economic Journal* **91**:375–388, June.

Sultan, F., Farley, J.U., and Lehmann, D.R., 1990, A meta-analysis of diffusion models, *Journal of Marketing Research* (forthcoming).

Takada, H. and Jain, D., 1988, *Cross-national Analysis of Diffusion of Consumer Durables*, Working Paper, Kellogg Graduate School of Business, Northwestern University, Evanston, IL, USA.

Tanny, S.M. and Derzko, N.A., 1988, Innovators and imitators in innovation diffusion modeling, *Journal of Forecasting* **7**(4):225–234.

Teng, J.-T. and Thompson, G.L., 1983, Oligopoly models for optimal advertising, *Management Science* **29**:1087–1101, September.

Teotia, A.P.S. and Raju, P.S., 1986, Forecasting the market penetration of new technologies using a combination of economic cost and diffusion models, *Journal of Product Innovation Management* **3**:225–237, December.

Thomas, R.J., 1985, Estimating market growth for new products: An analogical diffusion model approach, *Journal of Product Innovation Management* **2**:45–55, March.

Thompson, G.L. and Teng, J.-T., 1984, Optimal pricing and advertising policies for new product oligopoly models, *Marketing Science* **3**:148–168, Spring.

Tigert, D. and Farivar, B., 1981, The Bass new product growth model: A sensitivity analysis for a high technology product, *Journal of Marketing* **45**:81–90, Fall.

Tornatzky, L.G. and Klein, R.J., 1982, Innovation characteristics and innovation adoption-implementation: A meta-analysis of findings, *IEEE Transactions on Engineering Management* **EM-29**:28–45.

von Bertalanffy, L., 1957, Quantitative laws in metabolism and growth, *Quarterly Review of Biology* **32**:217–231.

Wildt, A.R. and Winer, R.S., 1978, Modeling structural shifts in market response: An overview, in S.C. Jain, ed., *Educator's Proceedings*, American Marketing Association, Chicago, IL, USA.

Wilson, L.O. and Norton, J.A., 1989, Optimal entry time for a product line extension, *Marketing Science* **8**:1–17, Winter.

Winer, R.S., 1985, A price vector model of demand for consumer durables: Preliminary developments, *Marketing Science* **4**:74–90, Winter.

Young, P. and Ord, J.K., 1985, The use of discounted least-squares in technological forecasting, *Technological Forecasting & Social Change* **28**:263–274.

Ziemer, D.R., 1988, *Marketplace Sizing and Timing Scenarios*, Working Paper, Management Development Center, Texas Instruments Incorporated, Dallas, TX, USA.

Chapter 7

The Research on Innovation Diffusion: An Assessment

Giovanni Dosi

7.1 The Role of Innovation Diffusion in Economic and Social Change

In recent years, innovation diffusion has attracted increasing attention within the economic discipline as well as from other social sciences, such as sociology, organization theory, economic geography, and political science.

Some of the reasons for the growing attention are likely to be found within the internal patterns of enquiry of the various disciplines themselves. Others are more obviously related to a general awareness of the importance of innovations and innovation diffusion – in products, processes of production, and forms of economic organization – for economic growth and, more generally, social change.

Certainly, the so-called *microelectronics revolution* has provided a powerful focus on the widespread economic and social consequences of major technological innovations. In this respect, fundamental, and still largely unanswered, questions concern, for example, its impact on employment and growth, the consequences for business organization, the induced changes in the patterns of consumption, the implications for educational requirements, the possible demand for new public policies, and many others. Moreover, in addition to microelectronics, other new technologies, such as bioengineering and new materials raise somewhat different but equally broad interpretative and normative issues.

The importance of changing patterns of innovation and diffusion has also emerged as a controversial issue in the interpretation of the trends in productivity growth within several OECD countries. Can the observed statistical slowdown be attributed to a parallel slowdown in the rates of innovation? Or, rather, to a slowdown in the rates of innovation *diffusion* and, thus, to an increasing gap between average and best-practice techniques? How can all this be reconciled with the intuitive evidence on the far-reaching productivity improvements which apparently the *microelectronics revolution* produces?

There is also an international aspect of innovation and diffusion. The most striking phenomenon has probably been the impressive Japanese capability of quickly adopting, improving, and – more recently – introducing new technologies. These capabilities, together with somewhat different organizational arrangements, have also meant a rapid growth of Japanese international competitiveness. More generally, many analysts suggest that the changing competitive strengths of the USA, the EEC countries, Japan and the newly industrializing countries, must also be attributed to a differential promptness of these countries in introducing and/or adopting technological and organizational innovations.

Another major issue, somewhat related to the previous one, concerns the relationship between innovativeness and the capability of appropriating economic benefits from the innovations themselves. This is the subject of great debat in the USA. To what extent is it necessary to be an innovator in order to enjoy relatively high per capita incomes? Under what circumstances is the innovator able to enjoy a quasi-rent on its technological achievements?

From a longer historical perspective, a view with respectable consensus holds that all the processes of economic growth and social change – at least since the English Industrial Revolution – cannot be explained without reference to the introduction and diffusion of major technological innovations – from the steam engine to electricity, the internal combustion engine, railroads, fertilizers, plastics, jet engines, and uncountable minor innovations.

Finally, very broad interpretative questions which are well beyond the domain of the economic discipline concern, for example, the relationship between new technologies and labor processes; the cultural and social structures which favor or hinder the introduction and diffusion of new technologies; the scientific and educational context within which innovation and innovation diffusion take place.

The list of questions could be much longer. Indeed, all the foregoing issues highlight the crucial importance of the phenomena of innovation and innovation diffusion in the interpretation of how economic and social structures keep together and change, sometimes in a rather orderly manner, and at other times with more abrupt discontinuities.

In a fundamental sense, the empirical evidence on the permanent process of technological and organizational change in contemporary societies confronts most social sciences such as, for example, economics, sociology, political science, and psychology, and demands theoretical explanations of its causes, patterns, and consequences.

The aim of this chapter is simply to sketch an overview of the state of the art, primarily in *the economics of innovation diffusion* highlighting some common themes and (often controversial) issues which underlie most other chapters of this book (see also, Arcangeli *et al.*, 1990, on whose introduction this chapter partly draws).

7.2 Invention, Innovation, and Diffusion

One of the contributions of J. Schumpeter's work that is often cited with reference to technological change concerns his distinction between invention, innovation, and diffusion. According to his definition, invention concerns the first development of a new artifact or process. Innovation entails its economic application. Diffusion describes its introduction by buyers or competitors. It is a rough and "heroic" conceptual distinction, which can hardly be found in practice, since the empirical processes are usually never quite like this. The *invention* is often introduced from the start as an *innovation* by economically-minded research establishments. *Diffusion* entails further innovation on the part of both developers and users. All three activities are often associated with changes in the characteristics of, and incentives for, potential innovators/adopters. However, Schumpeter's distinction between invention, innovation, and diffusion is still a useful theoretical point of departure. For example, *invention* is suggestive of some sort of exploited *potential* for technological progress, while *innovation* and *diffusion* hint at the economic, social and organizational incentives and impediments to the incorporation of technological advances into economic products and processes.

What progress has recently been made in the conceptualization of such phenomena? It is tempting to compare the contributions that follow with those presented almost thirty years ago at the Conferences of the National Bureau on Economic Research on *The Rate and Directions of Inventive Activities* (NBER, 1962). Significant elements of continuity as well as further developments appear.

7.2.1 Inventive opportunities

What is an *inventive opportunity*? Do all "new opportunities" emerge from apparently exogenous scientific progress? What shapes the dynamics of their actual exploitation? The analysis of *technology* as a quite specific sort of *information* characterized by indivisibilities and, at least some, *public-good* features in the sense of being able to be potentially transmitted and reused repeatedly without loss, certainly drew also from the seminal contributions of Arrow (*cf.*, for example, Arrow, 1962a and 1962b) and a few of the roots of such an approach are already witnessed by the cited NBER reading. That approach easily fostered widening streams of later analyses on "the economics of R&D" as a subset of the economics of imperfect and asymmetric information (for reviews of recent developments, see Stiglitz, 1985; Stoneman, 1990; Tirole, 1988). In an extreme synthesis, all the variegated contributions that can be joined under the "imperfect information" perspective have explored the properties of innovative worlds whereby *notional opportunities* are either given or are subject to the exogenous dynamics of scientific discoveries, but their *actual* exploitation depends on the particular incentive structures (related, e.g., to the forms of market competition, etc.) and possibly also on the *past* available information on which the agents can draw (*cf.* David, 1975; Atkinson and Stiglitz, 1969; Stiglitz, 1987).

Conversely, a somewhat different approach has drawn a sharper distinction between information and prior knowledge, the latter being the rather elusive set of cognitive structures, search rules, "tacit" capabilities guiding inventive activities (*cf.* Nelson and Winter, 1982; Pavitt, 1988; Dosi, 1988), all implicitly or explicitly linking with Simon's views on behaviors and decisions (see, e.g., Simon, 1965 and 1979). From such a perspective, *information* is still imperfect – indeed, *largely* imperfect – but, in addition, the rates and directions of inventive activities are shaped and constrained also by specific skills and heuristics of the searching agents, the activities in which "they are good", their past experiences, etc. The recent attempts to conceptualize the procedures and directions of innovative activities in terms of *technological paradigms* and *trajectories* clearly fits within this perspective (Dosi, 1988).

Relatedly, significant progress has been made in the empirical understanding of the varying balances between *private* and *public* aspects of technological knowledge, between new opportunities that are generated in non-profit institutions (such as universities, public research establishments, etc.) and those which are created within the business sector (in private R&D laboratories, but also through the more informal activities such as experimentation and learning) (on all this, see Nelson, 1988 and 1990).

7.2.2 Invention and innovation incentives

Irrespectively of the specific theoretical representations of inventive opportunities and their exploitation, the economic discipline has increasingly attempted to understand and conceptualize the effects of different economic incentives upon the actual rates of invention and innovation. Plainly, the issue goes back to the highly plausible fact that economically-motivated agents will undertake the costs and risks of innovating only in so far as there is, or they believe in, some differential economic returns from innovation. In other words, for whatever notional opportunities, the actual rates of innovation are going to be affected by the *appropriability conditions*. In fact, the recent economic literature has increasingly tried to explore the nature and effects of varying appropriability conditions (see Levin *et al.*, 1985 and 1987); a mainly empirical survey is in Dosi (1988); the review in Kamien and Schwartz (1982) concerns the relationship between market structures and incentives to innovate discussed within a relatively orthodox perspective; a discussion of some of the current theoretical literature is in Dasgupta (1988).

7.2.3 History and path-dependency

Both the theoretical and empirical literature reflect the growing recognition that *history counts*: past technological achievements influence future achievements via the specificity of knowledge that they entail, the development of specific infrastructures, the emergence of various sorts of increasing returns and non-convexities in the notional set of technological options. On theoretical grounds, this has led to the development of *path-dependent* models (*cf.* Arthur, 1988; David, 1985 and 1990); for some discussion of the general importance of this class of models for economic theory see Dosi and Orsenigo, 1988).

Certainly, over the last three decades, empirical analyses and theoretical modeling has made significant advances in the unfolding of the *black box* by which the economic discipline had traditionally represented technology and technological change (Rosenberg, 1982), by unfolding the determinants and nature of invention and innovation, and the driving forces, patterns and consequences of innovation diffusion (reviews and some evidence can be found in Dosi, 1988; Arcangeli *et al.*, 1990).

Below, I shall first sketch some empirical *stylized facts* and, second, try to present an overview of the diverse streams of analysis, organized around their methodological analogies, their basic assumptions on the diffusion process, and the characteristics of the adopters.

7.3 Some "Stylized Facts" on Innovation Diffusion

Even after a new product or production process or form of organization is developed, its economic and/or social significance is still going to depend on its acceptance amongst potential customers and the degrees to which it is imitated by competitors. The study of innovation diffusion concerns these phenomena. Not surprisingly these phenomena have been of interest to several social disciplines. For example, rural sociology has studied the circumstances which affect the pace of adoption of agricultural innovations. Other areas of sociology have investigated the diffusion of social innovations, such as particular forms of health care, pension funds, etc., and the social characteristics that influence the acceptance of new products. The latter have obviously also been the concern of marketing studies. Economic geography has studied innovation diffusion in its spatial dimension (somewhat overlapping with regional economics). A thorough review of these areas of research is in Rogers (1983). Innovation diffusion studies also have a relatively long tradition in economics, pioneered by the investigation on the diffusion of hybrid corn by Griliches (1957) and of a few industrial processes by Mansfield (1961). (For surveys, see Stoneman, 1983 and 1990; Metcalfe, 1988).

One can find in the literature different definitions of diffusion. However, whatever the definition, one of the basic *stylized facts* of the diffusion process is that it is never instantaneous. Innovation diffusion always takes time and occurs at rates that plausibly depend on the features of those technologies which are to be adopted; possibly on the features of those technologies which are to be substituted; on the incentives that the economic environment provides for adoption; on the characteristics of the would-be adopters; on the information available to them; on their technological competence; possibly, on their size. For example, the evidence from Mansfield (1968), Romeo (1975), Nasbeth and Ray (1974), von Tunzlemann (1978), and Ray (1984) show indeed quite a high inter-firm, inter-industry and inter-technology variance in the speed of diffusion (irrespective of the measures chosen).

7.3.1 Diffusion patterns

In general, as Rosenberg puts it,

> in the history of diffusion of many innovations, one cannot help being struck by two characteristics of the diffusion process: its apparent overall slowness on the one hand, and the wide variations in the rates of acceptance of different inventions, on the other [Rosenberg, 1976, p. 191].

Typically, one observes roughly S-shaped diffusion curves, whose precise form varies considerably across innovations (*cf.* Davies, 1979; and several contributions to this book). That the empirical curves are S-shaped should not be surprising: many time-dependent processes with some kind of asymptotic value present such a form. However, an important interpretative question concerns the determinants of particular diffusion patterns.

In one way or another diffusion analyses attempt to explain such empirical variety in the observed rates and patterns of adoption of new products, processes, and forms of organization:

- Why is adoption not instantaneous? Why is adoption distributed over time?
- What keeps the sequence of adoptions going forward through time, rather than stopping after the first or n-th firm or household has adopted?

At a more detailed level of empirical investigation, several studies, as mentioned, highlight the differences in the patterns of innovation diffusion and, also, in the origins of innovations themselves. For example, some innovations are embodied in specific artifacts produced somewhere else in the economic system (for example by machine or intermediate component manufacturers). Other innovations take the form of disembodied knowledge which diffuses via people's mobility and/or via competitive R&D.

In general, the *speed of diffusion* is inherently hard to judge because there is no precise way to define the ultimate scope of application or use of a new method of production or a new method of production or a new product (Rosenberg, 1976; Gold, 1981). In fact, whatever empirical definition of *potential adopters* one takes, their number tends to increase over a certain time after the introduction of the original innovation. At an early stage, new methods of production and new pieces of equipment "are of necessity, badly adapted to many of the ultimate uses to which they will eventually be put" (Rosenberg, 1976, p. 195). Diffusion is generally interlinked with more or less incremental improvements of the innovation itself which (a) enhance the technical/economic superiority of the new product or process *vis-à-vis* older ones, and (b) enlarge its scope of application (a detailed illustration of these processes for agricultural machinery is in David, 1975).

7.3.2 Potential adopters

The universe of potential adopters cannot be realistically assumed to be composed of identical units. One of the clearest cases is that discussed in David (1975) where the set of potential adopters of agricultural equipment were farms of different sizes and different configurations of the land. In turn, this affected the scope and the profitability of the mechanization of agricultural production. So, for example, the larger and on average flatter and regularly shaped American farms help to explain a faster rate of diffusion of agricultural mechanization in the USA as compared to, for example, the UK (of course in these international comparisons, differences in wage rates also determine differential incentives to the diffusion of mechanized equipment).

Another straightforward case is whenever the incentive of adopting an innovation is somewhat *scale-based*, or at least there is a minimum scale at which adoption is profitable (this general conjecture is argued in Sylos-Labini, 1967). Hence, the differences in size of potential adopters affect the incentive to adopt. A wealth of empirical evidence suggests, in fact, that the size of the firm is positively correlated to the speed of adoption (Mansfield, 1968; Metcalfe, 1970; Davies, 1979; see also Chapter 14 by Kelley and Brooks). This is not necessarily the case in inter-industrial, cross-innovation comparisons: that is, *ceteris paribus*, an innovation does not diffuse quicker simply because it is introduced into an industry where the average size of the firm is larger (Romeo, 1975; Davies, 1979). That the scale of firms matters, is also shown by the finding that the absolute cost of innovation affects, *ceteris paribus*, its rate of diffusion (Mansfield, 1968; Davies, 1979). Of course, from a theoretical point of view, one may attribute the phenomenon to various sorts of *imperfections* on the financial markets, etc. However, the expense of introducing and using an innovation does not include only the price tag on the new equipment, but also the costs of reorganization needed to take advantage of the new equipment or method. In turn, there is a fixed element in these costs, which can be spread, according to the size of the firm, over different volumes of production. More generally, a whole approach to innovation and diffusion studies would argue that it is often the case that the adopting firms differ in their technological capabilities, and that some of the potential adopters may not adopt because they do not have the technological and organizational capabilities to do so. To put it simply, they do not adopt since they lack the appropriate skills, internal knowledge, or managerial capabilities. Diffusion processes generally involve learning, modifications in the existing organization of production, and, sometimes, even modifications in the products, i.e., essentially, diffusion involves innovation for the user (Freeman, 1982; Rosenberg, 1976 and 1982; Gold, 1981; Lundvall, 1988). In brief, this second set of factors, which influence the patterns of diffusion, relates to the nature and distribution of *technological asymmetries* between firms. Conversely, each process of diffusion is matched by the development of skills among users, the solution of specific technical bottlenecks which hinder adoption, and the development of complementarities with other ancillary technologies (for detailed illustrations of these points, see Rosenberg, 1976, Chapter 11).

Over time, post-innovation changes in the price of the innovative goods (e.g., new machines) affect the incentive to adopt them, in general, and differentially for different sets of users (see again, Rosenberg, 1976; David, 1975).

7.3.3 Profitability and expectations

It is not surprising that there is robust evidence indicating that the rates of diffusion of innovation are influenced by the differential profitability of the new process of equipment as compared to the existing one (Mansfield, 1968; Mansfield *et al.*, 1977; Davies, 1979). It can plausibly be argued that this is indirect evidence of the *disequilibrium nature* (or at least a high degree of uncertainty) of many diffusion processes, since one would expect that according to most definitions of equilibrium, any firm for which the innovation is profitable – no matter *to what extent* profitable – would adopt it.

Certainly, part of the explanation of a discrete time lag in the diffusion of, e.g., a superior (i.e., unequivocally more profitable) machine is a "vintage effect" (Salter, 1969): given the general irreversibility of investment decisions, adoption decisions are to some extent scrapping decisions, which in turn depend on the "technological vintage" of the equipment currently in use.

Moreover, *technological expectations* matter, in the sense that the expected stream of future revenues from, e.g., a new type of machine of today's type, depends also on the expectations about the technical characteristics, productivity, and costs of future machines. In a way, the rate of adoption of innovation is implicitly influenced by expectations about future technological developments and also about the second order (how will that rate of change vary in the future). On empirical grounds, the evidence suggests very complex and often hardly formalizable processes of formation of these expectations: the decision rules of adoption may be highly *imperfect* in a neoclassical sense – heavily based on firm-specific and sector-specific institutional traits and *animal spirits* (for evidence, see Carter and Williams, 1957; Stoneman, 1976; Kleine, 1983).

7.3.4 Appropriability

The typical process of diffusion discussed so far, fits particularly well the description of the factors which drive or hinder the purchase and use of, say, a new type of machinery by a (changing) population of potential adopters. That is, it concerns primarily the introduction of a *process innovation* which is often a *product innovation* manufactured somewhere else in the economic system: for example, it can be the decision of a textile firm whether to introduce an automatic loom which in turn is a new product of a machine-building firm. However, the symmetric complement to this process is the diffusion of product innovations amongst potential suppliers (e.g., machine tool builders, etc.). After all, *product*-related R&D is estimated to account for 75 to 90% of total R&D expenditures in manufacturing (see Le Bas, 1981, for a discussion of the various sources of evidence). What affects the patterns of diffusion in supply (e.g., the number of machine tool builders which produce numerically controlled machines of a certain type or the number of drug firms which manufacture a new antibiotic)? Empirical research in this field is relatively young. However, some important findings emerge from the works of Gort and Klepper (1982) and Gort and Konakayama (1982) as well as from industrial case studies.

First, diffusion in supply – which implies more or less creative imitation of the original innovative product by other producers – relates directly to the *appropriability conditions* of innovations. Of course, a notional innovation with total appropriability would never be imitated by any other producers. Conversely, very low degrees of appropriability would allow, other things being equal, easy imitation and a quick diffusion in supply.

Second, note the double role that appropriability plays in diffusion in production. On the one hand, it acts as an incentive to imitation and entry since it is, *ceteris paribus*, correlated with a differential profitability in the production of the innovation good. On the other hand, it performs as an *entry barrier*, since appropriability is almost by definition based on some kind of appropriable asset, cumulated experience, differential technological capabilities or legal devices such as patents (see on these issues, Chesnais, 1990; Teece, 1986; Levin *et al.*, 1987; Philips, 1971; Gort and Klepper, 1982; Dosi, 1984). The net effect upon the rates of diffusion in production (that is, the rates of entry) are likely to depend on (a) the perceived opportunities of technical progress (a high opportunity with high appropriability conditions is likely to be a powerful incentive to enter and make a *better* product and/or innovate further along the same technological trajectory); (b) the nature of the knowledge-base on which a particular technology draws and, relatedly, the degree of specificity to the incumbent producers of their innovative capability (see Gort and Klepper, 1982, for empirical evidence).

7.4 Innovation Diffusion: Drawing Together Diverse Streams of Analysis

Let us first introduce a classification of diffusion models by means of some underlying dichotomies in the analytical hypotheses and *stylized facts* which they assume.

(1) *Heterogeneity versus uniformity of potential adopters of innovations*. Are all agents the same? Do they have similar *incentives* to innovation adoption? Or, conversely, do they differ in some structural characteristics, or in their *capabilities* of efficiently acquiring new products and processes of production, or in their technological *expectations*?

(2) *Perfect versus imperfect information*. Can the adopting agents be assumed to have adequate information – at least for interpretative purposes – about the nature and future developments of any one technology? Or,

rather, should we suppose that an essential determinant of innovation diffusion concerns information diffusion about the existence and attribute of particular innovations?

(3) *Non-increasing versus increasing returns in new technological developments.* Under what circumstances can we expect the *use* and/or the *production* of innovations to exhibit constant or decreasing returns? Conversely, are there factors which may yield size-related economies of scale, various sorts of learning processes and, generally, dynamic increasing returns?

(4) *The importance of history for the patterns of diffusion.* Clearly, the issue relates also to points (2) and (3). The higher the uncertainty about the technical and economic characteristics of innovation, the higher the importance of the *learning history* of individual agents is likely to be with respect to their adoption decisions. But does this also affect the general patterns of diffusion, or can one assume that the final "attractor" or stationary state of a diffusion process will still be independent of individual vicissitudes? Most is going to depend on the existence of dynamic increasing returns [point (3)] and on the feedback processes between the number of adopters of the technology, on the one hand, and the changing incentives to further adopt it, on the other. Whenever these circumstances occur we are clearly in the domain of path-dependent, non-ergodic processes, briefly recalled in Section 7.2 of this work. In all these cases, history counts, not only for individual patterns of behavior, but also in terms of the general long-term dynamics of the system.

(5) *The interaction between supply and demand of innovation.* When can one reasonably assume that the innovation to be adopted is supplied once-and-for-all, and conversely, when is it correct to assume a continuous process of improvement in its technical characteristics – which also make adoption easier and enlarges the set of potential users? How important are changes in supply conditions, *in primis* prices, for the changing pace of innovation diffusion?

(6) *Diffusion in demand versus diffusion in supply.* The way diffusion processes are often represented typically concerns a new good (say, a new type of production machinery) whose manufacturer is keen on selling to as many customers as possible. However, another side of the diffusion process concerns, as mentioned, the diffusion of the manufacturing capacity of this new good amongst the producers themselves. The theoretical representation of this kind of *diffusion in production* clearly relates to the *conditions of imitation* of an innovation and thus with the

theoretical analysis of technological appropriability, possibly entry- and mobility-barriers, "tacitness" versus "universality" of technical knowledge. Ultimately, it is an area where diffusion analysis joins with the economics of innovation and the economics of industrial dynamics.

(7) *The forces driving diffusion.* Are these forces mainly *exogenous* to the context in which the diffusion of a particular innovation takes place, such as general changes in relative prices and macro demand growth? Or rather do they mainly relate to factors that are *endogenous* to the supplying and adopting industries, such as for example, learning in the manufacturing of the innovation, learning by using, network externalities?

In addition to these basic dichotomies on the "stylized facts" that the analyses assume, some other fundamental alternatives concern directly the analytical methodology, and in particular:

(8) *Behaviors and choice processes of individuals or individual organizations.* At one extreme, one may represent decision processes about adoption/non-adoption of new technologies as a standard optimization exercise, whereby the agents explicitly form expectations about the returns on the new technologies, confronts the entire payoff matrix reachable through their actions, and choose by maximizing some objective function. Following an economist's convention, call this "rational" or "optimizing" behavior. At the other extreme, a few authors attribute much less "rationality" to individual choice processes, according to a methodological option grounded in the empirical observation and in some theoretical reasons for the impossibility of literally maximizing behaviors in environments that are sufficiently complex and nonstationary. Thus, in this other approach, behaviors are likely to be rather "routinized", influenced by specific "visions" and norms. Call this "institutionalized behavior".

(9) *Equilibrium versus disequilibrium dynamics of diffusion.* In the following, I shall use the convention that diffusion dynamics is an "equilibrium one" whenever micro decisions are postulated to be *reciprocally consistent* and "rational" microbehaviors all turn out to be fulfilled in their objectives. Conversely, I shall call "disequilibrium" diffusion processes all those dynamics wherein (a) the "attractors" of the process change themselves as a result of the very actions of the agents – such as when there are system-level increasing returns to technology adoption and/or (b) the diffusion process is explicitly represented in terms of the trial-

and-error efforts of the agents, which exhibit "disequilibrium behaviors" and deliver "disequilibrium signals" to other agents. (I refer here to equilibrium and disequilibrium diffusion dynamics as "macro level" analysis even when it refers to "macro behaviors" of single industries or groups of forms. The proper meaning of "macroeconomics" will be restated in the final section.)

As can be easily seen, the foregoing dichotomies in assumptions, in the postulated ***stylized facts***, and in the theoretical methodologies have a crucial importance well beyond the area of innovation diffusion. Indeed, issues like the diversity amongst agents, the access to information that the latter have, the consequences of increasing returns and non-convexities, the status of maximization and equilibrium assumptions, and the postulated processes which bring consistency among a multiplicity of agents, all raise challenging questions which are at the core of economic analysis in general. In fact, this is probably one of the reasons for the general importance of innovation diffusion: it does, after all, concern the processes by which the economy generates and accommodates "the new", and thus directly touches all those questions on coordination and change that have puzzled economists since the beginning of economics as a discipline.

7.5 Diffusion as Information Spread and Adaptation

Before adding some comments of my own, let me suggest a taxonomical guide through the state of the art in the field by mapping the various approaches found in the literature according to some of the earlier dichotomies. In *Table 7.1* diffusion analyses are grouped according to their methodological differences [points (8) and (9), above]. Four broad groups are given based on a two-dimensional classification based on whether the equilibrium or disequilibrium approach is used and on whether the particular model deals with optimizing or institutionalized behavior.

7.5.1 Equilibrium approaches

"Institutionalized" Behavior

The top-right corner of *Table 7.1* includes all equilibrium models with institutionalized behavior, developed with a strong *descriptive* emphasis, that

Table 7.1. Methodological classification of diffusion models.

	Micro behaviors	
Macro level	Optimizing behaviors	Institutionalized behaviors
Equilibrium *Steady-states*	Neoclassical models, e.g., David (1969); Stoneman (1983); Reinganum (1981)	*Traditional* models with adjustment lags, e.g., Griliches (1957); Mansfield (1968)
	Davies (1979)[a] *Marshallian* models in Metcalfe (1988)[a]	
Disequilibrium *Traverses* or *self-organization* processes	*Increasing returns* models of diffusion, cum. innovation, e.g., David and Olsen (1984, 1986); David (1985, 1986); Farrell-Saloner (1985)	Evolutionary models, e.g., Nelson and Winter (1982); self-organizational models, e.g., Silverberg, Dosi, and Orsenigo (1988)
	Arthur (1983, 1988)[a]	

[a]These models do not make explicit assumptions on microbehaviors and are, in principle, consistent with either hypothesis.

investigate the empirical relevance of various economic and social variables as favorable or retarding factors in the adoption of innovation. The starting point is generally the *empirical regularity*, mentioned earlier, on diffusion patterns often presenting an S-shaped profile. Indeed, it is generally found that time-patterns fit rather well rate equations of the generic form

$$\frac{dx_t}{dt} = f(N, x_t) \tag{7.1}$$

where x_t stands for the number of adopters at time t and N is the total number of potential adopters. Hence the analysis primarily concerns the factors determining the rates of change in adoption and the primary focus is about how people and organizations become exposed to novelty and react by rejecting or embracing change. The analysis of the diffusion of new technologies, in this approach, could in a way be considered as part of the study of more general behavioral patterns of humans and organizations.

Both in the economic and sociological literatures, primary attention is given to the propagation of information about novelties, which may come as *stimuli*, *surprises* or *threats* to individuals and corporations. The stimuli elicit responses, which come with varying lags, depending upon socio-psychological attributes of individuals, their position in the social system, or, somewhat analogously, on corporate cultures and organizational structures.

I classified this approach under "equilibrium" because it still generally treats diffusion as a process of convergence to some long-term, steady-state (say $X^* = N$ in equation 7.1) with non-instantaneous adoption primarily accounted by frictions, lack of information, response lags, and "out-of-equilibrium" behaviors of micro agents. The various studies that come under this heading are thoroughly reviewed in Rogers (1983). They have certainly provided rich insights into the empirical variety of socioeconomic determinants of diffusion patterns. However, if one looks for a *deeper explanation*, then it does not exist. In particular, economists often consider such a *deeper level* to rest in the *microfoundations* of any one aggregate process and most often are only satisfied when, in turn, micro behaviors can be grounded deductively into some "rational" choice procedure. This kind of theoretical diffidence toward simply descriptive models is probably also one of the explanations for the appeal of the class of models that we are going to consider next, those at the top-left of *Table 7.1*.

"Optimizing" Behavior

The top-left corner of *Table 7.1* includes "optimizing" equilibrium models. In many respects equilibrium diffusion models *cum* micro optimization represent the extension of neoclassical economic theory to diffusion phenomena. In this class of models diffusion is seen as the outcome of rational goal-directed choices, made by more or less fully informed firms and consumers, among the set of available technologies – particularly the choice between new technologies and those previously available. Here the focus is upon modeling the choice-process, and assimilating innovation-adoption into the larger corpus of (decision-theoretic or game-theoretic) microeconomic theory. This theory, when it is employed as a *positive* model (and not simply as a normative exercise) asserts a correspondence between the central (average) tendencies in individual behaviors and the (theoretically-derived) equilibrium.

Given the above, why, then do not all the agents adopt at the same time? One of the simplest answers is that the agents are "rational", but less-than-perfectly informed: there are objective costs in information acquisition, adjustment costs, etc. In a way this is the most straightforward "rationalization" (in terms of "rational agent" micro theory) of the evidence put forward by the studies reviewed in the previous section.

Another route is to assume that agents are "rational" *and different* in some structural characteristics. P.A. David long ago started exploring this route whereby thresholds to efficient innovation adoption are determined by the scale of output of adopters, and consequently, the benefits of adoption for each agent do not reach a maximum at the same time (David, 1969).

Consider the basic model:

- The users of the innovation belong to a competitive industry.
- They are heterogeneous with respect to output scale.
- There is perfect information about the characteristics and economic benefits of the innovation.
- The innovation is embodied in some fixed, lumpy piece of equipment (e.g., continuous rolling, annealing machines, tractors, etc.).
- Adopters are "myopic" in that they ignore the future acquisition costs of the technology.
- The scale of the firms is fixed and does not adjust in response to the innovations.

This *scale-constrained* world (as from David, 1969) is depicted in *Figure 7.1*, where $g(z)$ is the frequency distribution of some agents' characteristics (z, in our case, size). If we specify a time-path of the threshold level of the characteristics index $z^*(t)$ above which adoption becomes profitable, we re-map the characteristics distribution into the time domain. We can derive a so-called "Probit model" of diffusion.

Consider for example lognormal size distributions (as in David, 1969 and Davies, 1979) and some alternative time patterns of the threshold (for example, exponential relative decline in fixed-factor/variable factor price ratio, as in David, 1969, or linear time-paths with retardation in the decline of the threshold, as in Davies, 1969). Distributions of heterogeneous agents with a moving threshold for efficient adoption can generate a sigmoid diffusion curve. The time-profile of the threshold can obviously be given exogenously, but can also be derived from the diffusion-dynamics itself: for example, Stoneman and Ireland (1983) and David and Olsen (1984) derive it explicitly by specifying *learning effects* in the supply of the new technology.

Of course, size is by no means the only source of differentiation amongst adopters and equilibrium – "rational" models can accommodate other forms of heterogeneity, such as, for example, location and transport costs, regional wage levels, information costs, etc. For example, David and Olsen (1984)

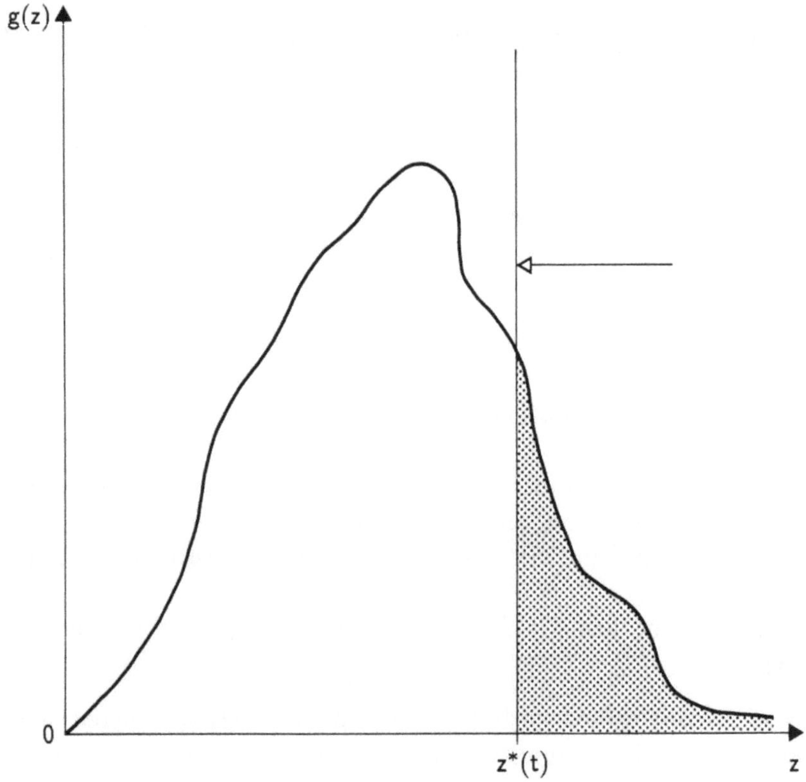

Objective characteristics distributions [g(z)] models:

- Vintage-distribution of fixed capital (Salter).
- Size-distribution of firms (output) (David, Davies).
- "Priors" on benefits (uncertain) (Stoneman, Jensen, Reinganum).
- Risk-aversion distribution.

Figure 7.1. "Equilibrium diffusion" with heterogeneous agents.

move beyond probit-type (or "threshold") models – in which the scale of output is not influenced by the new technology and the firm size distribution remains invariant with respect to the diffusion process – and allow for a competitive firm's choice of output jointly with its technology adoption decision. Hence, the heterogeneity which determines the time-distribution of adoptions must come from factors other than output size.

Equilibrium models with heterogeneous firms have typically assumed a competitive industry (with competitive or monopolistic suppliers of the innovation). Conversely, another stream of equilibrium analysis has explored diffusion in the context of *strategic oligopolistic* interactions *cum homogeneous agents* (e.g., Reinganum, 1981; Fudenberg and Tirole, 1985).

A closely related approach describes equilibrium diffusion with homogeneous, perfectly forward-looking agents *cum* non-appropriable learning (Jovanovic and Lach, 1989).

A taxonomy of equilibrium – "rational" models, categorized according to the heterogeneity versus homogeneity of the agents and endogeneity versus exogeneity of the *driving forces* of diffusion is presented in *Table 7.2.*

Table 7.2. A taxonomy of equilibrium models of technology diffusion.

Nature of dynamic	Objective characteristics of potential adopters	
driving process	Heterogeneous	Homonogeneous
Exogeneous	Competitive industry adoption, driven by: • market demand growth • output distribution changes • input price trends	Interdependent adopters with: pre-commitment (Reinganum, 1981) or strategic interaction (Fudenberg and Tirole, 1985) • exogeneous drivers
Endogeneous	Competitive industry adoption, driven by: • learning in innovation supply • learning in use • network externalities which are unbounded or bounded	Oligopolistic or duopolistic rivalry with *learning* or *system-scale* effects

One of the general questions addressed by equilibrium models with interdependent adopters is: can one conceive a *consistent* pattern of adoption whereby identical agents adopt at different times? Technically, an answer to such a question generally corresponds to the exploration of the existence of Nash equilibria in the domain of adoption decisions, thus extending the application of a methodology increasingly used by mainstream industrial economics with respect to, e.g., price and quantity variables in oligopolistic settings.

Certainly, these various versions of equilibrium-rational analyses have generated an increasing number of contributions, highlighting the implication for the diffusion process of different subsidiary assumptions (e.g., "myopic" versus "rational" expectations, strategic interactions with or without pre-committed choices, competitive versus monopolistic supply of the innovations, etc.). (A review of these developments is in Stoneman, 1990.) It is much harder to assess the heuristic value of this class of models for the *positive* interpretation of empirical phenomena of innovation diffusion. Any evaluation also involves a general judgement on the tool-box of the economic discipline as a whole, the status, and interpretative power of methodological assumptions such as equilibrium, rationality, etc. It is a judgement which certainly divides the economic discipline. However, irrespective of disciplinary preferences and the particular beliefs on "how much of the real world equilibrium-rational models can illuminate", there are phenomena which they can hardly handle, ultimately related to diffusion processes involving various forms of increasing returns, *circular feedbacks* between innovation and diffusion and environments where "rational behavior" is not only empirically unlikely but theoretically impossible to define. The models classified in the bottom half of *Table 7.1* try precisely to deal with some or all of these phenomena. I shall now turn to them.

7.5.2 Disequilibrium approaches

Increasing Returns, Path-dependency, and Evolutionary Processes in Innovation Diffusion

Many students of technology diffusion processes have arrived at the view, argued, for example, by Rosenberg (1972); Freeman (1982); Sahal (1981); Freeman and Perez (1988), that innovation and diffusion are continuously interlinked. Over time, technology adoption, and the creation of new technologies are mutually dependent. Moreover, the innovation-diffusion process must be seen as "macro-disequilibrium dynamics" (in the meaning of *macro* defined earlier), characterized by the exploration of various sorts of increasing returns along particular "trajectories" of technical progress (Nelson and Winter, 1977; Dosi, 1982), and the evolution of various institutions within the economy (including firms) along with the exploitation of the new technologies.

Of course, if *positive feedbacks* between adoption of innovations, their further improvements, and the cost of acquiring them are important enough, this implies, in technical terms, a source of non-convexity in the technological opportunities at the level of the firm, the industry or the whole economy. We leave the world of *convergence* to macro-level *solutions* and of equilibrium paths which can be defined independently of the *actual* technological and economic history of particular clusters of innovations. On the contrary, we are in the path-dependent world of Arthur (1983 and 1988) and David (1985), wherein the long-term positions of the system may well depend on even minor initial *fluctuations*, individual choices, institutions, and policy measures. (On these issues, *cf.*, in particular, the chapters by David; Amendola and Gaffard; Dosi *et al.* in Arcangeli *et al.*, 1990.)

In a different vein, all the models classified in the bottom half of *Figure 7.1* deal with diffusion processes that are implicitly or explicitly *historical* with multiple notional end-state and endogenously-generated opportunities (dynamic increasing returns). However, they differ in their approach to micromodeling.

Those on the bottom-left (e.g., David, 1985 and 1990; David and Olsen, 1990; Farell and Saloner, 1985) maintain a micro "rationality" assumption: agents still make their optimizing choices which collectively yield some sort of externality that in turn affects future returns and future options.

Conversely, the "evolutionary" and "self-organization" models on the bottom-right of *Figure 7.1* depart also from the standard micro-rationality and explicitly represent diffusion as the outcome of diverse behaviors of agents exhibiting "institutionalized" (so-called "bounded rationality") behaviors. Indeed, the supporters of this analytical approach do claim that the difference between "evolutionary" and "equilibrium" approaches to innovation diffusion is much deeper than the difference between the explicit (but somewhat unelegant) representation of a process and the description of the final state to which *that same process* is converging. Rather, evolutionary theory claims that (a) only under particular circumstances can "rational equilibria" be postulated to be the "attractor" or end-states of empirically more plausible "disequilibrium" trial-and-error processes, and relatedly, (b) actual end-states may well depend on non-average behaviors, so that an explicit account of the distributions of specific choice rules and of the specific mechanisms of competitive selection are theoretically required (these issues are discussed, in this perspective, in Nelson and Winter, 1982, and Dosi *et al.*, 1988, in particular the chapters therein by Silverberg, Allen, Dosi, and Orsenigo.

Broadly speaking, many "rational" and evolutionary/self-organization models of innovation diffusion overlap in the endeavor to *explain* theoretically *open-ended* (history-bound) processes. They do also sometimes overlap in the questions that they ask. For example: are there micro strategies that are stable in an evolutionary sense? How does dynamic mutual consistency come about? However, they depart in the relative faith that they place on the rational power of microagents as an "ordering factor". In a way, all micro-rational models look for order by investigating the conditions of existence of consistent rational strategies (whether pure or mixed strategies, in game-theoretic settings) that could support a diffusion process. Conversely, evolutionary/self-organization models rest their emphasis on diversity, learning and environmental selection as the main ordering factors, which *ex post, but not necessarily ex-ante* produce recognizable regularities in the diffusion process.

7.5.3 Optimizing equilibrium compared to institutionalized behaviors and disequilibrium

The reader may find a vivid illustration of the analogies and differences between the basic equilibrium and disequilibrium approaches by comparing Jovanovic and Lach (1989) and Silverberg *et al.* (1988). Loosely speaking, they could be considered as somewhat extreme archetypes of the two categories of models. In fact, they both start with quite similar technological assumptions: adoption of a capital-embodied innovation drives fixed and variable costs down and, therefore, the efficiency gains are a sort of collective result of adoption decisions. However, the analogies stop here. Jovanovic and Lach assume perfect micro technological forecasting and an *ex-ante* "equilibrium ordering" of entry decisions which, I must confess, I keep finding rather obscure in its microeconomic plausibility. (What is the mechanism generating the proper "queue"? How does one learn that he got an epsilon-wrong in this queue? What are the collective results of several epsilon mistakes?) Conversely, Silverberg *et al.* (1988) assume that people get it systematically wrong. This is not because they are "more stupid" than the Jovanovic-Lach (1989) agents. In principle, they form their expectations the best way they can. However, the end-results are truly collective phenomena, emerging from complex nonlinear interactions amongst all of them. What difference do the alternative analytical approaches make? In my view, one of the basic differences is that the "rational-equilibrium" approach loses interpretative significance the more the diffusion process is influenced by particular distributions of the expectational and technological characteristics of individual agents. In this sense, "equilibrium approaches" show the same limitations, and more so, than so-called rational expectation models in macroeconomics: "equilibrium paths" – whenever they exist – are not independent of the distribution of beliefs, technological capabilities, and learning processes of individual agents. In fact, one may simply check the robustness of the properties of a unique equilibrium diffusion path – from Jovanovic-Lach – allowing for different stochastic disturbances on, e.g., expectations: in general, one cannot presume the equilibrium path to be even *locally* stable; hence also the conclusions based on the properties of "perfect" equilibrium diffusion processes cannot be presumed to hold. In particular, it seems to me, equilibrium-"rational" analyses of diffusion must rule out *ex hypothesi* the possibility of (imperfect) out-of-equilibrium adjustment and learning from mistakes, which, I would argue, are an essential part of diffusion processes and affect also the specific path that is empirically observed.

It is indeed a major methodological difference which cuts across most fields of economic analysis: it is particularly evident with respect to innovation diffusion probably because, as mentioned, it is a crucial domain where novelties emerge in the economic system with an intrinsic tension between economic coordination and change. The supporters of methodological rationality believe that the calculating and planning powers of the agents can be stretched to also embrace highly complex, uncertain and non-stationary environments. Others suggest that for interpretative purposes this is a research program that despite unreasonable demands on individual calculating powers is bound to end up in rather indeterminate conclusions, and that institutions, together with more "impersonal" selection mechanisms can be part of a more promising microfoundation.

Moreover, innovation diffusion obviously bears major implications also in terms of macroeconomic growth of income, productivity, and employment (meaning here the proper "macro" level of the whole economy); industrial structures and organizational forms (including size, degrees of integration of business firms, etc.), and, finally, public policies. Let me now also briefly introduce also these broader issues.

7.6 Innovation Diffusion, Economic Dynamics, and Public Policies

Most diffusion models focus upon the determinants of innovation diffusion. However, a somewhat symmetric question concerns the *effects* of innovation diffusion upon changes in industrial structures and, more generally, on economic development.

In a genuinely dynamic framework, different capabilities, degrees of success or simply luck in innovation adoption by the various firms, affect firms' competitiveness, continuously generate *asymmetries* among them, and ultimately modify industrial structure (relative firm size, degrees of industrial concentration, etc.). It is probably a rather uncontroversial claim of "Schumpeterian"/"evolutionary" economics to have initiated this kind of investigation, albeit now paralleled by other approaches that rely more on optimizing microfoundations and equilibrium dynamics (see, for example, Dasgupta and Stiglitz, 1980a and 1980b).

Yet, irrespective of the specific analytical assumptions, any world where structural conditions and innovation diffusion are *dynamically-coupled*, that is wherein structural conditions influence innovation adoption and innovation adoption changes industrial structures, is going to exhibit nonlinear dynamics and path-dependency (hence, also multiplicity of dynamic paths and irreversibilities).

A crucial corollary of all this is that *institutions and policies matter in shaping economic dynamics*. Some of the contributions in Arcangeli *et al.* (1990) (especially Volume III) go further and conjecture some general "mappings" between (a) characteristics of diffusing technologies (or clusters of them), (b) institutions also shaping their adoption (and economic coordination, in general), and (c) patterns of economic growth. For example, Freeman and Perez (1988) argue that it is the "matching" or "mismatching" between broad "techno-economic paradigms" (such as electromechanic automation, microelectronics, etc.) and forms of socioeconomic organization which accounts for long-term historical regularities in economic growth intertwined by significant periods of instabilities and crises. On a somewhat similar level of analysis, contributions like Boyer (1988a and 1988b), Boyer and Coriat (1990), Coricelli *et al.* (1989), attempt some sort of explanation of the changing macrocoefficients (e.g., in the rates of change of income, employment, labor productivity, etc.) also in terms of plausible changes in the underlying trends in innovation and diffusion.

It is a formidable task, still at a very early stage of development, but it hints at the possibility of "microfounding" macrodynamics *cum nonstationarity* (at the very least in the available technologies, and most likely also in institutions and "preferences"). Some are keener on searching for these microfoundations through multiple equilibrium-rational dynamics. Others prefer to explore evolutionary/self-organization processes. Still, there seems a common quest for a micro-macro link that withholds (indeed, endogenously generates) non-stationarity in the so-called "fundamentals" and accommodates micro-diversity.

Another major consequence of differentiated patterns of innovation and diffusion concerns its regional and international dimensions. If regions and countries innovate and adopt at different rates, *first*, that very phenomenon must be explained, which in principle is a task of some sort of diffusion model *cum heterogeneous adopters. Second*, its implications for regional growth, international trade, and development must be explored. Several works have started to address these issues (see again, Volume III of Arcangeli *et al.*, 1990). Some increase the understanding of the "microcircularities" between technological advantages, patterns of international location, and international trade (such as the chapter by Cantwell and Dunning in Arcangeli *et al.*, 1990). Others (e.g., Cimoli *et al.*, 1990; Dosi *et al.*, 1990) link international diffusion with international institutional differences and "technology-gap" trade theories. Still a few others develop the tradition of regional economic studies in a fruitful parallelism between those models where *time* is the dimension of diffusion (such as, typically, in industrial economics), and those where it is *space*. Finally, all these complex and intertwined features of innovation diffusion have equally crucial normative (policy) dimensions.

Path-dependency and irreversibilities of diffusion patterns also imply the fundamental role of (plausibly less-than-perfectly informed) policies in the choice of major technological *trajectories*. But one side of the dilemma is that with *local* increasing returns, markets could most likely be more "myopic" and dynamically inefficient than approximate policy measures. However, somewhat symmetrically, it is hard to define the circumstances under which public policies can *plan* with more foresight toward notionally *superior* technologies. *Prima facie*, all this is a powerful argument for technological *pluralism* or at least for economic set-ups which allow for a sufficient diversity at the beginning of any new *technological paradigm* (Dosi, 1982 and 1988; David, 1990; Nelson, 1988). It remains also a major area of historical and theoretical exploration.

Micro-macro links add further complexity to the normative puzzles. If there are "multiple possible macro worlds" and institutions count in the determination of the one which finally emerges, then also very subtle questions come out concerning how individual agents, social groups, and public policies can influence such outcomes. Putting it another way, no matter whether one represents the world as multiple equilibrium dynamics *cum* non-convexities or as multiple *open-ended* evolutionary processes, one is still left with the questions on how and in which direction should the *initial conditions* be influenced – in one case – or how and in which form should micro diversities, selection processes, and economic institutions be shaped – in the other case.

All this, in our view, is another major frontier of research, which brings history, social disciplines, and policy analysis somewhat nearer to each other. Needless to say, it is an enormous task, but I believe, worth the endeavor.

References

Arcangeli, F., David, P.A., and Dosi, G., eds., 1990, *Frontiers on Innovation Diffusion*, Oxford University Press, Oxford, UK (forthcoming).

Arrow, K., 1962a, The economic implications of learning by doing, *Review of Economic Studies* **29**:155–173.

Arrow, K., 1962b, Economic welfare and the allocation of resources for invention, in NBER, *The Rate and Direction of Inventive Activity*, Princeton University Press, Princeton, NJ, USA.

Arthur, B., 1983, *Competing technologies and lock-in by historical events: the dynamic of allocation under increasing returns*, WP-83-90, International Institute for Applied Systems Analysis, Laxenburg, Austria.

Arthur, B., 1988, Competing technologies, in G. Dosi *et al.*, eds., *Technical Change and Economic Theory*, Pinter Publishers, London and New York.

Atkinson, A. and Stiglitz, J., 1969, A new view of technological change, *Economic Journal* **79**(315):573–578.

Boyer, R., 1988a, Technical change and the theory of regulation, in G. Dosi *et al.*, eds., *Technical Change and Economic Theory*, Pinter Publishers, London and New York.

Boyer, R., 1988b, Formalizing growth regimes, in Dosi *et al.* (*ibid*).

Boyer, R. and Coriat, B., 1990, Technical flexibility and macro stabilization: some preliminary steps, in F. Arcangeli *et al.*, eds., *Frontiers on Innovation Diffusion*, Oxford University Press, Oxford, UK (forthcoming).

Carter, C. and Williams, B., 1957, *Industry and Technical Progress*, Oxford University Press, Oxford, UK.

Cimoli, M., Dosi, G., and Soete, L., 1990, Technological gaps, institutional differences and patterns of trade: a North-South model, in F. Arcangeli *et al.*, eds., *Frontiers on Innovation Diffusion*, Oxford University Press, Oxford, UK (forthcoming).

Coricelli, F., Dosi, G., and Orsenigo, L., 1989, *Microeconomic dynamics and macro regularities: an "evolutionary" approach to technological and institutional change*, OECD-DSTI-SPR-89.7, Paris, France.

Chesnais, F., 1990, Some notes on technological cumulativeness, appropriation of technology and technological progressiveness in concentrated market structures, in F. Arcangeli *et al.*, eds., *Frontiers on Innovation Diffusion*, Oxford University Press, Oxford, UK (forthcoming).

Dasgupta, P., 1988, Patents, priority and imitation or, the economics of races and waiting games, *Economic Journal* **98**(389):66–80.

Dasgupta, P. and Stiglitz, J., 1980a, Industrial structure and the nature of innovative activity, *Economic Journal* **90**(358):266–293.

Dasgupta, P. and Stiglitz, J., 1980b, Uncertainty, industrial structure and the speed of R&D, *Bell Journal of Economics* **11**(1):1–28.

Dasgupta, P. and Stoneman, P., eds., 1987, *The Economic Theory of Technology Policy*, Cambridge University Press, Cambridge, UK.

David, P., 1969, *A Contribution to the Theory of Diffusion*, Memorandum No. 71, Stanford Center for Research in Economic Growth, Stanford University, Stanford, CA, USA.

David, P., 1975, *Technical Choice, Innovation and Economic Growth*, Cambridge University Press, Cambridge, UK.

David, P., 1985, Clio and the economics of QWERTY, *American Economic Review* **75**(2):332–337.

David, P., 1990, Narrow windows, blind giants and angry orphans: The dynamics of systems rivalries and dilemmas of technology policy, in F. Arcangeli *et al.*, eds., *Frontiers on Innovation Diffusion*, Oxford University Press, Oxford, UK (forthcoming).

David, P. and Olsen, T.A., 1984, *Anticipated Automation: A Rational Expectation Model of Technological Diffusion*, CEPR Working Paper No. 11, Stanford University, Stanford, CA, USA.

David, P. and Olsen, T.A., 1990, Equilibrium dynamics when incremental technological innovations are foreseen, in F. Arcangeli *et al.*, eds., *Frontiers on Innovation Diffusion*, Oxford University Press, Oxford, UK (forthcoming).

Davies, S., 1979, *The Diffusion of Process Innovations*, Cambridge University Press, Cambridge, UK.

Dosi, G., 1982, Technological paradigms and technological trajectories: a suggested interpretation of the determinants and directions of technical change, *Research Policy* **11**,(3):147–162.

Dosi, G., 1984, *Technical Change and Industrial Transformation*, Macmillan, London, UK and St. Martin's Press, New York, NY, USA.

Dosi, G., 1988, Sources, procedures and microeconomic effects of innovation, *Journal of Economic Literature* **26**(3):1120–1171.

Dosi, G. and Orsenigo, L., 1988, Coordination and transformation: An overview of structures, performance and change in evolutionary environments, in G. Dosi *et al.*, eds., *Technical Change and Economic Theory*, Pinter Publishers, London and New York.

Dosi, G., Orsenigo, L., and Silverberg, G., 1990, Innovation diffusion: A self-organization approach, in F. Arcangeli *et al.*, eds., *Frontiers on Innovation Diffusion*, Oxford University Press, Oxford, UK (forthcoming).

Dosi, G., Freeman, C., Nelson, R., Silverberg, G., and Soete, L., eds., 1988, *Technical Change and Economic Theory*, Pinter Publishers, London and New York.

Dosi, G., Pavitt, K., and Soete, L., 1990, *The Economics of Technological Change and International Trade*, Wheatsheaf, Brighton, UK.

Farrell, J. and Saloner, G., 1985, Standardization, compatibility and innovation, *Rand Journal of Economics* **16**(1):70–83, Spring.

Freeman, C., 1982, *The Economics of Industrial Innovation*, 2nd edition, Frances Pinter, London, UK.

Freeman, C. and Perez, C., 1988, Structural crisis of adjustment: business cycles and investment behavior, in G. Dosi *et al.*, eds., *Technical Change and Economic Theory*, Pinter Publishers, London and New York.

Fudenberg, D. and Tirole, J., 1985, Pre-emption and rent equalization in the adoption of new technology, *Review of Economic Studies* **52**(3):383–401.

Gort, M. and Klepper, S., 1982, Time paths in the diffusion of product innovations, *Economic Journal* **92**(367):630–653.

Gort, M. and Konakayama, A., 1982, A model of diffusion in the production of an innovation, *American Economic Review* **72**(5):1111–1120.

Gold, B., 1981, Technological diffusion in industry: Research needs and shortcomings, *Journal of Industrial Economics* **29**(3):247–269.

Griliches, Z., 1957, Hybrid corn: An exploration in the economics of technological change, *Econometrica* **25**:501–522.

Jensen, R.A., 1982, Adoption and diffusion of an innovation of uncertain profitability, *Journal of Economic Theory* **27**(1):182–193.

Jovanovic, B. and Lach, S., 1989, Entry, exit and diffusion with learning by doing, *American Economic Review* **79**:690–699.

Kamien, M. and Schwartz, N., 1982, *Market Structure and Innovation*, Cambridge University Press, Cambridge, UK.

Kleine, P., 1983, *Investitionsverhalten bei Prozeßinnovationen*, Campus Verlag, Frankfurt and New York.

Le Bas, C., 1981, *Economie des Innovations Techniques*, Economica, Paris, France.

Levin, R., Cohen, W.M., and Mowery, D.C., 1985, R&D appropriability, opportunity and market structure: New evidence on some Schumpeterian hypotheses, *American Economic Review* **75**(2):20–24.

Levin, R., Klevorick, A., Nelson, R., and Winter, S., 1987, Appropriating the returns from industrial research and development, *Brookings Papers on Economic Activity* (3):783–831.

Lundvall, B.A., 1988, Innovation as an interactive process: From user-producer interaction to the national system of innovation, in G. Dosi *et al.*, eds., *Technical Change and Economic Theory*, Pinter Publishers, London and New York.

Mansfield, E., 1961, Technical change and the rate of imitation, *Econometrica* **29**:741–766.

Mansfield, E., 1968, *Industrial Research and Technological Innovation: An Econometric Analysis*, Norton & Co., New York, NY, USA.

Mansfield, E., Rapoport, J., Romeo, A., Villani, E., Wagner, S., and Husic, F., 1977, *The Production and Application of New Industrial Technology*, Norton & Co., New York, NY, USA.

Metcalfe, S., 1970, Diffusion of innovation in the Lancashire textile industry, *The Manchester School of Economic and Social Studies* **38**(2):145–162.

Metcalfe, S., 1988, The diffusion of innovations: An interpretative survey, in G. Dosi *et al.*, eds., *Technical Change and Economic Theory*, Pinter Publishers, London and New York.

Nasbeth, L. and Ray, G.F., 1974, *The Diffusion of New Industrial Processes*, Cambridge University Press, Cambridge, UK.

NBER (National Bureau of Economic Research), 1962, *The Rate and Direction of Inventive Activity*, Princeton University Press, Princeton, NJ, USA.

Nelson, R., 1981, Research on productivity growth and productivity difference: dead ends and new departures, *Journal of Economic Literature* **19**(3):1029–1064.

Nelson, R., 1988, Institutions supporting technical change in the United States, in G. Dosi *et al.*, eds., *Technical Change and Economic Theory*, Pinter Publishers, London and New York.

Nelson, R., 1990, Institutions generating and diffusing new technology, in F. Arcangeli *et al.*, eds., *Frontiers on Innovation Diffusion*, Oxford University Press, Oxford, UK (forthcoming).

Nelson, R. and Winter, S., 1977, In search of a useful theory of innovation, *Research Policy* **6**(1):36–76.

Nelson, R. and Winter, S., 1982, *An Evolutionary Theory of Economic Change*, The Belknap Press of Harvard University Press, Cambridge, MA, USA.

Pavitt, K., 1988, International patterns of technological accumulation, in N. Hood and J.E. Vahlne, eds., *Strategies in Global Competition*, Croom Helm, London, UK.

Phillips, A., 1971, *Technology and Market Structure*, D.C. Heath, Lexington, MA, USA.

Ray, G., 1984, *The Diffusion of Mature Technologies*, Cambridge University Press, Cambridge, UK.

Reinganum, J.F., 1981, On the diffusion of new technology: A game theoretic approach, *Review of Economic Studies* **48**(3):395–405.

Rogers, E.M., 1983, *Diffusion of Innovations*, 3rd edition, The Free Press, New York, NY, USA.

Romeo, A.A., 1975, Interindustry and interfirm differences in the rate of diffusion of an innovation, *Review of Economics and Statistics* **57**(3):311–319.

Rosenberg, N., 1972, Factors affecting the diffusion of technology, *Explorations in Economic History* **10**:3–33, Fall.

Rosenberg, N., 1976, *Perspectives on Technology*, Cambridge University Press, Cambridge, UK.

Rosenberg, N., 1982, *Inside the Black Box*, Cambridge University Press, Cambridge, UK.

Sahal, D., 1981, *Patterns of Technological Innovation*, Addison-Wesley, Reading, MA, USA.

Salter, W.A.G., 1969, *Productivity and Technical Change*, 2nd edition, Cambridge University Press, Cambridge, UK.

Silverberg, G., Dosi, G., and Orsenigo, L., 1988, Innovation, diversity and diffusion: A self-organization model, *Economic Journal* **98**(393):1032–1054.

Simon, H., 1965, *Administrative Behavior*, 2nd edition, The Free Press, New York, NY, USA.

Simon, H., 1967, *Models of Man*, John Wiley & Sons Inc., New York, NY, USA.

Simon, H., 1979, Rational decision making in business organizations, *American Economic Review* **69**(4):493–513.

Stiglitz, J., 1985, Information and economic analysis: A perspective, *Economic Journal* **95**(5):21–41.

Stiglitz, J., 1987, Learning to learn, in P. Dasgupta and P. Stoneman, eds., *The Economic Theory of Technology Policy*, Cambridge University Press, Cambridge, UK.

Stoneman, P., 1976, *Technological Diffusion and the Computer Revolution*, Clarendon Press, Oxford, UK.

Stoneman, P., 1983, *The Economic Analysis of Technological Change*, Oxford University Press, Oxford, UK.

Stoneman, P., 1990, Technological diffusion: The viewpoint of economic theory, in F. Arcangeli *et al.*, eds., *Frontiers on Innovation Diffusion*, Oxford University Press, Oxford, UK (forthcoming).

Stoneman, P. and Ireland, N., 1983, The role of supply factors in the diffusion of new process technology, *Economic Journal Conference Papers* **93**:66–78.

Sylos-Labini, P., 1967, *Oligopoly and Technical Progress*, 2nd edition, Harvard University Press, Cambridge, MA, USA.

Teece, D., 1986, Profiting from technological innovation: implications for integration, collaboration, licensing and public policy, *Research Policy* **15**(6):285–305.

Tirole, J., 1988, *The Theory of Industrial Organization*, MIT Press, Cambridge, MA, USA.

Trachtenberg, M., 1987, *The Welfare Analysis of Product Innovations with an Application to CT Scanners*, Department of Economics, Tel-Aviv University, Tel-Aviv, Israel.

von Tunzlemann, N., 1978, *Steam Power and British Industrialization to 1860*, Clarendon Press, Oxford, UK.

Chapter 8

Adoption and Diffusion of Technology as a Collective Evolutionary Process

Gerald Silverberg

8.1 Introduction

Innovation diffusion occupies a special place in the economics of technological change. On the one hand it is empirically the best established and most intensively studied phenomenon in this area, and the logistic and other S-shaped curves have provided a sound mathematical, if somewhat phenomenological, inroad into a diverse range of applications. On the other, it still remains somewhat divorced from any microeconomically founded, overarching theory of the determinants of technological change which might constitute a central component of a general approach to economic dynamics and social evolution.

Why is this the case after many years of diffusion research and a vast accumulation of case studies, and a diversity of mostly *ad hoc* theoretical models? The reason appears to be twofold. First, most diffusion research has proceeded on the basis of a number of implicit assumptions which have inevitably narrowed the focus of inquiry and prevented a link-up with other aspects of the economics of technological change. Second, a number of features of the economic history of modern technology, which must be obvious to even the most casual observer, have failed to become part of the diffusion literature. As I shall argue below, these represent two sides of the same coin. Once they are brought together a new perspective opens up which, in my opinion, may lead to a fruitful generalization and integration of present work.

The study of technical change has been dominated in this century by what I call the linear model, which undoubtedly goes back to Schumpeter (1912): there is a linear progression from invention to innovation to imitation/diffusion. Each of these three stages is distinct, and a technology passes unidirectionally from one to the next, in the course of which the dominant economic factors and the nature of the actors change character significantly. The technology itself, however, remains more or less the same (once it has been invented) – it is simply passed along this pipeline (perhaps undergoing in the process some slight modifications and adaptations) to reveal its economic potential and be exploited until its innovative strength has been exhausted like a squeezed-out lemon. The diffusion literature has naturally embraced an analogous perspective: an innovation arrives at time zero as a consummated creation, like Venus from Zeus' brow. The problem is then to explain why it takes so long for it to be completely adopted by its (*a priori* clearly defined) potential adopters. Nonadoption is taken as self-evidently irrational, the (Schumpeterian) heroes being the inventors and innovators, the imitators being ambiguous figures obviously necessary to the process but somehow unseemly and undeserving, while a remnant remains of conservative stick-in-the-muds doomed to economic obscurity. The only questions to be examined are what determines the rate of diffusion, what distinguishes early from late adopters, and to what extent is forecasting (of the rates and ultimate level of saturation) possible. Implicit in this viewpoint is the notion that the faster the better, the earlier the timepoint of adoption the better, the higher the level of diffusion the better.

Another peculiar lacuna in the diffusion literature is the fact that, although the influence of profitability on adoption and diffusion rates has been extensively studied since Mansfield, the inverse – the effect of the timing of adoption on profitability, relative competitiveness and market structure – has surprisingly been largely neglected. In all of the concern to find an economic explanation for technological change, the economic implications of the latter have been left out of the picture, or relegated to the by now familiar implicit assumption that the earlier the better. It is almost as if decisions with respect to technology, although influenced by economic considerations, themselves had no economic repercussions, in particular in terms of the ability to compete. Thus studies which simply note the number or size of firms which adopt over time overlook the fact that the sizes, profitabilities, and strengths of these firms do in fact change, sometimes quite significantly, during the diffusion period, and in part at least precisely due to the adoption decisions implemented. In part this neglect is justified by the argument that most diffusion processes concern innovations which are themselves of secondary importance to the overall performance of a firm. But if this were so universally true, then there would seem to be little point in devoting so much scientific attention to questions of diffusion in the first place.

Of course this characterization is unfair to quite a number of writers on diffusion. The probit and other equilibrium traditions have argued that diffusion proceeds according to a pace consistent with some predetermined distribution of agents' characteristics (e.g., David, 1966; but see Olmstead, 1975, for a counterargument on precisely David's case study of the diffusion of reapers which reinforces the thesis of this chapter). Other writers (in particular Rosenberg, 1972; Sahal, 1981) have emphasized that the technical characteristics of an innovation develop simultaneously and interdependently with its diffusion. Yet none of these questions seem to have been dealt with systematically, and above all fitted into the overall pattern of technical change, of which diffusion is only a part, albeit an essential one.

The mirror image of these sins of omission are a number of observations which seems almost too obvious to have to be stated. The point of recollecting them here, however, is that they may serve as an entryway into a new perspective on the role of diffusion *per se* in the process of technical change, and thus bring into focus a number of previously disparate elements of the picture. The first observation is that technologies improve and develop considerably during the diffusion process. In fact, in many cases a new technology may actually be inferior in performance to an established one, and only overtake it later (Enos, 1962, e.g., gives a number of examples taken from petroleum refining). This is at variance with the implicit assumption noted above that a technology arrives full blown before diffusing, and that development and diffusion can be separated from each other. This adds an element to the adoption decision which until now only the probit models have incorporated, albeit in a very static sense: when to invest in a new technology given that its technical performance characteristics are both uncertain and changing, and that its future potential is poorly known. The fact that, in general, decision makers are conscious of the changing nature of technologies in their investment decisions and modify their criteria accordingly enables us to connect up with the considerable literature on durable investment under technical change and the role of technological expectations (e.g., Terborgh, 1949, on optimal replacement, and Rosenberg, 1976, specifically on expectations).

But we can go a step further with this observation to make the claim that one reason technologies develop considerably during the diffusion process is precisely due to this process itself. One obvious reason is learning by doing and learning by using on the part of producers and users, respectively (there is a large literature on this subject, but the classical references in economics are Arrow, 1962a, and Rosenberg, 1982). The more a technology is adopted and employed the faster producers can go down their learning curves due to increased cumulative investment and production, and the better users can exploit it due to their own accumulated experience. Moreover, user/producer interactions play a critical role in the development and adaptation process, as has been pointed out by Lundvall (1988) and von Hippel (1988) in particular. This process of incremental innovation, which quantitatively can be much more significant than the original *act* of invention, itself can take on the character of invention, not in the sense of the solitary inventor or pioneering firm still secretly cherished by most writers on technology, but what Robert C. Allen has labeled "collective invention" (Allen, 1983).

Allen defines collective invention as a fourth form of the generation of new practices and technological knowledge alongside non-profit university and governmental research, firm R&D, and individual inventors. In this process nonpatentable incremental *inventions*, i.e., explorations of possible technological practice, operating conditions, procedures, designs, etc. are more or less freely exchanged within an industry:

> The essential precondition for collective invention is the free exchange of information about new techniques and plant designs among firms in an industry .... Thus, if a firm constructed a new plant of novel design and that plant proved to have lower costs than other plants, these facts were made available to other firms in the industry and to potential entrants. The next firm constructing a new plant could build on the experience of the first by introducing and extending the design change that had proved profitable .... In this way fruitful lines of technical advance were identified and pursued [Allen, 1983, p. 2].

Whereas Allen restricts his concept of collective invention to a form of technical change in which firms, essentially noncollusively, build on each others' individual experience without committing resources to what we would now call R&D (and thus, he argues, was more significant in the nineteenth century), there are several aspects of the phenomenon which seem to be relevant to a more general evolutionary perspective on technical change in decentralized economies. First, much *incremental* technical change is incidental to more *normal* activities of firms such as investment, scaling up, and routine problem solving. Second, technology is to some extent a form of information and thus partakes of some of the peculiar economic properties of this elusive concept. It can be both private and public at different times, tacit or codifiable, reproducible, and transmissible at low cost but exploitable only if the recipient has already attained a certain technological level him/herself, and nonadditive. However, complementary pieces of information can be synergetic, allowing much larger advances to be achieved when combined than if they were used in isolation.

In dynamic terms these properties are the preconditions for the well-known Schumpeterian interaction between innovator and imitator. The implicit (linear) assumption of this picture is that the innovator enters with the new information, which eventually leaks out and is copied by others, enabling them to gradually catch up with his superior technological level and whittle away his superprofits. In view of our brief discussion of collective invention above, however, we may be emboldened to break with this one-way concept of innovation causality and advance the hypothesis that imitation, copying, and diffusion themselves contribute significantly to the further technological maturation of the original innovative idea through a complicated, two-way process of interaction and collective exploration. Almost any economic history of the development of technology which goes beyond the Samuel Smiles' exaltation of the heroic inventor will confirm that most innovations pass through many hands and strange byways before they attained the *dominant design* by which we now know them. Rare is the example of the Edison light bulb and electric power system which was single-handedly developed to market maturity by one research laboratory under one man's direction (*cf.* Friedel *et al.*, 1986). But even here, the original idea is much older, Edison's original inspiration (an electromechanical solution based on telegraph technology) was a dead-end, and crucial ancillary equipment (such as vacuum pump technology) came from outside. Furthermore, the important further advances on this technological trajectory such as high-voltage AC generation and transmission, came about through a multitude of partly complementary and partly competing international contributions totally outside of Edison's control and against which he actually fought a vain rearguard action for a time (see Hughes, 1983; David and Bunn, 1988).

Many of these factors and feedback channels have been recognized in the literature, though usually in isolation. From the economic point of view the discussion has focussed on their influence on the *appropriability* of technical change to the individual agent, that is, the agent's ability to capture a private economic reward for his inventive exertions and exclude free riders. The information character of technology is what puts this appropriability in doubt, of course. The classical solution is the legal institution of the patent. But it has become clear in recent years that patents are suitable protection for only certain classes of innovations. And in areas of very rapid technical change and concomitant rapid obsolescence of ideas, secrecy or first-mover advantages can be more cost effective forms of protection than the time-consuming and expensive process of patenting, which moreover mandates disclosure (*cf.* Levin *et al.*, 1985). Furthermore, patents can often be designed around, another example of the peculiar forms information *interchange* can take. The appropriability issue, even in a neoclassical setting, also leads to the well-known justification for government-sponsored R&D due to the discrepancy between social and private rates of return on R&D investment (Arrow, 1962b).

Appropriability is a fascinating issue for the economist because it is an example of an externality, and thus poses a challenge to the optimum welfare implications of those styles of general equilibrium analysis which, at least in theory, fully reconcile individual and social interests. This has led to a sophisticated literature on whether too much or too little R&D will be done compared to some posited social optimum, due to either the inadequate private incentive or the danger of redundancy and duplication of research efforts. From the perspective of this chapter I think these questions are somewhat besides the point. As I shall try to demonstrate in the following, both externalities and (near) duplication can be very useful, perhaps even necessary, components of technical change when seen as a collective evolutionary process. The key concept in this regard is *learning*, which can take place within the individual, the organization, and collectively through a network of feedbacks unfolding over time between both cooperative and competitive agents.

8.2 A Self-Organizational Approach to the Modeling of Technology Diffusion

I want to draw on the above observations to indicate some of the features of an appropriate modeling framework for the analysis of technical change, based on some of my own previous work (Silverberg, 1987; Silverberg *et al.*, 1988; Silverberg, 1990), and the literature reviewed in Silverberg (1988 and 1989). To begin with, it is clear that if technical change is an ongoing process, albeit proceeding in fits and starts, then the decision procedures of participants must eventually take this into account, and thus cannot be rooted in purely static analyses. This is one of the chief reasons (though certainly not the only one) in my opinion for discarding the production function approach for anything but rough indexes of total factor productivity (*cf.* Silverberg, 1990). In particular, investment and choice of technique decisions must be predicated on expectations about the future rate at which present commitments become technologically obsolete. Fortunately, there exists a considerable literature under the name of optimal replacement theory dealing with precisely these questions, if under certain simplifying assumptions, and insufficiently known to most students of technical change (Terborgh, 1949; Massé, 1962; Smith, 1961; Malcomson, 1975). The essence of a careful optimization treatment of the replacement decision is that the most common rule of thumb in practice, the cutoff (undiscounted) payback period, is theoretically valid, with the payback period determined by the longrun (expected) rate of future technical change and, to second order, the cost of capital.

This conclusion is only really valid, however, for fairly routine forms of capital-embodied incremental technical change and abstraction from problems of market power strategic rivalry. Given that entrepreneurs do employ the payback criterion (though with somewhat varying values of the required payback), one may ask how they collectively learn what the correct payback for any given historical epoch of technical change should be.[1] To this end I have constructed a simple evolutionary model of market competition and incremental technical change which is dynamic, rooted in bounded rationality and plausible decision rules, and displays the properties of cumulative causation and selection (Silverberg, 1987).

One pathway of diffusion is already incorporated in this kind of model: the vintage effect (whose pedigree goes back to Solow, 1959; Salter, 1960; Kaldor and Mirrlees, 1962). A rational investor will only gradually replace installed equipment once the gain in performance satisfies certain criteria.[2] This does not explain, however, why different agents first adopt a new technology at different times, only why diffusion through the aggregate stock of "machines" takes time. An obvious reason is that different entrepreneurs employ different payback periods, as many surveys of business practice have repeatedly confirmed. Moreover, their assessment of cost savings and the appropriateness of a technology to their line of business may differ (much as in the probit-type models). The variance in payback periods is not unexpected, given that these should depend in the first instance on expectations about the (necessarily uncertain) rate of increase of the cost advantage of new techniques over old ones, as well as the cost of capital. But it is also clear that systematic differences exist in the customary values in general use in different countries (see Silverberg, 1990), something which probably cannot be explained by reference to differences in the national costs of capital (about which there is considerable definitional and measurement confusion in any case) alone, as Flamm (1988) for example argues.

If we assume that the rate of change of best practice productivity is exogenously given, then the "optimal" payback period will allow a firm to track technical change by continually incorporating new equipment into its capital stock at a rate which just balances average unit cost advantages against the financial costs of capital turnover due to acquisition and scrapping. In contrast to the usual derivations of optimal replacement policy, however, this model critically turns on the dynamics of oligopolistic competition. Experience with simulation experiments seems to indicate that a unique "optimal" payback period may not exist independent of market structure and pricing strategies (as markups on units costs). In particular, collective effects may be at work which may lock an industry into alternative combinations of locally "evolutionarily" stable payback periods and markups. This is the focus of my current research and is a topic I cannot go into further here.

Since the thesis of most recent work in the economics of technical change is that the creation of best practice technology itself must be seen as endogenous to the system, this sort of model can only be a stepping stone to further work. This is particularly the case given that the thesis of the present chapter is that this process is endogenously intertwined with and inseparable from the process of diffusion as well. Thus the question is: How are we to model these feedbacks, and at what level must they be situated?

The concept of technological trajectory (Dosi, 1982; Nelson and Winter, 1982; Sahal, 1985) comes to our rescue at this point, enabling us to generalize the previous model in the direction of bona fide diffusion. As Grübler points out in Chapter 18 of this volume,

> ... should we not attempt to define the object of diffusion research prior to analysis by some sort of "evolutionary tree", spanning the whole diverse domain into which any particular case is embedded? Improvements and add-on innovations, which result that a particular technology would become competitive within a given market segment, could be represented as "branchings" along an evolutionary "tree" and allow for classification, taxonomy and rigorous definition of technologies in competition ...

A practical example of the utility of this *morphological* approach is also provided by Foray and Grübler's (1990) work on casting technologies (see Chapter 16). Technological trajectories enable us to distinguish, at some hierarchical level, between true branchings and incremental advances along well-defined technological "chreods" (to borrow a biological metaphor from Waddington, 1976). But what does a true branching, which seems to be what we really want to focus upon in diffusion studies, imply for the economist, as opposed to everyday technical change?

A change of technological trajectory entails, first of all, a quantum jump in uncertainty, not so much with respect to the relative merits of competing technologies at a given time, but rather concerning the rate and extent of future developments, since extrapolations from past experience, which may have been specific to the old trajectory, lose their validity. (In fact, even the further development of the old trajectory may be radically influenced by the advent of the new one, as the often remarked sailing ship effect seems to indicate.) Furthermore, the nature of these developments may well depend on the expectations and commitments of others, i.e., have a strong bandwagon element, such as has often been observed about the standards setting problem and network externalities (*cf.* Arthur, 1988).

We can go a step further, however, by recognizing that technologies are not just blueprints or collections of artifacts, but also require complementary skills, tacit and codifiable knowhow, and proficiencies of agents at various levels of involvement with them, from management and engineers to foremen, skilled, and unskilled workers. These skills, organizational structures and the like, may often be highly specific to a particular trajectory (one need only think of the myriad of organizational differences between job-shop skilled worker production and the Fordist assembly line). They are created in ways which are still poorly understood, as a generalized form of individual learning so to speak, but with specific team, organizational, and cultural components which still very much elude scientific analysis. Phenomenologically we can often represent them by means of the familiar learning-by-doing power law relationships. The other side of this coin is the fact that learning which takes place in one organization can *leak out* (or be stolen) to the benefit of another. Crucial in this regard is the dynamic aspect: the rates at which skills can be acquired through own activity versus the lags in their more general diffusion through formal and informal networks of communication and exchange.

Whereas the adoption/investment decision could be reduced to an application of the simple payback method in the previous model (albeit with some uncertainty as to the appropriate payback period to use), adoption decisions with respect to genuine changes of technological paradigm present some novel aspects which may generalize to other instances of the basic problem of innovation. First, decisions in the present have to be made on the basis of no more than hunches about the development potential of a technology in the future. This is as true of invention as of the adoption of a rapidly changing new technique [such as numerically-controlled (NC) machine tools or computer-aided design (CAD)]. We may even view these as being phases along the path from idea to actuality in which dimensions of uncertainty are reduced while the standard design(s) gradually comes into being. Second, the presence of both internal and interagent learning leads to the following dilemma. Should one adopt early/commit R&D to preemptory innovation, in order to secure a competitive lead (or, in the ideal but rare case, a watertight patent) via technological accumulation internal to the firm? Or should one keep one's powder dry, so to speak, and wait for the technological smoke to clear a bit (primarily due to the resource-intensive efforts of others in diffusing and developing) before entering the fray? Kleine (1983), Rosenberg (1976), and Stoneman (1976) provide good examples of this reasoning from different industries. Cohen and Levinthal (1989) go a step further and argue that much R&D itself is not undertaken primarily to realize own innovations so much as defensively, in order to allow the firm to keep up with the progress of others in the field.[3] This confirms once again my hypothesis that R&D and innovation on the one hand, and adoption/diffusion on the other have very many more structural features in common than has usually been remarked. The hard and fast Schumpeterian distinction between innovation and imitation may thus have outlived its usefulness.

In Silverberg *et al.* (1988) these stylized facts find expression in a simple but dynamic generalization of my model of incremental technical change. The productivity of a vintage of new technology at any point in time is posited to be the product of an underlying embodied maximum value and the efficiency at which it can currently be exploited within the economic organization due to cumulative learning. To each technological trajectory is associated a skill level internal to the firm and a *public* skill level available to new entrants, which lags behind the average of internal levels and results from the spillover externality. Firms express their assessment of the relative advantages of early entry into the new trajectory by augmenting their standard payback period by some factor, which I term the anticipation bonus.

On the assumption of a diversity of firm anticipations it can be shown that the *dynamic appropriability* of an innovative strategy is very much a function of the rates of learning, both internal and public. Thus for high enough values of the rate parameters governing each form of skill accumulation, first movers do derive the largest net benefit, in terms of ultimate gains in market share, from the introduction and diffusion of the new technology. This is of course the classical Schumpeterian picture. For intermediate values, however, a second-mover or imitator may be able to capture more of the benefits by letting early innovators bear an excessive share of the development costs. For even lower values of these parameters, an innovation will fail to reach maturity and its diffusion will spontaneously reverse, even though some firms are willing to commit to it and it is indeed potentially superior. Given that an unbiased survey of innovation prospects would seem to show that a large if not major share of all innovations fail to diffuse successfully, it is reassuring to find a theoretical model which can also yield this result, though this eventuality did not play a role in its original formulation.

Why then do firms differ in their assessment of and commitment to a new technological trajectory, and what implications does this fact have for the rate and direction of technical change? First, our results, while superficially paradoxical, are really self-consistent. Firms do not know whether in any particular case a first-in or a wait-and-see attitude is superior in terms of appropriating permanent gains from technological investments. Those that are sufficiently aggressive will secure first-in status, but at the risk of not having the staying power to see an innovation to maturity (on the assumption that it indeed does have potential). But even in this worst case for them, they may serve as the essential trigger to subsequent adoption decisions of other firms who do eventually see the innovation through to profitability. This is by no means uncommon in the history of technology, as Marx already remarked. The lure of the (Schumpeterian) innovative laurels is what sparks this process in sufficiently entrepreneurial economies, but the model provides no rational guarantee for the appropriateness of this stance. Finally, too

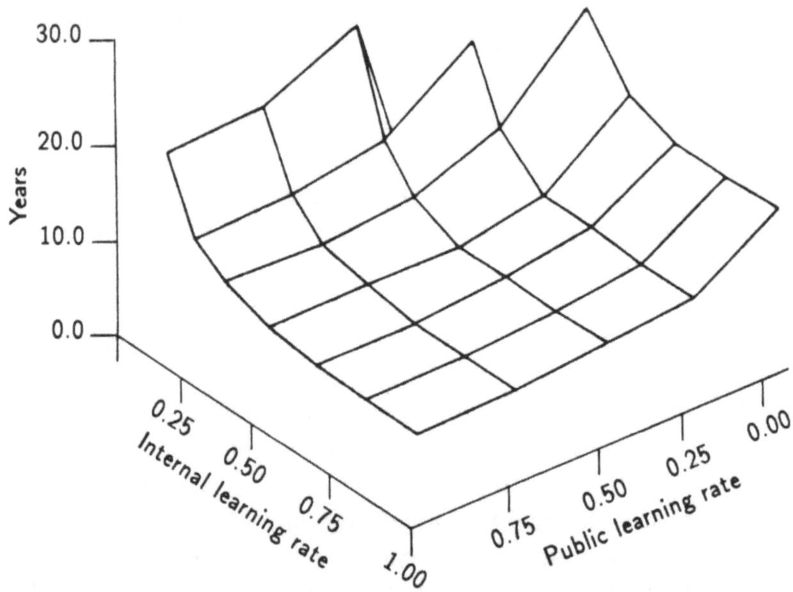

Figure 8.1. Time to diffuse from 10% to 90% of industry capacity as a function of the internal and public learning rates.

conservative an adoption strategy may lead to the inexorable elimination of the firm from the market and preclude any possibility of ever catching up, due to the cumulative causative forces at work in the model. Thus a certain amount of innovativeness becomes almost compulsory for everyone once it has become at all widespread among competitors.

The mesolevel manifestations of these contending considerations are also quite intriguing. If we plot diffusion speed (measured e.g., as the time to go from 10% to 90% of the saturation share) as a joint function of these two learning rates, then we find that they *both* contribute to more rapid diffusion (see *Figure 8.1*). Moreover, for a given rate of internal learning, an increase in the rate of "leakage" or public learning does not undermine the appropriability of an innovative strategy, as one might expect, provided one is indeed already in the Schumpeterian regime (see *Figure 8.2*). Thus higher rates of public learning are always socially desirable if the rate of internal learning is sufficiently high to guarantee economic incentives to innovation, if only risky ones.

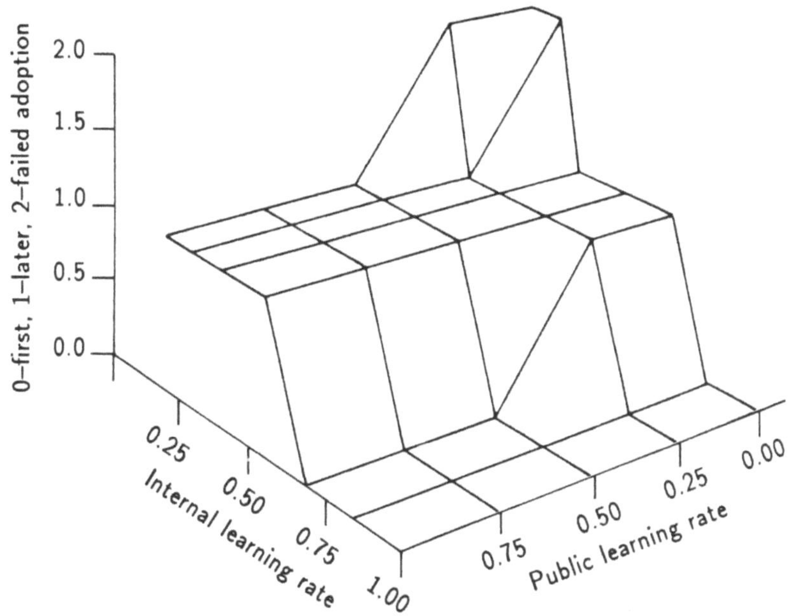

Figure 8.2. Diffusion winner: first versus later adopters, for a given distribution of strategies, as a function of the learning rates.

This state of affairs confronts innovation and industrial policy with a curious, two-handled instrumentarium. On the one hand, at least in a private capitalistic economy, it is necessary to ensure that technological investments are at least in part (and then only temporarily) appropriable. On the other it is highly desirable to encourage information flows, quick diffusion of experience, and formal and informal cooperation and sharing. The example of collective invention investigated by Allen reveals the spontaneous, one is tempted to say self-organizational, forms this can take which at first glance seems to run contrary to the principles of economic self-interest. An industrialist would gradually extend the technological frontier by building blast furnaces which were slightly taller, slightly larger, and operated at higher blast temperatures. This clearly produced fuel economies, but too large jumps along this technological trajectory often resulted in conspicuous failures. By pooling the accumulated experience of different furnaces it was possible to avoid mistakes and direct efforts in the right directions. This was possible because engineers quite willingly published and exchanged operations data on new furnaces. Allen rightly asks why firms would disclose this highly useful proprietary information to competitors and potential entrants. In this case they seemed to have little to lose from disclosure, a bit of prestige to gain from technical publicity, and significant mutual benefit to be derived from collective invention in terms of reaping production returns in relation to other iron-producing areas. Von Hippel (1988) also provides contemporary examples of engineers from different, sometimes competing, firms informally and unofficially exchanging information.[4] Saxonhouse (1974) relates how Japanese textile manufacturers pooled operating and equipment experience through their trade association during the early industrialization period. All of these examples hinge on the *quid pro quo* nature of the ongoing interaction between the agents which is reminiscent of the kind of emergent informal cooperation analyzed by Axelrod (1984). From a myopic rationality standpoint this may appear paradoxical, but seen in the long term when agents repeatedly interact and come to terms with one another, it may well be in the interests of competing agents to engage in certain forms of tacit cooperation and give and take. Industrial policy has begun to recognize this fact by encouraging and legalizing certain forms of precompetitive joint R&D, such as is exemplified by Sematech and other projects in the USA. The Japanese appear to have been singularly successful in reconciling economic competition with R&D cooperation and technology targeting.[5] Moreover, this synergy is potentiated when the rates of both forms of learning are high, for diffusion of knowledge and cooperation alone is relatively ineffectual. It is necessary for it to be accompanied by a correspondingly high ability to assimilate and internally storm the internal learning curves of the firms themselves in order to reap all the benefits. Thus the appropriability issue does not necessarily lead to a contradiction between public and private exploitation of technological innovations, for they can still be made socially and privately compatible if the learning rates are both sufficiently high.

If we want to descend from the Olympian heights of this sort of analysis to the actual determinants of what we have labeled public and internal learning then we are forced to confront a wide range of sociological, institutional, and even cultural issues. Thus a major difference between the innovation systems of the USA and Japan relates to the very different (wo)manpower, training, and employment traditions of the two countries. The turnover of skilled personnel in the USA is quite high, it being quite common for even top executives and engineers frequently to take jobs with rival organizations. This has been seen as a certain disincentive for firms to invest in human capital formation in the USA, since they are not assured of the *appropriability* of this investment in terms of employee loyalty. In the context of our model this could be interpreted as lowering the rate of internal learning. Japan, on the contrary, with its system of lifetime employment at least for the leading export firms, has evolved a unique system to build up internal skills, loyalty, and shop-floor cooperativeness. There are of course two sides to this issue, the high mobility of entrepreneurially-minded labor in the USA leading to frequent innovative company spinoffs (the semiconductor and computer industries being prime examples, where dissatisfaction with established companies' technology policies has lead to several generations of new foundings). While this may increase the rate of public learning, it may not guarantee a sufficiently high rate of internal learning (which, as we have seen, it quite crucial) to be socially desirable in terms of firm and national levels of technological appropriability. In particular, the Japanese may be enjoying the best of both worlds due to their very successful system of precompetitive cooperation and coordination.

Henry Ergas (1987) has introduced the distinction between innovation and diffusion oriented technology policies. While I do not quite find his choice of labels apposite (as Giovanni Dosi suggested at the conference, mission oriented versus generic broad-based might be more appropriate), this distinction is not without relevance for our purposes. A case in point is numerically controlled machine tools, a major innovation dating back to the early 1950s and the desire of the US Air Force to automate certain involved operations in aircraft manufacture.[6] The technology which resulted was indeed very sophisticated, in fact too sophisticated for anything but subsidized government work. It was not until over a decade later that the technology was increasingly tailored to the needs of general industry and became economically viable, but then American firms, which had specialized in the expensive and sophisticated varieties, were no longer in the forefront. This example reinforces my contention that innovation and diffusion are never entirely separable. The filling of different market niches, itself a form of diffusion, is very often contingent on a sequence of subsequent innovations, both major and incremental. The casting example analyzed by Foray and Grübler (1990), in which a new process eventually spread from the batch to the mass production segment of the industry precisely because of the learning economies realized during the initial diffusion phase, is also in this vein. A technology policy that does not take the interdependence of these two aspects into account will always be inherently flawed.

8.3 Theoretical Outlook and Conclusions

The concepts employed in our model have many points of tangency with other modeling approaches in the literature. The question of internal learning and spillovers in R&D and diffusion has also been examined in a neoclassical, equilibrium framework by Spence (1984) and Jovanovic and Lach (1989), for instance. While I cannot go into the technical issues which separate our approach from theirs here, I would only point out that we come to much less pessimistic conclusions regarding the disincentives and limitations to appropriability than they generally do for systems in the *high* internal learning regime. This is due primarily to the more dynamic, disequilibrium formulation we have opted for, which represents innovation advantages as being only temporary but potentially leading to cumulative divergences in development paths.

Our work of course is very much in the modeling tradition established by Nelson and Winter (1982) and further extended by Winter (1984) and Iwai (1984a and 1984b). An important distinction is the insistence on the embodied nature of much technical progress, i.e., that it can only be realized when new investment is taking place. This has not been emphasized sufficiently in the past in evolutionary modeling (although it should not be overemphasized, as in the pure vintage modeling tradition). Investment can influence the rate of technical progress through yet another pathway, however. If innovation is collective in the sense of Allen, then further exploration and extension of the technological frontier will only take place if new investment is kept up on a broad front. Allen argues that the demand stagnation of the last quarter of the nineteenth century in Britain brought investment, and thus this form of productivity-enhancing technical change, to a halt, which eventually led to an undermining of Britain's technological competitiveness and a further fall in her world market share.

The endogenization of the innovation frontier is the preeminent need in this area. I believe new analytical constructs will have to be introduced to deal with this problem, since, contrary to conventional wisdom, true invention and innovation are more analogous to a semi-blind, semi-stochastic exploration of a rugged landscape than an optimization problem with well-defined choice sets and payoff matrices. The application of analytical and simulational tools from mathematical biology and artificial intelligence, such as genetic algorithms, evolutionary stable strategies, and classifier systems, although bearing the danger of misplaced analogy, may hold the key to further progress.[7]

Notes

[1] This assumes of course that they have in fact discovered the *correct* value. One could argue, however, that these technological expectations might under certain circumstances be collectively self-fulfilling, so that a multitude of (possibly) suboptimal technological regimes could exist. This is the object of current research I am conducting which I cannot go into here.

[2] The simple criterion that total (current plus amortized capital) costs of the new investment equals current costs of the old is widely applied in the literature. Thus the very provocative discussion of the rationality of late Victorian entrepreneurs is usually couched in this framework (*cf.* McCloskey, 1974; Allen, 1981). This is all the more remarkable as this is definitely not the *rational* solution to this problem, as a glance at the optimal replacement literature would immediately reveal. We seem to be confronted here with another instance of academic economists presuming to have a claim to superior rationality over the practitioners (whose rationality, ironically, they are attempting to vindicate).

[3] Thus they write (Cohen and Levinthal, 1989, pp. 569–570): "... we argue that while R&D obviously generates innovations, it also develops the firm's ability to identify, assimilate, and exploit knowledge from the environment – what we call a firm's *learning* or *absorptive* capacity ... the exercise of absorptive capacity represents a sort of learning that differs from learning-by-doing ... . Learning-by-doing typically refers to the automatic process by which the firm becomes more practiced, and hence, more efficient at doing what it is already doing. In contrast, with absorptive capacity a firm may acquire outside knowledge that will permit it to do something quite different."

[4] I am very grateful to Harvey Brooks for bringing this example to my attention during the conference.

[5] Thus Flamm (1987, p. 151), for example, on computers: "Joint research, a major element in the rapid development of Japanese computer technology, has created a unique mix of cooperation and competition. In general, Japanese authorities have worked to preserve competition in *downstream* applications and commercialization of new products. But the results of more basic, precompetitive joint research have been shared quite widely to eliminate wasteful duplication and increase productivity of R&D spending."

[6] See Noble (1984) for a somewhat polemical and contradictory but valuable account.

[7] On genetic algorithms see Goldberg (1989) and Holland (1975), on classifier systems Holland, Holyoak, Nisbett and Thagard (1986). One application pathway to technology has been explored by Kwasnicka and Kwasnicki (1986) and Kwasnicki (1989). The original Nelson and Winter model of course can also be viewed as a kind of evolutionary algorithm which tries quite deliberately to argue by economic and not biological analogy.

References

Allen, R.C., 1981, Entrepreneurship and technical progress in the northeast coast pig iron industry: 1850–1913, in P. Uselding, ed., *Research in Economic History*, JAI Press, London and Greenwich, CT, USA.

Allen, R.C., 1983, Collective invention, *Journal of Economic Behavior and Organization* **4**:1–24.

Arrow, K., 1962a, The economic implications of learning by doing, *Review of Economic Studies* **29**:155–173.

Arrow, K., 1962b, Economic welfare and allocation of resources for invention, in R.R. Nelson, ed., *The Rate and Direction of Inventive Activity*, Princeton University Press, Princeton, NJ, USA.

Arthur, W.B., 1988, Self-reinforcing mechanisms in economics, in P.W. Anderson *et al.*, eds., *The Economy as an Evolving Complex System*, Addison-Wesley, Reading, MA, USA.

Axelrod, R., 1984, *The Evolution of Cooperation*, Basic Books, New York, NY, USA.

Cohen, W.M. and Levinthal, D.A., 1989, Innovation and learning: The two faces of R&D, *Economic Journal* **99**:569–596.

David, P., 1966, The mechanization of reaping in the ante-bellum midwest, in H. Rosovsky, ed., *Industrialization in Two Systems: Essays in Honor of Alexander Gerschenkron*, John Wiley & Sons Inc., New York, NY, USA.

David, P., and Bunn, J., 1988, *The Economics of Gateway Technologies and Network Evolution: Lessons from Electricity*, Paper presented at the International Schumpeter Conference, Sienna, Italy, May 24–28, 1988.

Dosi, G., 1982, Technological paradigms and technological trajectories, *Research Policy* **11**:147–162.

Enos, J.L., 1962, *Petroleum Progress and Profits*, MIT Press, Cambridge, MA, USA.

Ergas, H., 1987, Does technology policy matter? in B.R. Guile and H. Brooks, eds., *Technology and Global Industry: Companies and Nations in the World Economy*, National Academy Press, Washington, DC, USA.

Flamm, K., 1987, *Targeting the Computer: Government Support and International Competition*, Brookings Institution, Washington, DC, USA.

Flamm, K., 1988, *Differences in Robot Use in the United States and Japan*, Paper presented at the OECD workshop on Information Technology and New Economic Growth Opportunities for the 1990s, Tokyo, September 12–23, 1988.

Foray, D. and Grübler, A., 1990, Morphological analysis, diffusion and lock-out of technologies: Ferrous casting in France and the FRG, *Research Policy* **19**(6): 535–550.

Friedel, R. and Israel, P., with B.S. Finn, 1986, *Edison's Electric Light. Biography of an Invention*, Rutgers University Press, New Brunswick, NJ, USA.

Goldberg, D., 1989, *Genetic Algorithms in Search, Optimization, and Machine Learning*, Addison-Wesley, Reading, MA, USA.

Holland, J.H., 1975, *Adaptation in Natural and Artificial Systems*, University of Michigan Press, Ann Arbor, MI, USA.

Holland, J.H., Holyoak, K.J., Nisbett, R.E., and Thagard, P.R., 1986, *Induction: Processes of Inference, Learning, and Discovery*, MIT Press, Cambridge, MA, USA.

Hughes, T.P., 1983, *Networks of Power. Electrification in Western Society 1880–1930*, The Johns Hopkins University Press, Baltimore, MD, USA.

Iwai, K., 1984a, Schumpeterian dynamics: An evolutionary model of innovation and imitation, *Journal of Economic Behavior and Organization* **5**:159–190.

Iwai, K., 1984b, Schumpeterian dynamics II: Technological progress, firm growth and economic selection, *Journal of Economic Behavior and Organization* **5**:321–351.

Jovanovic, B., and Lach, S., 1989, Entry, exit, and diffusion with learning by doing, *American Economic Review* **79**:690–699.

Kaldor, N. and Mirrlees, J.A., 1962, A new model of economic growth, *Review of Economic Studies* **29**:174–192.

Kleine, J., 1983, *Investitionsverhalten bei Prozeßinnovationen*, Campus Verlag, Frankfurt and New York.

Kwasnicka, H. and Kwasnicki, W., 1986, Diversity and development: Tempo and mode of evolutionary processes, *Technological Forecasting & Social Change* **30**:223–243.

Kwasnicki, W., 1989, *Market, Innovation, Competition*, Discussion Paper, Institute of Engineering Cybernetics, Technical University of Wroclaw, Poland.

Levin, R.C., Cohen, W.M., and Mowery, D.C., 1985, R&D appropriability, opportunity and market structure: New evidence on some Schumpeterian hypotheses, *American Economic Review: Papers and Proceedings* **75**:20–24.

Lundvall, B.A., 1988, Innovation as an interactive process: From user-producer interaction to the national system of innovation, in G. Dosi *et al.*, eds., *Technical Change and Economic Theory*, Pinter Publishers, London and New York.

Malcomson, J.M., 1975, Replacement and the rental value of capital equipment subject to obsolescence, *Journal of Economic Theory* **10**:24–41.

Massé, P., 1962, *Optimal Investment Decisions*, Prentice-Hall, Englewood Cliffs, NJ, USA.

McCloskey, D.N., 1974, *Economic Maturity and Entrepreneurial Decline*, Cambridge University Press, Cambridge, MA, USA.

Nelson, R.R., and Winter, S.G., 1982, *An Evolutionary Theory of Economic Change*, The Belknap Press of Harvard University Press, Cambridge, MA, USA.

Noble, D.F., 1984, *Forces of Production: A Social History of Industrial Automation*, Alfred Knopf, New York, NY, USA.

Olmstead, A., 1975, The mechanization of reaping and mowing in American agriculture, 1833–1870, *Journal of Economic History* **33**:327–352.

Rosenberg, N., 1972, Factors affecting the diffusion of technology, *Explorations in Economic History* **10**:3–33, Fall.

Rosenberg, N., 1976, On technological expectations, *Economic Journal* **86**:523–535.

Rosenberg, N., 1982, *Inside the Black Box: Technology and Economics*, Cambridge University Press, Cambridge, MA, USA.

Sahal, D., 1981, *Patterns of Technological Innovation*, Addison-Wesley, Reading, MA, USA.

Sahal, D., 1985, Technology guide-posts and innovation avenues, *Research Policy* **14**:61–82.

Salter, W., 1960, *Productivity and Technical Change*, Cambridge University Press, Cambridge, MA, USA.

Saxonhouse, G., 1974, A tale of Japanese technological diffusion in the Meiji period, *Journal of Economic History* **34**:149–165.

Schumpeter, J., 1912, *Theorie der wirtschaftlichen Entwicklung*. English translation 1934:*The Theory of Economic Development*, Harvard University Press, Cambridge, MA, USA.

Silverberg, G., 1987, Technical progress, capital accumulation, and effective demand: A self-organization model, in D. Batten *et al.*, eds., *Economic Evolution and Structural Adjustment*, Springer-Verlag, Berlin, Heidelberg, New York.

Silverberg, G., 1988, Modelling economic dynamics and technical change: Mathematical approaches to self-organisation and evolution, in G. Dosi *et al.*, eds., *Technical Change and Economic Theory*, Pinter Publishers, London and New York.

Silverberg, G., 1989, Patterns of evolution and patterns of explanation in economic theory, in J. Casti and A. Karlquist, eds., *Newton to Aristotle: Toward a Theory of Models of Living Systems*, Birkhäuser, Boston, Basel, Berlin.

Silverberg, G., 1990, Dynamic vintage models with neo-Keynesian features, in OECD, *Technology and Productivity: The Challenge for Economic Policy*, OECD, Paris.

Silverberg, G., Dosi, G., and Orsenigo, L., 1988, Innovation, diversity and diffusion: A self-organisation model, *Economic Journal* **98**:1032–1054.

Smith, V., 1961, *Investment and Production*, Harvard University Press, Cambridge, MA, USA.

Solow, R., 1959, Investment and technical progress, in K.J. Arrow *et al.*, eds., *Mathematical Methods in the Social Sciences*, Stanford University Press, Stanford, CA, USA.

Spence, M., 1984, Cost reduction, competition, and industry performance, *Econometrica* **52**:101–121.

Stoneman, P., 1976, *Technological Diffusion and the Computer Revolution*, Clarendon Press, Oxford, UK.

Terborgh, G., 1949, *Dynamic Equipment Policy*, McGraw-Hill, New York, NY, USA.

von Hippel, E.A., 1988, *The Sources of Innovation*, Oxford University Press, Oxford, UK.

Waddington, C.H., 1976, Concluding remarks, in E. Jantsch and C.H. Waddington, eds., *Evolution and Consciousness*, Addison-Wesley, Reading, MA, USA.

Winter, S.G., 1984, Schumpeterian competition in alternative technological regimes, *Journal of Economic Behavior and Organization* **5**:137–158.

Chapter 9

Diffusion of Innovations Under Conditions of Uncertainty: A Stochastic Approach

Sergei Yu. Glaziev and Yuri M. Kaniovski

9.1 Introduction

The diffusion of innovations is at the core of the pattern of technological change. Many attempts to explain and describe this process have been undertaken during the last decade and a vast bibliography of publications on this subject is presented in Rogers, 1962 and 1983; and Rogers and Shoemaker, 1971. The theory of innovation is an important part of economic and social science, and is both conceptual and formal. Their unity is a necessary premise for the success of any scientific theory.

Currently, researchers are aware of some mismatch between the conceptual and formal sides of innovation theory. The conceptual part draws increasing attention to the hidden mechanisms of technological change. The problems of uncertainty and unevenness of innovations are at the center of current conceptual discussions. Economists argue about the relationships between ruptures and continuity in long-term technological change, and instability and consistency of technological trajectories during the different phases of an innovation's life cycle. A classification of innovations and some important new concepts, which reflect the technological pattern of change [technological and techno-economic paradigms, technological trajectories (Dosi, 1984; Perez, 1983; Freeman, 1987), radical, basic, incremental, process, and product innovations, having different diffusion regularities] were recently introduced into economic theory. These *conceptual innovations* have not yet been adopted by the formal side of innovation theory.

The majority of the present mathematical models treat the diffusion of innovations in a traditional way as a deterministic process, which can be described by means of differential equations or logistic curves. This approach has been quite successful as many studies have shown. Without questioning the usefulness of this approach, we must emphasize that the hypothesis about the deterministic character of innovation diffusion is appropriate only for the growth and maturity phases of the innovation life cycle under stable conditions. In this chapter we present another approach to innovation diffusion modeling which considers uncertainty and random fluctuations within the process. We consider a simple model that enables us to trace the influence of innovators and imitators on the final market share. It is worth mentioning that this approach for describing competing technologies, based on the generalized urn scheme, was proposed by Brian Arthur (1983).

We concentrate our analysis here on the early stage of innovation diffusion, when the costs and benefits of a new technology are not clear and the trajectory is fluctuating. This phase is not considered by the traditional deterministic approach because of the uncertainty and instability.

The early phases of radical innovation diffusion are characterized by the two important features which are often missed in diffusion models: (1) the instability of the present development and the uncertainty of the future evolution trajectory, and (2) the existence of different alternative technologies, which compete for the potential adopters. The random fluctuations play an important role in this phase and must be taken into consideration.

9.2 Formulation of the Problem

According to the Schumpeterian theory of innovation, innovation diffusion is a process of cumulative growth of imitators, which introduces the innovation into the market (after its exposure by entrepreneurs) with expectations of high profits (Schumpeter, 1939). We assume that several alternative (from the point of view of their expected profitability and possibilities for adoption) technologies were simultaneously introduced into the market by various entrepreneurs. The relative advantages of these technologies are not clear for the imitators, who must make their choice in order to survive in the changing economic environment.

For the sake of simplicity, we consider two new technologies (say, A and B), introduced into the market by a corresponding number of innovators.

The difference between our approach and the traditional approach concerning the classification of the participants of the diffusion process is apparent. According to the latter, all of the participants can be divided into the following groups: innovators, early adopters, early majority, later majority, and laggards. All groups except the first are considered to be imitators (Rogers, 1983; Bass, 1980). The difference between innovators and imitators is based on the characteristic features of their behavior, "imitators unlike innovators are influenced in the timing of adoption by the decisions of other members of the social system" (Bass, 1980). From our point of view, we aggregate these groups into two wider ones.

Starting with n_A A-technology innovators and n_B B-technology innovators we study how technologies are shared by the imitators in the market. We assume that for each time instant $t \geq 1$ one new imitator appears on the market (we consider a time scale connected with the appearance of new firms in the market). Technology A is chosen with probability $p_t(x_t)$ and technology B with probability $1 - p_t(x_t)$. Here x_t is the proportion (relative concentration) of the adopters that use technology A at time t:

$$x_t = \frac{n_t^A}{n_t^A + n_t^B} \ ,$$

where n_t^A is the number of adopters that use technology A at time t and n_t^B is the number of adopters that use technology B at time $t \geq 1$. The probabilities of technological choice are considered to be a function of the relative concentration of the alternative technologies in the market. According to the premises of the model we assumed that the number (and share) of adopters of this or that technology is the indicator of the accumulated experience of its utilization. Also p_t is a function which maps $R(0,1)$ on [0,1], where $R(0,1)$ is the set of rational numbers from the interval (0,1). As far as $n_t^A + n_t^B = n_A + n_B + t - 1$ and $n_t^A = (n_A + n_B + t - 1)x_t$ our probability of additions of new adopters depends on both the total number of units of the technologies in the market $n_t^A + n_t^B$ at time t and the number of the adopters that use technology A (n_t^A) and technology B (n_t^B).

We are interested in finding the final ratio of the adopters that use technology A and technology B under the assumption that the market has an infinite capacity. Formally speaking we shall study the limit behavior of the value x_t as $t \to \infty$.

Let us consider $\beta_n(x), n \geq 1, x \in R(0,1)$, independent with respect to n random values which have Bernoulli distributions. Assume that

$$P\{\beta_t(x) = 1\} = p_t(x).$$

Then the process x_t, $t \geq 1$, follows the dynamics (Arthur *et al.*, 1987):

$$\begin{aligned} x_{t+1} &= x_t + \frac{1}{n_A + n_B + t}\{\beta_t(x_t) - x_t\} = x_t + \\ &\quad + \frac{1}{n_A + n_B + t}\{p_t(x_t) - x_t\} + \frac{1}{n_A + n_B + t} z_t(x_t), \\ &\quad t \geq 1, x_1 = \frac{n_A}{n_A + n_B}. \end{aligned} \tag{9.1}$$

Consequently our process is driven on average by the term $p_t(x_t) - x_t$ (at time t).

The study of the asymptotic behavior of the process x_t, $t \geq 1$, may be done by means of the methods shown by Arthur *et al.* (1987 and 1988), but we will not consider it in detail. Here we are interested in the formation of probabilities p_t under different premises. We shall study the asymptotic behavior of the innovation diffusion process according to the different probability functions, inferred from the conceptual premises.

As was mentioned above participants of the real innovation diffusion process usually do not have sufficient information about the relative advantages of new technologies. According to the premises of the model, imitators when making their decisions, take into account the experience of earlier adopters. The information about this experience is not easily obtained because it is related to the competitive position of adopters (firms) in the market. As usual each firm can be acquainted with the experience of a limited sample, which is far less than the whole range. This is the main source of uncertainty in decision making and innovation diffusion that must be taken into consideration in a market economy. It can be eliminated only by accumulating experience about innovation adoption. But with decreasing uncertainty in the utilization of a new technology and the risk associated with its adoption, the profitability also decreases with the saturation of the market during innovation diffusion. The supernormal profitability of a successful innovation is temporal – it declines with the market shift towards a new equilibrium level while innovation diffuses according to well-known empirical laws.

We shall take into account both of the above-mentioned points. First, we shall consider the case of new technology uncertainty where imitators have no means to compare the expected profitabilities of the competing technologies, accompanied by information uncertainty about the real market situation (this case is typical of the early stages of radical innovation diffusion and for technological change during the turbulent phase of technological paradigm substitution). Second, we shall consider the case in which imitators have enough information to compare the expected profitability of competitive technologies, but they still do not have sufficient information about the market (it is typical for the growth phases of innovation diffusion and technological change within a consistent technological trajectory).

9.3 Diffusion of Innovations with Uncertain Probabilities (Imitative Behavior)

According to the above-described premises of the model, imitators make decisions to introduce a new technology according to the accumulated experience of its utilization by previous adopters. This is a traditional assumption made for diffusion innovation models (see Rogers, 1983). It is natural to suppose that among alternative, uncertain new technologies they will choose those that were successfully introduced by the majority of previous adopters from the known sample. In the case of two technologies this decision-making principle can be formulated strictly in the following way:

Rule 1. Ask an odd number r of the users of alternative technologies. If the majority of them use A, choose A. Otherwise choose B.

The probability of choosing technology A at time t under the above rule of decision making is given by the following formula:

$$\sum_{i=\frac{r+1}{2}}^{r} \frac{C^i_{n^A_t} C^{r-i}_{n^B_t}}{C^r_{n^A_t+n^B_t}}.$$

Here $C^p_q = \frac{q!}{p!(q-p)!}$ is the number of combinations from q to p. Also $q! = q(q-1)\dots 1$. Let us designate this probability $p^I_t(x_t)$, where x_t is the proportion of technology A in the market. Then $p_t^I(x)$ equals to $p^I(x)$ with the accuracy of the order $o(1)$ as $t \to \infty$ (uniformly with respect to $x \in [0,1]$). Here

$$p^I(x) = \sum_{k=\frac{r+1}{2}}^{r} C_r^k x^k (1-x)^{r-k}.$$

When $r = 1$ we have $p_n^I(x) = p^I(x) = x$ for all $n \geq 1$. The graphics of the function p^I for different r are given in *Figure 9.1*. The function f whose zeros determine all possible limits for values of x_t (see Arthur *et al.*, 1987) is given now by the following formula:

$$f(x) = p^I(x) - x.$$

Let us consider the case when $r > 1$. The corresponding set $B^f([0,1])$ of the zeros consists of three points: 0, $\frac{1}{2}$, 1. It may be shown that both 0 and 1 are attainable, but $\frac{1}{2}$ is unattainable (see Arthur *et al.*, 1988). Consequently x_t converges as $t \to \infty$ to 0 or to 1 (and to both points with positive probability). This means that finally we shall have only one of the alternative technologies in the market. But each of them has the probability of being the winner.

Let us study the relationship of the probability of being the winner $p_{n_A,n_B}(1)$ (starting with n_A innovators of A and n_B innovators of B) of technology A to the proportion of initial adopters. Then

$$p_{n_A,n_B}(0) + p_{n_A,n_B}(1) = 1. \tag{9.2}$$

As far as $p_n^I(x) = 1 - p_n^I(1-x)$, we have that

$$p_{n_A,n_B}(0) = p_{n_B,n_A}(1)$$

and

$$p_{n_A,n_B}(1) = p_{n_B,n_A}(0).$$

With equality (9.2) we obtain

$$p_{n_A,n_B}(0) + p_{n_B,n_A}(0) = 1 \tag{9.3}$$

and

$$p_{n_A,n_B}(1) + p_{n_B,n_A}(1) = 1. \tag{9.4}$$

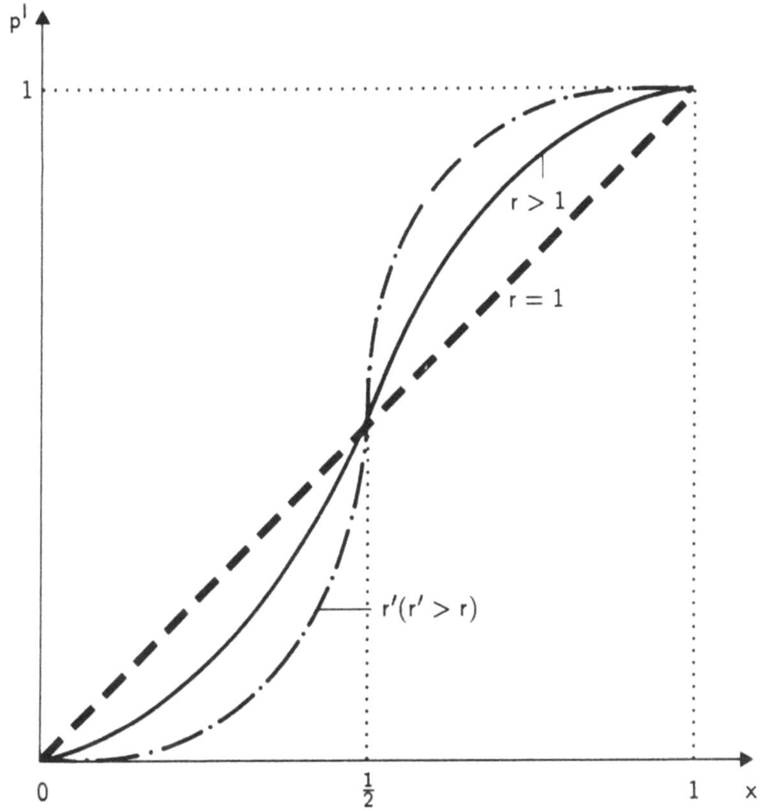

Figure 9.1. Probability of choosing (according to Rule 1) technology A as a function of its market proportion.

Combining equalities (9.3) and (9.4) with $n_A = n_B = n$ one has

$$p_{n,n}(0) = p_{n,n}(1) = \tfrac{1}{2}. \tag{9.5}$$

Let $n_A < n_B$. Then all trajectories $x_t, t \geq 1$, that lead to 1 should at least once exceed the value $\frac{1}{2}$. Let us show that the process $x_t, t \geq 1$, takes this value. Indeed, suppose that for some $k, m (1 \leq k < m)$ there will be

$$k/m < \tfrac{1}{2} \text{ and } \tfrac{k+1}{m+1} > \tfrac{1}{2} \tag{9.6}$$

(this means that the process does not take this value). If m is an odd number, i.e., $m = 2p + 1$ for some $p \geq 1$, then the smallest k that ensures the second one of the inequalities (9.6) is $k = p + 1$. So we have

$$k/m = \tfrac{p+1}{2p+1} > \tfrac{1}{2},$$

what contradicts inequalities (9.6). Similarly we obtain a contradiction in (9.6) when m is an even number. Consequently one of the inequalities (9.6) is indeed an equality. It means that the process $x_t, t \geq 1$, when crossing the middle of the segment [0,1] takes the value $\frac{1}{2}$. Taking into account the above and (9.5) we obtain

$$\begin{aligned} p_{n_A,n_B}(1) &= \sum_{k=n_B-n_A}^{\infty} P\Big\{x_t \to 1 | x_1 = \tfrac{k}{2k}\Big\} \\ & \quad P\Big\{x_1 = \frac{n_A}{n_A+n_B}, x_2 < \tfrac{1}{2}, \ldots, x_{k-1} < \tfrac{1}{2}, x_k = \tfrac{1}{2}\Big\} \\ & \stackrel{\text{def}}{=} \tfrac{1}{2}\, \gamma_{n_A,n_B}. \end{aligned} \tag{9.7}$$

As far as

$$P_{n_A,n_B}\Big\{x_t < \tfrac{1}{2}, t \geq 1\Big\} = 1 - \gamma_{n_A,n_B}$$

and

$$\begin{aligned} P_{n_A,n_B}\{x_t < \tfrac{1}{2},\quad t \geq 1\} &\geq P_{n_A,n_B}\{x_t = n_A(n_A+n_B+t-1)^{-1}, \\ t \geq 1\} &\geq \prod_{k \geq 1}^{\infty} \Big\{1 - p_k^I\Big(\frac{n_A}{n_A+n_B+t-1}\Big)\Big\} > 0 \end{aligned}$$

we have that $\gamma_{n_A,n_B} < 1$. Consequently [because of (9.7)] $p_{n_A,n_B}(1) < \frac{1}{2}$ and [because of (9.2)] $p_{n_A,n_B}(0) > \frac{1}{2}$ for $n_A < n_B$.

This implies that the probability of being the winner is greater for the technology with the larger number of innovators.

For $r = 1$ we can use the results of Polya (1931) and Athreya (1969) to find that the limit of x_t has a Beta distribution with parameters n_A and n_B. We designate this limit random variable $\bar{x}$. Then $\bar{x}$ has a density (with respect to the Lebesgue measure in R^1) of the following form

$$f_{\bar{x}}(y) = \begin{cases} \frac{(n_A+n_B-1)!}{(n_A-1)!(n_B-1)!} y^{n_A-1}(1-y)^{n_B-1} & \text{for } y \in [0,1], \\ 0 & \text{for } y \notin [0,1]. \end{cases}$$

This means that at the limit both of the technologies can exist in the market in all possible combinations. However, each individual combination has zero probability. We have for the following events non-zero probability "the final combination belongs to the interval (α, β)", where $(\alpha, \beta) \subseteq (0,1)$ is arbitrary. The case $r = 1$ corresponds to the situation in which each imitator only has sufficient information about a single previous introduction of an alternative technology. This reflects a situation of extreme high uncertainty in the market. A negligible share only is known by the followers. According to the results of this model, one can expect here instability in the final sharing of the market by alternative technologies.

To summarize the above argument we can make the following conclusion. When there is uncertainty about innovations, sharing the market depends on the strategies of the imitators (dependence of the final market share on the number r and numbers of innovators). In the case where followers make their choice according to the knowledge of their predecessors (if they know more than one case of previous innovation adoption) innovators can essentially affect the final sharing of the market. In particular the probability of dominating the market is greater for the technology with the larger number of innovators at the beginning of its life cycle. It is necessary also to mention that according to the premises and results of the model, innovators can only influence the market sharing tendency, but cannot predetermine (in a deterministic way) the domination of one of the alternative technologies. In addition, this model illustrates a very important regularity in the formation of new technological trajectories: the earlier phases of innovation diffusion play a relatively more important role in this process than later ones. As a result, the structure of the innovation diffusion process is formed in the very early stages.

9.4 Diffusion of Innovations with Expected Profitability and Uncertainty of Current Market Sharing

Now we shall consider the case in which the decisions of imitators are determined by the expected probability of alternative technologies. We assume that imitators have enough information about alternative innovations to estimate the expected dynamics of their relative profitabilities. This case corresponds to the diffusion of new technologies in stable environments in which the trajectories of evolution have already been formed and are stable. At the same time, as in the previous case, the main source of information for the followers of new technologies is the experience of earlier adopters. They make decisions based on their estimation of the dynamics of innovation profitability and the market situation. We assume that imitators interpret the apparent sharing of the market sample between alternative technologies as the sharing of the whole market. We must emphasize once more that according to the premises of our approach, imitators make their decisions based on limited information about market structure – it is an important source of uncertainty in the decision making and in the randomness of innovation diffusion.

As was mentioned above, the profitability of a new technology usually decreases with the increase in the number of adopters and later saturation of the market. Therefore it is natural to suppose that imitators will recognize the decreasing profitability of a technology as the number of its adopters increases.

Let us introduce positive functions g_A and g_B which describe the dependence of the proportions of the A and B technologies in the market sample and the imitators expected profitability. These principles of decision making can be formalized in the following way:

Rule 2. Ask an odd number r of the previous adopters. Let N of them use technology A. (Consequently $r - N$ use technology B.) Calculate the values $g_A(\frac{N}{r})$ and $g_B(1 - \frac{N}{r})$. If the first of these values is greater, choose A. Otherwise choose B.

According to the premise that a decrease in the proportion of a technology corresponds to an increase in the expected profitability, functions g_A and g_B should be nonincreasing. If they are decreasing there can be only one solution of the equation

$$g_A(x) = g_B(x), x \in [0,1]. \tag{9.8}$$

To ensure that the solution exists, one requires continuity of the functions, and that $g_A([0,1]) = g_B([0,1])$ for example. (This last condition means that values of the expected profitabilities change in the same interval.) Now we suppose that the solution of the equation (9.8) exists. *Figure 9.2* demonstrates one of the possible situations.

Here x corresponds to the proportion of technology A in the market. Functions g_A and $\tilde{g}_A$ demonstrate two possible ways in which the expected profitability can decrease. One can see that if $\tilde{g}_A(x) \leq g_A(x)$ for every x, then $\tilde{x}^* < x^*$.

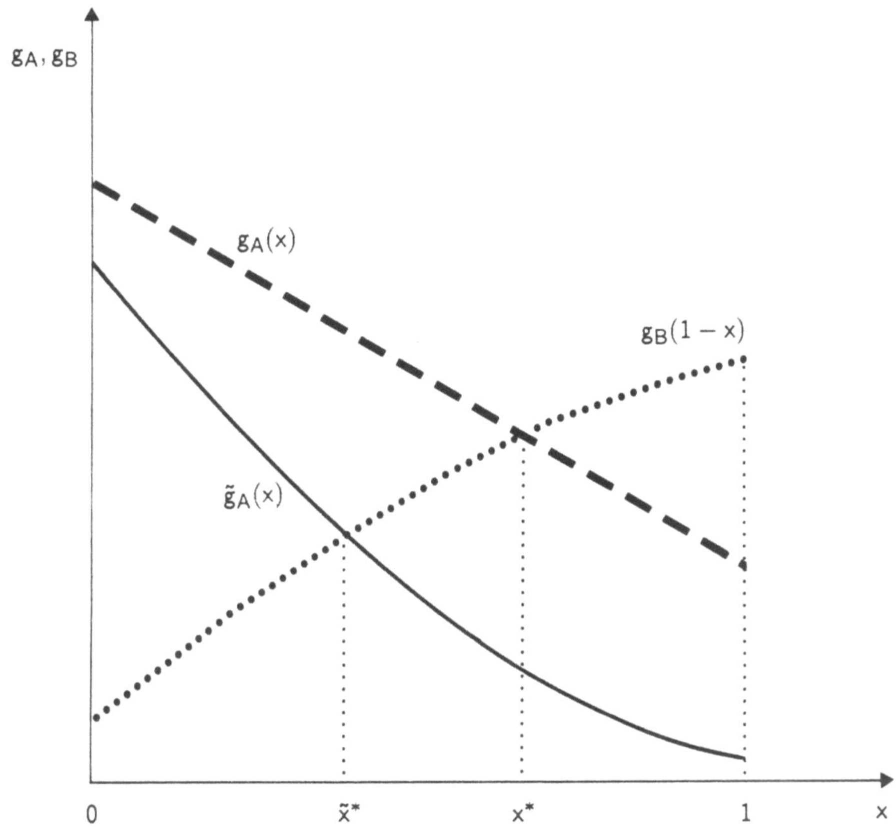

Figure 9.2. Expected profitabilities of the technologies as functions of a proportion of technology A.

Consider $r > 1$. Let $N(r)$ be the smallest N such that $N/r \geq x^*$. Then in the same manner as in the previous case, the probability of choosing technology A at time t is given by the formula

$$\sum_{i=0}^{N(r)-1} \frac{C^i_{n_t^A} C^{r-i}_{n_t^B}}{C^r_{n_t^A+n_t^B}}$$

and corresponding function $p_t^{II}(x)$ equals $p^{II}(x)$ with the accuracy of the order $o(1)$ as $t \to \infty$ (uniformly with respect to $x \in [0,1]$). Here

$$p^{II}(x) = \sum_{i=0}^{N(r)-1} C_r^i \, x^i (1-x)^{r-i}.$$

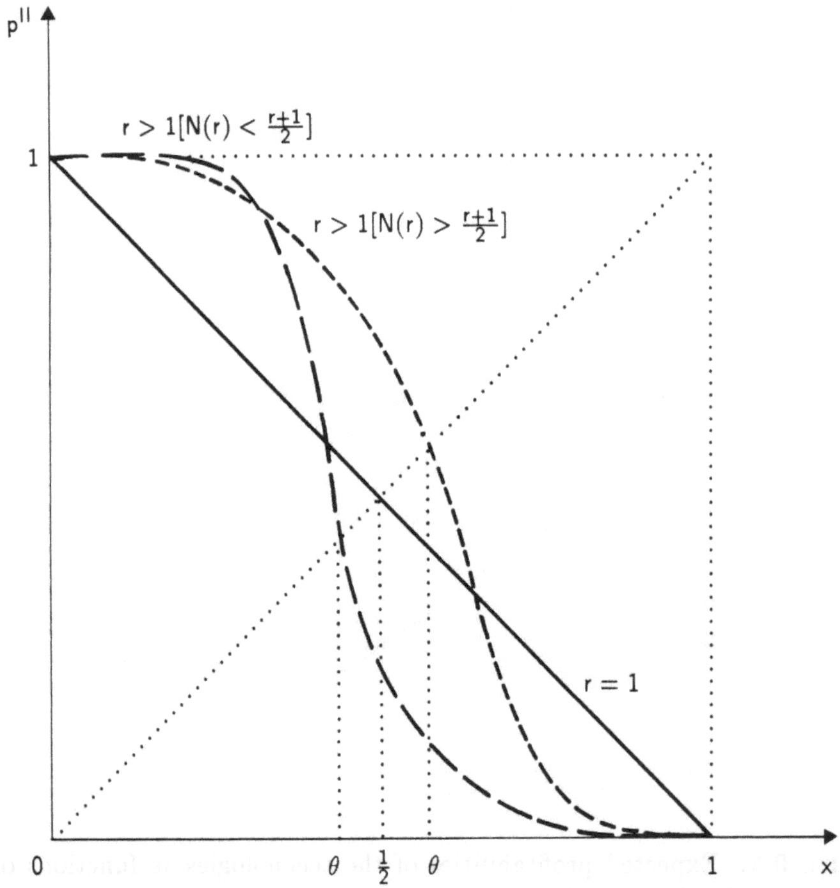

Figure 9.3. Probability of choosing (according to Rule 2) technology A as a function of its proportion on the market.

If $r = 1$ then $p_t^{II}(x) = p^{II}(x) = 1 - x$ for all $t \geq 1$. Function p^{II} is given in *Figure 9.3*.

The function f whose zeros determine all possible limit values for x_t (Arthur *et al.*, 1987) is given by the following formula:

$$f(x) = p^{II}(x) - x.$$

The corresponding set $B^f([0,1])$ of zeros is singleton. As it follows from Arthur *et al.* (1987) x_t goes to θ with probability 1 as $t \to \infty$. It is easy to see that $\theta = \frac{1}{2}$ for $r = 1$ and for $r > 1$ there will be

$\theta > \frac{1}{2}$ when $N(r) > \frac{r+1}{2}$
$\theta = \frac{1}{2}$ when $N(r) = \frac{r+1}{2}$
$\theta < \frac{1}{2}$ when $N(r) < \frac{r+1}{2}$.

These results then have the following conceptual interpretation:

(1) When imitators make their decisions according to the expected probability dynamics, final sharing (namely, the position of the point θ) of the market by the new alternative technologies is determined by the strategies of the imitators. In contrast with the previous case, imitators' strategies influence the domination of one of the alternative technologies in the market in a deterministic way.
(2) Under the given circumstances, both of the technologies will exist when the market reaches the limit.
(3) For the given function of the reduction in the expected profitability for technology B (say g_B) and two given functions (say g_A and $\tilde{g}_A$) of the reduction in the expected profitability for technology A, the limit share of technology A will be smaller for the faster decreasing function (this means that $\tilde{g}_A(x) \leq g_A(x)$ for all $x \in [0,1]$).

The results show that the final sharing of the market depends upon changes in the imitators' expectations of the profitability of new technologies. *Figures 9.2* and *9.3* illustrate the dependence between changes in the expectations of the imitators and shifts in the final structure of the market. The decrease in the rate of technology-expected profitability means that imitators estimate their chances of gaining profits by the introduction of technology A as greater than by the introduction of technology B. Therefore the changes in the expectations of imitators lead to a corresponding change in the limit structure of the market. Within this framework we can deal with a situation where one of the technologies (say B) is a conventional one. This means that $g_B(x) = \text{const}$ for all $x \in [0,1]$.

To illustrate this let us consider the simplest examples. Assume that $g_A(x) = a + b(1-x), g_B(1-x) = c + dx$. Here $g_A(0) = a + d$ is the expected profitability of technology A when nobody in the market sample uses it. Also $g(1) = a$ is the expected profitability of technology A when all adopters in the sample us it. Consequently $a \geq 0, b \geq 0$. Similarly $c \geq 0$ and $d \geq 0$. These functions are given in *Figure 9.4*.

If $a + b > c$ and $d + b > 0$ there is only one solution x^* of the equation $g_A(x) = g_B(x)$. It is easy to check that

$$x^* = \frac{a+b-c}{d+b}$$

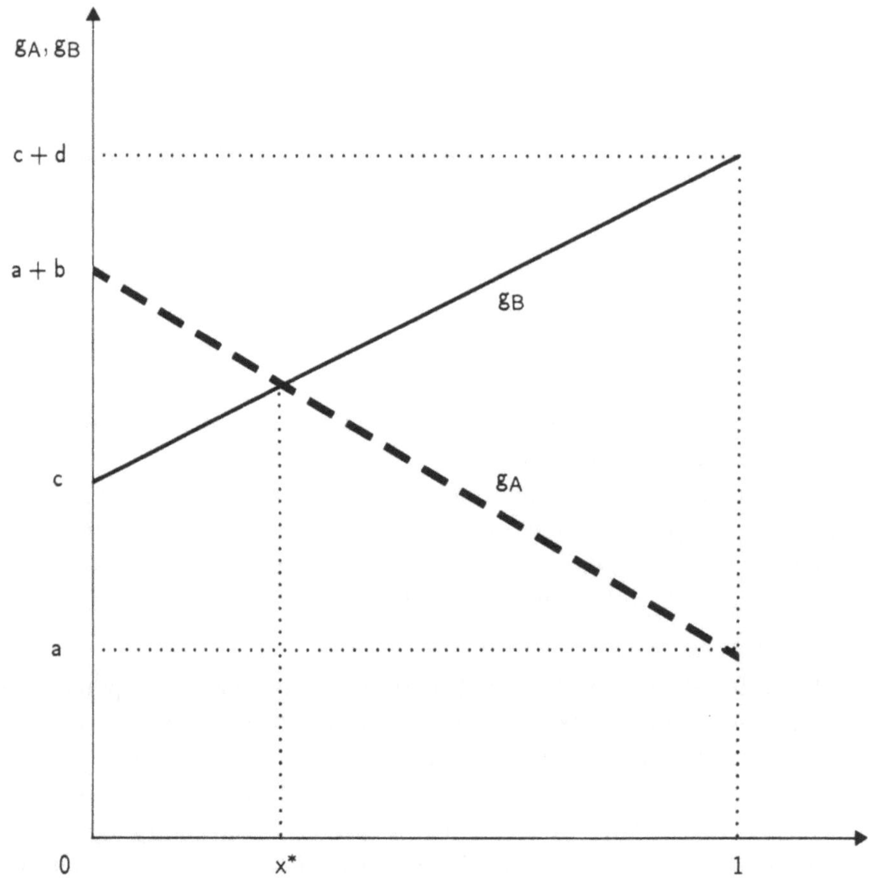

Figure 9.4. Linear (with respect to the proportions on the market) expected profitabilities of the technologies.

and x^* belongs to (0,1) if and only if $a + b > c$ and $a < c + d$. As one can see in *Figures 9.2* and *9.3*, if $\frac{a+b-c}{d+b} > \frac{1}{2}$ then technology A will dominate the market.

Thus, under this dependence of the expected probability and share of alternative technologies, the final structure of the market is determined by both the initial and final expected profitabilities of the alternative technologies or by the relation between initial expected profitabilities and rates of change. In the case when $g_A(0) = g_B(0)$, i.e., $a + b = c + d$, technology A will dominate in the market if, and only if, the rate at which its expected profitability will decrease is smaller (i.e., $b < d$). If the rates coincide (i.e.,

$b = d$) then $x^* = \frac{1}{2} + \frac{a-c}{2b} = \frac{1}{2} + \frac{(a+b)-(c+d)}{2b}$. Consequently technology A will dominate in the market in this case if and only if its initial profitability is larger (i.e., $a + b > c + d$ or $a > c$).

9.5 Conclusions

With the help of this model we can simulate the diffusion of alternative innovations under different assumptions about the influence of predecessors on the technological choice of followers so-called path-dependent processes of innovation diffusion. In this chapter we have considered the case of the diffusion of two new alternative technologies under different assumptions. It is not difficult to consider the general case with n alternative technologies. But some conceptual conclusions about the innovations diffusion can already be inferred from the results of this 2-technology model.

The interesting results concern the role of innovators (entrepreneurs) and imitators in innovation diffusion at the stage when the market share is decided. The model showed that imitators determine the trajectory of this process and the results of innovation competition. Entrepreneurs open up new technological possibilities, but their realization is determined by the imitators' choice of technologies. With uncertainty of technological choice, the probability of dominating the market is greater for the technology with the larger number of innovators. Of course, newcomers can change the situation. The result of technological competition is determined by the choice of all actors in the market. But the influence of earlier adopters on the formation of a technological trajectory is higher than those who adopt later.

These results describe important features of the alternative innovation diffusion. Both in market and centrally planned economies it is difficult to estimate the relative advantages of alternative innovations in the early phase of their diffusion or in the periods of technological paradigm substitution. In this case followers make their choice according to information about predecessor choices, and the trajectory of innovation diffusion is determined by the innovators. The technological trajectory is formed during the early phase of innovation diffusion.

The role of innovators become less important when imitators have enough information to estimate the dynamics of the expected profitability of new technologies. In this case followers make their choice according to their own estimations of future profits. These expectations determine the trajectory of innovation diffusion and the final share of alternative technologies. With the help of this approach one can simulate the formation of new technological trajectories under different types of imitator behavior.

References

Arthur, W.B., 1983, *On Competing Technologies and Historical Small Events: The Dynamics of Choice Under Increasing Returns*, WP-83-90, International Institute for Applied Systems Analysis, Laxenburg, Austria.

Arthur, W.B., Ermoliev, Yu.M., and Kaniovski, Yu.M., 1987, *Non-linear Urn Process: Asymptotic Behavior and Applications*, WP-87-85, International Institute for Applied Systems Analysis, Laxenburg, Austria.

Arthur, W.B., Ermoliev, Yu.M., and Kaniovski, Yu.M., 1988, *Non-linear Adaptive Process of Growth with General Increments: Attainable and Unattainable Components of Terminal Set*, WP-88-86, International Institute for Applied Systems Analysis, Laxenburg, Austria.

Athreya, K.B., 1969, On the characteristic property of Polya's urn, *Studia Scientiarum Mathematicarum Hungarica* 4:33–35.

Bass, F.M., 1980, The relationship between diffusion rates, experience curves, and demand elasticities for consumer durable technological innovations, *Journal of Business* **53**(3):51–67.

Dosi, G., 1984, *Technical Change and Industrial Transformation: An Application to the Semi-conductor Industry*, Macmillan, London, UK.

Freeman, C., 1987, *Technology Policy and Economic Performance*, Frances Pinter, London, UK.

Perez, C., 1983, Structural change and assimilation of new technologies in the economic and social system, in C. Freeman, ed., *Long Waves in the World Economy*, Butterworths, London, UK.

Polya, G., 1931, Sur quelques points de la theorie des probabilités, *Annales de l'Institut Henri Poincaré* **1**:117–161.

Rogers, E.M., 1962, *Diffusion of Innovations*, 1st edition, The Free Press, New York, NY, USA.

Rogers, E.M., 1983, *Diffusion of Innovations*, 3rd edition, The Free Press, New York, NY, USA.

Rogers, E.M. and Shoemaker, F., 1971, *Communication of Innovations: A Cross-cultural Approach*, 2nd edition, The Free Press, New York, NY, USA.

Schumpeter, J., 1939, *Business Cycles: A Theoretical, Historical and Statistical Analysis of the Capitalist Process*, McGraw-Hill, New York, NY, USA.

Chapter 10

Temporal Diffusion and Population Dynamics: A Systems Model

Ove Granstrand

10.1 Introduction

New technologies emerge, diffuse, and develop in many time- and space-related dimensions, thereby substituting for and/or coexisting with old technologies. This chapter presents a systems model of this process in terms of ordinary physical time, and of an *organizational space* rather than a physical space. In the organizational space, a new technology diffuses both among different organizations and among different locations or units within the same organization. In principle this could be looked upon as a special kind of spatial diffusion since the organizational space could be endowed with a topology, for instance a metric, that expresses the important characteristics of organizational neighborhoods and distances in terms of communication and decision-making structures.

This chapter considers two fundamentally separate populations of organizational units among which a new technology diffuses; organizational units as users of the new technology (with households as a special case), and organizational units as producers. It is important to note that a technological innovation (for example, as defined by Freeman *et al.*, 1982, p. 201) diffuses among both those who use it and those who produce it, regardless of whether or not there is a market that separates them.

These two dimensions of organizational diffusion apply to both product and process innovations, although in process innovations the using and producing units may sometimes be the same, implying that separability breaks down. In case any market mechanism separates users and producers, we will equate them in this chapter with buyers and sellers respectively, although in large organizations buyers and users may be quite distinct from each other, just as sellers and producers may be. For the sake of clarity, we can think of a new technology in the form of a product innovation that diffuses among buyers and sellers in a market.

An empirical example would be the electronic image-processing technology of the video cassette recorder (VCR) for consumer use, which was introduced into the Japanese market in 1975 by Sony and shortly afterwards by JVC. The VCR diffused rapidly among consumers, thereby substituting for movie cameras based on old chemical image-processing technology. At the same time more and more producers entered the market, such as Philips, Zenith, Hitachi, etc., which corresponded to a diffusion process among sellers/producers.

The rates of buyer and seller diffusion depend upon each other. For example, if the diffusion among buyers is profitable and rapid, diffusion among sellers will speed up (unless patent protection slows it down). Similarly, rapid diffusion among sellers may strengthen their marketing efforts and speed up the diffusion among buyers. The common factor behind the rates of diffusion is that the innovation generates an extra return over costs in both its use and its production. However, a slow seller diffusion does not necessarily slow down buyer diffusion unless capacity problems arise. If the innovator holds a strong patent and does not want to sell licenses, it is generally in his interest to promote buyer diffusion, while halting seller diffusion. On the other hand, a firm may sell licenses and thereby use rapid seller diffusion as a means to increase buyer diffusion in order to outcompete alternative subtechnologies. A case in point here is the licensing of JVC/Matsushita's VHS technology for video cassette recorders, which finally outcompeted Sony and its Betamax technology.

Thus the diffusion processes in the buyer and seller populations interact just as demand and supply factors interact in general. In fact it is a primary purpose of this chapter to present an explicit model of the interdependence between the demand and supply sides as characterized by buyer and seller diffusion processes generated by a stream of technological innovations. As will become clear, an equilibrium in the chosen state space will not be attained unless technological changes disappear.

10.2 What Happens to the New Product and Technology During the Diffusion Process?

The product and its technology continue to change and develop during its diffusion. On the buyer side, adaptations to different users are made, new applications are found and new ideas arise, often from the users themselves (von Hippel, 1976). On the seller side, imitations are rarely true copies, but both modifications and significant changes occur as a result of adaptations to the different production equipment of the makers, inventing around the patents of others, product differentiation and new ideas (Rosenberg, 1976). During the diffusion processes such changes often take the form of minor piecemeal improvements which accumulate, but radical changes also occur. Thus an innovation is never a one-shot affair, but triggers a swarm of mostly minor changes, occurring partly as a result of diffusion. Hence, the common innovation/imitation dichotomy easily becomes less useful for descriptive purposes.

Taken together, the subsequent changes and innovations mostly lead to gradual increases, with some jumps in the technical performance of the product (in a broad sense) along some of its performance parameters (weight, efficiency, durability, etc.). These increases in performance are sometimes correlated with cumulative production as well as with the cumulative stock of products in use, and thus can be interpreted as a result of learning – learning by producing and learning by using – as demonstrated by Sahal (1981). An important question relates to exactly what factors account for this learning; whether learning takes place predominantly at the buyer or at the seller side at different points in time, and who appropriates the benefits of learning. However, it may be argued that the really important point is not whether technological change, based on learning, is user-driven or producer-driven, but what makes the whole intra- and inter-organizational system of actors function as a learning system. Technological changes may take place among producers as well as among users with different applications, and among other producers and users connected to the producer-user environment.

All in all, the conceptual boundaries of a technology cannot be taken as fixed and the basic dichotomy between old and new technology becomes difficult to use empirically. Nevertheless, both empirical work and analytical tractability suggest that the distinction can be used, at least as a first approximation.

10.3 What Factors Govern the Buyer/Seller Diffusion Process?

Adoption decisions by users and imitation decisions by producers can be viewed as decisions to enter the new technology. The exit of an old technology is part of this process, although several technologies may coexist, at least temporarily, within a buying as well as a selling organization. These entry and exit decisions are largely governed by information about the new technology and its associated long-run profitability expectations among buyers and sellers. These expectations are formed on the basis of many more factors than short-run price signals. Each innovation means an increase in the technical performance of the product, which represents an increase in the value or utility to the buyer who is using the product in his production process (as an intermediate good – a new material, a new component, a new piece of machinery, a new ancillary product, etc.). The price of the new product distributes this value increase between the buyer and the seller. Thus, we can distinguish between two profitability measures, one for the buyer and one for the seller (*cf.* Metcalfe, 1981).

For each product innovation, profits are generated for buyers and sellers during the corresponding diffusion of the innovation among them. While the profit for an individual buyer essentially depends on the price and the product's technical performance (or the price-performance ratio), the profits for the innovator and the subsequent imitators depend on price, cost characteristics and the combined effects of the diffusion pattern among both buyers and sellers. Typically a high buyer-diffusion rate and a low seller-diffusion rate benefit the innovator. An efficient patent is sufficient to delay seller diffusion but it is not a necessary condition.

10.4 Quantitative Modeling

10.4.1 Notation and assumptions

Let us first briefly take a broader view and consider a techno-industrial system (K, A, R), composed of a time-dependent countable set of technologies $K(t)$, a time-dependent countable set of economic agents $A(t)$ and a set of relations R on $K \times A$. Typical relations between technologies would be *substituting*, *contingent upon* and *complementing*. Typical relations between economic agents would be *buying*, *selling*, *owning*, *competing*, and *cooperating*. Each economic agent is linked at each point in time to a subset of K, which represents the technology base of that agent, namely the set of technologies that agent has entered into but from which it has not yet made an exit. Empirically the multi-technological character of firms is of increasing importance (see Granstrand and Sjölander, 1990). Technology i is characterized by a performance vector $\mathbf{T}_i(t)$.

Now let the technology set K more specifically be composed of a stream of technologies, where each new technology is substituting for an older one, and the set of agents A be decomposed into one set B of agents buying and another set S of agents competing and selling. The state variables to be used to describe this system will be the number of buyers and sellers that have entered into each technology, the number of buyers and sellers that have exited from each technology, the price of the product based on each technology, and finally $\mathbf{T}_i(t)$ for $i = 1, 2, \ldots$.

The modeling is simple in the sense that only numbers of buyers and sellers are used as state variables; that is, the populations are assumed to be unstructured. (In more refined models, demand and supply variables could be used which incorporate intra-firm diffusion, and dynamic models could be built, based on the theory of structured populations.) We will first model the processes of entry of buyers and sellers into each new technology (diffusion), and next link these entry processes to exit processes (substitution). Then the interdependence between the buyer and seller diffusion processes which arise will be modeled, as well as technological development and price dynamics. Finally, the model will be assembled into an optimal control model, using a macro-performance criterion for the whole techno-industrial system.

Another way to look at it is as follows. Consider first the two primary populations of economic agents on a market: buyers/users (B) and sellers/producers (S) of a product, say a camera. (We assume that the populations are disjoint and finite, possibly also time-independent.) This product is subjected to a stream of technological innovations over time, each innovation causing, in each time period, some buyers to switch over to using the new camera and some sellers to switch over to producing the new camera. If we assume that pure switching takes place, i.e., the old camera is scrapped at the same time (you don't keep your old movie camera once you buy a video camera), we have at each time point four disjoint subpopulations: users of the old product, users of the new product, producers of the old product, and producers of the new product. Thus an innovation splits up the original two populations (B) and (S). In the case of two subsequent innovations we get six subpopulations, etc. What is of interest now is to model the dynamics of these (unstructured) subpopulations and their interactions which altogether result in diffusion, substitution, and technological and economic change.

Let us now use the following notations:

t	time variable, $t \geq 0$;
t_i	time point for innovation i (a simple arrival stream of radical innovations is assumed); $i = 1, 2, \ldots$;
$b_{ni}(t)$	cumulative number of buyers (firms) that have adopted the ith innovation by time t;
$b_{ei}(t)$	cumulative number of buyers that have made an exit from the market for the ith innovation as buyers by time t;
$s_{ni}(t)$	cumulative number of sellers (firms) that have taken up production of the ith innovation by time t;
$s_{ei}(t)$	cumulative number of sellers that have terminated production of the ith innovation, i.e., withdrawn from the market as producers;
$b_i(t)$	net number of active buyers on the market for the ith innovation at time t;
$s_i(t)$	net number of active sellers on the market for the ith innovation at time t;
$b(t)$	total number of actual and potential buyers at time t;
$s(t)$	total number of actual and potential sellers at time t;
$k_1, k_2, \ldots$	positive scalars or vectors incorporating profitability expectations and decisions at the firm level;
$\mathbf{T}_i(t)$	vector of technological performance parameters realized (for technology i) at time t;
$P_i(t)$	market (average) price of the product based on technology i at time t;
$C_{vi}(t)$	industry (average) variable unit cost at time t for the product based on technology i.

Continuous, differentiable variables are assumed to be adapted to the counting variables above, using the same notation.

The following relations (balance equations) now hold:

$$\begin{cases} b_i(t) = b_{ni}(t) - b_{ei}(t), \ i = 1, 2, \ldots \\ s_i(t) = s_{ni}(t) - s_{ei}(t), \ i = 1, 2, \ldots \end{cases} \tag{10.1}$$

If we assume that no "technological leap-frogging" occurs, that is, no buyer or seller can jump from technology $i-1$ to technology $i+1$ or further, the general dynamic laws of the system could be modeled as (with arbitrary functions f, $\mathbf{g}$ and h):

$$\begin{aligned} \dot{b}_{ni} &= f_{bn}(b_{ni}, s_{ni}, \mathbf{T}_i, P_i) \\ \dot{s}_{ni} &= f_{sn}(b_{ni}, s_{ni}, \mathbf{T}_i, P_i) \\ \dot{b}_{ei} &= f_{be}(b_{ni};\ b_{n,i+1};\ s_n;\ s_{n,i+1};\ \mathbf{T}_i, \mathbf{T}_{i+1};\ P_i, P_{i+1}) \\ \dot{s}_{ei} &= f_{se}(b_{ni};\ b_{n,i+1};\ s_n;\ s_{n,i+1};\ \mathbf{T}_i, \mathbf{T}_{i+1};\ P_i, P_{i+1}) \\ \dot{\mathbf{T}}_i &= \mathbf{g}(b_{ni}, b_{ei}, s_{ni}, s_{ei}) \\ \dot{P}_i &= h(s_{ni}, s_{ei}, P_i) \end{aligned} \tag{10.2}$$

In the following sections this general model will be further specified and simplified on empirical as well as on analytical grounds.

10.4.2 Entry processes

Buyer Side

Traditional research on diffusion processes (see, e.g., Mansfield, 1968; Mansfield *et al.*, 1977; Sharif and Ramanathan, 1984; Mahajan and Peterson, 1985) provides us with several models of $b_{ni}(t)$, e.g.:

$$\begin{aligned} \dot{b}_{ni}(t) &= k_1[b(t) - b_{ni}(t)], k_1 > 0, \quad t \geq t_i \\ b_{ni}(t) &= 0 \text{ for} \quad t \leq t_i \end{aligned} \tag{10.3}$$

(Linear diffusion model. Total market growth is proportional to the number of non-adopters.) or

$$\begin{aligned} \dot{b}_{ni}(t) &= k_2[b - b_{ni}(t)] \cdot b_{ni}(t)/b, k_2 > 0, \quad t \geq t_i \\ b_{ni}(t_i) &= 1;\ b_{ni}(t) = 0 \text{ for} \quad t < t_i \end{aligned} \tag{10.4}$$

(Logistic diffusion model, $b(t)$ = constant b. Total market growth is proportional to the number of non-adopters, times the fraction of adopters.)

Note that in these traditional models, technological diffusion among buyers is not explicitly dependent upon actions among sellers. Moreover, such models are sometimes applicable to diffusion among users in a planned economy, or to intra-organizational diffusion in a market economy (see Åstebro, 1989, for an example of the latter).

Seller Side

With regard to diffusion on the seller side, there does not seem to be much empirical research available. There is a great deal of research on innovation processes, but not on imitation processes and the aggregate diffusion process that various economic agents' imitation processes give rise to. There are strong reasons to believe that the dynamics of seller diffusion in many cases differ from buyer diffusion. For example, the patent system gives rise to a principal difference. However, in the absence of empirical research we will simply assume that we can model seller diffusion in a similar way to buyer diffusion. In case an efficient patent protection leads to a delay in the seller diffusion process, we could use a time lag. Thus, in simple logistic seller diffusion we have

$$\dot{s}_{ni}(t - L_{pi}) = k_3[s - s_{ni}(t - L_{pi})]s_{ni}(t - L_{pi})/s \text{ for } t \geq L_{pi} + t_i \qquad (10.5)$$

and

$$s_{ni}(t) = 1 \text{ for } t_i \leq t \leq L_{pi} + t_i; \; s_{ni}(t) = 0 \text{ for } t < t_i$$

where $L_{pi} \geq 0$ is the length of time the corresponding patent (if any) provides efficient protection of innovation i from imitation on the market.

In case the innovating firm sells patent licenses freely, the L_{pi} could be dropped (or modified). In this case there is a buyer diffusion process for the license market.

10.4.3 Exit processes

Regarding exit decisions at the firm level and exit processes on the market, there seems to be almost no empirical research available yet for modeling purposes. We then have to deal with special cases and assumptions to arrive at model specifications.

Buyer Side

One possibility is to assume that diffusion of a new technology $i+1$ instantly causes substitution for technology i:

$$b_{ei}(t) = b_{n,i+1}(t) \tag{10.6}$$

In other words the exit process is directly driven by the entry process into the next technology, i.e., we have a case of pure switching from the old to the next new technology, while neither technological leap-frogging nor technological coexistence is possible.

Seller Side

Similarly, we could bluntly assume that the exit process on the seller side could be modeled in the same way. However, as is well known, some manufacturers exit forever, while some new entrants are either entirely new firms or old firms outside the basic market, diversifying into it. Also, a seller mostly has to offer products based on both old and new technologies, since there are usually buyers who have not yet switched to the new technology. For example, in public telephone switching the old crossbar technology will be offered for more than a decade ahead by some sellers, that is, they will produce both crossbar and computerized switches in parallel for perhaps 20–25 years. This could be modeled by introducing a positive time lag, L_{si}, which in fact would partly depend on the pattern of buyer diffusion generally. Thus:

$$s_{ei}(t + L_{si}) = s_{n,i+1}(t) \tag{10.7}$$

10.4.4 Interdependence between buyer and seller diffusion

By viewing an innovation as giving rise to diffusion processes among both buyers and sellers, we are able to model that part of the interactions between buyers and sellers which results in interdependence between the buyer diffusion process and the seller diffusion process. As already mentioned, traditional diffusion research considers only buyer diffusion. Path-breaking exceptions are Metcalfe (1981), Stoneman and Ireland (1983), and Metcalfe (1988) who consider variants of simultaneous buyer and seller diffusion processes, although without any explicit modeling of their interdependence in terms of interactions between buyer and seller populations subjected to a stream of innovations. Several types of interdependences are conceivable and could be introduced without explicit reference to pricing and profitability considerations. For example, by viewing the sets of buyers and sellers as two interacting sets of subpopulations, some analogies with studies of population dynamics could be used. (The classic predator-prey interaction leading to the Lotka-Volterra type of equations is not applicable, however.)

In order to illustrate below, we may simply assume a logistic buyer diffusion model and a linear seller diffusion model with linear interdependences. Thus, we could assume that the entry rate at the buyer side is proportional not only to the number of non-adopters multiplied by the fraction of adopters but also to the number of sellers operating on the market, and that the contributions are additive. Similarly, the rate of entry of sellers could be assumed to have additive components, one proportional to the number of potential imitators and one proportional to the number of potential buyers on the market. Finally, there will be no buyers if there are no sellers and vice versa.

The rate of buyer diffusion could also be assumed to be additively reinforced by a factor proportional to the performance-to-price ratio (or more generally to a linear combination of such ratios in case several technological performance parameters are considered). Similarly the rate of seller diffusion could be assumed to be additively reinforced by a factor proportional to the price. As shown in the general model, equation (10.2), both performance and price depend on both buyer and seller diffusion and substitution, which creates a complex interdependence. This will be explicitly modeled in the following sections.

10.4.5 Technological development and substitution

It is far from clear how technological change in general should be represented. As mentioned, we will represent changes within technology i by a vector $\mathbf{T}_i(t)$ of realized best-practice levels of technological performance at time t (e.g., the number of elements per chip, chip size, and line width in the semiconductor field at time t). We then assume that the overall technological change is the combined result of a revolutionary process and an evolutionary process (*cf.* Nelson and Winter, 1982). The revolutionary process is represented by a stream of radical innovations occurring at time points t_i, producing large jumps in at least some component of the overall best-practice $\mathbf{T}$. The evolutionary process is represented by a continuous upgrading of $\mathbf{T}$ between the time points t_i. The rate of change in the evolutionary process is further assumed to depend on the amount of learning that takes place among buyers and sellers. Sahal (1981) has demonstrated empirically how evolutionary technological progress in certain areas could be interpreted as the combined result of learning by producing (measured as cumulative output) and learning by using (measured as the stock of products in use). Thus, it is reasonable as a first approximation to use a model as follows:

$$\dot{\mathbf{T}}_i = b_i(t) \cdot \mathbf{k}_5 + s_i(t) \cdot \mathbf{k}_6 \text{ for } t_i \leq t \tag{10.8}$$

The coefficient vectors $\mathbf{k}_5$ and $\mathbf{k}_6$ reflect the relative importance of learning by using and learning by producing. Generally speaking these coefficients depend on the industry and the technology. For some technologies learning by using is a dominant source of incremental technological advances (*cf.* von Hippel, 1976); in some technologies learning by producing (learning by doing) is more important (see Dosi, 1988).

Note that, modeled in this way, technological evolution is closely related to the integral of total sales over the product life cycle. The often observed S-shape of technological evolution then derives from the unimodality of the product life cycle curve.

10.4.6 Price and cost dynamics

Pricing of new products is an intricate decision on behalf of the innovator and could be done by using various strategies (e.g., skimming price or "price low – gain share"). For simplicity we will assume here that the innovator for technology i uses mark-up pricing and that the imitators and close-to-imitators are price takers. Thus $P_i = (1+k)C_{vi}(t)$, where C_{vi} is the industry average variable unit cost at time t for the product based on technology i and k is a constant.

According to empirical studies, learning by producing results in cost reductions which often takes the following form:

$$C_{vi}(t) = [C_{vi}(0) - C_{vi}(\infty)]\, exp\,[-k_9 \int_0^t s_i(u)du] + C_{vi}(\infty)$$

Here we have assumed that the rate of production is proportional to the number of producers and that learning effects are industry-specific rather than firm-specific.

Differentiation gives

$$\dot{C}_{vi}(t) = -k_9 s_i(t)[C_{vi}(t) - C_{vi}(\infty)]$$

and thus

$$\dot{P}_i(t) = -k_9 s_i(t)\, [P_i(t) - \check{P}_i] \text{ for } t \geq t_i, \text{ where } \check{P}_i = (1+k)C_{vi}(\infty) \tag{10.9}$$

Thus the rate of price reduction is proportional to the number of competitors times the price gap to a floor price $\check{P}$.

10.4.7 Optimal control of diffusion through investments

During buyer and seller diffusion, gross incomes and profits are generated at the seller side, while benefits through increased technological performance of the product are generated at the buyer side together with costs for the product equal to the gross incomes of the sellers. According to empirical studies of R&D budgeting practices (see, e.g., Granstrand, 1982), R&D investments at the firm level are often made as a fraction of gross income, the size of the fraction being determined by comparison with competitors. At each point in time, both a seller and a buyer face the options of investing in the old technology, investing in the new technology, investing in both the old and the new technology, or not investing at all (leap-frogging still excluded). Also, based on empirical studies, one can assume a strong (although lagged) correlation between R&D investments and inventive output. R&D investments moreover constitute an industry-specific proportion of the total investments needed in production and marketing for a new product (see, e.g., Eliasson, 1987). Through the feedback into R&D of resources that essentially are fractions of sales, technological change is speeded up and thereby buyer diffusion, thus increasing physical volume of sales, and possibly also monetary volume and profit unless prices are reduced and margins are competed away too swiftly.

From the above, one may conclude that $\mathbf{k}_5$ and $\mathbf{k}_6$ in Section 10.4.5 could also be considered as aggregate control variables reflecting the R&D investment behavior on the seller side, which is influenced by the buyer diffusion, and which through seller diffusion adds up to aggregate R&D investments, causing (actually with some delay) continued technological change. Thus equation (10.8) represents one feedback feature of the system.

It is also possible to at least partly endogenize the stream of radical innovations occurring at time points t_i by letting them be influenced by government supported R&D investments, typically basic research in universities. These investments are financed in turn through taxing the profits of buyers and sellers, profits which are influenced by the state variables of the buyer-seller system.

If a buyer's utility (flow) function $U(\mathbf{T}/P)$ of the performance-to-price ratios is finally introduced, a (macro) performance index J of the system could be specified in the following way (using continuous discounting with impatience):

$$J = \int_0^\infty \sum_{i=1}^{\infty} b_i(t)\, U\, [\mathbf{T}_i(t)/P_i(t)]\, e^{-rt} dt \tag{10.10}$$

where r = average rate of return, assumed > 0. The optimal control problem is then to maximize J for admissible $\mathbf{k}_5$ and $\mathbf{k}_6$ subject to equation (10.2) as specified above and below.

10.4.8 Example of a full model summarized

Thus, a full system of equations for innovation i ($i = 1, 2, \ldots$) in the case of only one technological performance parameter, logistic buyer diffusion, and linear seller diffusion with a linear coupling term looks like:

$$B \left[\begin{array}{lcll} \dot{b}_{ni} & = & k_1(b - b_{ni}) \cdot b_{ni}/b + k_2 s_{ni} + k_7 T_i/P_i, \text{ for } t \geq t_i, & \\ & & 0 \leq b_{ni} \leq b & \\ b_{ni} & = & 1 & \text{for } t = t_i \\ b_{ni} & = & 0 & \text{for } 0 \leq t < t_i \\ b_{ei} & = & b_{n,i+1} & \text{for } b_{n,i+1} < b_{ni}, \text{ otherwise} = b_{ni} \\ b_i & = & b_{ni} - b_{ei} & \text{for } s_i > 0, \text{ otherwise} = 0 \end{array} \right. \tag{10.11}$$

$$S \left[\begin{array}{lcll} \dot{s}_{ni} & = & k_3(s - s_{ni}) + k_4(b - b_{ni}) + k_8\, P_i, \text{ for } t \geq t_i, & \\ & & 0 \leq s_{ni} \leq s & \\ s_{ni} & = & 1 & \text{for } t = t_i \\ s_{ni} & = & 0 & \text{for } 0 \leq t < t_i \\ s_{ei} & = & s_{n,i+1} & \text{for } s_{n,i+1} < s_{ni}, \text{ otherwise} = s_{ni} \\ s_i & = & s_{ni} - s_{ei} & \text{for } b_i > 0, \text{ otherwise } = 0 \end{array} \right. \tag{10.12}$$

$$T \left[\begin{array}{lcl} \dot{T}_i & = & k_5 b_i + k_6 s_i \text{ for } t_i \leq t \\ T_1(t_1) & = & 1 \text{ (assumed w.l.o.g.)} \end{array} \right.$$

$$P \left[\begin{array}{lcl} \dot{P}_i & = & -k_9 s_i (P_i - \check{P}_i) \text{ for } t_i \leq t,\ \check{P}_i \text{ const.} > 0 \\ P_1(t_1) & = & 1 \text{ (assumed w.l.o.g.)} \end{array} \right.$$

System performance index:

$$J = \int_0^{\infty} \sum_{i=1}^{\infty} b_i(t)\, U\, [T_i(t)/P_i(t)]\, e^{-rt} dt$$

where U = the utility (flow) of performance-to-price for a buyer, and r = average rate of return.

Control variables: k_5 and k_6.

Unfortunately, as soon as nonlinearities are introduced, solutions rapidly become difficult to find or know if they even exist. In principle the solutions for different i give a family of trajectories in the state space, describing in this framework the continuous evolution of market structures driven by innovation processes. As new technologies arrive and entries into them drive exit processes (possibly lagged) from old technologies, the corresponding trajectories will eventually return to the origin. A dynamic (periodic) equilibrium could be imagined but not a static one, unless technological developments stagnate. An illustrative simulation run is given below.

10.4.9 An illustrative simulation run

For the sake of illustration, a simulation of the preceding model has been run in the absence of sufficient data and analytical tractability. ***Figures 10.1*** to ***10.3*** present the run for the case of a stream with four radical innovations.

The parameters in the model are:

Number of innovations = 4; total number of buyers b = 100; total number of sellers s = 50.
Constant k_1=0.3, k_2=0.05, k_3=0.15, k_4=0.005, k_5=2, k_6=2, k_7=0.0002, k_8=0.02, k_9=0.002.
Time t_1=0, t_2=20, t_3=40, t_4=45.
Time lag L_{p1}=5, L_{p2}=8, L_{p3}=8, L_{p4}=12.
(All time constants in years.)

The T_i-curves have been linked together to the curve $T(t)$ through defining $T(t) = T_i(t)$, $t_i \le t < t_{i+1}$ and $T_{i+1}(t_{i+1}) = T_i(t_{i+1} - 0)$.

Similarly the P_i-curves have been linked together to the curve $P(t)$ through defining

$$\begin{aligned} P(t) &= P_i(t),\ t_i \le t < t_{i+1} \\ & \quad P_{i+1}\,(t_{i+1}) = P_i\,(t_{i+1} - 0). \end{aligned}$$

Finally $\check{P}_1 = 50, \check{P}_2 = 45, \check{P}_3 = 10, \check{P}_4 = 5$.

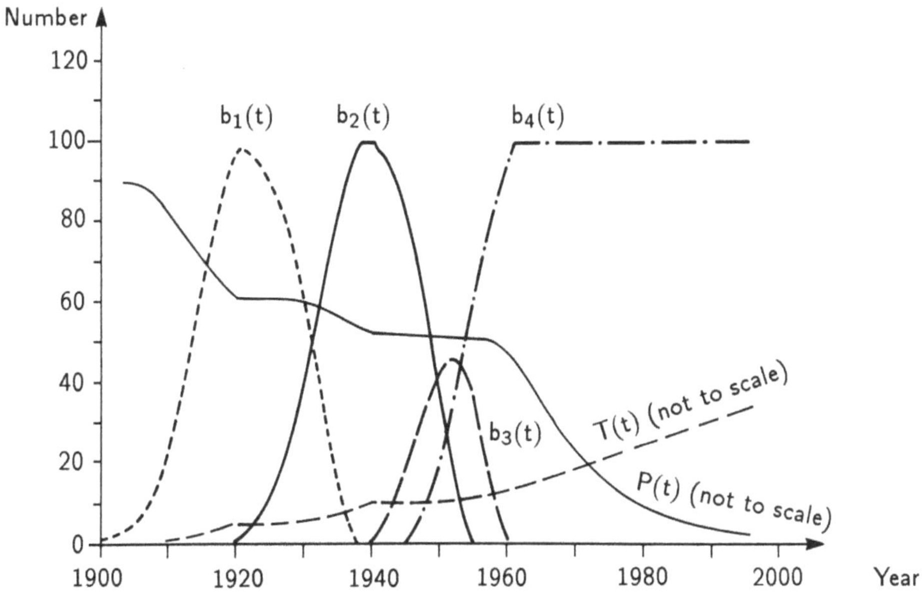

Figure 10.1. Buyer diffusion $b(t)$, technological performance $T(t)$, and price development $P(t)$.

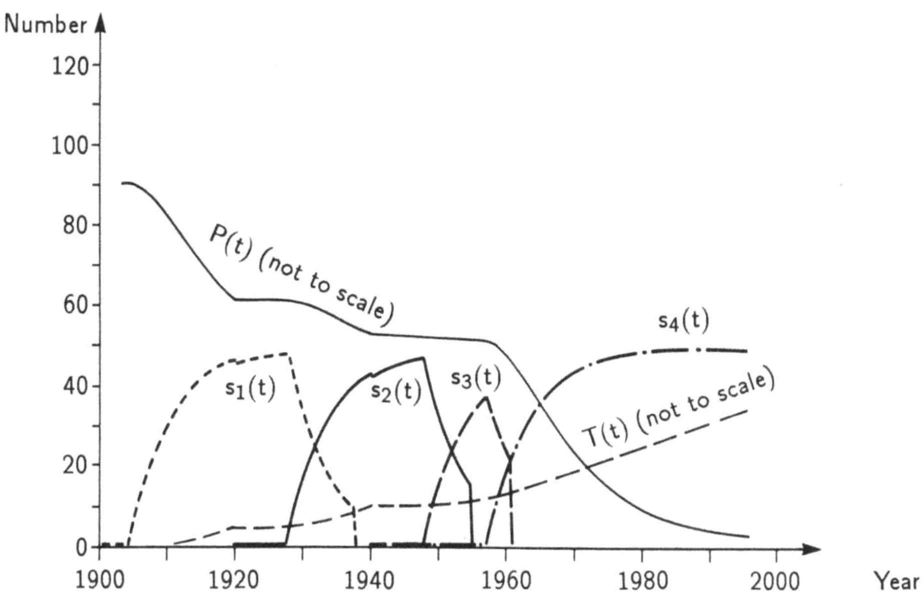

Figure 10.2. Seller diffusion $s(t)$, technological performance $T(t)$, and price development $P(t)$.

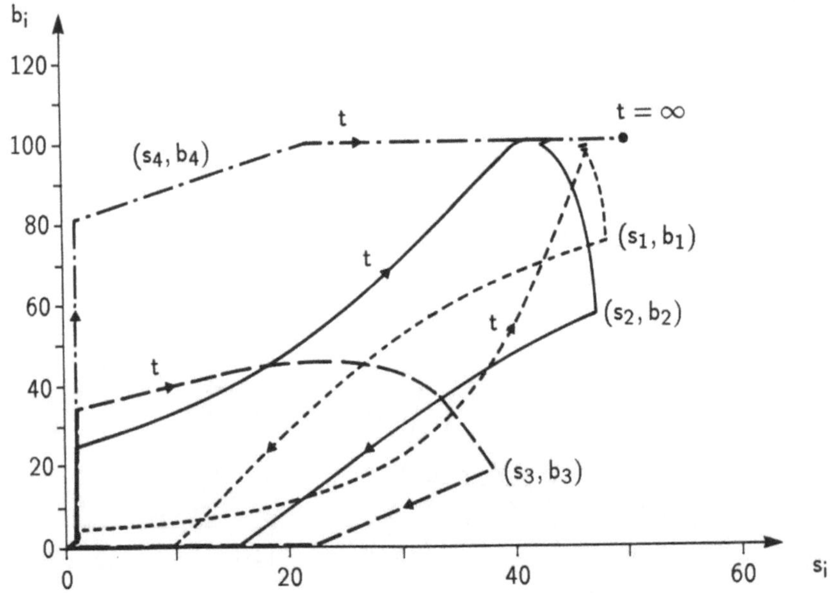

Figure 10.3. Buyer and seller diffusion (phase diagram, time t parameterized). Note: $(s_i, b_i) = [s_i(t), b_i(t)]$ as above.

Acknowledgments

The author wishes to thank Professor Mats Gyllenberg for many valuable suggestions about population dynamics, Mr. Magnus Svernlöv for assistance in simulations, and Ms. Ulla Larsson for editorial assistance.

The research on which this chapter is based is part of an ongoing research program, titled "Economics and Management of Technology" at Chalmers University of Technology, and carried out under the auspices of Institute for Management of Innovation and Technology at Chalmers University of Technology, Göteborg, Sweden.

References

Åstebro, T., 1989, *Intra-Firm Diffusion Patterns of Electronic Mail Systems*, Paper presented at the IIASA Conference on Diffusion of Technologies and Social Behavior, 14–16 June 1989, International Institute for Applied Systems Analysis, Laxenburg, Austria.

Eliasson, G., 1987, *The Dynamics of Supply and Economic Growth: How Industrial Knowledge Accumulation Drives a Path Dependent Economic Process*, IUI, Working Paper No. 182, The Industrial Institute for Economic and Social Research, Stockholm, Sweden.

Freeman, C., Clark, J., and Soete, L., 1982, *Unemployment and Technical Innovation*, Frances Pinter, London, UK.

Dosi, G., 1988, Sources, procedures, and microeconomic effects of innovation, *Journal of Economic Literature* **26**:1120–1171.

Granstrand, O., 1982, *Technology, Management and Markets*, Frances Pinter, London, UK.

Granstrand, O. and Sjölander, S., 1990, Managing innovations in multi-technology corporations, *Research Policy* **19**(1):35–60.

Mahajan, V. and Peterson, R.A., 1985, *Models for Innovation Diffusion*, Sage Publications Ltd., London, UK.

Mansfield, E., 1968, *Industrial Research and Technological Innovation*, Norton & Co., New York, NY, USA.

Mansfield, E., Rapoport, J., Romeo, A., Villani, E., Wagner, S., and Husic, F., 1977, *The Production and Application of New Industrial Technology*, Norton & Co., New York, NY, USA.

Metcalfe, J.S., 1981, Impulse and diffusion in the study of technical change, *Futures* **13**(5):347–359.

Metcalfe, J.S., 1988, The diffusion of innovation: an interpretative survey, in G. Dosi *et al.*, eds., *Technical Change and Economic Theory*, Pinter Publishers, London and New York.

Nelson, R.R. and Winter, S., 1982, *An Evolutionary Theory of Economic Change*, The Belknap Press of Harvard University Press, Cambridge, MA, USA.

Rosenberg, N., 1976, *Perspectives on Technology*, Cambridge University Press, Cambridge, UK.

Sahal, D., 1981, *Patterns of Technological Innovation*, Addison-Wesley, Reading, MA, USA.

Sharif, M.N. and Ramanathan, K., 1984, *Temporal Models of Innovation Diffusion*, IEEE Transactions on Engineering Management, Vol. **EM-31**, No. 2:76–86.

Stoneman, P. and Ireland, N.J., 1983, The role of supply factors in the diffusion of new process technology, *Economic Journal Conference Papers* **93**:66–78.

von Hippel, E., 1976, The dominant role of users in the scientific instrument innovation process, *Research Policy* **5**:212–239.

Chapter 11

Diffusion Through Interfirm Cooperation: A Case Study[1]

Gerhard Rosegger

11.1 Introduction

The last two decades have seen a remarkable growth in the number of cooperative agreements among large, multinational enterprises (MNE). Why, and under what conditions, would these vertically and horizontally integrated organizations choose to give up a part of what has often been cited as their main *raison d'être*, i.e., secure control over firm-specific technical and market information, in favor of sharing such information with rivals?

This chapter examines how the interplay between international rivalry and cooperation has affected the diffusion of technology in the automobile industry, with the focus on manufacturers' reactions to changed conditions in the American market and on their active efforts to shape these conditions. For this purpose, technology is defined as including not only knowledge embodied in capital equipment and products but also knowledge concerning the organization of production, management systems, and operational controls. The rapid spread of interfirm, cooperative arrangements can of course itself be interpreted as the diffusion of an institutional innovation, although it would be difficult, conceptually and empirically, to describe it in terms of standard diffusion curves!

While the consensus among observers seems to be that the trend toward cooperation has been *technology-driven*, and that it has accelerated the global diffusion of product and process innovations, a more specific answer to the question posed above would have to distinguish among at least three different types of arrangements, each deriving from a different set of motives:

- Multifirm consortia concerned primarily with the development of generic (pre-competitive) knowledge about newly emerging technologies.
- Joint projects, frequently encouraged by government policies and government financing, aimed at major technological innovation.
- Bilateral cooperation ("strategic partnering") between MNE in established industries with mature basic technologies.

Of these, strategic partnering is especially widespread in the motor vehicle industry. As *Figure 11.1* shows, the world's major manufacturers are linked in a large number of bilateral relationships, ranging from formal joint ventures and the acquisition of equity positions by one partner in the other, to contractual agreements covering specific aspects of their operations.[2] Whatever their form and content, these arrangements occupy the middle ground between arm's length transactions and the full internalization of an activity within a single firm's administrative structure, which may of course include an outright merger between two corporate entities.

In addition to the connections shown in the figure, many of the large MNE also have partnership links with minor manufacturers as well as with suppliers of parts and components (see, for example, Sekaly, 1981; Womack, 1988). Thus, even though agreements to cooperate are typically bilateral, networks of partners have in fact evolved around the major producers. The American Big Three (General Motors, Ford, and Chrysler) are among the leaders in the cooperation game, even while maintaining a higher degree of integration in their worldwide operations than most of their foreign competitors.

In the next section, data are presented reflecting changes in the American market, still the world's largest; these changes were triggered by import competition but have assumed a momentum of their own. As a consequence,

	GM	Ford	Toyota	VW/Audi	Nissan	PSA	Renault	Chrysler	FIAT	Honda	Mazda	Mitsubishi	Suzuki	Subaru	Daihatsu	Isuzu	Rover GP	Volvo	Hyundai	Kia	Daewoo	*
GM													5%			42%					50%	6
Ford											24%									10%		5
Toyota	JV														15%							8
VW/Audi		JV	JV																			4
Nissan		JV		CA										6%								7
PSA		CA		CA																		5
Renault				CA		JV																8
Chrysler			CA	CA		CA	JV					22%										3
FIAT		JV		CA		JV	CA															11
Honda		CA																				4
Mazda		JV			CA	CA														8%		5
Mitsubishi	CA							JV			CA								15%			6
Suzuki	CA																					3
Subaru					CA	CA																1
Daihatsu			CA																			
Isuzu	JV												CA	JV								6
Rover GP				CA		CA	CA		CA	JV		CA										
Volvo	JV			CA		JV	JV		CA													
Hyundai								CA														
Alfa Romeo					JV			JV														
SAAB																						
Kia		JV																				
Daewoo	JV																					

JV = joint venture; CA = other forms of cooperative arrangement. * = number of equity holdings in smaller manufacturers.

Figure 11.1. Equity positions, joint ventures, and other cooperative arrangements among major manufacturers.

a technology gap developed between the Big Three and their rivals from Japan and Western Europe. Drawing on some, admittedly anecdotal, empirical evidence for the gap, the US producers' and their partners' motives in establishing cooperative relationships are examined. In a concluding section, the more general lessons concerning the diffusion of innovations in mature industries that might be drawn from the American experience are speculated upon.

11.2 The Transformation of the US Market

Had the American manufacturers continued to dominate their home market as they did in the 1950s and through the middle 1960s, they might have had little reason to engage in major product innovation. After a flurry of entries (and more or less rapid exits) induced by the pent-up demand of the immediate postwar decade, the domestic industry had settled into a three-firm oligopoly with a small competitive fringe (Rosegger and Baird, 1987). The "standard American car", huge and unwieldy by the standards of other markets, had become the dominant product. What little demand there existed for vehicles deviating from the preferred size and configuration was met by imports amounting to a small fraction of total sales.

Figure 11.2(a) indicates how this situation changed, starting in the late 1960s. Imports began to capture increasing shares of what had become an essentially stagnant, albeit cyclically fluctuating, market. Subsequently, Japanese producers took over from the Europeans as major suppliers of foreign-made cars, especially at the so-called *entry level*, in the process eventually forcing the exit of some large rivals (like Fiat and Renault) from the American market.

The effects of these developments on the market shares of the Big Three can be traced in *Figure 11.2(b)*: General Motors saw its lead position eroding steadily, as did Ford. Chrysler experienced a dramatic drop in sales and entered the last decade on the brink of bankruptcy; the company was saved only through massive infusions of government-guaranteed loans. By the middle 1980s, all of the domestic fringe producers had left the market. Their places were taken by a rising number of foreign-owned operations in the United States and Canada.[3]

Explanations of the domestic industry's (relative) decline have focused on virtually every aspect of market structure, managerial behavior, and technology. Anecdotal evidence suggests that there is at least a kernel of truth in

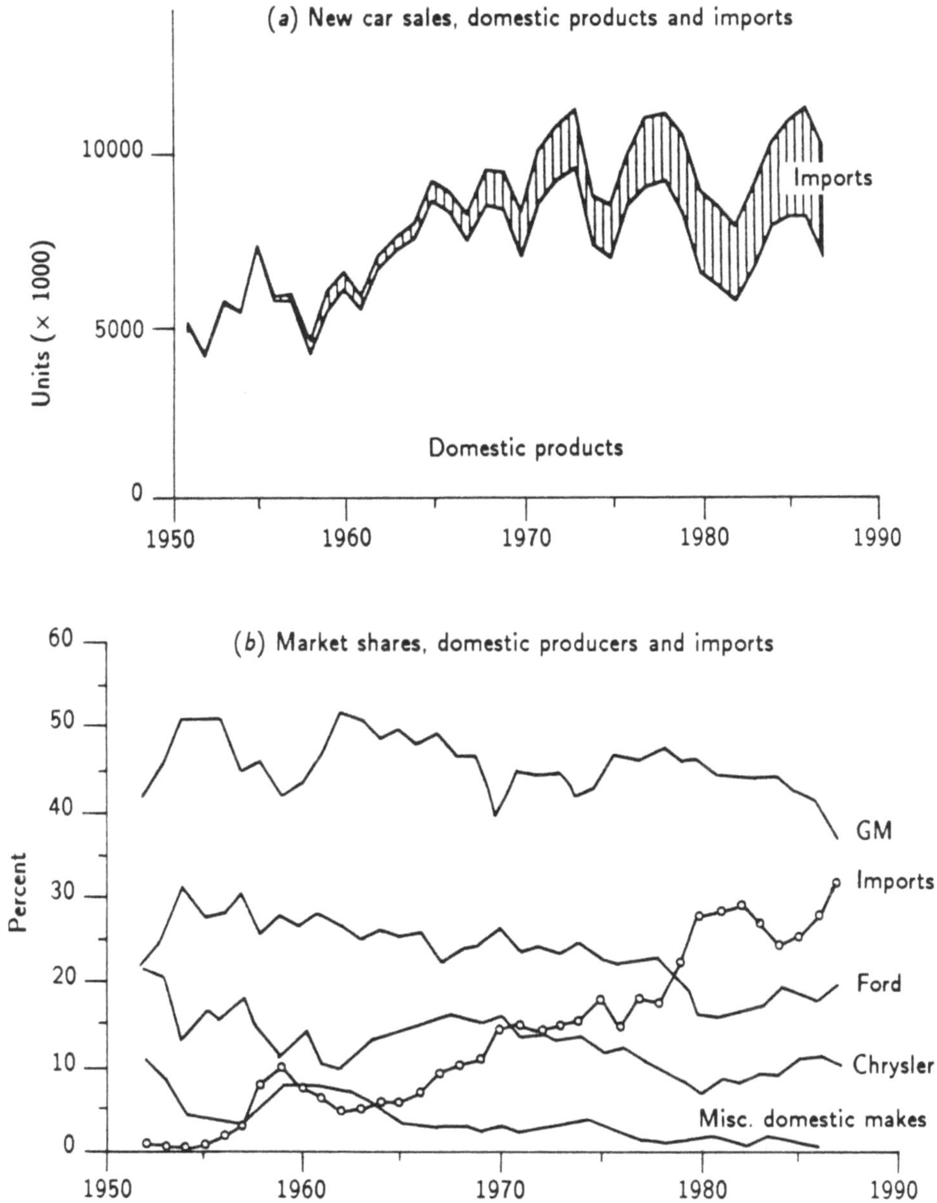

Figure 11.2. Total sales of domestic products and imports; market shares of major domestic producers and imports, 1951–1987. Panel (a) – new car sales. Panel (b) – market shares. (Source: Motor Vehicle Manufacturers Association, *Annual Statistical Report*, var. vols.)

most of the accusations levelled at Big-Three strategists as well as at wrongheaded government policies. The list of failures is irrelevant for the purposes of this chapter. What matters is that product characteristics and consumer tastes were changing, that a number of innovations made abroad were diffused through imported vehicles, and that the American manufacturers' capacities for product and process innovation apparently proved inadequate to the task of meeting these challenges. Not surprisingly, they turned to the government for relief and received it in the form of a "voluntary" export restraint program negotiated with the Japanese.[4] Quotas did little to improve conditions for the Big Three; in fact they probably increased the range of competitive pressures by motivating Japanese firms to de-emphasize the entry-level market and to meet their American and European rivals head-on in the higher priced market segments.

It is worth noting that the contours of a changing market had emerged well before the first *oil crisis* of 1973–1974. Although the rise in fuel prices triggered a burst of consumer interest in energy-efficient cars, as well as the promulgation of government-mandated *corporate average fuel economy* (CAFE) standards, these developments merely accelerated the trend towards vehicles quite different from the standard American car.

As a consequence, competitive strategy called for the offering of a much wider range of models by each of the mass-market competitors. The resulting increases in the spread between the lowest-priced and highest-priced products are illustrated in *Figure 11.3*. Also clearly observable is the fact that two of the European competitors, Audi and SAAB, pursued a different strategy, moving up market rather than facing the rising pressures of rivalry in the medium-price range.

11.2.1 Changing product characteristics

The effects of the developments sketched above on the US firms have been dramatic. Most significantly, domestic manufacturers have had to adjust their production programs to accommodate the trend away from the standard car and its luxury variants. *Figure 11.4* shows the decline in these mainstays of their business and the concomitant rise in the shares of subcompact, compact, and intermediate-size cars in Big-Three output between 1971 and 1987; the market share of all imported cars is indicated as reference.

The implications of these changes were manifold, ranging from the need to develop entirely new capabilities in development, design, and engineering

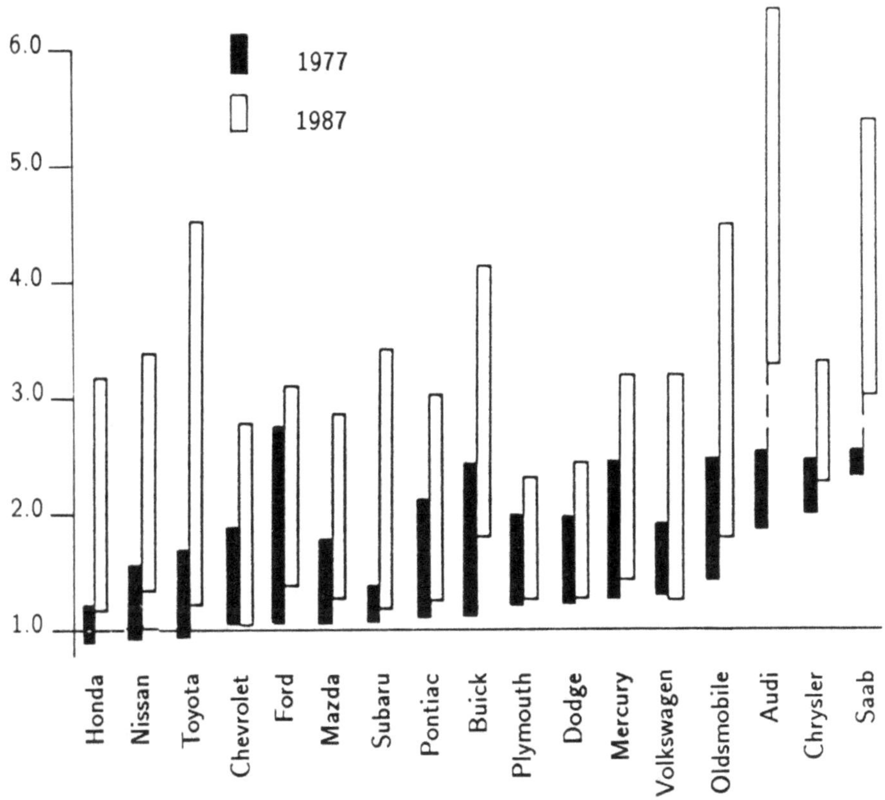

Figure 11.3. Price ranges of models of major sellers in the United States, as multiples of the lowest-priced domestically produced passenger car for 1977 and 1987. (Source: Rosegger, 1989.)

to the redirection of marketing efforts that for several decades had emphasized the alleged advantages of standard-size vehicles. The greatest difficulties, however, stemmed from the fact that the new production programs required new types of capital equipment as well as new approaches to the organization and control of manufacturing operations.

This section examines briefly how successful the Big Three were in bringing off what amounted to a catch-up strategy on the product side. While it is impossible to determine the extent to which their success is attributable to cooperative agreements with their foreign rivals, there can be no doubt that the existence of an obvious technology gap helped to motivate their pursuit of selective cooperation.

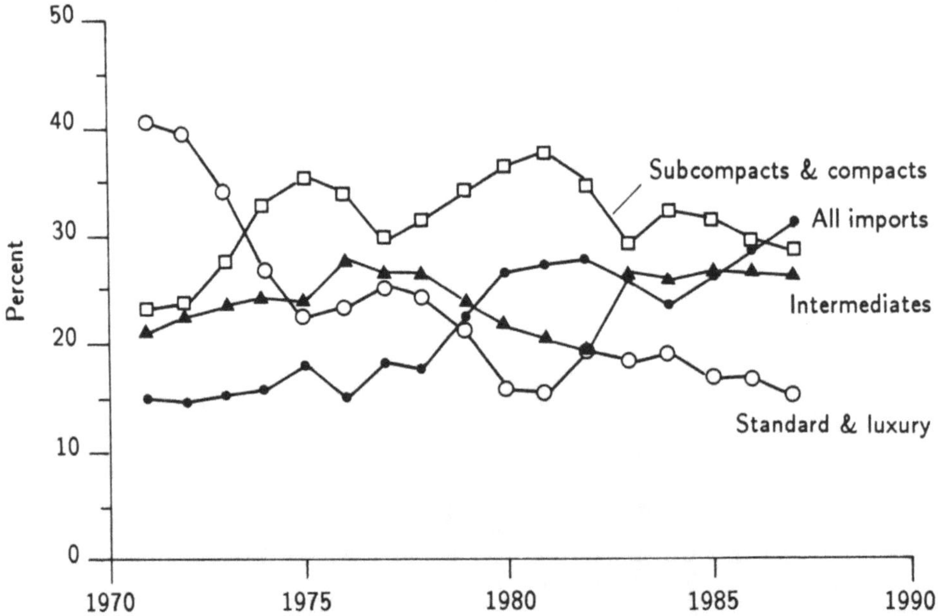

Figure 11.4. Market shares of US-made automobiles, by size class, and market share of all imports, 1971–1987. (Source: Motor Vehicle Manufacturers Association, *Annual Statistical Report*, var. vols.)

There are numerous dimensions along which one could, in theory, compare the objective, technical characteristics of American-made and imported vehicles.[5] For purposes of this chapter it is sufficient to concentrate on arguably the most important parameter, engine performance. Just how revolutionary the required adjustments turned out to be is indicated, first and foremost, by the shift in engine types produced (*Figure 11.5*). In the early 1970s, over 80 percent of all American-produced cars were equipped with V-8 engines. Note, however, that in 1960 the 6-cylinder engine had still been the dominant type and that the V-8 more than doubled its share during the subsequent decade, when the Big Three were engaged in what became known as as *horsepower race*. This was a period of booming demand for big engines, and it would be difficult to claim that the US producers made a strategic error in meeting this, highly profitable, demand. From 1970 onward, cars with 4-cylinder engines became increasingly popular. By the early 1980s they had become the most common type.

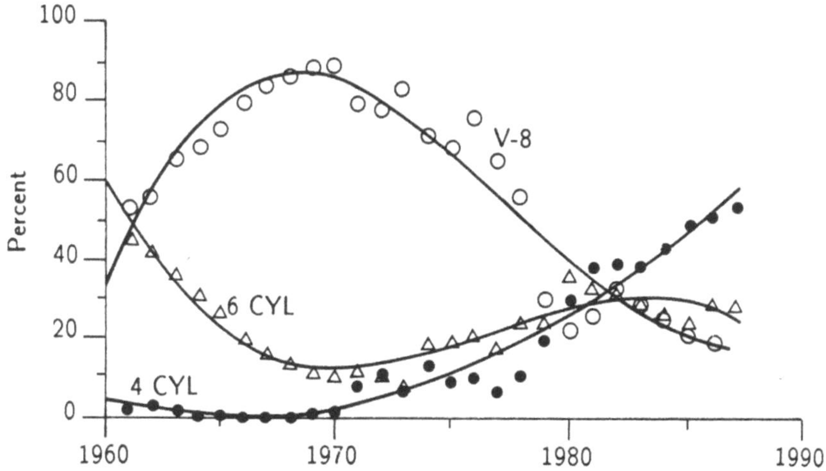

Figure 11.5. Shares of engine types in US-made automobiles, 1960–1987. (Source: Motor Vehicle Manufacturers Association, *Annual Statistical Report*, var. vols.)

As the mix of American-made engines began to approach that of the imports (though few of the latter were offered with engines of eight or more cylinders), their average displacement declined. As *Figure 11.6* demonstrates, convergence of average sizes lasted until 1982, when US engines once again increase in displacement. Fuel prices were dropping, and buyers of domestic cars once more were beginning to show a preference for larger and more powerful vehicles. The average displacement of import engines increased only slightly from 1976 to 1989; however, this average is the result of two offsetting trends: on the one hand, small, entry-level cars (supplied by the Japanese as well as by new manufacturers in South Korea and Yugoslavia) had once again managed to increase their market shares; on the other hand, several Japanese firms raised their engine sizes as they moved into the higher-priced end of the market (see also Rosegger, 1986).

With the convergence of American and foreign engine types, one would expect measurable performance to follow. As panels (*a*) and (*b*) in *Figure 11.7* suggest, however, a substantial gap remained. The average horsepower of domestic cars dropped sharply with the decline in engine displacement and then rose again during the 1980s. Nevertheless, the average power output of imports surpassed that of domestic cars in the same period. Demonstrably, many of the foreign-made cars had successfully shed their "economy" image

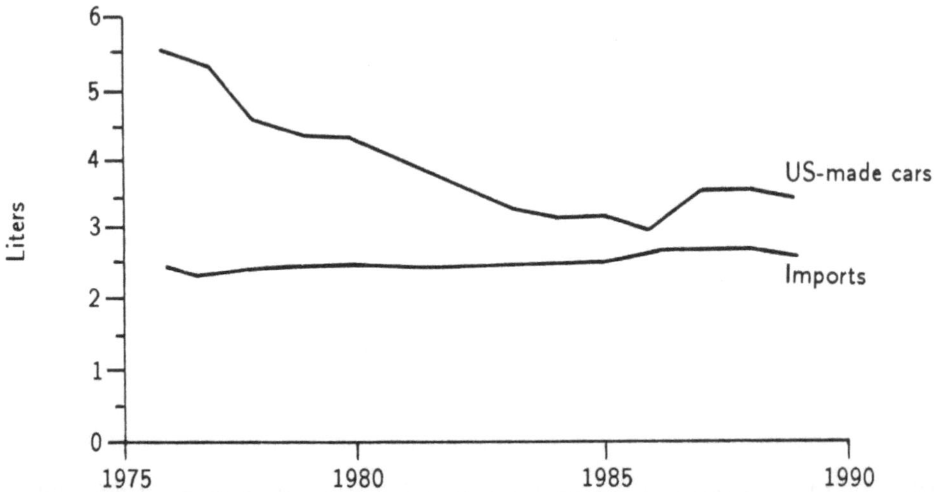

Figure 11.6. Average displacement, engines of US-made and imported automobiles, in liters, 1976–1989. (Source: *Automotive Industries*, 1989.)

and had stolen much of the "performance" thunder from the American-made products.

In specific power output (brake horsepower per liter of displacement), the average 1989 American-made model had reached roughly the performance level of the 1976 imports. In the meantime, performance of the latter had increased substantially, so that, by this measure, there was little in the way of convergence. A Japanese study (Marumo, 1984) ranked American cars third, behind Japanese and European products, in acceleration, top speed, and weight/horsepower ratio. There are long-standing historical reasons for the European emphasis on the development of low-displacement, high-horsepower engines, but the same does not hold for the Japanese producers. In fact, their first foray into the American market during the early 1960s had failed mainly on account of inadequate engines. This handicap was more than overcome in about fifteen years of intensive technical development.

Despite their higher average performance, imported cars also continue to lead in fuel consumption (see *Figure 11.8*). American efforts have tracked the government's changing CAFE (corporate average fuel economy) standards. This should be no surprise to students of the effects of regulation on technical performance; nor is it surprising that intensive lobbying by General Motors and Ford (though not by Chrysler, which had managed to "beat"

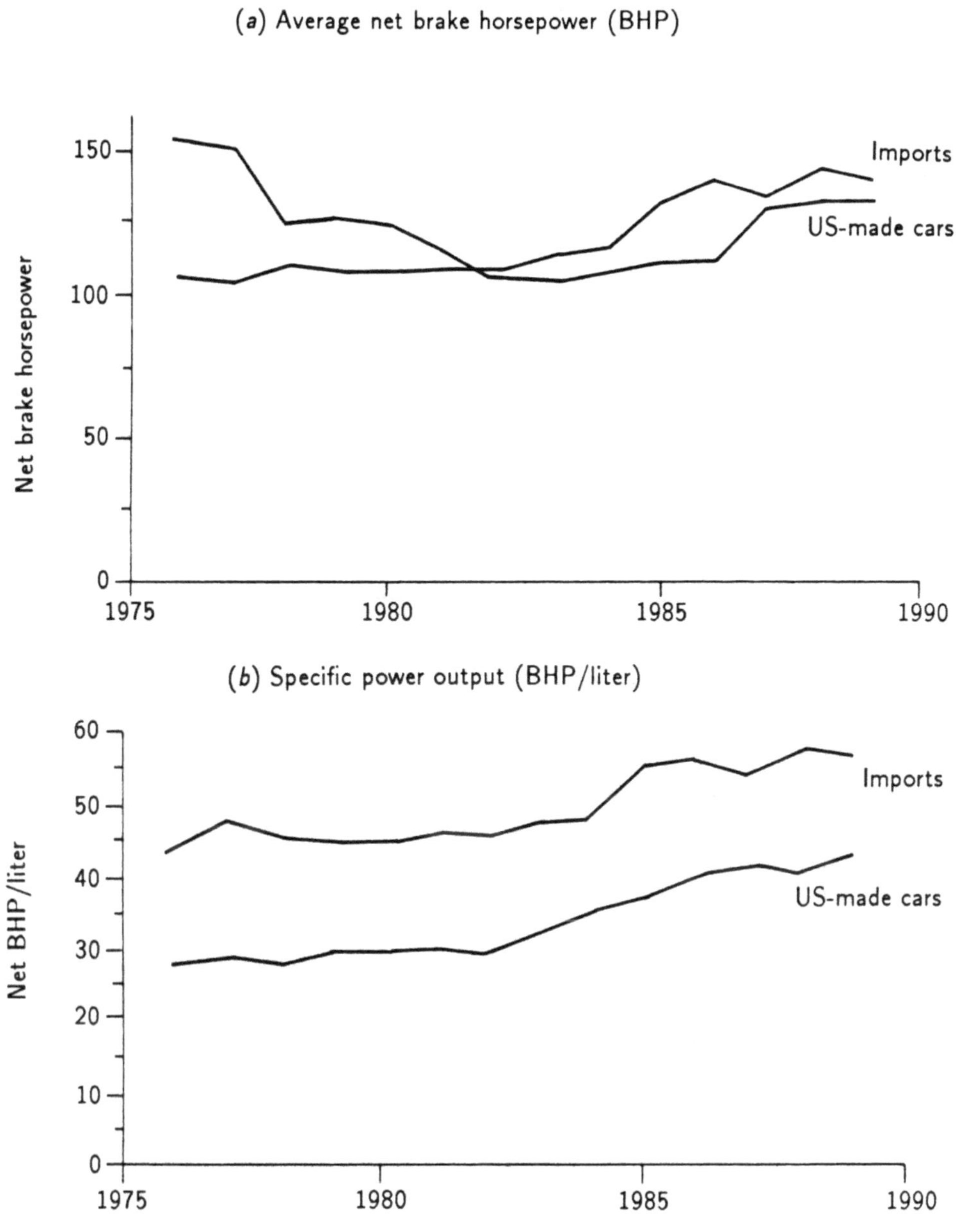

Figure 11.7. Comparative average engine performance, US-made automobiles and imports, 1976–1989. Panel (a) – average net brake horsepower (BHP). Panel (b) – specific power output (BHP/liter). (Source: *Automotive Industries*, 1989.)

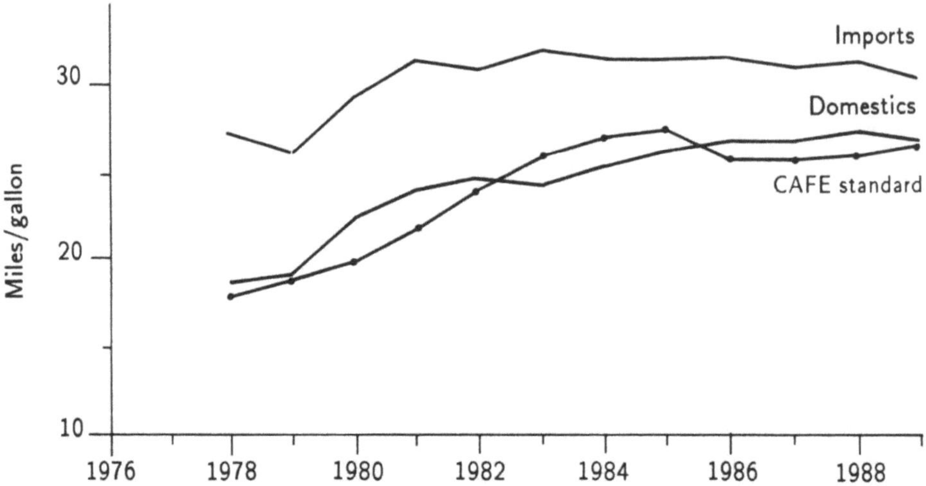

Figure 11.8. Average fuel consumption, US-made automobiles and imports, and corporate average fuel economy standard, 1978–1989.

the standard) resulted in a lowering of fuel-economy requirements from 1986 onward. Of course, certain imported luxury and sports vehicles have also failed to meet the CAFE standard by wide margins and have had to pay substantial *gas guzzler* taxes.

All of these data suggest that American manufacturers' strategies, whether they involved technical cooperation with foreign rivals or stand-alone efforts, have not yet succeeded in closing the performance gap. According to mostly impressionistic evidence, the same appears to be true with respect to product quality, an issue that has received public attention since the years when US manufacturers had fallen far behind their European and Japanese competitors (see, for example, Subcommittee on Trade, 1980). Here again, however, there is general agreement that the Big Three have made substantial strides toward improvement.

11.2.2 Process technology and manufacturing capability

Why has it been so difficult for American manufacturers to match the objective and subjective advantages of foreign producers? On the face of it, any company with long-standing experience in a mature-technology industry should be able to adopt and adapt incremental innovations readily. Yet, the Big Three appear to lag behind their competitors in the technical arena, despite having returned to profitable operations in the middle 1980s.

As has been pointed out in a number of studies (e.g., Womack, 1988; Albernathy, 1978; Altshuler, 1984; Cusumano, 1985), a good part of the answer has to be sought in the realm of process technology, i.e., in the American firms' plant and equipment, in the degree of vertical integration, and in operating techniques.

Equally significant, however, is a set of attitudes and ways of doing things that is often subsumed under the vague notion of *corporate culture.* Nelson and Winter (1982) have endowed this notion with some technical and economic substance by introducing the concepts of *routines* and *organizational capabilities.*

The relevant point here is that gaps in process technology may be closed through the adoption of innovations generally available via the conventional mechanisms of diffusion, albeit with the technological specificity of existing facilities as a constraint on their effectiveness in particular situations (Rosegger, 1976). On the other hand, it seems obvious that a transfer of attitudes, routines, and standard operating practices requires channels other than those through which individual, capital-embodied techniques are diffused. Langenfeld and Scheffman (1986) make a useful distinction between *discrete* innovations and *systems innovations*, defining the latter as collections of disembodied advances yielding major performance gains even within the context of a given set of *hardware.* As a consequence, one would expect to be able to observe and measure gaps in discrete process technology, while systems technology clearly escapes such neat quantification.

Nothing reflects the interacting effects of all these factors more persuasively than the standard comparisons of *productivity* (*Figure 11.9*). First and foremost, data on output per employee are determined by the extent of a firm's vertical integration, which of course has nothing to do with conventional concepts of productivity but is the result of past strategic decisions. Thus, for example, approximately 75 percent of a General Motors vehicle's value is added by internal operations, with 25 percent coming from outside suppliers; these proportions are almost exactly the reverse for Toyota, the largest Japanese manufacturer. One would expect the degree of integration to change only very gradually, even in response to a radical reorientation of strategies. Second, productivity comparisons have to take into account differences in capital intensity at various stages of production as well as structural relationships among these stages in geographically dispersed multiplant firms. And third, differences are no doubt explained in part by the

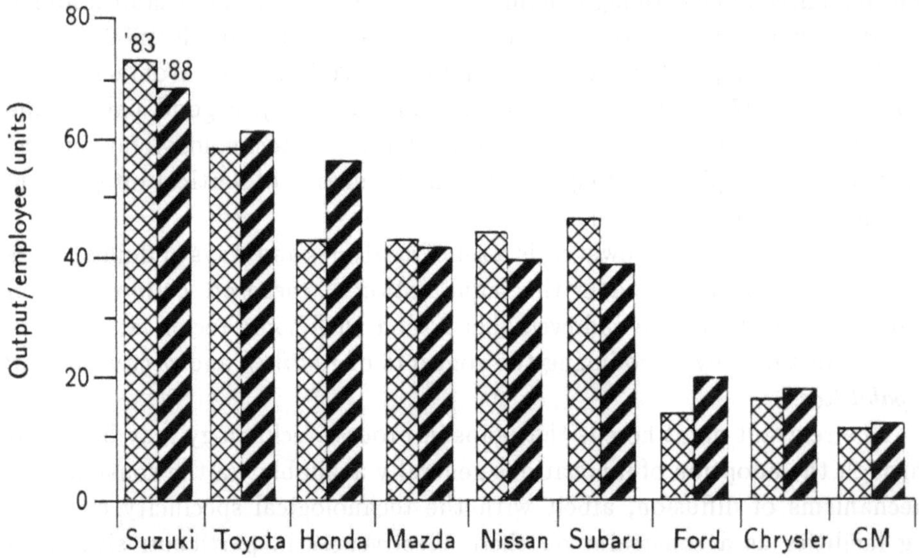

Figure 11.9. Output per employee, selected Japanese and US firms, 1983 and 1988. (Source: *Automotive Industry*, 1989.)

sizes and role of various staff functions within enterprises and by the work rules governing the assignment of factory-floor labor.

When all these differences are taken into account, however, experts still agree that Japanese firms produce vehicles more efficiently and more effectively than do their American competitors. Hard data on the matter are difficult to come by. *Table 11.1* shows the results of a Japanese evaluation, made several years ago.

While average estimates of this kind should not be taken too seriously, one striking observation emerges: the degree of automation, usually taken to be the prime mover toward higher labor productivity, apparently cannot explain differences in output per man-hour. One may surmise that advantages in "systems technology" give Japanese producers an edge even in situations where embodied techniques are similar.

Some aspects of this advantage have been widely discussed: labor relations, worker attitudes, quality circles, just-in-time inventory systems, lifetime employment, and so on. Their transfer to the American setting has been equally widely advocated, and their adoption has brought some beneficial

Table 11.1. Comparison of process technology, Japan and USA (production of passenger cars, 1800–2500 cc.).

Criterion	Unit of evaluation	Japan	USA
Capital intensity	10,000 Yen/worker	70	30
Automation	Percent of output		
Casting		80	80
Machining		90	85
Stamping		90	70
Welding		95	50
Painting		80	80
Productivity	Units/man-hour		
Stamping		0.91	0.40
Welding		0.40	0.15
Painting		0.52	0.25
Assembly		0.20	0.07

Source: Marumo (1984).

results. Yet there remains a conviction that there is more to "the Japanese approach to manufacturing", and that this approach cannot be copied but must be learned in context. If this is so, then neither the licensing of techniques nor a straightforward know-how transfer is likely to be sufficient for a catch-up program.

11.2.3 Changes in research, development, design, and engineering

The fragmentation of the automobile market referred to earlier has controlled all mass producers with new problems in the research, development, design, and engineering (RDD&E) work required to bring new models into production. One reason is that large firms have to offer a full range of products, from entry-level cars to luxury and sports vehicles, or they run the risk of having competitors take over profitable segments. A second reason is that the same strategy requires them rapidly to adopt new technical features, such as turbocharging, multivalve engines, four-wheel drive, and automatic braking systems.

Two developments have affected the cost of RDD&E:

- Changes in the technical complexity of motor vehicles, which increasingly require a "systems integration" approach to design.
- There is a more intensive link between RDD&E and manufacturing in an effort to minimize production problems at the design stage instead of having to solve them later on. According to the estimate of one expert, this means that between 50 and 70 percent of a vehicle's manufacturing costs are already determined when the "layout" of the car (i.e., the first complete cross-section drawings) is completed (Eaton, 1987).

For these and other reasons related to the increase in capital intensity of RDD&E work, current estimates for the cost of developing and putting into production an entirely new model go as high as US$3 billion. As a consequence, RDD&E also has the potential for considerable economies of scale and scope. Thus, while the last three decades have seen reductions in the efficient scale of assembly operations for mass-market cars from around 600,000 to 250,000 units per year, minimum efficient scales for RDD&E have been rising steadily. One recent (and probably exaggerated) estimate puts them at 5 million units/year (Rhys, 1988). But even if this figure were too high by a factor of two, it would still exceed minimum unit cost capacities in all other distinctive operations that go into the production of an automobile.

To some degree, mass producers have always been able to extricate themselves by the practice of "badge engineering", i.e., equipping identical basic models with differentiating add-on features and selling them under different brand names. But there are limits to this approach, especially since it is constrained by its nature to one particular size class. Besides, as systems integration requires increasing cooperation at the RDD&E stage between manufacturers and component suppliers, badge engineering tends to run into diminishing returns unless these suppliers can serve several assemblers with essentially identical products.

These problems are accentuated to the extent that innovations from other fields of technology, with which auto makers have little experience, have begun to be diffused into mass-produced vehicles. The growing application of sophisticated electronic devices and the substitution of more or less exotic materials for ferrous and other metals are but two examples of this trend.

Here, too, the Japanese are generally conceded to have a lead over their American and European competitors. While the latter typically require five years to develop a new model and bring it to market, Japanese do the job in less than four years (*Autoweek*, 1984). They are reported to accomplish this by dealing coherently with all design alternatives before beginning actual development work, instead of making modifications after that work has started, and by keeping senior management from interfering with the process once certain milestones have been passed.

Between 1982 and 1988, US producers introduced one-third fewer new models than the Japanese. Some of their efforts at radically new designs, like the Pontiac Fiero, proved technically flawed and were withdrawn from the market after a few years of disappointing sales (and, it has been claimed, at about the time when most technical problems had been corrected). General Motors' Saturn project, which was intended to incorporate all of best practice in RDD&E, plant design, and production operations, has been modified several times, its ambitions scaled down, and completion dates pushed forward.

Advantages are not entirely one-sided, however. American manufacturers are generally conceded to have enjoyed a substantial lead in the design and production of automatic transmissions, air-conditioning systems, and other passenger-comfort features. In addition, during the last decade the Big Three, led by Chrysler, have managed to pioneer in the introduction of an entirely new product, the mini-van, which has proved to be a major market success. Foreign producers, including the Japanese, have rushed to market their own versions, but so far the Americans have maintained their lead positions. They registered similar successes in the manufacture of innovative light trucks, which increasingly have become substitutes for conventional passenger vehicles, especially among young and rural buyers.

11.3 Cooperating to Compete

Although the acquisition or exchange of technological knowledge (in the broad sense outlined earlier) are not the only motives for forming strategic partnerships, the evidence adduced in the preceding sections offers ample reasons why American manufacturers should have been interested in cooperating with foreign firms. More generally, the global network of partnerships shown in *Figure 11.1*, above, suggests that such arrangements are regarded as advantageous whenever firms possess asymmetric capabilities for dealing with rapidly changing technologies and markets.

Historically, the Big Three have been reluctant to establish permanent, cooperative relationships with other firms, domestic or foreign. In their home market, they were no doubt sensitive to the antitrust implications of any kind of formal cooperation among themselves or with large component manufacturers. Fusfeld (1986) attributes the failure of the Carter administration's Cooperative Automotive Research Program (CARP) to this sensitivity. Only in 1988 did they break the pattern of (domestic) independence by forming the Automotive Composites Consortium, within the new, permissive legal framework provided by the National Cooperative Research Act of 1984. Aimed at hastening the development of generic automotive plastics technology, this is the first such venture among American auto makers (*Automotive News*, 1989).

In foreign markets, the Big Three relied heavily on their subsidiaries.[6] There seemed to be little point in impairing the decision-making autonomy of these operating divisions by establishing long-term relationships with foreign partners.

11.3.1 The development of cooperative networks

Attitudes toward international cooperation changed radically under the impact of accelerating import competition. Chrysler pioneered the new strategy with its 1971 acquisition of a 15 percent equity stake in Mitsubishi. Although the initial arrangement was narrow in scope, involving primarily the marketing of a Mitsubishi model (the Colt) under Chrysler brand names, it has grown to full-fledged technical and production cooperation, culminating in the creation of Diamond-Star Motors, a fifty-fifty joint venture for the manufacture of a new range of models to be sold under Chrysler and Mitsubishi labels.

The now-defunct alliance between American Motors and Renault followed in 1978. This venture, for production of Renault models in the United States, collapsed mainly because of the French firm's image (it had failed and been withdrawn from the American market once before) and its American partner's old-fashioned plant and inadequate distribution network. American Motors' exit from the industry was eased by Chrysler's purchase of all remaining assets, including the small but profitable Jeep operations.

In 1979, Ford acquired its interest in Toyo Kogyo (Mazda). Again, cooperation evolved gradually, leading to the recent formation of an assembly joint venture in the United States. In this, as in the Chrysler case, RDD&E and plant operations were left largely to the Japanese partner; however, the arrangement has resulted in many other forms of cooperation, such as the joint development and use of components across a range of models.

In 1980, General Motors began discussions with Toyota that led to the 1983 formation of a joint venture for the production of cars to be sold under the Toyota and Chevrolet names. The New United Motor Manufacturing Inc. (NUMMI) was established in an abandoned GM plant in California. This case is of particular interest because from the outset it was the announced objective of the American partner to facilitate the acquisition of Japanese "systems technology" through the undertaking. A program for rotating American engineers and managers through the NUMMI plant was set up, and other mechanisms for transferring technical knowledge were put in place. It was reported that, after four years of operation, "running the plant with a Japanese management system and American workers has resulted in the highest quality of any GM-sold car" (Langenfeld and Scheffman, 1986).

In the meantime, each of the Big Three has extended its network of partnerships, to include not only Japanese and South Korean companies but also a number of European firms. *Figure 11.10* shows these direct linkages between American and foreign producers. Not included are the numerous connections among European producers, nor is it possible to trace the complex networks of manufacturer-supplier agreements that have grown up around the core groups. Development of the latter has been hastened by the establishment in the United States of subsidiaries of Japanese suppliers to Toyota, Nissan, Mazda, and Honda operations in this country.

11.3.2 Strategies and motives

In contrast to the joint benefits expected from multifirm cooperation in the development of generic technologies, the establishment of a strategic partnership involves a direct *quid pro quo*. Transferring technical knowledge to a competitor without some expected compensation is not a viable objective for a profit-oriented firm! The point would be trivial, were it not for the fact that some supporters and critics of international, interfirm cooperation (see, for example, Reich and Mankin, 1986), have focussed entirely on the advantages and disadvantages of these arrangements as though they involved a *one-way* flow of benefits.

From the foregoing it seems clear that the American manufacturers' strategic objective was to move toward parity with their foreign rivals in product and process technology, and to match, if not surpass them in the range of models offered. It seems equally clear, however, that partnering is only one of a number of strategies whereby they could achieve this objective. *Table 11.2* presents a classification of these strategies which is intended to suggest the options available to mass-producers in their four main areas of

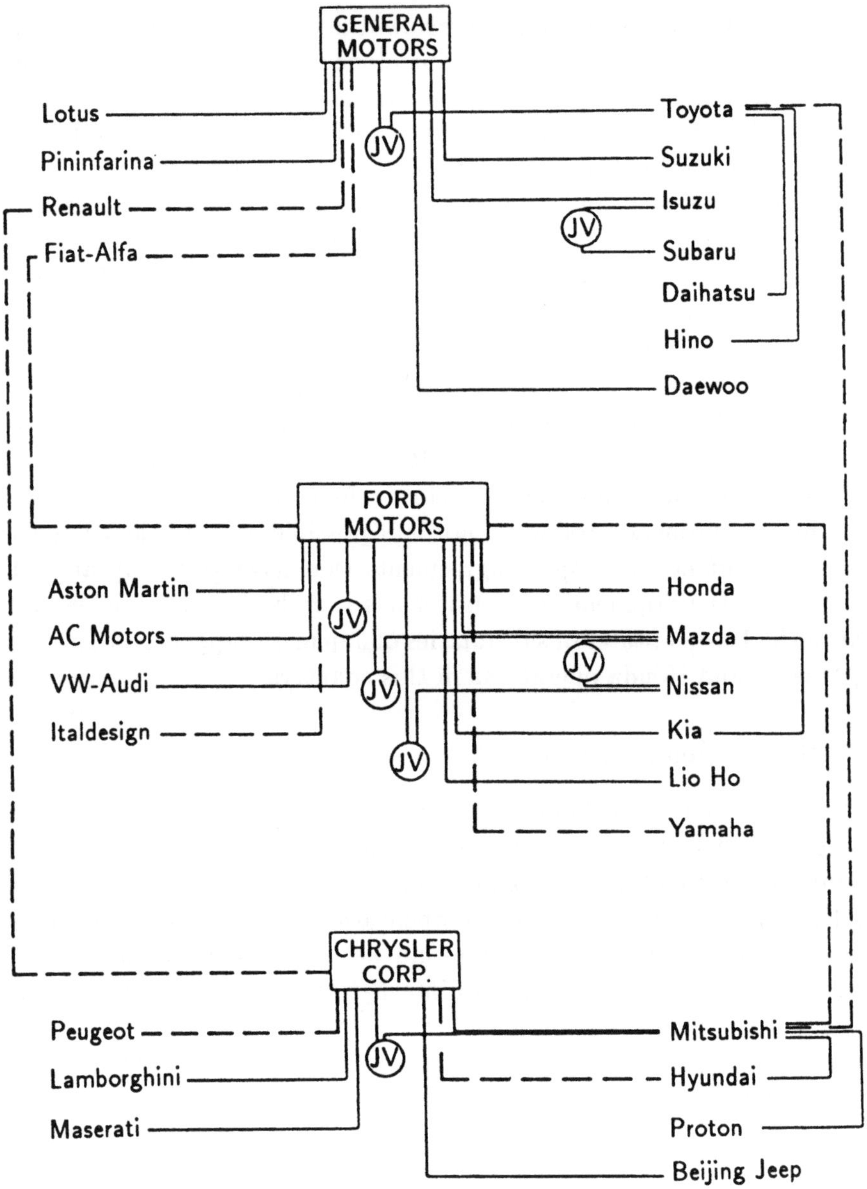

Figure 11.10. Major cooperative relationships of US manufacturers. (Source: see [2].)

Table 11.2. Taxonomy of automotive mass producers' strategies.

Research, development, design, and engineering	Acquisition of parts and components	Manufacturing and assembly	Marketing and distribution
Stand-alone RDD&E	Vertical integration	Production in and for home market	Sales under own name, through own dealers
Arm's length acquisition of technology (licensing)	Spot transactions with outside suppliers	Production in home country, exporting	Sales of partner's products through own dealer network
Long-term agreements for interfirm exchanges of technical information	Long-term contractual relations with suppliers	Stand-alone DFI in country of sale or third country	Sale under own name of partner's products, through own dealers
Joint support, with other manufacturers of third party RDD&E	Acquisition of equity interest in suppliers	Joint ventures for production, in home country or third country	Joint distribution system with partner
Joint ventures in RDD&E, joint projects for development of models	Joint supply ventures with other manufacturers. Exchange of components with partners	Merger with domestic or foreign manufacturer	Merger of existing distribution systems

Source: Rosegger (1989).

activity: RDD&E, the acquisition of parts and components, manufacturing and assembly, and marketing and distribution.

Given these options, it is not surprising that large MNE like the Big Three would pursue a mix of strategies, of which cooperation is at present far from the most important aspect. The tradition, after all, has been to "go it alone" whenever possible. There is no better indicator of the American firms' ambivalence than the fact that, even while establishing an increasing number of partnerships with their major rivals, they also continued their advocacy of protectionism. But this would be a subject for a separate investigation.

Consideration of the reasons for the Big Three establishing each of the partnerships shown in *Figure 11.10* would go well beyond the scope of this survey. One can, however, identify three basic sets of motives:

- The partnerships with the Japanese mass producers obviously have as their objective the acquisition of all the knowledge that seems to give the latter their technical and commercial advantages. Given the nature of systems technology and know-how transfers, joint RDD&E and production joint ventures are the optimal arrangements.
- Partnerships with smaller manufacturers are motivated mainly by the desire to extend the range of products offered under one of the Big Three's nameplates. At one end of the market, firms like Suzuki and Isuzu provide small cars for GM to sell as "Geo" models, while the South Korean partner, Daewoo, actually manufactures a car (the Pontiac LeMans, derived from the German subsidiary's Opel Kadett) for GM. Kia performs a similar function for Ford, supplying it with the entry-level Festiva. At the other end of the market, European specialist producers like Aston Martin, Maserati, and Pininfarina enable their American partners to offer luxury and sports vehicles at a cost lower than that of a stand-alone effort at developing such products. In addition, small specialists possess the technical resources and the flexibility required for the execution of major innovation projects. The work of Lotus on an active suspension system, financed by GM, is an example.
- Some linkages are established not so much for technical reasons as by opportunities for the joint development of emerging markets. Autolatina, the Brazilian joint venture between Ford and Volkswagen, as well as several partnerships in the Far East can be interpreted in light of this objective.

The history of strategic partnering is still relatively young, but every indication seems to be that, whatever the motives which triggered their initial establishment, they inevitably evolve in the direction of closer technical cooperation, frequently accompanied by a division of labor between the partners. A further consequence seems to be a growing *vertical disintegration* of production. To the extent that joint product development involves increasing reliance on long-term cooperative arrangements with outside suppliers, common parts are no longer being produced by the final assemblers themselves. As suggested above, however, in the case of the Big Three this trend has slowed down because of the traditional high degree of vertical integration, as well as by the now-fading convention of selecting suppliers annually on the basis of competitive bids.

On the whole, then, the move of American producers toward strategic partnering can be understood in terms of some reasonably well-defined short-term and medium-term gains. Whether the same holds true for the long run, is a question that will be addressed in the concluding section.

What are the motives for the foreign partners? Here the answers are more complex; nevertheless, one can discern some general patterns:

- The fact that technological advantages were not tilted entirely in favor of the foreign firms has already been alluded to. The acquisition of knowledge concerning the production of automatic transmissions, comfort features, and certain electrical systems was a *sine qua non* for success in the American market at all but the entry level.[7] In a similar vein, the Japanese manufacturers were aware of the fact that their American partners were more attuned to the special requirements of designing products for this market. The need to pay attention to these requirements was certainly demonstrated by the disastrous results of the first Japanese foray into the United States, in the early 1960s.
- Given their strategy of setting up independent, direct foreign investment (DFI) operations in the United States, joint-venture production enabled Japanese managements to acquire experience in dealing with uniquely American conditions, including labor relations, supplier systems, and governmental affairs. Perhaps the first stand-alone DFI venture in this country, the Westmoreland, Pennsylvania, plant of Volkswagen, served as an example of how not to proceed: it failed to reach profitable levels of operation and was shut down.
- To the smaller foreign manufacturers, American partners had something to offer that they themselves could not obtain, given their resources: market knowledge and a marketing and distribution network that provided countrywide access to potential customers. While the concept of the minimum efficient size of a dealer system is elusive, it seems clear that the entry of smaller-scale and specialty producers is eased greatly by the opportunity to rely on a partner with an established network. Here again, as in the case of joint-venture production, some firms – like Suzuki and Isuzu – followed up on partnering by beginning to build up their own dealerships.

- There can be little doubt that the major Japanese producers also sought ties with American partners in order to mitigate the possible effects of further protectionist legislation. In particular, their experience with the local-content controversies that have hampered their European-market strategies would convince them that cooperation in the United States might avoid some of these troubles, should such local-content requirements be imposed here.
- In the case of the up-market European specialty producers, partnerships with the Big Three provided welcome infusions of capital. Although they were successful developers of *high-technology* products, most of these firms had faced continuous financial difficulties. Partnering brought them a measure of stability they had not previously known.
- Finally, firms in the newly industrializing countries, chief among them South Korea, were motivated by the need to acquire manufacturing know-how in an industry with which they had scant experience. While there is little in the way of cooperation in RDD&E at this time, for these companies partnering represents a first step toward independent entry into the global automobile market. It is interesting in this connection to compare the experience of two recent, entry-level "independents" in the American market: the South Korean Hyundai, relying heavily on design and production cooperation with Mitsubishi and Chrysler, has been successful as a stand-alone exporter to the United States and Canada, whereas the Yugo, derived from an obsolete Fiat model and produced independently by the Yugoslav Zastava firm, failed in the American market, mainly on account of quality and service problems.

Although any kind of conclusive assessment of partnering would be premature, one may surmise from the evolution of these arrangements that, on the whole, they have met partners' expectations of strategic gains. Except for the failure of the American Motors-Renault venture, firms appear to have reaped sufficient benefits to continue support and expansion of their cooperative managements. In particular, there can be little doubt that new foreign product and process technologies have been diffused more rapidly among American manufacturers through the medium of the strategic partnership.

11.4 Concluding Observations

Little more than a quarter-century ago, visions of a "world car" dominated forecasts of the global automobile industry's technological evolution. This was predicted to be a relatively unsophisticated vehicle, incorporating state-of-the-art technology in some highly standardized form, with but minor adaptations to the requirements of different markets. Prices would be low because of the resulting economies of scale in production. No doubt the remarkable, worldwide success of the Volkswagen *Beetle* stood at the cradle of these forecasts, which were based on an essentially technocratic notion of how efficiently to satisfy growing demand for private transportation.

Actual developments went precisely in the opposite direction: in order to remain competitive, mass producers had to cater to the market's increasing preference for product diversity, and a succession of innovations put into question the concept of a "standard technology" sufficiently stable to build a mass-production system around it. As a consequence, rapid changes in process technology and in management systems resulted in a sharp reduction in the benefits of large-scale assembly, conventionally assumed in strategic thinking about plant design and production programs. On the other hand, minimum efficient scales in RDD&E increased as vehicles became integrated technical systems, rather than assemblies of more or less standard components, and as the closer linking of design and manufacturing made greater demands on all functions involved in bringing a new product to market.

Although diversity and flexibility became essential if firms wanted to capitalize on first-come or fast-second advantages, the conditions outlined above also produced a growing interest in interfirm cooperation. In the case of the American manufacturers this interest was motivated primarily by the recognition of a technology gap between themselves and their foreign, especially Japanese, competitors. Revolutionary as these cooperative arrangements may be in terms of the global industry's organization, it is important to recognize that they did no more than to *accelerate* the diffusion of new technology by providing yet another channel. Convergence of technical capabilities in an industry is inevitable, unless the lead firms can sustain a long-run rate of innovation that is higher than the rate at which these innovations are diffused to the industry's other members. To the best of my knowledge, the history of technology life cycles offers no examples of this possibility.

In this view, then, interfirm cooperation represents an institutional innovation induced by fast-changing technical and market conditions. It would be difficult to argue that, whatever may be meant by the often-cited *globalization* of the auto industry, the existence of strategic partnerships has changed the underlying pattern of rivalry. This observation gains additional weight if one recognizes the extent to which the struggles by national industries for a share in the world market have been supported by their governments' policies.

There exists, of course, another interpretation of strategic partnering: that it represents but a first move toward the reconstitution of the international manufacturers' cartels which, in the interwar period, were notorious for the "... restricting of production, withholding of new products, and fencing in and blocking off new developments" (Berge, 1944). While conceding that, so far, cooperation has not produced any of these results, proponents of what might be called the "conventional industrial-organization view" could point out that the true test will come only during a prolonged slump in the world market for motor vehicles. Will the networks of partners be tempted to allocate market shares, fix prices, or retard the introduction of innovations?

Two arguments speak against a repetition of this pattern, one encouraging and the other less so. First, virtually all students of cooperative arrangements in mature industries agree that these arrangements have not only accelerated the diffusion of technology among existing firms but also lowered entry costs and thus contributed to a worldwide dispersion of manufacturing capabilities. In the absence of radical governmental intervention, the resulting, irreversible interpenetration of markets has created a situation in which successful collusive action has become highly improbable.

The second argument would recognize that, for better or for worse, many of the control functions that may have been exercised by cartel-like arrangements in the past, have been taken over by governments in the name of *industrial policy.* Whatever restrictions on entry, on the spread of innovations, and on pricing are likely to inhibit global competition today stem from governmental or intergovernmental policies rather than from the efforts of an industry's member firms. In the strategies of the latter, rent-seeking has taken the place of collusion.

Even here, however, globalization of technology may have salutary effects. In order to survive in worldwide competition, firms have to find the proper mix in the integration of activities, arm's length transactions, and cooperation. National boundaries may become increasingly irrelevant as the members of strategic alliances draw on information, physical assets, and production systems that are dispersed across the globe. Once an industry has reached that stage, unequivocal definitions of a *national interest* on which to orient industrial policies may become difficult, if not impossible.

Against this sanguine view we must set one final question: what are the likely long-run consequences of cooperation for an individual firm's technological capabilities? There can be little doubt that firms have benefitted from the more rapid transfer of technology and the enhancement of know-how resulting from the strategic partnering. *If* the principles of comparative advantage and specialization apply to the production of new technical and market knowledge, long-run efficiency will also be enhanced by the inevitable division of labor among partners. If, on the other hand, a firm's success in all forms of innovation hinges on a full range of capabilities, from widely-dispersed basic and applied research to more narrowly-focussed work involving all products in the firm's output mix, cooperation could in the long run impose a considerable cost.

Given the developments outlined in this chapter, such skepticism would seem out of place; however, the history of technology suggests that the well-established concept of learning by doing may well have an opposite, "forgetting by not doing". Capabilities given up instead of honed, and redundancies of technological effort eliminated in the interest of short-run efficiency, may be hard to regain in the more distant future. This is a truly strategic issue, to which neither business decision makers nor academic observers seem to have paid much attention.

Acknowledgments

I want to thank Asim Erdilek (Case Western Reserve University), Albert N. Link (University of North Carolina at Greensboro), Nicholas W. Balabkins (Lehigh University), and Samuel J. Mantel, Jr. (University of Cincinnati) for their active support at various stages of my research. Special expressions of gratitude go to Heinz Hübner, Head of the Department of Management Science and Innovation Research, University of Kassel, FRG, for providing me with a congenial and stimulating setting in which to prepare the version of the chapter presented at the International Conference on Diffusion of Technologies and Social Behavior, June 1989, International Institute for Applied Systems Analysis, Laxenburg, Austria.

Notes

[1] This chapter is an extension of work reported in Rosegger (1986 and 1989). The points raised in the "Concluding Observations" have since been elaborated in "The Benefits and Costs of International Technical Cooperation in Mature Industries: An American Perspective, *International Journal of Technology Management* (in press, 1990).

[2] There exists no single, authoritative source on cooperative arrangements. The data presented in *Figure 11.1* were assembled from the trade and technical press as well as from business publications; therefore, they can make no claim to completeness.

[3] As of 1989, direct foreign investment operations in the United States included the following stand-alone facilities: Honda (Marysville, Ohio), Nissan (Smyrna, Tennessee), Toyota (Georgetown, Kentucky); in addition, Honda and Toyota had assembly subsidiaries in Canada. Production joint ventures included: General Motors/Toyota (Fremont, California), Ford/Mazda (Fiat Rock, Michigan), Chrysler/Mitsubishi (Bloomington, Illinois), General Motors/Suzuki (Canada), Subaru/Isuzu (projected).

[4] The initial (1981) agreement restricted Japanese export to the United States to approximately 1.7 million units/year; this limit was subsequently raised to 2.4 million units.

[5] The emphasis on *objective* characteristics is necessary because less easily documented images of quality, reliability, and pleasing design also played a major role in the success of imports in the American market.

[6] One of the puzzling aspects of the case is that the European subsidiaries of Ford and GM had consistently managed to match their competitors in technical features, design, and quality. What factors retarded (or prevented) the transfer of this knowledge to their American parent organizations?

[7] In this connection, it is interesting to note that one of the earliest joint ventures was formed in the 1960s, between Borg Warner, an American transmission producer, and Aisin Seiki, a transmission supplier to Toyota (Womack, 1988).

References

Albernathy, W.J., 1978, *The Productivity Dilemma*, The Johns Hopkins University Press, Baltimore, MD, USA.

Altshuler, A., Anderson, M., Jones, D., Roos, D., and Womack, J., 1984, *The Future of the Automobile*, MIT Press, Cambridge, MA, USA.

Automotive Industries, 1989, **169**(4), April.

Automotive News, 1989, Composites Cooperative, *Automotive News* **E10**, April 24.

Autoweek, 1984, Development game belongs to the Japanese, study says, *Autoweek* **4**, May 1.

Berge, W., 1944, *Cartels: Challenge to a Free World*, Public Affairs Press, Washington, DC, USA.

Cusumano, M., 1985, *The Japanese Automobile Industry: Technology and Management at Toyota and Nissan*, Harvard University Press, Cambridge, MA, USA.

Eaton, R.J., 1987, Product planning in a rapidly changing world, *International Journal of Technology Management* **2**(2):183–189.

Fusfeld, H.I., 1986, *The Technical Enterprise: Present and Future Patterns*, Ballinger, Cambridge, MA, USA.

Langenfeld, J., and D. Scheffman, 1986, *Innovation and US Competition Policy*, US Federal Trade Commission, Washington, DC, USA, mimeo.

Marumo, K., ed., 1984, *Japanische Industrietechnologie im internationalen Vergleich*, mimeo., German translation of a study prepared by the Agency of Industrial Science and Technology of the Ministry of International Trade and Industry, I.S.I., Karlsruhe, FRG.

Motor Vehicle Manufacturers Association, *Annual Statistical Report*, various volumes, Detroit, MI, USA.

Nelson, R.R. and S.G. Winter, 1982, *An Evolutionary Theory of Economic Change*, The Belknap Press of Harvard University Press, Cambridge, MA, USA.

Reich, R.B. and E.D. Mankin, 1986, Joint ventures with Japan give away our future, *Harvard Business Review* **86**(2):78–86.

Rhys, G., 1988, Economics of the motor industry, *Economics* **34**(4):158–168.

Rosegger, G., 1976, Diffusion research in the industrial setting: Some conceptual clarifications, *Technological Forecasting & Social Change* **9**:401–410.

Rosegger, G., 1986, *Rivalry and Cooperation in the World's Automobile Industry*, Paper presented at the Colloquium on International Marketing Cooperation between Rival Trading Nations, University of Pisa, San Miniato, Italy, May 29–31.

Rosegger, G., 1989, Cooperative research in the automobile industry: A multinational perspective, in A.N. Link and G. Tassey, eds., *Cooperative Research and Development: The Industry-University-Government Relationship*, Kluwer Academic Publishers, Norwell, MA, USA.

Rosegger, G. and R.N. Baird, 1987, Entry and exit of makes in the automobile industry, 1895–1960: An international comparison, *OMEGA* **15**(2):93–102.

Sekaly, R.R., 1981, *Transnationalization of the Automobile Industry*, University of Ottawa Press, Ottawa, Ontario, Canada.

Subcommittee on Trade, US House of Representatives, Committee on Ways and Means, 1980, *Auto Situation 1980*, US Government Printing Office, Washington, DC, USA.

Womack, J.P., 1988, Multinational joint ventures in motor vehicles, in D.C. Mowery, ed., *International Collaborative Ventures in US Manufacturing*, Ballinger Publishing Co., Cambridge, MA, USA.

Chapter 12

Some General Regularities of Techno-Economic Evolution*

Sergei Yu. Glaziev

12.1 Introduction

One of the most exciting phenomena of the modern world is the fundamentally homogeneous direction of the overall techno-economic development trajectory in practically all regions of the world. The existing economic systems in different countries are collapsing one after another under the pressure of an expanding *industrial culture* and are becoming drawn into the international division of labor. Simultaneously, their economic development is influenced by the general regularities of the world techno-economic system, the rhythm of which is set by the industrially developed countries. These general regularities of long-term techno-economic development, invariant under different sociopolitical systems, are the subject of this chapter.

*We gratefully acknowledge permission to use parts of the original version of this paper which appeared in *Communist Economies*, Volume 2, No. 2 (1990),

12.2 Regularities of Technological Change

The key factor that directs the overall techno-economic development trajectory is technological change. During the last few decades the problems of technological change have been the focus of interest in economics. A new concept has evolved which views economic dynamics as an uneven and uncertain process of evolutionary development (Dosi *et al.*, 1988). From this point of view technological change is a complex interaction of various technological alternatives, carried on by competing and collaborating economic agents in similar institutional environments. The selection of techno-economic development alternatives and their implementation in technological shifts and structural changes take place as a result of learning processes (which are determined by various nonlinear feedbacks – positive and negative) that influence the dynamics of the interaction of technological and social change.

The concept of economic growth as a complex, nonlinear and uncertain process involving permanent changes, allows us to develop a new approach in studying the regularities in long-term techno-economic development and the management of technological change. Feedbacks stipulate the interaction of various elements in the socioeconomic system in the course of technological shifts and determine the directions and rates of evolution of the economy. The modeling of these feedbacks becomes a priority task of economics.

This new approach predetermines a new vision of economic structure. It is important to select a view of the economic system that can ensure the stability of its components and of interrelations between them in the process of technological change. Such a vision of economic structure assumes a corresponding choice of its primary *element.* This element should not only preserve integrity in the process of technological change, but it should be a carrier of corresponding innovations, i.e., it need not necessarily be disaggregated to describe and measure them.

The changing driving forces of the form and direction or of the overall techno-economic development trajectory have recently been summarized by Freeman and Perez (1988). In a Schumpeterian tradition, they distinguish four successive modes of growth (or *techno-economic paradigms*) since the onset of the industrial revolution. These modes of techno-economic development are driven by the growth of *leading branches* and growth sectors which involves a synergetic aggregate of key factor industries, technologies, and infrastructures.

Thus, as an element for the analysis of the overall techno-economic development we suggest a totality of technologically connected production processes preserving its unity in its evolution. Such totalities are united into a stable self-producing unity – a technological wave (TW). The latter covers a macroeconomic production cycle – from the extraction of natural resources and labor force education to final consumption.

The production sections contained in a single TW develop more or less simultaneously because of their technological interconnection. Any change in one of the elements of the TW initiates a corresponding change in other elements in the chain of production units. The broadening of each technological process is conditional on the development of the whole group of interconnected production systems.

The primary element of technological change is the innovation. The development of any technological system begins with the introduction of a basic innovation which is then followed by others. Basic innovations are usually radically different from their traditional technological surroundings. The effective functioning of a new formation can be achieved only through an adequate technological chain. The inclusion of new technological systems in traditional technological chains does not, as a rule, lead to significant technical progress, because their advantages are not always recognized. Techno-economic development takes place through the establishment of new technological chains based on clusters of technologically connected innovations which are combined in new TWs. Technological change does not proceed with the more or less smooth introduction of new technologies, the elimination of the old, and the gradual raising of the overall technological level of production. It is, in essence, a sequential process of long periods of evolutionary development of TWs, broken up by occasional revolutionary changes in the technological base of the economy through replacement of the dominant TWs. The replacement of one dominant TW by another is accompanied by the reconstruction and elimination of a large number of technological processes which form the base of the previous TW, and by the disintegration of the corresponding production links with the rapid diffusion of new technological processes and production systems.

It makes no sense to ask what is the main driving force in the genesis of each TW – technological possibilities or consumer needs – or in which sector of the economy does the original impulse for the technological breakthrough arise. Research shows that innovations and shifts in different sectors (machinery, raw materials, energy sources, construction, transport, and communications) are interdependent. They stimulate and complement each other. Radical discoveries and inventions relating to one of these sectors are not put into practice, or at the very least do not receive adequate diffusion, without the appearance of corresponding innovations in other sectors.

Each period of domination of a particular TW is characterized by a set of base technologies, a group of leading branches, a type of consumption and labor qualification, dominant energy sources, materials, and modes of transport and communications. There is a continuity between the consecutive stages of development and the corresponding TWs, i.e., the material and technical base for a particular stage emanate from the development of the previous stage. The birth of a new TW occurs within the old one, and in further development it adapts the production units that existed in the previous stage to the needs of its evolution.

The results of the research done in recent years into the long-term trends of technical and economic evolution in the industrially developed countries (see, for example, Dosi, 1983; Freeman, 1983; Grübler, 1990; Mensch, 1979; Piatier, 1984; Nakićenović, 1986; Vasko, 1987) allow us to conclude that since the first industrial revolution five technological waves have been observed, based successively on the mechanization of the textile industry (first TW), machine technologies connected with the use of the steam engine and railways (second TW), electrotechnical machinery and the electrification of the economy (third TW), comprehensive development and the use of chemicals and motor vehicles (fourth TW), and telecommunications and microelectronic technologies (fifth TW).

The evolution of each technological wave can be schematically represented by two pulsations. The first pulsation (the formation phase) takes place in an unfavorable economic and social environment which is determined by the previous technological wave (still predominating at that time). In the second stage the technological wave expands rapidly after it overcomes the social and economic barriers as a result of institutional changes. In this stage the TW becomes the main carrier of economic growth. With the exhaustion of technical improvement possibilities in the industries that constitute the predominant TW, and with the saturation of corresponding consumer needs, the predominant TW expansion becomes inefficient and the period of institutional change comes round again. The successive substitution of predominant TWs generates low-frequency variations (long waves) in world economic development.

12.3 Technological Change in Different Economic Systems

In a market economy the discontinuities of economic growth associated with the TW substitution appear empirically to have become more established, albeit interpreted from a number of different theoretical perspectives (see, for example, Delbeke, 1983; Freeman, 1983; Mensch, 1979; Vasko, 1987, among others). Such discontinuities in economic development can be explained with the help of known regularities of capital reproduction. With saturation of the corresponding type of consumption demand and exhaustion of the possibilities of further technological improvement in the life cycle of each TW, a phase of decay sets in, when the marginal efficiency of investment in that group of technologies and industries falls sharply. Further growth of national consumption and production, in addition to maintaining the profitability of the latter, requires new investment to be channeled to the radically new technologies of the next TW. The latter, as a rule, already exist in the form of inventions, R&D results, and design documentation. However, their pervasive diffusion is restricted by inadequate socioeconomic conditions: the common interests of workers (unwilling to lose their jobs and see their skills devalued), shareholders (interested in a return on capital invested in traditional TW industries), and corporations (interested in the expansion of conventional technologies). A high level of inertia in socioeconomic institutions leads to prolonged depression. During this period a large part of the available capital can find no profitable investment application and is partially lost in speculative operations. Simultaneously, the growth rates of all macroeconomic indicators decrease, including the real income of the population. It is only with the implementation of corresponding institutional changes and organizational and social innovations (it takes time to realize that they are necessary) that possibilities for the rapid growth of the new TW industries appear. Surplus capital is gradually spent, large-scale redistribution of resources takes place, and the economy embarks on a new expansion path of economic growth.

During the TW growth phase in a market economy stable circuits of resource distribution are formed, which determine the expansion of production. Private and public organizations, official regulatory bodies, and other economic institutions become connected with the technologies of the given TW. This prevents the redistribution of resources into new technologies, leading to overaccumulation of capital in traditional TW industries which are saturating. The economy enters a stage of long-term recession, characterized by low investment, underused capacity, and high unemployment. With the aggravation of social tension a search for new directions of economic activity begins. Government and corporations start to introduce social, organizational, and technological innovations, making way for the new TW.

In a world market economy the above mechanism works on a global scale. Facing falling demand in the local market, corporations start to export goods and capital, thus diffusing the corresponding technological wave throughout the world economic system. It reaches the limits of its diffusion in a number of countries, connected by intensive trade flows, more or less simultaneously, leading to worldwide overaccumulation of capital. The country that is the first to introduce the necessary social and technological innovations in order to make way for the next TW initiates a new long growth phase, frequently attaining a leading position in the world.

The substitution of technological waves in market economies is determined by the regularities in the way the capital market functions. In modern centrally planned economies there is no capital market, therefore technological wave substitution takes other forms.

Centrally planned economies are characterized by a high degree of stability in industrial relations (i.e., relations between enterprises, ministries, state authorities, etc.) and institutions that determine the flow of resources. The result of this is an inertia in technological evolution which leads to difficulties for structural change in the economy. The reasons for this great stability and inertia in centrally planned economies can be found in the regularity in the functioning of large economic organizations which form the basis for its economic mechanism.

The natural tendency of any organization to strive for stability, both internally and externally, produces an internal resistance to innovation. Any serious innovation will inevitably disrupt existing communications and management processes, which obviously results in resistance among those affected.

Usually management introduces a radical innovation only if it is necessary for its survival in a changing environment. Consequently, in a stable external environment one can hardly expect a high degree of innovative activity. The reverse side of a high degree of stability is the absence of independence. Enterprises concerned with introducing significant innovations do not as a rule have the capacity to do so independently. Decisions on important innovations are made at the top level of the economic hierarchy. The enormous scale of its *subordinate estates* and the solid management power of the ministry apparatus give a high degree of stability to its work and consequently result in a low degree of innovation.

The stability of the enterprises' economic environment under a certain ministry is complemented by a system of relevant interbranch relationships and central planning institutions. Its basic components are: the practice of planning by means of extrapolation, the allocation of capital investment according to production growth objectives, the provision of supplies to enterprises according to schedules set in advance by Gossnab (the State Supply Administration), and the preparation of plans on the basis of the *national economic need* as set by Gossnab – which in practice is a simple compilation of the orders of enterprises, usually calculated (with annual planning and a lengthy interval between putting in orders and receiving plan tasks) by proportionally increasing the orders of the previous year. Yet other components include the budget financing of capital investment plus a *soft* credit system, cost-based prices which guarantee the covering of any production costs regardless of the social utility of the resulting product, and the activity of Gosstandart (the State Standards Administration) which prevents any changes in the technology of production, whether for better or for worse. The organization of R&D also forms part of this mechanism of stable extensive production across different (unchanged) technological trajectories. The subordination of most scientific organizations to the ministries and the management of science by mass production methods promote an incremental style of technological changes.

Thus extremely stable industrial relations arose in the centrally planned economy, providing stable conditions for the extensive reproduction of existing technological processes. Its basic features were formed before the Second World War, serving as a powerful means for the rapid expansion of production typical of the third TW. Today it continues to provide unchanging flows of resources and products. The technologies of a new TW are introduced along with the old, by the formation of new branches and sub-branches, which receive resources for expanded production. As a consequence of their technological incompatibility with conventional technologies, however, the management bodies responsible for their development try as much as possible to create their own technological base. Thus, the new TW develops autonomously, but such reproduction, based on internal accumulation, proceeds extremely slowly. The expansion of the new TW requires redistribution of resources and the adaptation of conventional technologies.

This presupposes the breakdown of old technological chains and the corresponding economic information flows, which in turn is impossible without the reorganization or liquidation of many institutions serving the traditional technological systems. Until this happens the new technological structures develop parallel to the expanding reproduction of the old. With time a number of autonomous TWs form, functioning within a stable régime of expanded production.

In the system of management of the centrally planned economy there was no mechanism of automatic redistribution of resources from obsolete industries to new ones. Resources were distributed by stable circuits of industrial relations, according to the interest of ministries. For that reason development of new industries was very slow, with simultaneous excessive expansion of the established ones. The result is a specific situation of simultaneous diffusion of the consecutive and parallel existence of multiple TWs. It is followed by a number of negative consequences; overproduction of obsolete products, superfluous economic activity, the overexpansion of resource-producing industries, considerable national economic losses, and generally low efficiency of production.

12.4 Measurement of Technological Wave Substitution

In order to obtain evidence supporting the above hypotheses, we have undertaken empirical research on the formation and substitution of TWs during the last century. During this period one can see three successive TWs or techno-economic paradigms replacing their predecessors in industrialized countries (see, in particular, Freeman and Perez, 1988, for a detailed account).

The formation of the first, which we refer to here as the third TW, began in Russia in the last quarter of the 19th century, slightly later than in the leading industrialized countries. Its nucleus was electric power and electrical machinery. The diffusion of the technologies of this TW was accompanied by the mechanization of the basic technological processes and corresponding changes in the quantity and skill level of labor. The most important industrial material was steel, including rolled steel. The energy source was coal, while the main form of land transport was rail. Production was oriented toward high-volume resources, universally applicable machinery, and labor with low qualifications (by today's standards). The development of this TW was accompanied by rapid urbanization and radical changes, not only in the structure of consumer demand, but also in the lifestyle of the population.

By the beginning of the 1930s the mechanization of production, the improvement in the qualifications of the labor force, shifts in the fuel-energy balance toward liquid fuel, the growth in total energy consumption, as well as the establishment of new systems of mass communications and a new transport infrastructure (roads), created the conditions for the growth of a new (the fourth) TW in the developed market economies. Research into the rate of change in the use of traditional industrial materials, as well as the production of types of products characteristic of the third TW, shows that the technologies contained in it continued to be used in the leading market economy countries right up to the mid-1960s.

However, the basic engine of technical and economic development in the post-war years was the fourth TW. Its nucleus was the chemical industry and associated machine-tool production industries (chemical engineering) and the motor industry, which underwent further development, while road vehicles became the major mode of transport. Characteristics of this stage were the mechanization and automation of many basic technological processes, the growth in the specialization of production, and its reorientation from the use of high-volume resources and universal machinery to quality raw materials and specialized equipment. Electric power consumption grew at very high rates, while the shift in the fuel-energy balance toward oil continued until it finally replaced coal as the leading source of energy. The use of new industrial materials greatly increased, including plastics and high-quality steels. Secondary education became the norm, while the skills of the labor force and the production culture moved to qualitatively higher levels. In the mid-1970s the fourth TW reached the limits of its growth in the developed capitalist countries, which can be seen in the dynamics of the indicator (see discussion below) of its life cycle (*Figure 12.1*). At this stage in the major capitalist countries, the relative consumption of basic materials, energy sources, and non-production consumption items for this TW stabilized, while the diffusion of its basic technologies peaked. Further technical and economic development is instead connected with a new stage of technological change, based on microelectronics and telecommunication technologies (the fifth TW).

The nucleus of the fifth TW includes electronics, robotics, and microprocessor technology. This TW is characterized by computer integrated manufacturing technologies, new systems of mass communication based on computer networks and satellite links, among others.

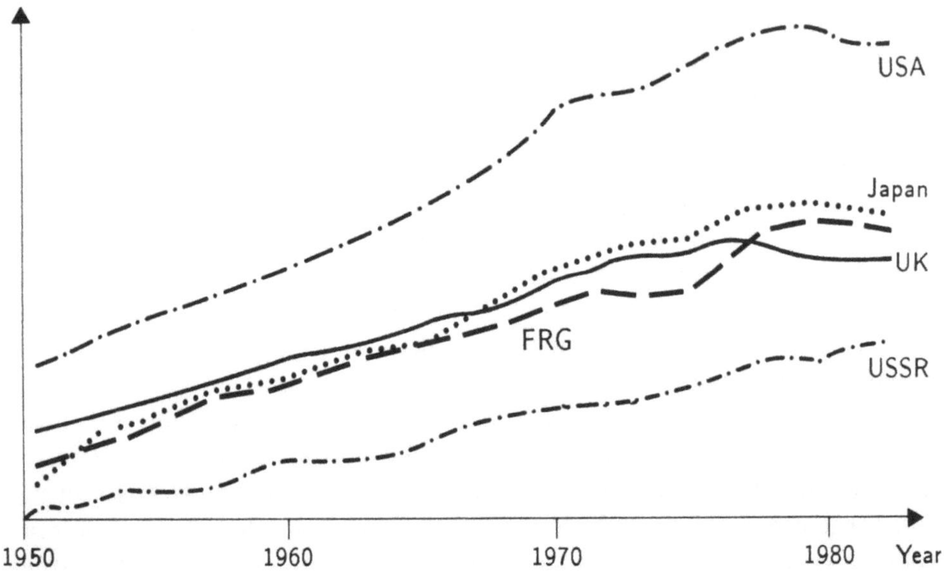

Figure 12.1. General indicator of the evolution and intensity of the fourth TW based on a two-step analysis of principal components of 50 indicators of technological change.

A conceptual analysis of the technological waves successively predominating in the technological structure of the developed countries in this century will make it possible to build systems of indicators reflecting the evolution of these waves. Work with available statistical data revealed about 50 initial indicators. They were presented as time series for the period 1951–1986, reflecting technological changes connected with the diffusion of the fourth TW in the following economies: Bulgaria, Czechoslovakia, Federal Republic of Germany (FRG), German Democratic Republic (GDR), Hungary, Japan, Poland, Romania, USSR, UK, and USA.

The information sources used in this work were publications of the UNO statistical services, as well as those of CMEA, OECD, UNESCO and other international organizations, data from national statistical sources, and separate research works. A list of the initial indicators is given in the *Tables 12.A1* to *12.A8* of the Appendix. Principal component analysis was used for aggregating a large number of initial indicators into general indicators. The justification for its use was shown by the high correlation of the initial indicators (see Appendix), reflecting a single process – the diffusion of the fourth TW.

The initial indicators used in this research reflect technological change connected with the diffusion of the fourth TW in seven economic spheres: in construction materials, the chemical industry, electric power generation, the fuel-energy complex, the agricultural complex, transportation, and final consumption. Different economic sectors are represented in the system of initial indicators by a different number of indicators. Direct application of principal component analysis to them would give, however, a distorted picture of techno-economic development in the general index, because separate economic spheres would then bear weights proportional to the number of indicators which reflect them, but not to their actual economic value. Therefore the initial data were aggregated step-by-step. First, general indicators of technological change in each economic sphere were constructed by principal component analysis (*Tables 12.A1* to *12.A7* in the Appendix), then these general indicators themselves were subject to component analysis (*Table 12.A8* in the Appendix).

The structure of the corresponding principal components is given in the Appendix. As can be seen from the coefficients of factor utilization with the first principal components, the latter not only aggregate the overwhelming majority of information contained in the initial indicators, but reflect the main direction of mutual variability of the initial indicators of techno-economic development. Therefore, they may be used as general indicators of the technical evolution of the corresponding economic spheres. Similarly, as the results of component analysis show, the principal component of the second level may be used as a general indicator of the technical evolution of the economy as a whole.

The results of the principal component analysis of the second level are presented in *Figure 12.1*. The period under consideration covers the maturity stage of the fourth TW in developed market economies and its growth stage in the USSR. The results of measurements show that the TW is more *prolonged* in a centrally planned economy in comparison with market economies: technological shifts are more even and slower. This phenomenon reflects the less intensive redistribution of resources from conventional technologies to new ones in a centrally planned economy. As a result, its technological development is more inert.

As one can see from the results of empirical research, the general direction and intensity of techno-economic development can vary depending on the peculiarities of economic relations. Despite the fact that the general direction and historical succession of technological shifts in both socialist and capitalist countries were the same, their rates, methods of realization, and manifestations were essentially different. Certain differences are also characteristic of countries within the same socioeconomic system. Thus, rates of technological change were substantially higher in Japan than in other developed capitalist countries. Thanks to an efficient national policy of long-term techno-economic development, the fourth TW life cycle was reduced to a quarter of a century, and growth in production capacities of the third TW were curtailed promptly as soon as they proved unpromising from the point of view of long-term trends in technological change. Simultaneously, at the growth stage of the fourth TW the government and corporations in Japan made great efforts to form basic industries for the fifth TW, which had been identified in the long-term forecasting of Japan's techno-economic developments. As a result, along with the growth of the industries in the fourth TW, large-scale resource redistribution into the industries of the fifth TW took place in Japan. The consequence was rapid advancement with Japan becoming one of the leaders in technological development during the process of technological wave substitution in the world economy.

In the USSR the third TW was formed during the industrialization of the 1930s. The formation of the fourth TW began in the second half of the 1950s. At this stage production units, typical of the third TW, had not yet reached their limits of diffusion and the material and technical base needed for the fourth TW had not been formed. The dynamics of the relative consumption of traditional industrial materials, the production of universal metal-cutting machinery, and a number of other indicators of the life cycle of the third TW, indicate its continued growth right up to the mid-1970s and its reproduction up to the present day in the USSR. The allocation of an enormous amount of economic resources to the continuing reproduction of the third TW resulted in an inadequate allocation of resources to the development of the fourth TW. Therefore the diffusion of its technological systems has occurred at rates and levels considerably below those demonstrated by a range of industrially developed market economies.

At present the formation of the fifth TW is beginning. Research shows a comparatively high rate of diffusion of robotics, computer-controlled machine tools, and other CIM technologies in the USSR. However, it must not be forgotten that as yet this TW is still in the initial phase of its life cycle. Therefore high rates of diffusion of technologies can be achieved without large allocations of resources, because of their insignificant weight in the technological structure and in the economy as a whole. For these high rates to be maintained in the longer term an ever-increasing redistribution will be necessary. The problems of organizing such a redistribution are apparent from the experience of the development of the fourth TW. High rates of growth at the beginning of the life cycle of the fifth TW declined sharply after expansion began. Our measurements revealed a decline in diffusion of the fifth TW-technology over that last decade in the USSR and an increase in the technological gap between the Soviet and Western economies. A qualitative analysis of the reproduction of the technological structure of the economy in the USSR, and individual calculations, indicate the parallel presence of three TWs, all at different stages of their life cycles (we refer to this phenomenon here as *technological multi-modeness*).

Technological multi-modeness is normal in periods of TW substitution. In the course of interwave interactions that take place at that time, the obsolete TW is destroyed, forming the prerequisites for the growth of the new one. Coordinated actions of government and industrial and social organizations can significantly accelerate this process and minimize social costs of technological wave substitution. However, when there are no incentives for resource redistribution from obsolete industries into new ones, the growth of the new TW may take place at the same time as the extended reproduction of the obsolete TW. If this happens, the possibilities of the new TW are restricted by the previous TW, which is still expanding and if structural policy is inadequate the parallel existence of several TWs may easily be reproduced. This is accompanied by a stratification in the economic structure.

Reproduction of each TW is autonomous. Interwave interactions are not directed at the replacement of obsolete TWs in order to meet the demands of new ones, but at the elimination of bottlenecks which restrict their simultaneous reproduction. Under these circumstances the tightest bottleneck is satisfying the demand for primary resources. Consequently, excessive efforts are concentrated in mining industries and the resource base of the economy becomes overloaded. With the prolongation of the obsolete TW, the formation of a new TW is restricted by general resource limitations – and missing links are provided through foreign trade. Owing to the deficit of high-quality resources, mass-produced resources are exported in exchange for high-quality imports. As a rule, the exports are raw materials that have undergone a low degree of processing, so that this exchange is non-equivalent and leads to an intensification in the structural crises.

One more negative consequence of *technological multi-modeness* is the stratification of the system of economic values, reflecting the peculiar structure of the economy. As a result, economic estimations cease to act as reliable reference points in making economic decisions – new technologies look inefficient in systems reflecting the conditions of the obsolete TW and economic agents turn out to be interested in the preservation of an obsolete technological structure. The reproduction of different TWs in the economy is supported by a multitude of positive feedbacks, exacerbating the losses connected with its preservation. An inadequate technical policy thus brings the economy into a permanent structural crisis with the country becoming increasingly impoverished.

12.5 International Comparisons of Techno-Economic Evolution

One can assess the technological structure of a country by comparing variations in the structural evolution of the technological and infrastructural base to the international level or to that of the leading country in a particular TW development. Below we discuss some long-term technological changes in the transport and energy systems of the USSR in comparison with a similar analysis performed for the USA (Marchetti and Nakićenović, 1979; Nakićenović, 1986 and 1987). In the case of the changing mix of transport infrastructures in the USA and USSR, the comparison shows that technological shifts in the Soviet transport infrastructure took place a quarter of a century later than in the USA. They also occurred more slowly, which is due mostly to slower curtailing of obsolete transport technologies (see Chapter 19 by Nakićenović). For instance, the slow decrease in the share of railways in the general transport infrastructure and also in intercity passenger transport are explained by the continuing reproduction of the third TW, characterized not only by active railway construction (as opposed to decommissioning in the USA), but also by a strong demand for rail transport. The high resource consumption of third TW industries means high demand for raw materials, which places great pressure on the transport system. This pressure is substantially enhanced by the demand of the fourth TW industries for oil and energy in general, and also by the fifth TW demand for imported equipment (satisfied by exporting ores and fuels). The result of this is a hypertrophied transport structure in which the role of rail transport is exceptionally high.

Comparative analysis of the structure of energy consumption in the USSR and the USA yields interesting conclusions (note that the dynamics of structural change in the US economy are close to those of the world economy). After the protracted predominance of coal in the consumption of primary energy carriers in the 1950s (compared to the peak in market dominance of coal in the USA some forty years earlier) technological shifts in the fuel-energy complex in the USSR speed up. The period from when the share of coal in primary energy consumption was at its maximum up to its replacement as the predominant energy carrier by oil took more than two decades in the USSR, compared with three-and-a-half in the USA. However, the peak market share of oil in the primary fuel mix of the USSR attained a significantly lower level than in the USA (where the market share of oil peaks above 50 percent of primary energy consumption). Even more rapid was the development of gas technology, which took only a decade and a half to become predominant in the USSR, starting from a 10 percent share in energy consumption.

The relatively rapid development of oil and gas technologies in the Soviet economy are explained by three major interrelated reasons. First, the rapid formation of the fourth TW from 1950 to 1960 caused a sharp increase in demand for oil, which had held a relatively large share in energy consumption even before (for a long period coal consumption was complemented by oil consumption – oil technology was separated from coal technology later). Second, a large share of oil consumption was a substitute for coal in industries of the third TW, the expansion of which had exerted equal pressure upon oil and coal consumption. Its predominance in the technological structure of the economy up to the period 1960–1970 also contributed to the prolongation of the coal technology life cycle. Third, the development of the fourth TW in the Soviet economy involved considerable use of imports for the production process, which in turn required the export of energy resources. This was an important incentive for the rapid expansion of oil production, and, also partially, for gas production, and this in turn influenced the structure of internal energy consumption.

The appreciation of the long-term trajectories of technology diffusion and their distinctive variations in different socioeconomic conditions makes it possible to put interstate comparisons on a more scientific basis. At the same time it is necessary to note that one must be very careful when using interstate comparisons in scientific research, and, even more, in practical economic decisions, in view of structural peculiarities and goals of techno-economic development. In particular, in the calculation of technical lags and technological gaps one must not rely on *absolute* levels of techno-economic indicators. As a rule, these values are determined not only by the level of techno-economic development but also by a number of social, geographical, political, and other factors specific to each country. What is far more important is the dynamics of such indicators and their structural evolution over time. Thus, the gap in the level of production diffusion of the fourth TW between the USA and West European countries is largely explained by geographical peculiarities such as population and consumption. More important is the almost parallel movement of the general indicator of the fourth TW development in the USA, the FRG, the UK, and Japan. This is confirmed by the synchronization of the saturation level of the given indicator (see *Figure 12.1*), despite substantial differences in the absolute value of this level. The fact that West European countries reach saturation of the given indicator only two to three years later than the USA reflects the small technological lag.

It must be noted also that, as a rule, the leading country in the development of a particular TW develops on a somewhat larger scale than the countries that follow it. This is explained by the relatively high stability of economic structures developed over a relatively long period. Such structures tend to promote the reproduction of the technologies connected with the country's earlier successes in the world market.

12.6 Economic Policy Implications

To concentrate on the absolute levels of US techno-economic indicators, as is usual in measurements of national techno-economic development, is scarcely justified. By orienting the economic policy of a country to the technological level and structure of a leading country condemns that country to permanent *lagging behind* and to the reproduction of the leading economy's development trajectory. Such an orientation does not allow the country concerned to benefit from the *advantage of the backward* in the organization of *overtaking* techno-economic development.

The *advantage of the backward* is the opportunity to use the experience of advanced countries, and forecasts for their future techno-economic development, in determining the optimal strategy for closing the technological gap. This advantage becomes most acute during the periods of large-scale structural change in the world economy connected with the substitution of overall TWs. At such a period in time the *backward* country has an opportunity to *take a short cut*, establishing the principal directions of technological shifts and concentrating resources in the key industries of a new technological wave.

The economic structure of the countries that were leading during the life cycle of the previous TW is closely connected with obsolete technologies and this leads to a high degree of sluggishness in their economic systems. Backward countries, on the other hand, find themselves in a comparatively better position, having no need to break the powerful old production machinery and to overcome the resistance of the people and organizations involved. This opens up possibilities for making a technological push with the aim of leap frogging or *surpassing without overtaking.* This was exactly how Japan made its remarkable push during the last two decades (see Freeman, 1987), followed by the new industrial countries of southeast Asia. They did not develop industries of the third and fourth TWs (and still less the second and first TWs) but put all available resources into the fifth TW.

Appendix

Indicators and factor matrices used for principal compound analysis of techno-economic change in selected market and centrally planned economies for the period 1951 to 1986.

Table 12.A1. Factor matrix for construction materials.

	Factor coefficients		
	I	II	III
Share of steel in consumption of construction materials	−0.3505	0.1739	0.0558
Share of plastics in consumption of construction materials	0.3602	−0.2342	0.0875
Share of aluminum in consumption of construction materials	0.2371	0.5553	−0.3332
Share of copper in consumption of construction materials	−0.3492	−0.0150	0.0723
Consumption of steel per unit of national income	−0.1751	0.2558	0.8396
Consumption of plastics per unit of national income	0.3540	−0.1389	0.2654
Consumption of aluminum per unit of national income	0.3214	0.3750	0.0772
Consumption of plastics per capita	0.3693	0.0014	0.0191
Consumption of steel per capita	0.2550	0.5083	0.1575
Consumption of aluminum per capita	0.3290	0.3497	−0.2574
Dispersion of factors	65%	15%	10%

Table 12.A2. Factor matrix for electric energy.

	Factor coefficients		
	I	II	III
Share of primary electric energy in consumption of energy resources	0.2338	−0.5816	0.3606
Share of electric energy in consumption of energy	0.4181	−0.2802	−0.0309
Consumption of electric energy per unit of national income	0.3861	0.2580	−0.1424
Consumption of electric energy per capita	0.3964	0.3334	−0.0770
Consumption of electric energy for lighting and household needs per capita	0.3932	0.3408	−0.1120
Consumption of fuel by combined heat and power electricity plants per unit of national income	−0.3634	0.0496	0.5329
Electricity consumption per worker in industry	0.2635	−0.5165	−0.1964
Share of atomic electricity plants in production of electric energy	0.3335	0.1411	0.6721
Dispersion of factors	59%	17%	8%

Table 12.A3. Matrix structure of energy consumption.

	Factor coefficients		
	I	II	III
Share of coal[a]	−0.4897	0.2537	0.1172
Share of oil[a]	0.4678	−0.0458	−0.2260
Share of natural gas[a]	0.3117	−0.6952	−0.0712
Share of primary energy[a]	0.3273	0.6032	−0.3277
Share of electric energy in energy consumption	0.4707	0.2376	0.0098
Share of atomic electricity plants in production of electric energy	0.3395	0.1065	0.9014
Dispersion of factors	61%	18%	10%

[a]Shares of coal, oil, natural gas, and primary energy are the shares of corresponding energy resources in the total consumption of energy resources for energy production.

Table 12.A4. Factor matrix for the chemical industry.

	Factor coefficients		
	I	II	III
Consumption of plastics per unit of national income	0.4991	0.5189	−0.5338
Consumption of synthetic fibres and yarns per unit of national income	0.4921	−0.5814	−0.5461
Consumption of plastics per capita	0.5039	0.4693	0.3570
Share of synthetic fibres and yarns in consumption of chemical fibres and yarns	0.5046	−0.4150	0.6145
Dispersion of factors	85%	8%	2%

Table 12.A5. Factor matrix for agriculture and related industries.

	Factor coefficients		
	I	II	III
Share of employees in agriculture	0.4288	0.4818	0.5827
Consumption of mineral fertilizers per 1,000 hectares of arable land	−0.4609	0.2067	0.4438
Number of tractors per 1,000 hectares of arable land	−0.3887	0.7897	−0.4261
Milk yield per cow	−0.4825	−0.2987	−0.0788
Grain harvest	−0.0685	−0.1099	0.5257
Dispersion of factors	72%	12%	9%

Table 12.A6. Factor matrix for transport.

	Factor coefficients		
	I	II	III
Length of railways	0.4198	−2.2871	0.1444
Length of roads	0.4396	−0.1151	−0.3227
Length of oil pipelines	0.4380	−0.1459	−0.3048
Length of natural gas pipelines	0.4425	−0.0075	−0.2859
Number of cars	0.2660	−0.0376	−0.8365
Share of containers in goods turnover	0.2660	0.9392	−0.0205
Dispersion of factors	83%	16%	3%

Table 12.A7. Factor matrix for private consumption.

	Factor coefficients		
	I	II	III
Consumption of national income used for non-productive consumption per capita	0.4008	0.1494	−0.2883
Consumption of energy used for lighting and household needs per capita	0.4005	−0.2324	−0.1490
Number of students per 1,000 inhabitants	0.2477	0.9136	0.1961
Number of TV sets per 100 inhabitants	0.4073	−0.0138	−0.2079
Consumption of paper per capita	0.4016	−0.1548	−0.0496
Share of employees in public services	0.3523	−0.2299	0.8794
Number of telephones per 100 inhabitants	0.4075	−0.1086	−0.1924
Dispersion of factors	80%	11%	5%

Table 12.A8. Factor matrix of the second level (seven main components of the first level).

	Factor coefficients		
	I	II	III
Construction materials	0.3295	0.3327	−0.0193
Chemical industry	0.3772	0.4648	−0.3885
Energy consumption	0.4402	0.3416	0.9000
Electric energy	0.3467	−0.3437	0.6347
Agriculture and related industries	0.0335	−0.4216	−0.1486
Private consumption	0.4749	−0.0621	−0.3239
Transport	0.4378	0.5616	−0.3026
Dispersion of factors	64%	32%	2%

References

Delbeke, J., 1984, *Long Wave Research: The State of the Art, Anno 1983*, CP-84-53, International Institute for Applied Systems Analysis, Laxenburg, Austria.

Dosi, G., 1983, Technological paradigms and technological trajectories: The determinants and directions of technical change and the transformation of the economy, in C. Freeman, ed., *Long Waves in the World Economy*, Butterworths, London, UK.

Dosi, G., Freeman, C., Nelson, R., Silverberg, G., and Soete, L., eds., 1988, *Technical Change and Economic Theory*, Pinter Publishers, London and New York.

Freeman, C., ed., 1983, *Long Waves in the World Economy*, Butterworths, London, UK.

Freeman, C., 1987, *Technology Policy and Economic Performance: Lessons from Japan*, Pinter Publishers, London and New York.

Freeman, C. and Perez, C., 1988, Structural crisis of adjustment: Business cycles and investment behaviour, in G. Dosi *et al.*, eds., *Technical Change and Economic Theory*, Pinter Publishers, London and New York.

Grübler, A., 1990, Technology diffusion in a long wave context: the case of the steel and coal industries, in T. Vasko *et al.*, eds., *Life Cycles and Long Waves*, Lecture Notes in Economics and Mathematical Systems 340, Springer-Verlag, Berlin, Heidelberg, New York.

Marchetti, C., and Nakićenović, N., 1979, *The Dynamics of Energy Systems and the Logistic Substitution Model*, RR-79-13, International Institute for Applied Systems Analysis, Laxenburg, Austria.

Mensch, G., 1979, *Stalemate in Technology*, Ballinger Publishing Co., Cambridge, MA, USA.

Nakićenović, N., 1986, *Patterns of Change: Technological Substitution and Long Waves in the United States*, WP-86-13, International Institute for Applied Systems Analysis, Laxenburg, Austria.

Nakićenović, N., 1987, *Transportation and Energy Systems in the US*, WP-87-01, International Institute for Applied Systems Analysis, Laxenburg, Austria.

Piatier, A., 1984, *Long Waves and Industrial Revolution*, CP-84-54, International Institute for Applied Systems Analysis, Laxenburg, Austria.

Vasko, T., ed., 1987, *The Long-Wave Debate*, Springer-Verlag, Berlin, Heidelberg, New York.

Chapter 13

The Adoption of Applications of Information Technology for Process Development: A Regional Case Study

Charlie Karlsson

13.1 Introduction

This chapter concerns the determinants of innovative activities at the plant level in the engineering industry in two peripheral regions in Sweden. More precisely, its purpose is to attempt to determine the characteristics of those establishments in these regions that are early users of applications of information technology, i.e., innovations, for the development of their production processes. In general, profit-seeking enterprises will plausibly allocate resources to the adoption of innovations if

- they know or believe in the existence of some as yet unexploited technical opportunities;

- if they have at their disposal the human skills necessary for adopting the innovations; and
- if they expect some economic benefit, net of the incurred costs, deriving from the adoption of the innovations (*cf.* Dosi, 1988).

Thus, we postulate that enterprises possessing certain characteristics at the time a new technique becomes available will introduce that technique fastest. Of course, this idea is not new but has been the basis of a number of studies by economists and other social scientists. One method employed in the literature, which ever since the pioneering analysis by Mansfield (1963) has become the dominant approach among economists, is to present and empirically test a number of separate *a priori* hypotheses regarding the economic characteristics which the authors expect early adopters of innovations to possess. This exercise has usually been performed on an *ad hoc* basis. The characteristics considered in the economics literature can be organized into four groups (Karlsson, 1988):

- The size of enterprises.
- The economic and financial characteristics of enterprises.
- The human capital characteristics of enterprises.
- The internal and external communication networks of enterprises.

A rather limited number of empirical studies analyze the characteristics of those enterprises which are relatively quick to introduce new techniques. It is no easy task to evaluate the empirical results not only because they are often contradictory or non-significant, but also because:

- The theoretical foundation is often weak or even non-existent.
- The data used often contain few observations.
- The samples are sometimes non-representative, and may, for example, include only large enterprises, and countries.
- Questions can be raised concerning the statistical treatment of the data in at least some of the studies.
- The variance explained in these studies is often quite small (*cf.* Karlsson, 1988).

Hopefully, the results presented here will provide another small piece of evidence that together with the results from earlier studies will give us a better understanding of the characteristics of those enterprises that adopt innovations early and also of how an efficient technology and innovation policy at the regional level might be formulated.

The chapter is organized as follows: In Section 13.2 a conceptual framework for analyzing process development by means of the adoption of innovations is presented. On the basis of this framework a number of hypotheses concerning the characteristics of those enterprises that tend to adopt innovations early is then derived. These hypotheses are presented in Section 13.3. In Section 13.4 the data is presented and the methods used in this study for characterizing early adopters of innovations are discussed. The empirical results are presented in Section 13.5 and the main empirical conclusions as well as the policy conclusions can be found in Section 13.6.

13.2 A Conceptual Framework for Analyzing Process Development

The empirical part of this chapter deals with the adoption of new techniques (innovations) in process development at the level of the individual establishment (plant). We define an establishment as a production system consisting of a number of cooperating but partially separable subsystems with great differences in durability. Examples of such subsystems are the employees with their technical, management, marketing, etc. skills; building capital; machine equipment and production techniques; a set of products; distribution systems and market organization; and administrative systems.

It is expected that every establishment goes through a life cycle that can be summarized in the following way (Johansson and Strömqvist, 1980): Under theoretically ideal market conditions a new establishment will be *born* with a set of products that is well adapted to demand conditions, and a production technique that gives the best possible conditions for competition during the lifetime of the investment.

However, market conditions change over time and the production technique available for new production systems becomes more efficient. Hence, a production system ages much more rapidly in the economic and technical senses than in a physical sense. The productive skills of labor and the capital equipment often have a durability that is longer than the economic lifetime of the production system. Aging in the economic sense is caused, among other things, by the fact that the *technique* that is determined by the existing capital equipment and the production skills of the employees successively becomes more and more inferior to that in new production systems.

The different subsystems of an establishment can also be thought of as production systems that develop according to a life cycle with an introductory phase, a longer period of increasing efficiency due to learning economies and an increasing volume of production, i.e., a growth phase, a maturity phase, and lastly, a phase of decline when productivity decreases in comparison with the productivity in new production systems. By means of investment, the starting point for the decline phase can be delayed. This means that an old production system could be renewed. As the different subsystems within an establishment age at different speeds, certain subsystems will have to be renewed faster than others. The result of such a gradual renewal process will be that every establishment, except the newest, will have a mixed vintage structure. One observation to note is that the life cycles of the different subsystems are interdependent, which means, on the one hand, that the situation in one subsystem may be an impediment to renewal in other subsystems and, on the other hand, renewal in one subsystem may necessitate or encourage renewal in other subsystems. Thus, the interdependence may delay the adoption of a given technique first but then, after adoption, increase the speed of adoption of the same or other techniques in other subsystems of the establishment.

The life cycle position of an establishment as a whole as well as of its different subsystems, together with its competitive situation and the competitive strategy chosen, determines the need for the establishment to engage in the renewal of its different subsystems, such as, for example, its machine equipment and production technique, and thus its need to adopt new techniques. Its capacity to meet this need is determined by:

- What it has learned in the past about the use of new techniques in the relevant subsystems.
- The skill profile of its employees.
- The level of affinity between techniques already in use and the technique that is to be adopted.
- The amount of resources available for investments, including R&D, etc.

In deciding whether to engage in development activities or not, the establishment will weigh its need against its capacity in the light of information about available techniques and their characteristics, and about the actions of competitors.

The interdependence between different subsystems within an establishment means that the exploitation of most new techniques requires changes in many subsystems of the establishment. The introduction of a new technique into the production system of an enterprise produces imbalances that must be redressed dynamically. Irrespective of the subsystem into which the new technique is introduced, it may nonetheless lead to a dynamic connective reorganization of several (all?) other subsystems. The micro units in our study are machines with their attached software which, by means of investments, are used to renew the technique used within the production processes in establishments, which are our units of observation.

By means of the framework presented here, we want to emphasize the active role played by the establishments themselves in the adoption of new techniques. They spend R&D resources on the specific adaptation of the new technique itself to the specific characteristics of the subsystem(s) in question or on the specific adaptation of the subsystem(s) to the specific characteristics of the new technique. In-house R&D seems in many cases to be a necessary ingredient in the process of adopting new techniques made available by suppliers.

However, an establishment does not act in a vacuum. Rather, it acts as part of a complicated network system. It is the network system that conveys information about those changes in the market situation that determine the need to engage in renewal activities, about available techniques and their characteristics, about the possibilities of recruiting skilled personnel, etc.

13.3 Formulation of Hypotheses

In this section, we develop the hypotheses that will be tested later in the chapter. The first hypothesis concerns the role of the knowledge level, i.e., human skills, for the adoption of applications of information technology (IT) in the production process, i.e., for process development. We expect the early utilization of new IT-applications, i.e., innovations, in an establishment to be critically dependent upon the knowledge level within the establishment.

H 1: The capability of an establishment to adopt new applications of information technology early for process development is positively dependent on the knowledge level of its employees.

It seems, however, possible to qualify this hypothesis. In Karlsson (1988) it is emphasized that it is the ability to model production activities and production systems theoretically that matters. This suggests that the level of theoretical education is the skill factor or central knowledge.

H 2: The capability of an establishment to adopt new applications of information technology early for process development is positively dependent on the number of university graduates and, in particular, the number of graduates from technical universities, it employs.

The employment of a large number of university graduates may, of course, merely reflect the total size of the establishment. We suggest, however, that the number of university graduates employed has a significant effect on the propensity to adopt innovations for process development and also when we take the size of the establishments into account.

H 3: Considering the size of establishments and the number of university graduates employed simultaneously it is still expected that the number of university graduates employed has a positive influence on the propensity to adopt new applications of information technology early for process development.

In Section 13.2 above, we assumed that the information and communication networks of the individual establishments have a central role in the innovation adoption process. Here we make a distinction between specialized information channels, and nonspecialized information channels. Specialized information channels imply channels to machine producers and to agents working as intermediaries between the machine producers and the potential adopters.

H 4: Regarding the role of specialized information channels in the adoption of innovations for process development, we expect adopters and, in particular early adopters to have more developed specialized information networks than non-adopters.

H 5: Well-developed specialized information channels are expected to have a positive influence on the propensity to adopt innovations early for process development for establishments of all sizes.

H 6: Considering the influence of specialized information channels and the knowledge level simultaneously we expect both to have a positive influence on the propensity to adopt innovations early for process development.

Nonspecialized information channels imply channels to major owner(s)/head office and to the market. In a spatial perspective, we expect the location of the major owner(s)/head office function to be a variable that is critical for an early adoption of innovations. The location of major owner(s) should indicate whether or not an establishment belongs to an enterprise (corporation) with a rich network of information channels while the location of head office is a factor that merely reflects intra-enterprise information networks.

H 7: The location of major owner(s)/head office outside the region favors process development by means of the adoption of innovations.

H 8: Considering the simultaneous effects of the location of major owner(s)/head office and the knowledge level, we expect both to exert a positive influence on the propensity to adopt innovations early for process development.

The market networks of an establishment are reflected in its trade patterns. The characteristic market extension measured by the share of total sales going to exports is an important indicator of the flow of information that reaches an establishment from different markets. Customer dependence, measured by the share of total sales going to the four largest customers, on the other hand, indicates the number of extensions in the market information network.

H 9: Market extension, i.e., a large export share, is expected to be positively related to an early adoption of innovations for process development.

H 10: Market extension is expected to have a positive influence on the propensity to adopt innovations early for process development in enterprises of all sizes.

H 11: Considering market extension and the knowledge level simultaneously, we expect both to have a positive influence on the propensity to adopt innovations early for process development.

H 12: High customer dependence, i.e., a large share of total sales going to the four largest customers is expected to have a negative influence on the propensity to adopt innovations early for process development.

13.4 Statistical Methods for Characterizing (Early) Adopters of Innovations

In an earlier study (Karlsson, 1988), the results from a number of empirical studies that had tried to simultaneously assess the characteristics of (early) adopters of innovations were surveyed. The use of regression analysis to characterize early adopters of new techniques dates back to Mansfield (1963). The Mansfield approach has been used in several more recent studies. In some of these studies the authors had to deal with a problem – the censored data problem – which Mansfield did not have to contend with, namely that the innovation(s) had not diffused to 100%, i.e., their material contained enterprises which were classified as potential adopters but which had not yet become adopters at the time of investigation (Nabseth, 1973; Håkansson, 1974; Smith, 1974). In this situation there are at least four options available (Karlsson, 1988):

(a) To ascribe a hypothetical adoption date to the non-adopters (the Nabseth approach) (Nabseth, 1973).
(b) To base the assessment on adopters only and thus leave out the information available about non-adopters.
(c) To use logit analysis (Oster and Quigley, 1977; Oster, 1982).
(d) To use tobit analysis (Oster, 1982).

Each method is afflicted with certain weaknesses. We have chosen here a modified Nabseth approach as our main approach. The *best* explanations found by the Nabseth approach are then tested on the group of adopters only [option (b) above].

Our modified Nabseth approach can be described as follows: We start by constructing an introduction index or speed indicator. To do this, we estimate for each individual type of application a function with the following logistic form:

$$\begin{aligned} \ln Z(t) &= a_0 + a_1 t + \epsilon; t = 1, 2, \ldots \ a_1 > 0 \\ Z(t) &= f(t)/[1 - f(t)] \end{aligned} \tag{13.1}$$

where $f(t)$ is the share of all production units which have introduced a given type of application in year t and ϵ is a random error term, which is assumed to be normally distributed. Now, let $\mathbf{T}$ be the year of observation and consider the variable Δf:

$$\Delta f = [1 - f(\mathbf{T})]/2 \tag{13.2}$$

where $f(\mathbf{T})$ is the share value obtained from the estimated equation in (13.1).

To all units which at year **T** have not adopted the application studied we assign an artificial introduction date t^* such that

$$f(t^*) = f(\mathbf{T}) + \Delta f \tag{13.3}$$

Using equations (13.1) and (13.3) we can, for the actual type of IT-application i and production unit k, attach an introduction year t_i^k. Let $k = 1$ be the first adopter of application i, and let $t_i^1 = 1$. Then we can define the following introduction index, y_i^k, for each type of application.

$$y_i^k = 1/t_i^k. \tag{13.4}$$

From equation (13.4) we can construct a compound or average introduction index, y^k, based upon the introduction index y_i^k, for the n different applications of a given category.

$$y^k = \sum_{i=1}^{n} y_i^k/n. \tag{13.5}$$

Consider now a set of characteristics of establishments, $x_1, x_2 \ldots$, and let y signify the introduction index of IT-applications in the production process. Then we can use OLS to estimate regression equations of the following form, where β_1 is an estimate of the elasticity, b_j can be transformed into an estimate of the elasticity and ϵ is a random error term, with an assumed normal distribution:

$$\ln y = a + \sum_i \beta_i \ln x_i + \sum_j b_j x_j + \epsilon. \tag{13.6}$$

The main reason for choosing this as our main approach was that we felt it important to be able to make a distinction between early and late adopters. We employ the logistic function as the simplest and most widely used of a number of possible alternatives. As Hernes (1976) and Sahal (1981) note, it is not possible to discriminate between alternative models on empirical grounds. Hence we have not found it worthwhile for our purposes to provide a theoretical basis for specifying a precise functional form.

We also use equation (13.6) for making OLS estimations on adopters only. In that case we use the above procedure for constructing our introduction index except we do not assign any artificial introduction year to non-adopters.

There are several examples in the literature of how contingency table analysis has been used to investigate the connection between adoption rates or speed of introduction and establishment characteristics (Rees *et al.*, 1983). We also use contingency analysis to investigate how the speed of introduction is associated with single characteristics as well as pairs of characteristics of individual establishments. Unfortunately our material is too small to enable us to include more than two characteristics at a time in the contingency analysis. Of course, the contingency analysis only serves as a simple hypothesis test of somewhat limited value. We have, however, found it valuable to include the contingency analysis in the empirical part of our study because it also serves to describe our data.

In the contingency analysis, an introduction index is obtained simply by dividing users of applications of information technology into two equally large groups: early users and late users. If we include the non-users, we can then form the following simple introduction index: Z^0 = non-users, Z^1 = late users, and Z^2 = early users. Forming analogous classes of the characteristics of each establishment, we construct contingency tables in which the Z-classification is matched with those other classifications. For each table we also calculate the pertinent χ^2-value. The contingency analysis has obvious relations to the logit model. In particular, the contingency investigations describe the statistical material without using any assumptions about functional forms (Goodman, 1971; Bishop and Fienberg, 1969).

13.5 Empirical Results

The purpose of this section is to try to determine the characteristics of those establishments that are early users of applications of information technology for process development. The statistical investigation in this section is a study of the adoption of applications of information technology (IT) in the engineering industry in two Swedish counties – Värmland and Älvsborg. The industries included in the study are engaged in the manufacture of metal products, machinery and equipment, electrical equipment and components, transport equipment, and instruments.

The study is based on a postal survey of all engineering establishments with more than four employees. Some information has been collected from approximately 95% of the total number of establishments and full information exists for about 75% of the total number of establishments. The coverage of the total number of employees in the pertinent industries is approximately the same. The results presented here are based on the statistical analyses of approximately 350 establishments.

This section is organized as follows: In Subsection 13.5.1 we illustrate the process of diffusion of IT-applications for process development. Subsection 13.5.2 is devoted to a contingency analysis of the connection between human skills and the adoption of IT-applications for process development. In Subsection 13.5.3 we investigate, by means of contingency analysis, the importance of different specialized information channels for the adoption of IT-applications for process development and in Subsection 13.5.4 we make a similar investigation as regards nonspecialized information channels. The simultaneous effect of human skills and specialized and nonspecialized information channels, respectively, on the speed of introduction of IT-applications for process development is examined in Subsection 13.5.5. Finally, we present the results of a simultaneous assessment of the influence of different variables on the speed of introduction by means of regression analysis.

13.5.1 The diffusion process

To illustrate the process of the diffusion of IT-applications for process development, we have estimated the logistic curve of equation (13.1). In formula (13.7) we present the estimated equation for process development. The t-values of the estimated parameters are given in brackets.

$$\begin{array}{rcccl} \ln Z(t) & = & \underset{(-40.9)}{-21.12} & + & \underset{(35.8)}{0.245t}, \; \mathrm{R}^2(\text{adj.}) = 98.5. \end{array} \qquad (13.7)$$

We can see that the logistic function approximates the diffusion process quite well (*Figure 13.1*). However, our main interest is not attached to the diffusion process as such but to the characteristics of those establishments that are early adopters. In the next section we investigate those characteristics.

13.5.2 Human skills and the adoption of IT-applications

In this section we will examine the role of the knowledge level, i.e., human skills, for the adoption of IT-applications in production processes. We measure the knowledge level by means of the basic educational variables we have

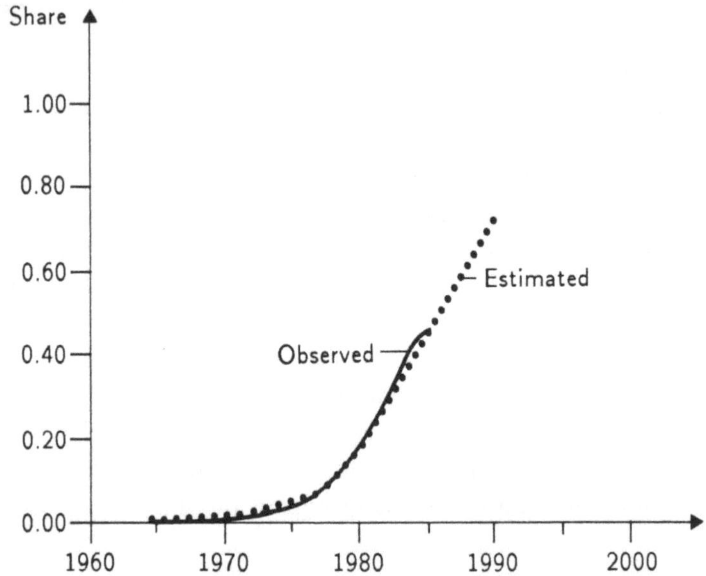

Figure 13.1. Introduction of IT-applications in the production process.

distinguished. In *Table 13.1*, we present a contingency table for the variables Z, the introduction index (defined in Section 13.4); and E, the number of university engineers employed, showing that for production processes the speed of introduction of IT-applications seems to be influenced by the size of E.

The influence factor in *Table 13.1* is the most significant of the basic education variables shown in *Table 13.2*, where we summarize the significance of the influence of three variables as regards the introduction of IT-applications in the production process.

Table 13.1. Employment of university engineers and IT-applications in the production process (%).

Speed of introduction	E=0	$1 \leq E \leq 4$	E>4
Non-adopters (Z^0)	60	31	11
Late adopters (Z^1)	26	31	26
Early adopters (Z^2)	14	38	63
Total	100	100	100

N = 319; $\chi^2 = 50$; df = 4; significance level: 0.001.

Table 13.2. Basic education variables and the introduction of IT-applications in the production process.

Speed of introduction against	χ^2	df	Significance level
University engineers (E)	50	4	0.001
College engineers (CE)	45	4	0.001
Other university graduates (U)	41	4	0.001

Table 13.3. Size, basic education, and the introduction of IT-applications in the production process (%).

	4<S<20		20≤S<50		S≥50	
Speed of introduction	E=0	E≥1	E=0	E≥1	E=0	E≥1
Non-adopters (Z^0)	68	33	52	53	33	16
Late adopters (Z^1)	22	48	34	20	29	26
Early adopters (Z^2)	10	19	14	27	38	58
Total	100	100	100	100	100	100

N = 319; χ^2 = 77; df = 10; significance level: 0.001.

We now turn to a discussion of the combined effect of size and knowledge level upon the speed of introduction. *Table 13.3* gives a contingency table in which the size S, measured in terms of the number of employees, and basic education effects are examined simultaneously for IT-applications in the production process. We see here that for establishments not employing any university engineers, the probability of adoption of IT-applications increases as the size of the establishment increases. For establishments employing university engineers, we find, somewhat unexpectedly, the lowest adoption rate among the medium-sized establishments. Thus, we have clear (but somewhat mixed) size effects. This is quite in line with the product life cycle theory, which postulates a relationship between process development and increasing scale of production. However, the dependence on the employment of university engineers for an *early* adoption of IT-applications in the production process increases in proportion to the size of the establishment.

13.5.3 Specialized information channels and the adoption of IT-applications

In this section we continue our study by investigating the role of specialized information channels in the adoption of IT-applications. In the survey a distinction was made between the following six types of specialized information channels:

C = consultants;
CC = courses and conferences;
F = fairs and exhibitions;
MP = producers and sellers of IT-equipment;
J = journals;
OI = other information channels.

In the case of IT-applications in production processes we found that consultants together with courses and conferences ($\chi^2 = 5.4$; df = 2, in both cases) are the only specialized information channels which seem to influence the introduction speed significantly. If we reduce the introduction speed variable to a simple yes/no variable, we obtain the results shown in *Table 13.4*. The table shows that while adopters have consultants, courses and conferences, and other information sources as important information channels, non-adopters seem to read journals, visit fairs and exhibitions, and have contacts with machine producers and sellers as frequently as adopters.

Table 13.4. Specialized information channels and the adoption of IT-applications in the production process.

Adoption (yes/no) against	χ^2	df	Significance level
Consultants (C)	5.4	1	0.025
Courses and conferences (CC)	4.5	1	0.05
Other information channels (O1)	3.3	1	0.10
Journals (J)	2.2	1	–
Fairs and exhibitions (F)	1.8	1	–
Machine producers and sellers (MP)	0.0	1	–

Note: C, CC, OI, F, and MP are coded yes and no.

We now focus on the simultaneous influence of size and specialized information channels. In *Table 13.5* we present a contingency table in which we examine the simultaneous effects of the strongest specialized information channel, consultants, and size on the introduction speed of IT-applications in the production process. We notice that consultants as an information channel seem to have a definite effect on the adoption propensity in all three size classes.

Table 13.5. Consultants as an information channel, establishment size, and the introduction of IT-applications in the production process (%).

	4<S<20		20≤S<50		S≥50	
Speed of introduction	C=No	C=Yes	C=No	C=Yes	C=No	C=Yes
Non-adopters (Z^0)	43	31	32	17	15	9
Adopters (Z^1 + Z^2)	57	69	68	83	85	91
Total	100	100	100	100	100	100

N = 220; χ^2 = 45; df = 5; significance level: 0.001.

13.5.4 Nonspecialized information channels and the adoption of IT-applications

In this section we investigate the role of nonspecialized information channels in the adoption of IT-applications in the production process. A general observation is that the strongest influence on the speed of introduction seems to come from market extension (M) and the location of the major owner (K).

We start by examining the influence of market extension on the speed of introduction of IT-applications in production processes. *Table 13.6* is a contingency table for M and Z showing that the speed of introduction of IT-applications in the production process seems to be strongly influenced by the market extension of the establishments. There is a clear positive effect on the speed of introduction in those cases where exports account for both a high- and a medium-sized share of total sales. This is in line with what we have reason to expect. In the product life cycle theory, process development is connected with extended markets. Customer dependence, on the other hand, does not seem to have any significant influence on the speed of introduction (χ^2 = 6; df = 4).

Table 13.6. Market extension and IT-applications in the production process (%).

Speed of introduction	M_1	M_2	M_3
Non-adopters (Z^0)	63	45	29
Late adopters (Z^1)	26	35	19
Early adopters (Z^2)	11	20	52
Total	100	100	100

N = 313; χ^2 = 52; df = 4; significance level: 0.001.

Table 13.7. Market extension, size, and IT-applications in the production process.

	4<S<20			20≤S<50			S≥50		
Speed of introduction	M_1	M_2	M_3	M_1	M_2	M_3	M_1	M_2	M_3
Non-adopters (Z^0)	65	67	55	63	48	44	46	19	10
Late adopters (Z^1)	24	28	25	27	35	28	39	44	12
Early adopters (Z^2)	11	5	20	10	17	28	15	37	78
Total	100	100	100	100	100	100	100	100	100

N = 313; χ^2 = 104; df = 16; significance level: 0.001.

Table 13.8. Location of major owner(s) and IT-applications in the production process (%).

Speed of introduction	K_1	K_2+K_3
Non-adopters (Z^0)	58	27
Late adopters (Z^1)	26	27
Early adopters (Z^2)	16	46
Total	100	100

N = 328; χ^2 = 38; df = 2; significance level: 0.001.

We now turn to a simultaneous examination of the effects of market extension and size on the introduction speed (see *Table 13.7*). We see that the influence of market extension is very clear in the case of the large establishments. At the same time it is clear that size alone has an important influence on the speed of introduction.

We continue by examining how the speed of introduction is influenced by the location of the major owner(s) of the establishment. The location of the major owner(s) indicates mainly whether or not an establishment belongs to an enterprise with a rich network of information channels. Being part of a larger enterprise means that an establishment can take advantage of information gathered or generated by other establishments within the enterprise. We also expect that establishments located in a peripheral region like the Värmland-Älvsborg region while having their major owner(s) located in another, more central region, will be able to adopt new techniques earlier than locally owned establishments, as a result of the process of relocation when products become standardized.

The contingency table presented in *Table 13.8* illustrates the effects of the location of the major owner(s) on the introduction of IT-applications in the production process. We see that the speed of introduction is positively stimulated when the major owner(s) is (are) located outside the region but within the country (K_2) or abroad (K_3). Establishments where the major owner(s) is located within the region (county) (K_1) lag behind in the adoption process. The location of the head office shows the same influence pattern as the location of the major owner(s) but the significance level is lower ($\chi^2 = 20$; df = 2).

With regard to the influence of the location of the major owner(s) on the speed of introduction, we see a clear influence in *Table 13.8*. This can be interpreted within the framework of the spatial product life cycle theory, according to which the relocation of production and the introduction of new process techniques can be seen as simultaneous processes. In this context, we assume that the existence of a major non-local owner indicates that the production is likely to have been spatially relocated at an earlier stage.

13.5.5 Human skills, information channels, and the adoption of IT-applications

In this section we examine the simultaneous effect of skills and specialized and nonspecialized information channels, respectively, on the speed of introduction of IT-applications. In the case of specialized information channels, the main purpose is to examine the relationship between flows of external information and the capability to generate internal information. In particular, we are interested in seeing whether the two kinds of information sources substitute or complement each other. In *Table 13.9* we present a contingency table showing how the speed of introduction of IT-applications in the production process is influenced by the simultaneous effects of the strongest basic education variable, the number of university engineers, and the strongest of the specialized information channels, consultants. The table shows that both variables seem to have a powerful influence on the introduction speed but that the influence of consultants is limited to those enterprises which employ no university engineers. Thus, we may assume that, to a certain extent, consultants function as a substitute for staff skills and internal knowledge in the enterprise.

According to the spatial version of the product life cycle, we expect the introduction of IT-applications for process development to be related to a relocation of production in later phases. At this stage, production mainly takes place in large-scale establishments which need extended markets to be able to sell their total output. Extended markets provide a lot of information about the IT-applications for process development used by other establishments.

Table 13.9. Skills, information, and IT-applications in the production process (%).

	E= 0		E≥ 1	
Speed of introduction	C=No	C=Yes	C=No	C=Yes
Non-adopters (Z^0)	41	16	15	18
Late adopters (Z^1)	37	60	30	29
Early adopters (Z^2)	22	24	55	53
Total	100	100	100	100

N = 220; $\chi^2 = 31$; df = 6; significance level: 0.001.

Table 13.10. Education, market extension, and IT-applications in the production process.

	E=0			E≥ 1		
Speed of introduction	M_1	M_2	M_3	M_1	M_2	M_3
Non-adopters (Z^0)	67	57	42	37	24	20
Late adopters (Z^1)	24	33	15	44	36	22
Early adopters (Z^2)	9	10	43	19	40	58
Total	100	100	100	100	100	100

N = 309; $\chi^2 = 104$; df = 10; significance level: 0.001.

However, we also expect the early adoption of IT-applications by an establishment to depend upon its knowledge stock. Thus, we expect both market extension and knowledge stock to have a positive influence on the speed of introduction of IT-applications in the production process.

In *Table 13.10* we examine the simultaneous effects of skill and market extension on the speed of introduction of IT-applications in the production process. The table indicates that both the employment of university engineers and market extension have clear effects on the speed of introduction. Thus, our expectations are confirmed.

For establishments located in a peripheral region like the Värmland-Älvsborg region we expect a major owner(s) located in another region to be a major source of information on new techniques for process development. However, at the same time, we expect that the adoption of new product techniques by an establishment, irrespective of where its major owner(s) is (are) located, to depend upon its knowledge stock. Thus, we expect the location of the major owner(s) outside the region and skill level to have a powerful influence on the speed of introduction of IT-applications for process development. *Table 13.11* is a contingency table which illustrates the simultaneous

Table 13.11. Education, location of major owner(s), and IT-applications in the production process (%).

	E=0		E≥ 1	
Speed of introduction	K_1	K_2+K_3	K_1	K_2+K_3
Non-adopters (Z^0)	65	36	31	20
Late adopters (Z^1)	23	38	42	18
Early adopters (Z^2)	12	26	27	62
Total	100	100	100	100

N = 319; χ^2 = 72; df = 16; significance level: 0.001.

influence of the number of university engineers and the location of a major owner(s) on the speed of introduction of IT-applications in the production process. We see how the education variable exerts strong influence on the speed of introduction, as does the owner variable.

13.5.6 Establishment characteristics and the adoption of IT-applications

Here we try to make a simultaneous assessment of the influence of different establishment characteristics on the speed of introduction of IT-applications in the production process. As far as possible, we will also try to interpret our results within the framework provided by the product life cycle theory.

The empirical results are presented in *Tables 13.12* and *13.13*. In both tables, we present different estimations of equation (13.6). Equation (1) in *Table 13.12* is presented here because this model had the *best* explanatory power measured by the R^2 (adj.)-value. We may note that early use of IT-applications in the production process is positively associated with a large number of employees and with the major owner(s) located within Sweden but outside the region. This fits very neatly with the product life cycle theory, which stresses that process development is associated with an increasing scale of production and a relocation of production to more peripheral locations. Actually, we cannot prove that there has been a relocation of production. It might be the case here that we have the effect of production units being bought by external owners but we still have a significant positive effect on the propensity to adopt IT-applications for process development within the production processes. We also see that early use is positively associated with the appreciation of fairs and exhibitions as an important information channel. Early involvement with IT-applications in the production process is

Table 13.12. Establishment characteristics and IT-applications in the production process.

	(1)		(2)	
Characteristic	Parameter	t-value	Parameter	t-value
Intercept	*−3.4330*	(−265.0)	*−3.2556*	(−89.5)
IT-investment share (HP)	*0.0043*	(12.0)	*0.0021*	(3.8)
Number of employees (S)	*0.0012*	(13.1)	*0.0010*	(8.4)
Number of employees with IT-training for administrative applications (AE)	*−0.0032*	(−4.2)	−0.0018	(−1.8)
Fairs and exhibitions are important information channels (F)[a]	*0.0924*	(4.7)	0.0561	(1.6)
Major owner(s) located in Sweden but outside the region (K_2)[a]	*0.0804*	(3.3)	0.0514	(1.3)
N	308		143	
R^2 (adj.)	74.8		68.2	

[a]Dummy variable.
Italic type indicates significance at the 5% level.

also positively associated with a high share of total investments in machines and equipment going to IT-equipment for the production process. There is also a negative association between our speed indicator and the number of employees who have received IT-training for administrative applications.

Equation (2) in *Table 13.12* shows the result when model (1) is applied to the set of adopters only. We can see that the parameters for total employment and the IT-investment share are still significant while the parameters for our three other variables have now become insignificant. In the models presented in *Table 13.12* there is no sign of the positive influence of any skill variable. This is not in line with our expectations so we now go on to test the influence of one skill variable – the number of university engineers employed. The results of this test are presented in *Table 13.13*. We still use equation (13.6) for our estimations.

Equation (1) in *Table 13.13* is a reestimation of equation (1) in *Table 13.12* with the number of employees replaced by the number of university engineers employed. We can see that the R^2 (adj.)-value only decreases slightly and that our skill variable is highly significant. In equation (2) in *Table 13.13* we present the result of estimating model (1) for adopters only. Our skill variable continues to be highly significant. In equation (3) we present the model which showed the highest explanatory power when estimated for adopters

Table 13.13. The introduction of IT-applications in the production process with special emphasis on the role of university engineers.

Characteristic	(1) Parameter	(1) t-value	(2) Parameter	(2) t-value	(3) Parameter	(3) t-value
Intercept	*−3.4046*	(−256.9)	*−3.1897*	(−98.4)	*−2.7402*	(−20.0)
IT-investment share (HP)	*0.0043*	(11.4)	*0.0016*	(3.2)	*0.0014*	(2.8)
Number of university engineers employed (E)	*0.0199*	(12.1)	*0.0196*	(10.7)	*0.0154*	(15.5)
Number of employees with IT-training for administrative applications (AE)	*−0.0015*	(−2.1)	−0.0014	(−1.8)	–	–
Fairs and exhibitions are important information channel (F)[a]	*0.0889*	(4.4)	0.0295	(0.9)	–	–
Major owner(s) located in Sweden but outside the region (K_2)[a]	*0.1548*	(6.5)	*0.1143*	(3.3)	–	–
Speed of introduction of IT-applications in administration	–	–	–	–	*0.1377*	(3.2)
Export share (M)	–	–	–	–	*0.0019*	(3.1)
Head office located outside Sweden (L_3)[1]	–	–	–	–	*−0.1468*	(−2.1)
N	305		142		137	
R^2 (adj.)	73.4		73.6		76.2	

[a]Dummy variable.

Italic type indicates significance at the 5% level.

only. We may note once again that the skill variable is highly significant. Given the results presented in *Table 13.13*, we feel that we have shown that our skill variable, the number of university engineers employed, is certainly not unimportant. Model (3) in *Table 13.13* indicates that the early use of IT-applications in the production process among adopters is positively associated with an early use of IT-applications in administration and a large export share, but negatively associated with a head office location outside the country. As before, we see that early users devote a large share of their total investments in machines and equipment to machines and equipment containing IT-applications.

13.6 Conclusions

In this chapter we have determined by means of analysis what establishment characteristics are associated with an early adoption of new applications of information technology (IT), i.e., innovations, in production processes, i.e., for process development, in two peripheral regions in Sweden. The analysis has been conducted by means of contingency and regression analysis. In particular, a number of hypotheses concerning the role of human skills and different types of information channels have been tested. By means of regression analysis we have shown that an early use of IT-applications for process development is positively associated, among other things, with the following:

- The establishment is large in terms of number of employees.
- Fairs and exhibitions are an important information channel.
- Having the major owner(s) located in Sweden but outside the region.

The number of university engineers employed also seemed to be an important explanatory variable, in particular, if we consider the characteristics of early adopters within the group of adopters only.

What relevance do these results have for a regional technology and innovation policy? Our research has, in this respect, been based upon the idea that the renewal of the technology used within the manufacturing industry at a general level can be related to three categories of capital formation:

- Investments in production units (including investments in new production units).
- Investments that lead to the transformation and/or expansion of the production milieu.

- Investments in transformation conditions such as R&D systems including educational systems for production, reception, and distribution of new knowledge; and marketing systems for the establishment of new buying and selling channels.

If we consider the decision systems for these three types of investments, it may be observed, that decisions can be taken within the single production unit but also in the surrounding system and at different regional levels. This means, for example, that some of these investment decisions are taken by governmental bodies at different levels. Our empirical results above indicated the importance of direct and indirect information channels as well as the knowledge level of the labor force for an early adoption of innovations – in our case applications of information technology to be used within the production processes in the engineering industry. The existence, capacity, and efficiency of information channels as well as the knowledge level of the labor force in a region are to a large extent a function of governmental decisions to invest in the production milieu and in transformation conditions, i.e., to invest in the infrastructure of the region. Hence, we suggest that our results indicate that a regional technology and innovation policy should have the character of infrastructure investments, such as airports, roads, universities, R&D-institutions, fair centers, etc. This may also mean that the traditional regional technology and innovation policy with information campaigns, isolated technology centers, casual advice campaigns, etc. is not the most efficient way to achieve a sustained and continuous improvement of the technology level in a region.

References

Bishop, Y.M.M. and Fienberg, S.E., 1969, Incomplete two-dimensional contingency tables, *Biometrics* **25**:119–128.

Dosi, G., 1988, Sources, procedures, and microeconomics effects of innovations, *Journal of Economic Literature* **26**:1120–1171.

Goodman, L.A., 1971, The analysis of multidimensional contingency tables, stepwise procedures and direct estimation methods for building models for multiple classification, *Technometrics* **13**:33–61.

Hernes, G., 1976, Diffusion and growth – the non-homogeneous case, *Scandinavian Journal of Economics* **78**:427–436.

Håkansson, S., 1974, Special presses in paper-making, in L. Nabseth and G.F. Ray, eds., *The Diffusion of New Industrial Processes: An International Study*, Cambridge University Press, Cambridge, UK.

Johansson, B. and Strömqvist, U., 1980, *Vinster och sysselsättning i svensk industri* (Profits and employment in Swedish industry), SIND 2, Liber förlag, Stockholm, Sweden.

Karlsson, C., 1988, *Innovation Adoption and the Product Life Cycle*, Umeå Economic Studies No. 185, Umeå University, Umeå, Sweden.

Mansfield, E., 1963, The speed of response of firms of new techniques, *The Quarterly Journal of Economics* **77**:290–311.

Nabseth, L., 1973, The diffusion of innovations in Swedish industry, in B.R. Williams, ed., *Science and Technology in Economic Growth*, Macmillan, London, UK.

Oster, S., 1982, The diffusion of innovation among steel firms: The basic oxygen furnace, *The Bell Journal of Economics* **13**:45-56.

Oster, S. and Quigley, J., 1977, Regulatory barriers to the diffusion of innovation: Some evidence from building codes, *The Bell Journal of Economics* **8**:361–377.

Rees, J., Briggs, R., and Oakey, R., 1983, *The Adoption of New Technology in the American Machinery Industry*, Occasional Paper, No. 71, Metropolitan Studies Program, Maxwell School of Citizenship and Public Affairs, Syracuse University, New York, NY, USA.

Sahal, D., 1981, *Patterns of Technological Innovation*, Addison-Wesley, Reading, MA, USA.

Smith, R.J., 1974, Shuttleless looms, in L. Nabseth and G.F. Ray, eds., *The Diffusion of New Industrial Processes: An International Study*, Cambridge University Press, Cambridge, UK.

Chapter 14

External Learning Opportunities and the Diffusion of Process Innovations to Small Firms: The Case of Programmable Automation

Maryellen R. Kelley and Harvey Brooks

14.1 Introduction

In this chapter, we are concerned with explaining which types of firms have failed to adopt well-known improvements in process technology. This problem has, of course, been the underlying concern of all studies of diffusion *"to rationalize why, if a new technology is superior, it is not taken up by all potential adopters"* (Stoneman, 1983). Drawing on various theoretical perspectives, we identify a number of different barriers to adoption. With data collected from a 1987 nationally representative sample of US establishments in 21 metal-working and machinery manufacturing industries, we then construct a multivariate logistic regression model to empirically test for the effects of these factors on the likelihood of adoption of a particular process innovation, namely programmable automation (PA) machine tools.

A widely accepted tenet of contemporary analyses of the diffusion of innovations is that certain types of organizations are better positioned than others to generate and to adopt innovations (David, 1969 and 1975; Mansfield, 1968; Mansfield *et al.*, 1977; Nabseth and Ray, 1974; Stoneman, 1980; Utterback, 1988). With respect to process innovations in particular, economic research on technology diffusion has demonstrated the importance of differences, or *heterogeneity* in what Dosi (1989) has termed the *incentive structures* of firms to explain why some firms are quick to adopt a process innovation while others fail to do so. For example, some firms are price leaders in labor markets, willing to pay a premium in order to attract the best quality labor; other firms are willing to accept somewhat lower quality labor in order to keep their wages at or below the average paid by their competitors. Unless the expected labor savings from a new technology are greater than the capital costs of purchasing the equipment, a firm is apt to delay making that investment (Metcalfe, 1990; Salter, 1960). Thus, at any one point in time, high-wage firms are apt to have a greater incentive than low-wage firms to adopt a labor-saving technology. Moreover, there may be some minimum threshold scale (i.e., volume of output), below which the labor savings are too small for it to be profitable for the small firm to invest (David, 1975). In addition, there may be scale requirements that make it technically infeasible for small firms to adopt it. For example, Mansfield (1968) found that for certain innovations, there is a minimum scale at which a technology can be profitably used in particular industries. Hence, where the scale of investment necessary for a new process technology is very large, it can only be undertaken by large firms; small firms will simply lack the financial resources or size of revenue stream to make such an investment. From this body of research, we learn that the failure of small firms to adopt an innovation may be attributable to the heterogeneity of firms with respect to relative factor prices (of labor and technology), profitability, and the lumpiness of capital investment.

A second stream of research on the economics of innovation emphasizes differences in firms' technological and organizational competencies, which develop or accumulate over time (Cohen and Levinthal, 1990; Dosi, 1988; Freeman, 1988; Nelson and Winter, 1977 and 1982; Rosenberg, 1972 and 1982). In this line of inquiry, the problem of imperfect information for learning about new technologies and the importance of accumulated knowledge and expertise are given prominence in explaining why some firms are more likely to adopt a new technology or to be sources of innovation themselves. Since information about the possible uses and relevance of a new technology to the firm is difficult to assess (Rosenberg, 1972), firms with more resources to devote to scanning the technological environment are likely to be better positioned than less well-endowed organizations to identify and exploit a new technology (Cohen and Levinthal, 1990). Moreover, firms differ in their experience with related technologies. These technological competencies can be expected to enhance a firm's capability to make use of other related innovations. With respect to process innovations, we would therefore expect to find adoption rates to be higher among firms that have demonstrably greater technological competencies and resources for scanning the external environment.

A third set of factors identified in some studies are special features of economic institutions and inter-firm relationships which explain why – in some regions, nations, or among certain groups of firms – the pace of diffusion was found to be more rapid and the rates of adoption much higher. For example, with respect to the adoption of hybrid corn among American farmers during the 1940s and 1950s, Griliches (1960) observes that the more rapid pace of diffusion in certain regions could be attributable to agricultural extension services in a number of different states which were a source of innovation for additional improvements. As Nelson and Winter (1977) point out, extension service agents have been an especially reliable source of information to farmers in these regions. As such, they may very well have contributed to the faster speed of adoption observed by Griliches. Similarly, Saxonhouse (1974) attributes the rapid rate of diffusion of new techniques among Japanese textile manufacturers both to the importance of business trade associations serving as a conduit for information and to the accepted practice of sharing technical know-how among these firms. The importance of information exchange or *know-how trading* for achieving improvements in utilizing a new technology among steel mini-mill producers has also been demonstrated by von Hippel (1988). In a world of imperfect information and considerable uncertainty about whether and how best to deploy a new technology, these extra-firm economic institutions and networks of relationships among firms can be expected to be particularly important for explaining differential rates of adoption.

In this chapter, we attempt a synthesis of these various theoretical perspectives, taking into account the heterogeneity of firms with respect to the cost incentives or profitability of adopting a particular process innovation, their organizational capacity for learning and technological competencies, and their external linkages to resources for learning about new technological developments. With detailed survey data on the technical, economic, and organizational characteristics of a large sample of US manufacturing establishments – all of which are potential adopters of the new technology – we are able to operationalize a model for predicting the likelihood of adoption of this new technology that simultaneously takes into account all three types of influences.

After accounting for the influence of differences in cost incentives and organizational capabilities, we find that a small firm's propensity to adopt a process innovation is particularly enhanced by the nature of its linkages to external resources for learning about technological developments. These results suggest that the well-known scale and size disadvantages of small firms for engaging in the risky learning-by-doing process necessary to the adoption of new productivity-enhancing technologies may, at least in part, be overcome when there are well-developed social networks for sharing expertise and acquiring new knowledge among economic actors and institutions. In regions or sectors where linkages to such external learning opportunities are particularly well-developed, we would expect to find a more rapid rate of diffusion of productivity-enhancing process innovations to small firms.

14.2 The Implications for Small and Large Firms of Radical Shifts in the Technological Trajectory

Productivity increases arise both from radical shifts to a new, more efficient technology, and from continued, incremental improvement in the way in which an existing technology is utilized (Dewar and Dutton, 1986; Ettlie *et al.*, 1984). Indeed, for some period of time when both emerging and mature process technologies coexist, additional improvements in the mature techniques also occur and frequently accelerate (Harley, 1973). Whether emerging or mature, every technology has its own associated trajectory (Dosi, 1982). Incremental learning about how best to use a particular configuration of equipment is the basis for productivity improvements which proceed under the same technological regime. Moreover, each firm has its own associated learning curve. The knowledge derived from marginal adaptations of the organization and the technology accumulate over time and becomes part of the informal or *tacit* know-how – the craft *art* recognized by many observers as a key ingredient that distinguishes high from low productivity operations employing the same technology (Bohn and Jaikumar, 1986; Kusterer, 1978; Pavitt and Patel, 1988; Skinner, 1986).

New process technologies always involve a change in the ways in which products are made – a change in the allocation of tasks, a change in machinery, a change in work methods which may imply retraining, or a change in organization. For the firm, there is always some uncertainty about how much new knowledge will be necessary and how drastic a change the new configuration of equipment and people will entail (Bohn, 1987; Rogers, 1983). If these changes require substantially new skills and expertise, then a displacement of the learning curve results, i.e., a discontinuity arises between the organizational learning accumulated under the previous production regime and that which is needed for the new technology. This could even result in a short-term decline in productivity until a certain portion of the new learning curve has been traversed as the organization develops the additional expertise needed to more fully exploit the potential advantages inherent in the new technological trajectory.

Certain changes in technology involve such a radical shift *away* from existing techniques that traditional competencies and skills are made obsolete (Abernathy and Clark, 1985; Tushman and Anderson, 1986). In the case of information technology's application to manufacturing in the form of programmable automated (PA) machine tools, the shift to this new technological regime requires the integration of new science-based knowledge of electronics, computers, and software engineering with the accumulated tacit knowledge of metal-cutting practices acquired through years of practical experience; PA also makes some traditional skills obsolete (Kelley, 1989a, 1989b, 1989c, and 1990a). In order to make this shift to the new trajectory successfully, firms have to buy, borrow, or somehow internally develop that expertise and integrate it with the relevant traditional practices to match the requirements of the emerging system. Because of their size, and hence very limited base of resources available to absorb mistakes, small firms are likely to face more severe consequences from underestimating the displacement of their learning curves.

Small firms have little organizational slack and, over the long term, are more vulnerable to business failure than large firms (Hage, 1980; Hage *et al.*, 1989; Scott, 1987). In small manufacturing companies, engineering and management resources are limited to a few individuals per plant. For small firms to engage in an experimental learning-by-doing process requires the diversion of existing resources from production activities. That may have a high opportunity cost. The time that one engineer spends assisting with the break-in of any new piece of equipment is time spent away from solving other design or production problems. The diversion of this scarce resource can cause delays in production which increase costs and may lead to delays in shipment and possibly lost orders from customers, contributing to lower profits and possibly lower sales.

By contrast, large, relatively resource-rich organizations can afford to embark on a number of experiments with process innovations, only some of which may turn out to be successful, without risk to the firm's survival and profitability (March, 1981). Related to size is the tendency of large firms to have developed specialized capabilities in production engineering and management. By devoting some specialized resources to improving production techniques, large firms have an experience advantage in the kind of "learning by doing" that Arrow (1962) identified as a key generator of continued productivity improvements under any technological regime. Moreover, because of their size advantage, large firms are less vulnerable to any severe consequences (such as their own demise) from making a strategic error (such as underestimating start-up time and cost or training time) in deploying any single piece of new equipment that does not achieve its expected savings.

14.3 Economic Limitations on the Technology Choices of Small Manufacturing Firms

As part of a strategy to attain or maintain leadership in one market, the large firm may seek to develop proprietary technology which provides a unique cost or quality advantage over its competitors. Moreover, when a firm operates in a number of markets for which there is a shared technical basis and sufficient scale of operations, there is the possibility of achieving a greater synergy from exploiting advances in process technology. Hence, being relatively quick to use new production techniques can provide such a firm with multiple cost or quality advantages over its competitors in several markets at once.

By contrast, because of the small scale at which they tend to operate, small firms have less opportunity to achieve and exploit such technical synergies. Moreover, even when the incentive to cut costs is great – as is likely to be the case with the small manufacturing firm that has expertise in a mature technology operating in industries where the prospects for sales growth are poor (i.e., where sales trends are flat or only growing slowly), and profit margins are slim – the small business owner/manager is of necessity focused on the short-term. To him/her, the adoption of a new process technology is viewed as being outside the realm of rational choices.[1] Instead, as March and Simon (1958) have suggested, management is likely to focus its attention on familiar problems, attempting to adapt by gaining greater control over variable production costs in order to make more efficient use of existing equipment and labor within the declining technological paradigm.

In the short run, such small manufacturing firms' aspirations are modest, being concerned simply with survival. Management may forestall wage increases or actually reduce wages. Equipment may be operated more continuously, sacrificing downtime for preventative maintenance and further depleting the useful life of the capital stock. The manager/owner may be willing to accept lower revenues and even profits in order to be more certain of staying in business.[2]

We might further plausibly assume that there are barriers to exit. Small manufacturing firms whose prior success depended on their capacity to exploit their accumulated experience within an increasing obsolete technological paradigm may lack the human or financial resources to absorb the one-time effort and cost of entering a new line of business. At the same time poor growth prospects in their present market niche may not attract entry of technologically more advanced firms. Thus, technologically backward firms may survive in narrow market niches for protracted periods of time by lowering wages and deferring investment, thus retarding the diffusion of new production technology and productivity growth.[3]

14.4 The Importance of External Learning Opportunities to the Diffusion of New Process Technologies

Whether or not a firm will adopt a new technology is generally believed to be determined by some combination of the relative importance of economic incentives for doing so (e.g., to lower costs) and its internal capability to undertake an experimental learning-by-doing process. However, with the exception of Nelson (1990) and von Hippel (1988), little attention has been given to examining how various kinds of linkages with other economic organizations and institutions matter to the adoption and implementation of innovations. The conventional wisdom evident in economic models of diffusion is that late adopters learn about the experience of early adopters through *osmosis*, that is, through informal contact and exchange of know-how among managers and engineers employed in different firms. The importance of the social context, or the set of linkages the firm has (or has somehow developed) to external learning opportunities, has hardly been considered in these models.

The proposition that the economic actions undertaken by management of a particular firm need to be understood as being affected by its network of relations with other firms and economic institutions has long been recognized by sociologists, political scientists, anthropologists, and historians (*cf.* Granovetter, 1984). Membership in trade associations, relationships with equipment vendors and customers are part of a social nexus in which the economic decisions of individual firms are embedded. Moreover, even among competitors, inter-firm relationships may take on a special character that is of particular importance to the success of a region or to the diffusion of innovations. For example, a number of studies on the industrial districts in Northern Italy attribute the success of these regional agglomerations in large measure to long-established relationships of trust and cooperation among technically inter-dependent small and medium sized firms whose economic ties are sometimes rivalrous and at other times collaborative – as suppliers to or customers of one another (Becattini, 1987 and 1989; Bellandi, 1989; Brusco, 1982 and 1986; Lorenz, 1989; Piore and Sabel, 1984; Sabel, 1989).

With respect to the learning opportunities that such external resources present, some firms are better connected than others, belonging to active, service-oriented trade associations, having collaborative relationships with their customers, and being part of an industrial community in which know-how trading with other firms is an accepted practice. We would expect these opportunities to be unevenly distributed among firms of different sizes, industries, and locales. Some firms are poorly linked to external resources as a matter of management policy. More commonly, we believe, the presence or absence of such linkages reflects historical differences in the evolution of economic institutions in particular locales (as seems to be the case of the Italian industrial districts) or particular sectors, and in the relative importance of leading firms in shaping orderly relationships with large networks of supplier-firms.[4]

If learning is the *product of experience* as Arrow (1962) has argued, then for firms to learn about the capabilities of *new* technologies, they must have access to opportunities for gaining trustworthy information about others' experience with them. For this to happen, there must be trustworthy institutions or forums which facilitate the exchange of useful information, help filter out erroneous or irrelevant news, and promote the accumulation of tacit know-how within the firm. External resources on which a firm may depend to learn about new advances and the experience of others with technology include: informal contacts with production managers or engineers in other firms; direct contacts with sales representatives of equipment vendors and distributors; participation in trade and professional associations; sharing information with the firm's customers; and reading trade journals, marketing newsletters, and brochures for general knowledge about the potential of new technologies. Firms may not be persuaded by such published information, however, because they have no way of assessing its trustworthiness, but gatherings at professional associations or industry trade shows may enable users and potential users of a new technology to meet and examine the latest equipment offered by vendors and exchange practical tips with other users about a new technology's limitations as well as its capabilities. Such linkages may be particularly important for explaining why some small firms adopt a process innovation while others with apparently similar characteristics do not.

Through their service activities, capital equipment manufacturers and distributors are also an agency through which best-practice techniques for utilizing a new technology may be *taught* (Ettlie and Rubenstein, 1980; Leonard-Barton, 1988).[5] These equipment manufacturers have been known to sometimes customize the design of new systems for lead users, adapting the innovation to the customer's specific production requirements and providing intensive follow-up support services during the initial implementation phase (Collis, 1988; von Hippel, 1988). They do so in anticipation of winning a large, loyal customer or as part of an experimental developmental effort which will result in improvements in the design of future generations of the technology. When this user dedicates some of its own organizational resources toward that collaborative effort, then it is also likely to engender organizational expertise within the user-firm as a result of close interactions of key personnel involved in such working relationships. These types of contacts are known to occur particularly in the early phases of the development of a new technology, are sometimes reserved for customers that purchase expensive systems, or are made available to large users from which the vendor expects a hefty order.[6]

Another important learning opportunity may arise from a firm's relationships to the businesses that purchase its products. Kelley and Harrison's (1990) research on subcontracting relationships suggest that many firms choose suppliers because of their specialized capabilities. Such business customers with which a small firm has some special relationship could be an important source both for learning about new technological developments and about how to use an innovation.

Previous studies have documented the ebb and flow of relations of trust and collaboration between firms and their business customers (Cusamano, 1985; Dore, 1986; Kenney and Florida, 1989; Minato, 1986; Sato, 1983; Trevor and Christie, 1988). Close relations between a firm and a few customers can be both beneficial and inhibiting to the adoption of new production technology. On the one hand, a close collaborative relationship to one or a few customers may open up the possibility of gaining favored status, and the benefits of technical and financial assistance that flow from that relationship. On the other hand, too close a dependence on a few customers may make small firms more vulnerable to price-cutting pressures and to fluctuations in customers' demands. Under pressure to cut costs, or faced with greater volatility in orders for its products, firms caught in such close relationships could have such small profit margins as to lack the resources and incentive to invest in new technology.

14.5 Programmable Automation: A Comparison of Adopters and Non-adopters of New Process Technology

Programmable automation (PA) in the form of numerically controlled (NC) and computerized numerically controlled (CNC) machine tools and flexible manufacturing systems (FMS), interconnecting such tools by automatic transfer, has been hailed as signalling a fundamental techno-economic paradigm shift which promises to greatly reduce the economies of scale that have driven the design and organization of manufacturing since the beginning of the industrial revolution (Freeman and Perez, 1986; Hirschhorn, 1984; Kaplinsky, 1984; Perez, 1986; Piore and Sabel, 1984). Since instructions controlling the operation of programmable machines can be incorporated into easily altered software rather than unalterable hardware, one piece of manufacturing hardware is adaptable to many different products which can be made in both small and large volume in a wide variety of industries that require the shaping and cutting of metal parts.

To date, PA has been applied mainly to the precision metal-cutting operations of turning, milling, grinding, and boring – operations important in the manufacture of a diverse range of products, from aircraft engines and industrial machinery to coffee grinders and lawn mowers. Machining is a "batch" production process, in which products are made in small to medium-size lots – too small to benefit from the earlier form of fixed-cycle *hard* automation. By 1982, the combined output of NC/CNC machines from six of the major machine-tool producing countries (USA, Japan, Federal Republic of Germany (FRG), France, Italy, and the UK) comprised two-thirds of the total value of production of all metal-cutting machines (Edquist and Jacobsson, 1988, p. 26).

This technology is not unique to particular product lines and appears to be equally technically and economically feasible for large and small firms. Single machines can be installed one at a time and used alongside conventional, non-programmable machines. Since the mid-1970s, improvements in the technology – particularly the incorporation of microprocessors – have made it easier to use, while machine productivity has increased at the same time as purchase costs have come down (Edquist and Jacobsson, 1988).

In what follows, we draw on a national survey of US manufacturing establishments completed in 1987 to examine adoption rates of PA in a range of establishment and firm sizes and to evaluate the significance of three sets of factors for distinguishing how adopters differ from non-adopters of the technology: cost and profitability incentives, organizational resources and technical capabilities, and linkages to external resources and sources for learning about technological developments.

14.5.1 Data description and methods

The data we employ come from a national sample of establishments belonging to twenty-one manufacturing industries, at the 3-digit level of the US Standard Industrial Classification scheme. Following the convention of the US Census of Manufactures, establishments in the sampling frame were grouped into the following size categories: fewer than 20, 20 to 49, 50 to 99, 100 to 249, and 250 or more, employees. In order to ensure a sufficient number of cases within each size stratum, establishments were disproportionately randomly sampled by strata in order to yield a data set with an equal number of establishments from each stratum. Since the distribution of establishments by employment size is highly skewed, with fewer than 10 percent of all plants employing 100 or more workers in the industries studied, this procedure guarantees a sufficient number of large size plants to allow for variation among them in the use of technology, type of product market.

All population estimates are weighted averages, which were constructed by weighting each observation by the inverse of the probability of selection from its sampling stratum. The production managers in these plants were surveyed between October of 1986 and March of 1987 (Kelley and Brooks, 1988). All told, 1,015 plant managers were successfully interviewed by mail, yielding a 50 percent response rate. Half of the non-respondents were then contacted by telephone and asked questions which, apart from their substantive value, confirmed the absence of response bias in the mail survey.

The twenty-one industries were chosen because they account for the great majority of machining activity in the US economy.[7] Twenty-five percent of all US manufacturing workers in 1986 were employed in these industries. The data base includes information on the size of the parent company (as measured by corporate-wide employment in the USA) and considerable detail on organizational, technical, and economic characteristics of each plant.[8] All sample establishments use machine tools for some aspect of production operations in their plants. Hence, they are all *potential* adopters of PA technology.

At the time of the survey, fifty-seven percent of the sample plants had not yet installed even one programmable machine. As of 1987, most firms that had failed to adopt programmable automation seem unlikely candidates for doing so at some time in the near future. Despite improvements in the technology that have made it easier to use and less costly to install, two-thirds of non-adopters perceive the payback period associated with the introduction of PA as being too long to justify any investment.

These firms' unwillingness to invest in programmable automation seems to be related to a general reluctance (and possibly lack of financial resources) for making investments to improve their capital stock. The average investment in new equipment of any kind for firms that had not purchased any PA at the time of our survey was less than one-third the amount invested per employee by PA users in the same year.[9] Over the previous five year period (from 1982-1986), during which time more than half of all programmable machine tool installations presently in use in the United States were purchased, less than one-third of the enterprises that made no such purchases ever even considered that alternative. Moreover, in 1987, only 18 percent of those that had not invested in PA said that they had any plans for purchasing this equipment in 1988 or 1989.

In the previous section of the chapter we described how three different sets of factors could be expected to affect the likelihood of a firm adopting a new process technology such as PA: cost and profitability incentives, organizational resources and technical competencies, and external linkages. The majority of PA users (71 percent) have some programmable machines that were first installed more than five years ago, with 10 percent of PA-users having at least some programmable machines still in use that were purchased more than 10 years ago. However, eighty-five percent of PA-using plants have at least one machine from the latest micro-processor generation of the technology – with computerized numerical controls – which was first introduced only in the late 1970s. We cannot know from these data the *changes* in firm and establishment characteristics that may have accompanied or followed the adoption of programmable machines.[10] Our analysis is thus limited to a comparison of the ways in which PA adopters differ from non-adopters at what must be understood to be an intermediate stage in the diffusion of this technology.

We estimated a binomial logistic regression model, with the dependent variable, PA, defined equal to 1 if the production manager at the establishment reported there were any programmable machines in use, and equal to 0 if no programmable tools were present. Technical definitions for all independent variables can be found in Appendix *Table 14.A1*. Complete data on all variables needed to estimate the model were available for 75 percent of the cases in the sample.

14.5.2 Factors distinguishing PA adopters from non-adopters

Cost and profitability

Five variables representing cost and profitability factors were hypothesized to be important inducements to the firm to adopt programmable automation. Relative labor costs, the degree to which production operations at a plant were dependent on machining skills, the presence of a union, the scale of machining operations at a plant, and product markets with a high requirement for "flexibility" are all expected to be important economic factors favoring the adoption of PA.

In previous research on programmable automation, managers have reported that reductions in direct labor costs or increased productivity are the major expected gains from adopting this process innovation (Ayres and Miller, 1982; Hicks, 1983; Parsons *et al.*, 1984; Rosenthal, 1984). As a form of automation that is expected to lead to productivity gains (i.e., increases in output per person hour), we would expect the cost-cutting impetus for adopting PA to be the greatest in establishments where machining labor costs are relatively high. Thus, PA is more likely to have been adopted in plants with relatively high wages for machining occupations.

Another economic factor expected to favor PA adoption is the relative importance of machining skills to the overall production activities of the establishment. The greater the share of all production workers in occupations requiring these skills, the more likely it is that management will seek ways of reducing costs and improving productivity within that operation. Where machining skills are of minor importance to the overall production activities, management is less likely to make an effort to automate that process with computer-controlled machinery.

There is some disagreement in the literature as to how we might expect unionization to affect investment in a labor-saving process innovation such as programmable automation. Freeman and Medoff (1984) argue that when a plant is unionized, management is more likely to pay attention to weeding out inefficient practices and to streamline production, suggesting that PA might be adopted more rapidly in unionized establishments. Similarly, Clark's comparison (1980) of union and non-union firms in the cement industry suggests that management is more likely to introduce labor-saving technology in order to improve productivity when collective-bargaining governs the firm's relationship to its employees. However, research by Kochan (1985) and Schmenner (1982) suggests that when a unionized plant is part of a multi-plant enterprise, *corporate* management's investment decisions are apt to be informed by other industrial relations policy considerations. In his analysis of Conference Board data on corporate patterns of investment in plant and equipment, Kochan finds that unionized establishments received much less investment from the parent company than their non-union sister plants (controlling for age of the plant). Schmenner's research on the Fortune 500 shows a similar pattern of withholding investment from unionized plants. If corporate (multi-plant) strategic considerations are found to dominate the union effect, we would expect to find the adoption of PA technology to be less likely to occur when a plant is unionized, or for there to be no significant impact from unionization at all.

The concept and measurement of *scale* is somewhat ambiguous in studies of innovation.[11] In this analysis, we include a variable that controls for the effect of scale in the sense of *size* of machining operations at a plant. With respect to machine tools, we would argue that the smaller the number of tools in use at a plant (and thus the smaller the scale of machining operations), the more risky it would be for the firm to adopt even one programmable machine. The potentially disruptive consequences from introducing the new technology are greater, the fewer the number of machines in use. Moreover, for firms with a plant that has few tools, the cost of replacing any one of them represents a much larger share of its total capital investment in machine tools than would be the case for firms having plants with larger stocks of machines. Similar to David (1975), we would argue that the cost of adopting a single programmable machine may be too high for firms operating at too small a scale of machining. Thus, for smaller scale machining operations, management may find the adoption of PA to be a much more risky and *lumpy* investment and therefore be unwilling to adopt it, even though it may be profitable to do so. For that reason, we would expect the chances that management will have adopted PA to be less, the smaller the scale of machining operations at a plant.

A number of students of industrial change (Carlsson, 1989; Piore and Sabel, 1984; Kern and Schumann, 1984 and 1987) have argued that manufacturing firms face increasingly volatile and uncertain product markets for their goods. In discussions of the advantages of programmable technologies for small firms, PA has been touted as being especially well-suited to meet the high *technical flexibility* requirements of firms operating in such environments, specializing in manufacturing a diverse array of parts or products in very small batches (Dosi, 1988; Kern and Schumann, 1984; Piore and Sabel, 1984). Because programmable machines can be re-instructed for each change in product, a firm that operates in such product markets will have a greater incentive to adopt the technology since it lowers the costs of switching from one product to another. Hence, establishments with high flexibility requirements should find PA a more attractive investment than plants without such high switching costs.[12]

Internal resources and technical competencies

Firm size is a proxy measure for the extent to which an organization may be said to be resource-rich. Large firms have multiple production sites and a larger base of experience with various technologies that can be brought to bear in adopting any particular process innovation. The larger the firm, the more likely it is to employ professional staff at the corporate level with responsibility for providing technical expertise to production managers at any one of its plants. At any one point in time, larger firms can marshal greater resources more rapidly than smaller firms to deal with unexpected problems in implementing a new technology. Because of this superior adaptive capacity, we would expect that the larger the firm the more capable it is (in terms of resource capacity) to make the necessary adjustments in its operations to accommodate to a process innovation such as PA, and the more likely it will be to have adopted PA in plants for which the technology is suitable, i.e., those using machine tools in production operations.

Information technology has many applications. In manufacturing, computers are used for monitoring and planning functions, such as process planning and production scheduling, quality control and materials flow monitoring, and inventory control. That a plant has adopted any of these information technology applications would indicate greater formalization of management systems of information and control. Such a change may involve only a shift from written record-keeping and inventory procedures to computerization, or it may involve a more radical shift from an informal organic organizational structure to one with a more formal structure and system of control. In either case, the use of information technology in these functions suggests a greater technological sophistication (and by implication, an enhanced organizational capacity) to exploit PA. Although there may be no technical interdependence between the use of IT in these applications and PA (as there would be with computer-aided design), it is a complementary technological competency that should facilitate the adoption of programmable machines. We would therefore expect an establishment with such advanced information technology capabilities in monitoring and planning functions to be more likely to have adopted PA.

Linkages to external resources

As mentioned earlier, there are multiple external resources by which a firm can acquire knowledge about a new technology's capabilities. These can be distinguished by the type of linkage (whether it has a social or interpersonal dimension or not) and by the source of information.

Stories in trade journals and mailings from equipment vendors and their distributors are external sources for learning about technological developments in the form of written media. Thirty-eight percent of all production managers surveyed considered linkage to this channel of information flow to be very important for learning about new technology. Nevertheless, by itself, we do not expect this kind of linkage to external resources to be a sufficiently reliable means for learning about technology. We would therefore not expect to find any difference between PA users and non-adopters with respect to their reliance on such written sources of communication.

Know-how trading in the form of informal exchanges of information through conversations with production managers and engineers outside of the establishment (i.e., in other companies or in other plants of the same company) was the most common external resource cited by respondents as a very important way to learn about new technology. This type of exchange is of course recognizable as the casual individualized form of learning by osmosis which Stoneman (1980) assumes to be the major channel through which new technological expertise diffuses. In our formulation of the set of inter-firm linkages that distinguish PA adopters from non-adopters, we have the opportunity to test how important this kind of informal exchange is relative to other, more structured, collective forms of information exchange. Following Stoneman and von Hippel, we do expect to find a positive effect of know-how trading on the likelihood of adoption of PA, *ceteris paribus.*

Trade associations and professional technical societies provide an avenue for managers and engineers from member organizations to meet and discuss problems of common concern. In contrast to individualized know-how trading, demonstrations of new equipment at meetings of such associations is a highly structured, collective way of learning about technological problems and capabilities. At such events, groups of managers and engineers with common problems and issues have an opportunity to exchange information with each other and to compare the features of equipment offered by different vendors. We would expect that being connected to such organizations and participating in such events affords members a more intensive and comparative learning experience than may occur through individualized know-how trading. Managers and engineers of firms who participate in such activities are likely to be better informed and to have a broader set of knowledgeable contacts to whom they can turn to discuss technological issues. For these reasons, we would expect firms with such linkages to have a higher propensity to adopt advanced manufacturing technologies such as PA.

Direct contact with sales representatives from equipment vendors or their distributors (independent of contact through trade shows) is another structured way of learning about new technology. Unlike the case of consumer products, where sales persons are not expected to know much about the products they sell, sales representatives for industrial equipment products such as PA are expected to have some expertise with the technology, if only to be capable of explaining how it can be used and what the expected benefits are from using it. Hence, they may be an additional source of expertise that managers can draw upon in deciding to adopt and use new technology.

Another way in which expertise is transferred from one firm to another is through a firm's contacts with its customers. Eighty-eight percent of all US metal-working establishments make products using machine tools for sale to other manufacturing firms. More than three-fourths of these say that they have business customers for whom they make machined parts or products on special order and from whom they receive technical information and engineering assistance in making these special parts or products. Such a transfer of technical information suggests a degree of dependency between the two firms that could provide the occasion for other exchanges of technical expertise that may be particularly important in augmenting the capabilities of small firms to adopt a new technology. More generally, we hypothesize that firms with such close connections to their business customers are also more likely to be able to draw on these customers for assistance in implementing a new technology. For these reasons, firms that have such an information sharing relationship with their customers will (we expect) be more likely to adopt PA.

14.6 Findings

The results of our estimating procedure are shown in *Table 14.1.* Ten of the thirteen variables in the model are significant in predicting which establishments are likely to have adopted PA by 1987. As expected, cost and profitability incentives are important. In addition, plants that are part of firms with greater technical and organizational resources are much more likely to be PA users. Finally, establishments with certain kinds of linkages to external learning opportunities have an enhanced chance of adopting PA technology that can permit the small firm to overcome the liabilities of small scale and its lack of adequate internal resources.

Cost incentives

Both the cost of labor and the degree to which the overall production process at a plant is dependent on machining skills are significant predictors of PA adoption. Independent of wages, we do not find that unionization has significantly affected management's deployment of programmable machines.

Figure 14.1 shows a simulation of the predicted probabilities of adopting PA for the typical US machining establishment for different wage rates, with all other variables in the model held constant at their sample means.[13] At more than twice the average wage for machining occupations (about

Table 14.1. Logistic regression of the likelihood of adopting programmable automation.

Variable name	Mean	Standard deviation	Predicted sign	Regression coefficient	Standard error
Programmable Automation	0.46	0.50			
Incentives: Cost & Profitability					
Hourly wage[a]	9.83	3.46	+	1.078[b]	0.310
Dependency on machining skills	76.90	32.15	+	0.027[b]	0.004
Union	0.125	0.33	?	–0.223	0.321
Machining scale[a]	27.00	83.39	+	0.239[c]	0.105
High flexibility requirements	0.34	0.47	+	0.185	0.197
Internal Resources/Competencies					
Firm size[a]	2,590.00	3,516.00	+	0.568[b]	0.088
Complementary IT application	0.46	0.50	+	0.373[d]	0.212
External Resources/Information Sources					
Brochures, ads and newsletters	0.38	0.49	+	–0.337[d]	0.195
Equipment vendors' sales reps.	0.34	0.47	+	0.547[b]	0.186
Informal know-how exchange	0.53	0.50	+	0.176	0.186
Trade & prof. association meetings	0.34	0.47	+	0.490[b]	0.193
Customers	0.88	0.33	?	–0.410	0.356
Customer-provided technical assistance	0.69	0.46	+	0.563[a]	0.248
Intercept				–7.598	0.842
–2 Log likelihood				813.97	
χ^2 (df=13)				236.76	
C				0.808	
N	761.00				

[a]Natural logarithm of variable entered in regression model. Untransformed variable's mean and standard deviation shown in this table.
[b] Probability ≤ 0.01. [c] Probability ≤ 0.05. [d] Probability ≤ 0.10.

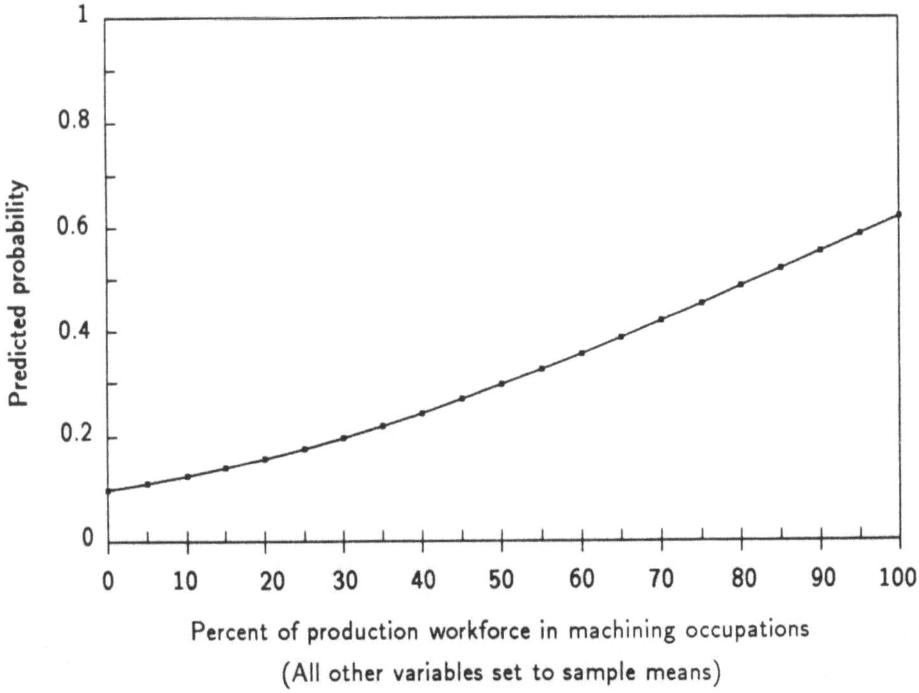

Figure 14.1. Probability of PA adoption by average hourly wage for machining occupations at the plant.

US$20.00 per hour), the probability that there will be at least some PA tools in a plant is quite high ($p > 0.66$). At less than one-half the average hourly wage (about US$5.00 an hour), fewer than 3 out of 10 such low-wage employers will have introduced any PA.

The degree of machining skill dependency is a technical attribute of production that is expected to vary with the kinds and mix of products being manufactured at a plant and the extent to which machining operations needed for these products are performed in their entirety within the plant or, in part, contracted out to other firms.[14] The typical manufacturing establishment engaged in machining activity is very dependent on the skills of workers specializing in these operations. On average, about 77 percent of all production workers in the manufacturing plants studied are employed in machining occupations. In *Figure 14.2*, we see that the less the manufacturing process at a plant depends on a work force with specialized machining skills the lower the probability of PA adoption. For example, as the degree

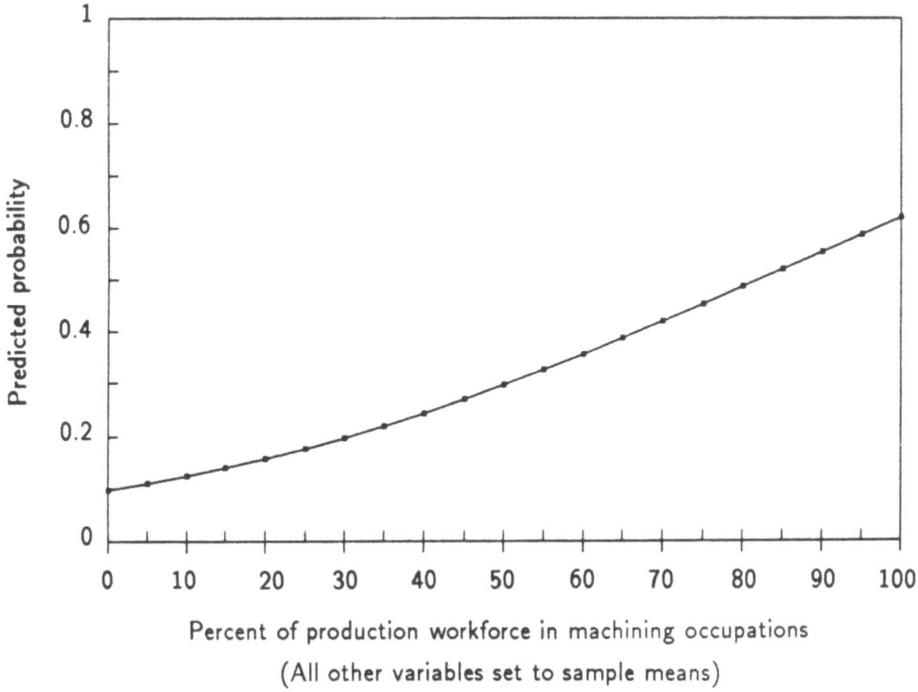

Figure 14.2. Probability of PA adoption by degree of machining skills dependency at the plant.

of skill dependency falls from 75 to 50 percent of all production workers, the predicted probability of PA adoption is reduced at about the same rate, by one-third, from p = 0.45 to p = 0.30 (again with all other variables held constant at their sample means).

These results are consistent with the hypothesis that when management faces higher labor costs and has a greater dependency on particular skills, there will be a greater incentive to adopt such a labor-saving technology as programmable automation. The higher wages we find associated with PA use may in part reflect an increase in the skill demands of machining occupations associated with the introduction of the new technology. Such cause and effect relationships cannot be sorted out with the available data. Hence, we cannot determine the magnitude of the incentive effect that high labor costs may have served to initially induce management to adopt PA.

More than one-third of all machining establishments have high flexibility requirements, making a diverse (50 or more) array of parts/products in very small batch sizes (with 50 percent or more of total machining output in batches of fewer than 10 units). Although the direction of the effect of high flexibility requirements is as predicted (indicating a tendency to adopt PA), these demands are not sufficiently strong, in and of themselves, to significantly affect the firm's propensity to use PA tools.

Scale (or the size) of machining operations across the wide range of industries studied varies from plants that employ only one machine tool to those utilizing up to 1,500 different machines. Our results show that plants with smaller scale machining operations (as measured by the number of tools) have thus far been deterred from adopting PA. This is illustrated in *Figure 14.3*, which also shows that the marginal effect of differences in scale on the probability of PA adoption are relatively small. For example, establishments with 50 tools have 5 times the scale of machining operations as plants in which only 10 tools are deployed, yet the change in predicted probabilities of adopting PA increases from $p = 0.44$ for plants with 10 tools to only $p = 0.54$ for plants with 50 tools, an increase of less than 25 percent. Even for very small scale machining operations, the probability of PA adoption is quite high, *ceteris paribus*. A plant with only 20 tools is nearly as likely to have at least one PA tool as it is to rely exclusively on non-programmable machine tool technology ($p = 0.49$).

Internal resources: the importance of firm size

At plants where information technology is utilized to support a system of control and production planning, there is a significantly greater probability that PA will also be deployed. Independent of the scale of machining operations, cost incentives, and the technological sophistication in related IT applications at a given plant, we find the size of the parent company's organizational resources to be a significant predictor of PA adoption.

As shown in *Figure 14.4*, for the very small single-plant firm with fewer than 20 employees, the chances that there will be even one programmable machine are no better than about 40 percent. For firms with more than 10,000 employees nationwide, we are practically certain ($p \geq 0.96$) of finding at least one programmable machine in its plants (all other things being equal). Even for the plants belonging to moderately large firms with about 500 employees throughout the United States, the chances of there being *no* PA tools at the plant are quite low ($p = 0.17$).

Small firms may be invariably small scale, but the converse is not always true: large firms do not invariably have large scale operations.[15]

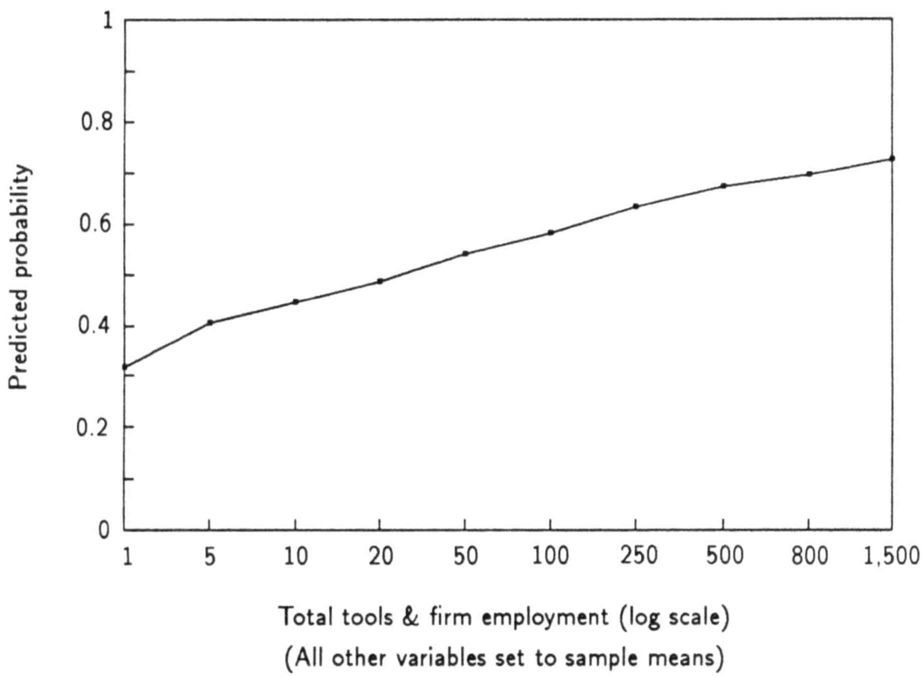

Figure 14.3. Probability of PA adoption by scale of machining operations at the plant.

When we compare the predicted probabilities for the marginal effects of firm size to that of machining scale at a plant (*Figure 14.5*), it is apparent that differences in firm size (taken as indicative of differences in organizational capabilities) are far more important than differences in machining scale *per se* in explaining differential adoption rates among establishments. Small firms with invariably small scale machining operations are, as expected, least likely to have adopted any PA tools.

How much do external learning opportunities affect the chances of PA adoption for small firms?

When the production manager reports that he has linkages – which support learning about new technology – to equipment vendors, to industry or trade associations, and to his customers, we are more likely to find some programmable machines in use at the plant he manages. But the causality is more complex than we had imagined.[16]

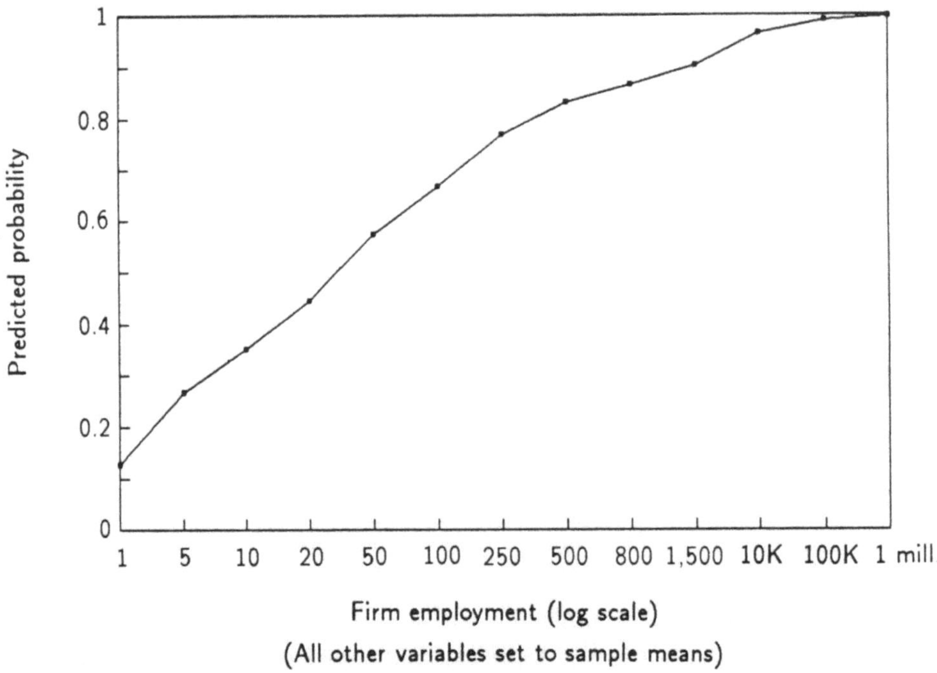

Figure 14.4. Probability of PA adoption by size of parent company.

Informal, individualized know-how trading may be the most common sort of linkage to external resources, but it is not a statistically significant predictor of PA adoption. Moreover, relying solely on written sources of information independent of any personal contact with equipment vendors, distributors, other users, or in isolation from trade shows and demonstrations at professional associations actually *reduces* the likelihood of PA adoption. In addition, when a firm sells its machining output to other firms but has no special order customers willing to share technical information, it's chances of adopting PA are significantly reduced. Linkages that actually reduce the chances of PA adoption have a passive and asocial character to them.

The kind of highly structured and social linkage to other potential adopters and new technology manufacturers that occurs in meetings of industry and professional associations is a significant predictor of PA adoption. Moreover, independent of the contacts that are made at such group settings, further direct contacts with sales representatives from equipment vendors and their distributors are another external linkage that significantly increases the

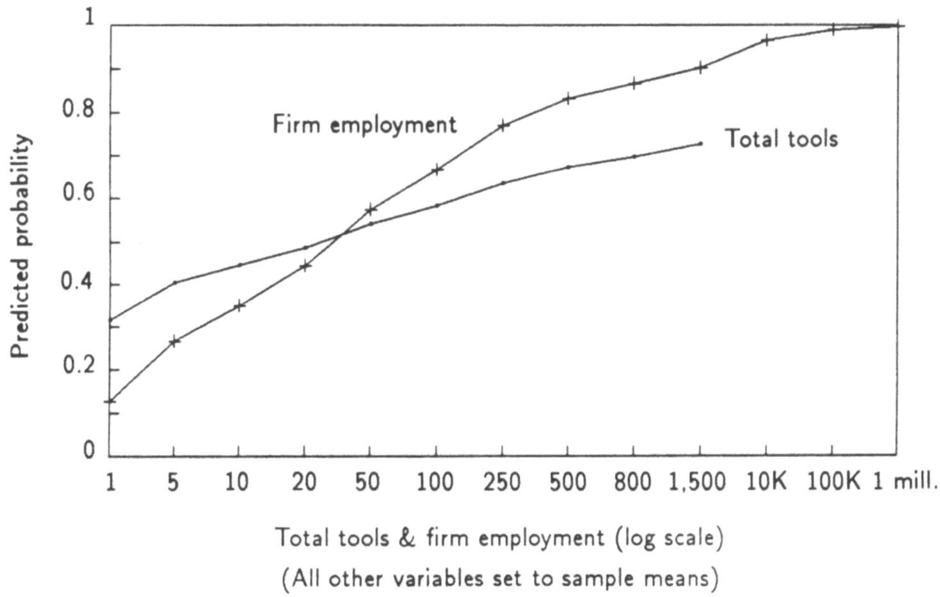

Figure 14.5. Probability of PA adoption by machining scale at plant and by size of parent company.

likelihood of PA adoption. Firms with special-order customers who provide detailed technical information about their orders have significantly greater chance of adopting PA than firms without such close relationships to their customers. Along with informal know-how trading – with it's positive but insignificant impact on the likelihood of adoption – this set of linkages can be best described as having an essentially social and active quality, involving direct interpersonal interchange or contact with outside resources.

In order to evaluate the *combined* effects of those various external linkages that we have identified as *active* and *social* from those we have identified as largely *passive* and *asocial* in nature on the likelihood of PA adoption for firms of different sizes (and hence different internal resources and capabilities for undertaking technological change), we computed the predicted probabilities for the average plant in each of four different firm size categories: firms with fewer than 20; 20 to 99; 100 to 499; and over 500 employees. The scenario identified as indicative of a "passive/asocial" set of external linkages refers to the case in which reliance on external resources is limited to written media, i.e., brochures, ads, and newsletters, and to firms that sell their machining products to some other firm but have no special order

Table 14.2. Estimates of the importance of external learning opportunities to the probability of PA adoption in plants of different size firms.

	Estimated probability of PA adoption		
Size of parent company	Passive/asocial linkages	Active/social linkages	
Firms with <20 employees	0.18	0.65	+256.5%
Firms with 20–99 employees	0.29	0.77	+167.7%
Firms with 100–499 employees	0.42	0.83	+105.4%
Firms with 500 or more employees	0.94	0.99	+5.6%

Notes:
(1) For both scenarios, the plants of these firms are assumed to sell output from the machining process to some other firm.
(2) "Passive/asocial linkages" refers to those situations in which management depends *only* on written media (newsletters, brochures) as an outside source of information about technological developments, is *not* an active participant in industry or professional association meetings, does *not* rely on know-how trading with managers or engineers outside the plant, does *not* depend on contacts with sales representatives from equipment vendors or their distributors to learn about new technology, and does *not* have business customers who share any technical information or expertise.
(3) A plant with "Active/social linkages" is one where management *is* an active participant in industry or professional association meetings, relies on know-how trading with managers or engineers outside the plant, depends on sales representatives from equipment vendors or distributors to learn about new technology, and has special order customers who provide technical information and expertise.
(4) Probabilities are estimated by setting variables measuring one type of linkage (e.g., "active/social") equal to one and those measuring the other type of linkage (e.g., "passive/asocial") equal to zero. For each estimate, all other variables in the model are set equal to the group mean for establishments in that firm size category.

customers who share technical information with them. To have "active" and "social" linkages means that plant management does not rely solely on written media independent of various personal contacts with sales representatives from equipment vendors or their distributors, and with other users. Moreover, management with "active/social" linkages participates in meetings of trade and professional associations where demonstrations of new technology occur, has additional direct contacts with vendors, and has sufficiently close relationships to some of its special order customers, such that some technical information sharing regularly occurs between the two.

The results of these simulation are shown in *Table 14.2.*[17] For all plants but those attached to the largest firms (≥ 500 employees), we find that when management has "active/social" linkages to external resources for learning about new technological developments the chances of PA adoption are substantially higher than those plants with only "passive/asocial" types of linkages. "Active/social" linkages raise the chances of PA adoption the most among plants belonging to the smallest size firms. For the very smallest firms with fewer than 20 employees that have only "passive/asocial" linkages to various external resources, the chances of PA adoption are quite low, less than 1 in 5 ($p = 0.18$). With "active/social" linkages through which the exchange of technical expertise and learning among firms is facilitated, the conditional probability of PA adoption increases by more than 250 percent (to $p = 0.65$). For small firms with 20 to 99 employees, "active/social" linkages to external resources outside the firm increase the chances of PA adoption over firms with "passive/asocial" linkages from less than 3 in 10 ($p = 0.29$) to more than 3 in 4 ($p = 0.77$). Even for medium-sized firms with 100 to 499 employees, we find that such "active/social" linkages to resources external to the firm augment the chances of PA adoption by a substantial margin.

When the very smallest size firms (with fewer than 20 employees) are very well-connected to all four of the "active/social" linkages for which we have identified a positive impact on PA adoption, these external economic advantages compensate for much of the diseconomies of small size and scale. Indeed, the chances of PA adoption for such well-connected small firms actually exceed those estimated for the typical plant of medium-sized firms that are nearly 10 times as large (with between 100 and 499 employees) but have only "passive/asocial" linkages to external sources of expertise.

14.7 Conclusions

In this chapter, we have argued that a combination of three sets of factors explains which establishments are likely to have adopted programmable automation. Cost and profitability incentives (or deterrents), the internal resources and accumulated technical competencies of the firm, and the firm's linkages to external sources of expertise for learning about the new technology's capabilities and limitations are all important in predicting PA use. As of 1987, we find that small-scale, small-size firms are the least likely to have introduced any PA tools. Small firms lack the internal resources and operate at too small a scale of production to generate the kind of internal synergies across a number of product markets, that is, the economies of scope, enjoyed by large diversified companies. Yet as we have shown, the deterring effect of small scale and the lack of internal resources (small size) for engaging in the kind of experimental learning-by-doing process necessary to make the shift to the new technology's learning curve can be overcome when certain external resources are available to supplement the small firm's limited capabilities. Without the kinds of connections to trade associations, to equipment manufacturers, and to special order customers – which help to transfer technical know-how and therefore underwrite the risks of adopting a new process technology – our results would suggest that the isolated small firm (i.e., characterized by passive/asocial linkages to external resources) that relies wholly on traditional techniques is not likely to even attempt the necessary retooling of machines and people. Instead, we would expect such firms to pursue a strategy of lower wages and more intensive use of aging capital for the short to intermediate term. That strategy may permit these firms to continue to exist for a while longer as long as they do not face much of a threat from new entrants to their markets who are more technologically advanced.

There are two ways to consider the implications of these findings for policy. One could argue that the greatest obstacle to diffusion is simply the tradition-bound firms' tenacity, i.e., their stubborn commitment to continuing to do business as usual as long as they can. Hence, a policy designed to more aggressively drive them out of business would presumably hasten the process of diffusion. If there is no alternative use for the capital of such firms and if the work force would need to be substantially retrained in order to be productively employed elsewhere in the economy, a policy designed to force the closure of these plants could result in a net social welfare loss. Even though the firms that remained in business would be more efficient than those that were encouraged to close, such a policy may be more costly to society than allowing these firms to continue to produce, albeit less efficiently. Alternatively, one could argue that the absence of strong "active/social" linkages to external resources for learning about new technology developments among small firms in different sectors and locales is what is limiting the more widespread and rapid diffusion of new manufacturing technologies such as programmable automation.

In certain other national economies, the most well-known example being that of Japan, these linkages are reported to be far more common, helping to diminish the disparities in technological sophistication between large and small firms that might otherwise prevail.[18] National economies with institutional arrangements that are generally supportive of such ties among many manufacturing firms may thus be more successful in sustaining their technological leadership in various markets in this period of transition to a new techno-economic paradigm.

In the United States, technology assistance programs to manufacturing firms now operate in more than thirty states (Shapira, 1990). Such programs can provide a trustworthy source for connecting potential PA users to equipment vendors or their distributors. But technology assistance programs that emphasize one-on-one assistance to small companies are in danger of ignoring the problem of weak linkages. These activities cannot substitute for the collective learning experience that is more likely to occur through interactions among members of the same trade or professional association. Nor can these agencies substitute for the absence of special customers with whom a small business can develop a collaborative relationship that not only is a resource for learning about technology but may also provide some degree of stability of demand for its products. The creation of these kinds of strong, supportive linkages of an "active/social" character should be an objective of public policies designed to foster modernization among small manufacturers.

Acknowledgments

Earlier versions of this chapter were presented at a research seminar in the Science Policy Research Unit at the University of Sussex and at the International Conference on Diffusion of Technologies and Social Behaviour in the International Institute for Applied Systems Analysis. We are grateful to Christopher Freeman and Robert Ayres for providing us with the opportunities to try out our ideas in these forums. For their careful criticisms and many helpful suggestions for revisions, we would especially like to acknowledge Alice H. Amsden, Kim B. Clark, Paul A. David, Peter Doeringer, Bennett Harrison, Susan Helper, Steve Klepper, F. Michael Sherer, and Todd Watkins. Research assistance was ably provided by Lan Xue and Todd Watkins. For Japanese translations, we would like to thank Toshihiro Toyoshima. The original data collection effort for this research was supported by grants from the National Science Foundation (Award No. SES-8520174) and the US Congressional Office of Technology Assessment (Contract No. 633-2470-0).

Notes

[1] For the small firm, the economic environment so conditions these choices, that as Simon (1957) put it, the firm's "planning horizon" becomes "sharply limit[ed]."

[2] Howland's research (1988) on plant closings provides some support for our view of small manufacturing firms as facing a severely constrained set of investment alternatives and strategic choices. For example, she finds that when faced with the same poor economic prospects, small, single-plant firms are much less likely

than branch plants of large firms to close, preferring instead to stay in business at a reduced level of operations and profits.

[3] Insofar as large firms that may themselves be quite sophisticated in their use of advanced manufacturing technologies continue to rely on suppliers with these characteristics, their ability to compete against firms with a technologically advanced supplier chain is also diminished.

[4] For example, nearly all researchers see the collaborative production networks in Japan as being led by great industrial conglomerates (Cusamano, 1985; Dore, 1986; Florida and Kenney, 1989; Freeman, 1988; Johnson, 1982).

[5] See Guile (1986), for a discussion of the possibility that distributors, rather than manufacturers of the new technology themselves, will play an increasing role in disseminating such knowledge and the possibility of market failure when the conditions for appropriating returns from such marketing activity are very weak.

[6] Such close connections to equipment vendors may not be equally available to all potential adopters of a process innovation. There is little incentive for vendors or distributors to provide individualized tutoring to the myriad of small firms that have not adopted any new technology and are individually likely to make only a small investment. Services available from makers of new equipment or their distributors may thus be a very imperfect mechanism for accommodating many small, weakly linked firms to a new technological trajectory.

[7] The industries surveyed include: nonferrous foundries (SIC 336), cutlery, hand tools, and hardware (SIC 342), heating equipment and plumbing fixtures (SIC 343), screw machine products (SIC 345), metal forgings and stampings (SIC 346), ordnance and accessories, not elsewhere classified (SIC 348), miscellaneous fabricated metal products (SIC 349), engines and turbines (SIC 351), farm and garden machinery and equipment (SIC 352), construction and related machinery (SIC 353), metalworking machinery and equipment (SIC 354), special industrial machinery, excluding metalworking (SIC 355), general industrial machinery and equipment (SIC 356), miscellaneous machinery, excluding electrical (SIC 359), electrical industrial apparatus (SIC 362), motor vehicles and equipment (SIC 371), aircraft and parts (SIC 372), guided missiles and space vehicles (SIC 376), engineering and scientific instruments (SIC 381), measuring and controlling instruments (SIC 382), jewelry, silverware, and plateware (SIC 391). Fifty percent of the sample establishments were in SIC 354 and SIC 359.

[8] These data have been analyzed to examine the morphology of subcontracting relations (Kelley and Harrison, 1990; Harrison and Kelley, 1990), productivity in the use of PA (Kelley, 1990b; Kelley and Xue, 1990), and the determinants of skill-upgrading approaches to job design and training opportunities (Kelley, 1989a, 1989b, 1989c, and 1990a).

[9] In 1986, an average of US$6,265.51 per employee was invested in new equipment among establishments with programmable machines, compared to the US$1,972.40 per employee in establishments that had no PA technology.

[10] However, from responses to questions asked of those who were planning to install PA in 1988-89, it seems clear that the introduction of this new technology is associated with an expectation of greater requirements for flexibility, that is, an expansion in the line of products the firm plans to manufacture. In seventy five percent of the plants where managers say they plan to introduce PA in the next two years, they expect to increase the variety of products being manufactured, which suggests that plans to *diversify* away from product lines, or new customers, may be an important stimulus to technology adoption.

[11] To David (1975), for example, scale implies both size and volume of production. Scale economies in the use of the reaper permitted larger size farms (in terms of acreage and hence volume of output) to amortize the cost of purchasing a reaper over its greater expected use in harvesting more acreage. With respect to scale as size, he finds that for farmers operating at a very small scale, only when the relative costs of labor rose above some threshold, were new harvesting machines likely to be purchased. For a detailed review of the different measures and concepts, see Gold (1981).

[12] Appendix *Table 14.A2* shows averages of selected characteristics of establishments, grouped by size of firm and by whether or not any programmable machine tools have been installed as of 1987. Larger firms (and larger scale facilities) are often assumed to specialize in fewer products and longer production runs, and hence to be less *flexible*. We do not find support for this assumption. As shown in the table, establishments with high technical flexibility requirements are as prevalent in plants of large as in plants of small-size firms.

[13] The estimates of the predicted probabilities of PA adoption displayed in *Figures 14.1* to *14.5* and *Table 14.2* were derived using the following method: For each variable X_j of value V,

$$(1)\quad Z_j = \sum_{i=1}^{13} \hat{b}_i \bar{X}_i + \hat{b}_j(X_j); \qquad i \neq j$$

$$(2)\quad \text{Prob (Adopt PA} = 1 | X_j = V) = (e^{Zj})/(1 + e^{Zj}).$$

[14] For an analysis of subcontracting behavior, see Harrison and Kelley (1990) and Kelley and Harrison (1990).

[15] Appendix *Table 14.A2* shows the mean values of selected characteristics of establishments belonging to firms of different sizes. An inspection of the group means shows that among large firms (with greater than 500 employees) there are great differences in the scale of machining operations related to PA use, but for the smallest size firms (with fewer than 20 employees) there are no differences in scale and skill dependency related to PA use.

[16] We cannot know from these data whether some of these linkages – especially to equipment vendors – existed before or were developed after purchasing the technology. However, when we compare the differences among non-adopters between those firms that are planning to purchase PA in the next two years and those that have no such investment plans, we find that well-linked firms

are more likely to have plans to introduce PA. Reliance on other users, active involvement in trade associations, and ties to business customers willing to share technical expertise are all significant factors, positively related to plans to purchase PA. However, firms with a tendency to rely more on equipment vendors or distributors as a source of information, independent of their contacts through trade shows and professional association meetings, are no more likely to be planning to introduce the new technology. These findings provide additional evidence for our contention that in political-economic environments where networks of relationships among economic actors for transferring technical know-how flourish, there will be a more rapid diffusion of new technologies to small firms.

[17] The details on average characteristics by PA use for plants grouped by these firm size categories can be found in Appendix *Table 14.A2*.

[18] In a recent report of the US Congressional Office of Technology Assessment (Gorte, 1990), entitled *Making Things Better: Competing in Manufacturing*, firms in Japan were described as having business customers that are nearly 1.5 times as likely to provide engineering support as we find among US firms. The higher incidence of such close relationships among Japanese firms may explain the higher rate of adoption of PA technology by very small Japanese subcontractors (to that of US firms of similar size) reported in a recent unpublished survey of such firms undertaken by the Shoko Chukin Bank in 1988.

Appendix Table 14.A1. Variable definitions.

Variable table	*Definition*
Hourly wage	$=\log_e$ average hourly wage of workers employed in machining occupations at the plant.
Dependency on machining skills	=% of all production workers in the plant employed in machining occupations.
Union	=1, if production workforce is unionized; =0, if non-union
Machining scale	$=\log_e$ (total number of machine tools).
High flexibility requirements	=1, if 50 or more different parts or products manufactured with machine tools *and* if 50 percent or more of that machining output is produced in batch sizes smaller than 10 units; =0, otherwise.
Firm size	$=\log_e$ (total company employment in the USA).
Complementary IT application	=1, if computers are used for any of the following purposes: process planning / scheduling, quality assurance, process monitoring, materials flow/inventory control, materials/parts planning; =0, otherwise.
Brochures, ads, newsletters	=1, if plant manager considered any of the following sources of information to be very important for learning about new developments in machining technology: advertisements, direct mail (catalogues/brochures), articles in publications; =0, otherwise.
Equipment vendors' sales rep.	=1, if plant manager considered either of the following sources of information to be very important for learning about new developments in marketing technology: sales representatives from manufacturers or sales representatives from distributors; =0, otherwise.
Informal know-how exchange	=1, if plant manager considered conversations with individuals at other companies or conversations with individuals at other plants of this company to be a very important source of information about new developments in machining technology; =0, otherwise.
Trade & prof. association mtgs.	=1, if plant manager considered presentations at technical society meetings or exhibits at trade shows to be very important sources of information for learning about machining technologies; =0, otherwise.
Customers	=1, if plant manager reported that the output of machining operations was sold to customers outside this company; =0, if output of the machining process used solely internally by the firm.
Customer-provided tech. assistance	=1, if plant managers reported that products or parts were made to special order for customers who provided any of the following technical information: blueprints or drawings, written specification sheets detailing how each operation is to be performed, direct assistance from the customer's own manufacturing engineering or programming staff or other types of technical assistance; =0, if no technical assistance provided by special order customers.

Appendix Table 14.A2. Selected establishment characteristics[a] by firm size and by PA use.

Variable names	Firm size: <20 No PA	Any PA	Total
Total machine tools (scale)	13.00	16.00	14.00
Plant employment	8.00	9.00	8.00
Company employment (firm size)	8.00	10.00	9.00
Company/plant employment	1.09	1.11	1.10
Hourly wage (US$)	8.92	11.14	9.69
Dependency on machining skills (%)	82.67	97.07	87.66
Union	0.048	0.057	0.051
Complementary IT application	0.205	0.336	0.250
Customers	0.885	0.992	0.922
Customer-provided technical assistance	0.679	0.914	0.760
High flexibility requirements	0.295	0.430	0.342
Brochures, ads, newsletters	0.348	0.373	0.357
Equipment vendors' sales reps.	0.219	0.406	0.284
Informal know-how exchange	0.527	0.619	0.559
Trade & prof. association mtgs.	0.256	0.357	0.291
% of total	65.40	34.60	100.00
Unweighted N	96.00	56.00	152.00

Appendix Table 14.A2. (continued)

Variable names	Firm size: 20–99 No PA	Any PA	Total
Total machine tools (scale)	30.00	30.00	30.00
Plant employment	35.00	37.00	36.00
Company employment (firm size)	41.00	42.00	41.00
Company/plant employment	1.28	1.22	1.25
Hourly wage (US$)	9.94	9.46	9.66
Dependency on machining skills (%)	51.95	78.57	67.74
Union	0.131	0.113	0.120
Complementary IT application	0.801	0.671	0.724
Customers	0.841	0.818	0.827
Customer provided technical assistance	0.625	0.689	0.663
High flexibility requirements	0.381	0.350	0.363
Brochures, ads, newsletters	0.345	0.433	0.397
Equipment vendors' sales reps.	0.305	0.493	0.416
Informal know-how exchange	0.396	0.550	0.487
Trade & prof. association mtgs.	0.227	0.491	0.383
% of total	40.70	59.30	100.00
Unweighted N	90.00	145.00	235.00

[a]Means weighted by the reciprocal of the probability of selection in the sample stratum.

Appendix Table 14.A2. (continued)

Variable names	Firm size: 100–499 No PA	Any PA	Total
Total machine tools (scale)	22.00	46.00	38.00
Plant employment	126.00	115.00	119.00
Company employment (firm size)	211.00	190.00	197.00
Company/plant employment	4.33	3.13	3.55
Hourly wage (US$)	10.22	10.45	10.37
Dependency on machining skills (%)	24.24	61.38	48.54
Union	0.340	0.317	0.325
Complementary IT application	0.862	0.853	0.856
Customers	0.736	0.851	0.811
Customer-provided technical assistance	0.512	0.607	0.574
High flexibility requirements	0.245	0.327	0.289
Brochures, ads, newsletters	0.270	0.434	0.377
Equipment vendors' sales reps.	0.219	0.511	0.410
Informal know-how exchange	0.283	0.446	0.390
Trade & prof. association mtgs.	0.318	0.447	0.402
% of total	34.60	65.40	100.00
Unweighted N	60.00	114.00	174.00

Appendix Table 14.A2. (continued)

Variable names	Firm size: ≥500 No PA	Any PA	Total
Total machine tools (scale)	19.00	121.00	102.00
Plant employment	230.00	595.00	530.00
Company employment (firm size)	57,862.00	23,627.00	29,807.00
Company/plant employment	673.19	85.60	191.67
Hourly wage (US$)	10.91	10.74	10.77
Dependency of machining skills (%)	16.17	53.67	46.90
Union	0.492	0.493	0.493
Complementary IT application	0.881	0.946	0.934
Customers	0.533	0.778	0.734
Customer-provided technical assistance	0.260	0.445	0.412
High flexibility requirements	0.414	0.298	0.319
Brochures, ads, newsletters	0.304	0.560	0.513
Equipment vendors' sales reps.	0.371	0.483	0.462
Informal know-how exchange	0.444	0.528	0.513
Trade & prof. association mtgs.	0.356	0.615	0.568
% of total	18.10	81.90	100.00
Unweighted N	38.00	162.00	200.00

References

Abernathy, W.J. and Clark, K.B., 1985, Innovation: Mapping the winds of creative destruction, *Research Policy* **14**:3–22.

Arrow, K., 1962, The economic implications of learning by doing, *Review of Economic Studies* **29**:155–173.

Ayres, R.U. and Miller, S.M., 1982, Robotics and conservation of human resources, *Technology in Society* **4**:181–197.

Becattini, G., ed., 1987, *Mercato e Forze Locali: Il Distretto Industriale*, Il Mulino, Bologna, Italy.

Becattini, G., 1989, Sectors and/or districts: Some remarks on the conceptual foundations of industrial economics, in E. Goodman *et al.*, eds., *Small Firms and Industrial Districts in Italy*, Routledge and Kegan Paul, London and New York.

Bellandi, M., 1989, The industrial district in Marshall, in E. Goodman *et al.* (*ibid*).

Bohn, R.E., 1987, *Learning Experimentation in Manufacturing*, Harvard Business School Working Paper No. 88-001, Harvard University, Cambridge, MA, USA.

Bohn, R.E. and Jaikumar, R., 1986, The development of intelligent systems for industrial use: A conceptual framework, *Research on Technological Innovation, Management and Policy* **3**:169–211.

Brusco, S., 1982, The Emilian model: Productive decentralization and social integration, *Cambridge Journal of Economics* **6**(2):167–184.

Brusco, S., 1986, Small firms and industrial districts: The experience of Italy, in D. Keeble and F. Weever, eds., *New Firms and Regional Development*, Croom Helm, London, UK.

Carlsson, B., 1989, The evolution of manufacturing technology and its impact on industrial structure: An international study, *Small Business Economics* **1**:21–37.

Clark, K.B., 1980, Unionization and productivity: Micro-economic evidence, *Quarterly Journal of Economics*:613–639, December.

Cohen, W.M. and Levinthal, D.A., 1990, Absorptive capacity: A new perspective on learning and innovation, *Administrative Science Quarterly* **35**:128–152.

Collis, D.J., 1988, The Machine tool industry and industrial policy, 1955–82, in M.A. Spence and H.A. Hazard, eds., *International Competitiveness*, Ballinger Publishing Co., Cambridge, MA, USA.

Cusamano, M., 1985, *The Japanese Automobile Industry: Technology and Management at Nissan and Toyota*, Harvard University Press, Cambridge, MA, USA.

David, P.A., 1969, *A Contribution to the Theory of Diffusion*, Stanford Center for Research in Economic Growth, Memorandum No. 71, Stanford University, Stanford, CA, USA.

David, P.A., 1975, *Technical Choice, Innovation and Economic Growth*, Cambridge University Press, Cambridge, MA, USA.

Dewar, R.D. and Dutton, J.E., 1986, The adoption of radical and incremental innovations: An empirical analysis, *Management Science* **12**:1422–1433, November.

Dore, R., 1986, *Flexible Rigidities: Industrial Policy and Structural Adjustment in the Japanese Economy 1970–1980*, Stanford University Press, Stanford, CA, USA.

Dosi, G., 1982, Technological paradigm and technological trajectories, *Research Policy* **11**(3):147–162.

Dosi, G., 1988, Sources, procedures, and microeconomic effects of innovation, *Journal of Economic Literature* **26**:1120–1171.

Dosi, G., 1989, The research on innovation diffusion: An assessment, in A. Del Monte, ed., *Recent Developments in Industrial Organization*, MacMillan, London, UK.

Edquist, C. and Jacobsson, S., 1988, *Flexible Automation: The Global Diffusion of New Technology in the Engineering Industry*, Basil Blackwell, Oxford and New York.

Ettlie, J.E., Bridges, W.P., and O'Keefe, R.D., 1984, Organization strategy and structural differences for radical vs. incremental innovation, *Management Science* **30**:682–695.

Ettlie, J.E. and Rubenstein, A.H., 1980, Social learning theory and the implementation of production innovations, *Decision Sciences* **11**:648–668.

Florida, R. and Kenney, M., 1989, High technology restructuring in the USA and Japan, *Environment and Planning A* **22**(2):233–252.

Freeman, C., 1988, *Technology Policy and Economic Performance: Lessons From Japan*, Pinter Publishers, London and New York.

Freeman, C. and Perez, C., 1986, *The Diffusion of Technical Innovations and Changes in Techno-economic Paradigm*, Science Policy Research Unit, University of Sussex, UK.

Freeman, R.B. and Medoff, J.L., 1984, *What Do Unions Do?* Basic Books, New York, NY, USA.

Gold, B., 1981, Changing perspectives on size, scale, and returns: An interpretive survey, *Journal of Economic Literature* **19**:5–33.

Gorte, J.F., 1990, *Making Things Better: Competing in Manufacturing*, OTA-ITE Report No. 443, US Government Printing Office, Office of Technology Assessment of the US Congress, Washington, DC, USA.

Granovetter, M., 1985, Economic action and social structure: The problem of embeddedness, *American Journal of Sociology* **91**:481–510, November.

Griliches, Z., 1960, Hybrid corn and the economics of innovation, *Science* **3422**: 275–280.

Guile, B., 1986, *Investigation of a Transaction-cost Approach to Market Failures in the Development and Diffusion of Manufacturing Technologies*, Paper presented at the Annual Meeting of the Association for Public Policy Analysis and Management (October), Austin, TX, USA.

Hage, J., 1980, *Theories of Organizations: Form, Process and Transformation*, John Wiley & Sons Inc., New York, NY, USA.

Hage, J., Collins, P.D., Hull, F., and Teachman, J., 1989, *Liabilities of Organizational Form in a Family of Populations: The Case of Capital Intensive American Manufacturing in 1973–87*, Working Paper, Center for Innovation and Department of Sociology (July), University of Maryland, College Park, MD, USA.

Harley, C.K., 1973, On the persistence of old techniques: The case of North American shipbuilding, *Journal of Economic History* **33**:372–398.

Harrison, B. and Kelley, M.R., 1991, Outsourcing and the search for flexibility: The morphology of "Contracting Out" in US manufacturing, in M. Storper and A.J. Scott, eds., *Pathways to Industrialization and Regional Development in the 1990s*, Unwin Hyman, Boston and London.

Hicks, D.A., 1983, *Automation Technology and Industrial Renewal: Adjustment Dynamics in the US Metalworking Industry*, American Enterprise Institute, Washington, DC, USA.

Hirschhorn, L.J., 1984, *Beyond Mechanization: Work and Technology in a Postindustrial Age*, MIT Press, Cambridge, MA, USA.

Howland, M., 1988, *Plant Closing and Worker Displacement: The Regional Issues*, W.E. Upjohn Institute, Kalamazoo, MI, USA.

Johnson, C., 1982, *MITI and the Japanese Miracle: The Growth of Industrial Policy, 1925–1975*, Stanford University Press, Stanford, CA, USA.

Kaplinsky, R., 1984, *Automation: The Technology and Society*, Longman, Harlow, UK.

Kelley, M.R., 1989a, An assessment of the skill-upgrading and training opportunities for blue-Collar workers under programmable automation, *Industrial Relations Research Association: Proceedings of the Forty-First Annual Meeting*, pp. 301–308, July.

Kelley, M.R., 1989b, Alternative forms of work organization under programmable automation, in S. Wood, ed., *The Transformation of Work?* Unwin Hyman, London, UK.

Kelley, M.R., 1989c, Unionization and job design under programmable automation, *Industrial Relations* **28**:174–187.

Kelley, M.R., 1990a, New process technology, job design, and work organization: A contingency model, *American Sociological Review* **55**:191–208.

Kelley, M.R., 1990b, *Plant, Process, and Product-Specific Determinants of Efficiency in the Use of New Technology: The Case of Programmable Automation*, Working Paper No. 90–26, School of Urban and Public Affairs, Carnegie-Mellon University, Pittsburgh, PA, USA.

Kelley, M.R. and Brooks, H., 1988, *The State of Computerized Automation in US Manufacturing*, Center for Business and Government, John F. Kennedy School of Government, Harvard University, Cambridge, MA, USA.

Kelley, M.R. and Harrison, B., 1990, The subcontracting behavior of single vs. multi-plant enterprises in US manufacturing: Implications for economic development, *World Development*, September.

Kelley, M.R. and Xue, L., 1990, Does decentralization of programming responsibilities increase efficiency? An empirical test, in W. Karwowski and M. Rahimi, eds., *Ergonomics of Advanced Manufacturing and Hybrid Automated Systems II*, Elsevier Science Publishers, New York, NY, USA.

Kenney, M. and Florida, R., 1989, *Reindustrialization Within Deindustrialization: Japanese Steel, Rubber and Automobile Production in the United States*, Working Paper No. 89–58, School of Urban and Public Affairs, Carnegie-Mellon University, Pittsburgh, PA, USA.

Kern, H. and Schumann, M., 1984, *Das Ende der Arbeitsteilung?* Beck, Munich, FRG.

Kern, H. and Schumann, M., 1987, The limits of the division of labor: New production and employment concepts in West German industry, *Economic and Social Democracy* **8**(2).

Kochan, T.A., 1985, Worker participation and American unions, in T. Kochan, ed., *Challenges and Choices Facing American Labor*, MIT Press, Cambridge, MA, USA.

Kusterer, K., 1978, *Know-How on the Job: The Important Working Knowledge of "Unskilled" Workers*, Westview Press, Boulder, CO, USA.

Leonard-Barton, D., 1988, Implementation as mutual adaptation of technology and organization, *Research Policy* **17**:251–267.

Lorenz, E., 1989, The search for flexibility: Subcontracting networks in French and British engineering, in P. Hirst and J. Zeitlin, eds., *Reversing Industrial Decline?* Berg, Leamington Spa, UK.

Mansfield, E., 1968, *Industrial Research and Technological Innovation*, Norton & Co., New York, NY, USA.

Mansfield, E., Rapoport, J., Romeo, A., Villani, E., Wagner, S., and Husic, F., 1977, *The Production and Application of New Industrial Technology*, Norton & Co., New York, NY, USA.

March, J.G., 1981, Footnotes to organizational change, *Administrative Science Quarterly* **26**:563–577.

March, J.G. and Simon, H.A., 1958, *Organizations*, John Wiley & Sons Inc., New York, NY, USA.

Metcalfe, J.S., 1990, On Diffusion, Investment and the Process of Technological Change, paper presented at the conference on *Technology and Investment*, sponsored by the Royal Swedish Academy of Engineering Sciences, the Swedish Ministry of Industry, and the Organization for Economic Co-operation and Development, Stockholm, Sweden, January 21–24.

Minato, T., 1986, *A Comparison of Japanese and American Interfirm Production Systems*, unpublished mimeo, School of International Politics, Economics, and Business, Aoyama Gakuin University, Tokyo, Japan.

Nabseth, L. and Ray, G.F., eds., 1974, *The Diffusion of New Industrial Processes: An International Study*, Cambridge University Press, London, UK.

Nelson, R.R., 1990, Capitalism as an Engine of Progress, *Research Policy* **19**: 193–214.

Nelson, R.R. and Winter, S.G., 1977, In search of a useful theory of innovation, *Research Policy* **6**:36–76.

Nelson, R.R. and Winter, S.G., 1982, *An Evolutionary Theory of Economic Change*, The Belknap Press of Harvard University Press, Cambridge, MA, USA.

Parsons, C., Scott, R., Crozier, P., and Guile, B., 1984, *The Development of Programmable Automation Systems in Discrete Parts Manufacturing Industries: Agricultural Machinery; Auto Parts; and Pumps and Compressors*, unpublished paper of the Berkeley Roundtable on International Economy, University of California, Berkeley, CA, USA.

Pavitt, K. and Pari, P., 1988, The international distribution and determinants of technological activities, *Oxford Review of Economic Policy* **4**:35–55.

Perez, C., 1986, The new technologies: An integrated view (translation of Las Nuevas Tecnologias: Una Vios de Conjunto), in C. Ominami, ed., *La Tercera Revolucion Industrial: Impactos Internacionales del Actual Viraje Tecnologico*, Grupo Editor Latinoamericano, Buenos Aires, Argentina.

Piore, M. and Sabel, C., 1984, *The Second Industrial Divide*, Basic Books, New York, NY, USA.

Rogers, E., 1983, *Diffusion of Innovations*, The Free Press, New York, NY, USA.

Rosenberg, N., 1972, Factors affecting the diffusion of technology, *Explorations in Economic History* **10**:3–33, Fall.

Rosenberg, N., 1982, *Inside the Black Box: Technology and Economics*, Cambridge University Press, Cambridge, MA, USA.

Rosenthal, S.R., 1984, *Managerial Perspectives on Factory Automation in the USA*, Manufacturing Roundtable Research Report Series, Boston University, MA, USA.

Sabel, C., 1989, Flexible specialization and the reemergence of regional economies, in B. Hirst and J. Zeitlin, eds., *Reversing Industrial Decline?* Berg, Leamington Spa, UK.

Salter, W.E.G., 1960, *Productivity and Technical Change*, Cambridge University Press, Cambridge, MA, USA.

Sato, Y., 1983, The subcontracting production (Shitauke) system in Japan, *Keio Business Review*, Keio University, Tokyo, No. 21:1–25.

Saxonhouse, G., 1974, A tale of Japanese technological diffusion in the Meiji period, *Journal of Economic History* **34**:149–165.

Schmenner, R., 1982, *Making Business Location Decisions*, Prentice-Hall, Englewood Cliffs, NJ, USA.

Scott, W.R., 1987, *Organizations: Rational, Natural and Open Systems*, 2nd edition, Prentice-Hall, Englewood Cliffs, NJ, USA.

Shapira, P., 1990., *Modernizing Manufacturing: New Policies to Build Industrial Extension Services*, Economic Policy Institute, Washington, DC, USA.

Shoko Chukin Bank, 1988, *Fact-Finding Survey on Subcontractors in the Machinery Industries*, Tokyo (available only in Japanese; privately translated for us).

Simon, H.A., 1957, Rational Choice and the Structure of the Environment, *Models of Man: Social and Rational*, John Wiley & Sons Inc., New York, NY, USA.

Skinner, W., 1986, The productivity paradox, *Harvard Business Review*:55–59, July–August.

Stoneman, P., 1980, The rate of imitation, learning, and profitability, *Economic Letters* **6**:179–183.

Stoneman, P., 1983, *The Economic Analysis of Technological Change*, Oxford University Press, London, UK.

Trevor, M., and Christie, I., 1988, *Manufacturers and Suppliers in Britain and Japan*, Policy Studies Institute, London, UK.

Tushman, M.L. and Anderson, P., 1986, Technological discontinuities and organizational environments, *Administrative Science Quarterly* **31**:439–465.

Utterback, J.M., 1988, Innovation and industrial evolution in manufacturing industries, in B.R. Guile and H. Brooks, eds., *Technology and Global Industry: Companies and Nations in the World Economy*, National Academy Press, Washington, DC, USA.

von Hippel, E., 1988, *The Sources of Innovation*, Oxford University Press, New York and Oxford.

Chapter 15

Material Substitution: The Role of New Technology

John E. Tilton

15.1 Introduction

Material substitution has for centuries reduced the costs and enhanced the quality of the tools, weapons, shelter, and other material objects of use to society. It has pushed back the cost-increasing effects of resource depletion while expanding man's horizon of opportunities. In recent years, the pace of material substitution has quickened. Many new materials have been introduced, and old ones improved.

In analyzing material substitution, and the forces behind this important activity, economists and others have traditionally emphasized the role of relative material prices. In recent years, however, an alternative view has surfaced suggesting that technological change is the dominant force behind material substitution.

This study examines these two views – the traditional view and the alternative view – of material substitution. It also attempts to determine which is the more useful in explaining changes over time in the mix of materials used in the production of particular goods.

Since it is not feasible to analyze substitution in all material applications, we will focus here on one industry, the US beverage container industry. The industry, as is well known, has experienced intense material competition over the last several decades. In addition, a particularly thorough investigation of material substitution in this industry has been carried out by Demler (1980 and 1983), which we can draw on.

The results indicate that new technology is the primary cause of change in the composition of materials consumed, at least in the US beverage container industry, and so support the alternative rather than traditional view of material substitution. The implications of this finding for the nature of material substitution and for our understanding of this phenomenon are examined at the end of the study.

15.2 The Traditional View

Economic theory considers the substitution of one material for another to be part of the broader more general process by which firms determine the particular factors of production they use and the quantities of each. According to the theory of the firm, that branch of microeconomics concerned with this process, the potential for one material to substitute for another – or more generally for one factor of production, such as capital, to substitute for another, such as labor – is captured by isoquant curves.

These curves indicate the various combinations of any two factors of production required to produce a certain output of final product given existing technology. For example, the isoquant curves Q_1, Q_2, and Q_3 shown in *Figure 15.1* could reflect the various tonnages of steel and aluminum required to produce one, two, and three thousand automobiles of a particular model.

Normally, isoquant curves have a negative slope, indicating that the two factors are substitutes and that increasing the amount of one reduces the need for the other. Isoquant curves are also usually drawn convex to the origin, as in *Figure 15.1*, on the assumption that as more of one factor is used in place of the other, greater quantities of the first are required per unit of the second. In our automobile example, for instance, this assumption implies the first substitutions of aluminum for steel involve trim or other applications where a little aluminum can replace a pound of steel. As the use of aluminum is increased, however, steel parts that carry heavier loads and rely on the greater strength of steel have to be replaced. For this and

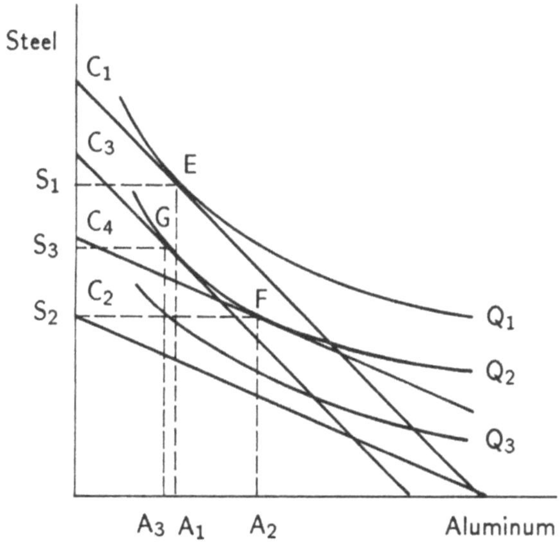

Figure 15.1. Material demand determined by conventional isoquant and isocost curves.

other reasons, these substitutions require much more aluminum per pound of steel replaced.

After the firm determines its desired level of output (that is, the particular isoquant curve on which it wishes to operate), it must choose the factor input mix it will use. This entails selecting a particular point on the desired isoquant curve, which according to the theory will be the point that minimizes the cost. It is determined by imposing on the relevant isoquant curve a series or map of isocost curves.

Isocost curves, such as the curves C_1, C_2, C_3, and C_4 in *Figure 15.1*, indicate the various combinations of two factors of production, such as steel and aluminum, that can be acquired for a given cost or expenditure. Their negative slope reflects relative factor prices, or the amount of one factor that must be given up to purchase an additional unit of the other. Normally isocost curves are assumed to be linear and are drawn with invariant slopes, on the assumption that the firm can vary its factor purchases over the range of quantities covered by the curve without affecting factor prices. The further an isocost curve is from the origin, the larger the quantities of factor inputs represented and the greater the costs. Consequently, to minimize production costs, a firm must operate at that point on its desired isoquant curve that intersects the lowest isocost curve. In *Figure 15.1*, for example, if the isoquant curve Q_1 reflects the desired output, production at point E where the isocost curve C_1 is just tangent to Q_1 results in the lowest possible production costs. Any other point on Q_1 requires a mix of steel and aluminum with higher combined costs since it lies on an isocost curve further from the origin. For this reason, the firm will produce its desired level of output Q_1 with A_1 tons of aluminum and S_1 tons of steel.

Now let the price of steel increase, for simplicity, by 100 percent. The quantity of steel the firm could purchase for any level of expenditure would fall by half, causing the isocost curve C_1 to rotate to C_2. Such an increase in steel costs would likely cause the firm to raise the price it charges for its final product, automobiles, which in turn presumably would reduce demand and the desired level of output. If the isoquant curve Q_2 reflects the new desired level of production, point F on that curve represents the firms' new demand for steel and aluminum. As expected the demand for steel has fallen, from S_1 to S_2 tons.

This decline in steel demand can be separated into two parts. The first, the output effect, is the reduction in steel demand that occurs because the production of automobiles is now lower. Had the price of steel remained the same and automobile production fallen for some other reason, the slope of the isocost curves would not have changed and the firm would have minimized its cost of aluminum and steel by operating at point G (where the isocost curve C_3 with the same slope as C_1 is just tangent to the isoquant curve Q_2). Consequently, the reduction in automobile output is responsible for the fall in steel demand from S_1 to S_3.

The second, the substitution effect, is the result of the shift in relative factor prices. The rise in the price of steel causes the slope of the isocost curve to rotate encouraging the firm to move along the new isoquant curve from point G to F. This in turn causes a further decline in steel demand, from S_3 to S_2.

For aluminum, the output effect also reduces demand, from A_1 to A_3. The substitution effect, however, increases demand, as one would expect since the price of aluminum has fallen relative to that of steel, from A_3 to A_2. In *Figure 15.1*, the substitution effect more than offsets the output effect, so the rise in the price of steel produces an increase in aluminum demand. This, though, does not necessarily have to be the case.

The output effect may be substantial, or it may be negligible. The latter is the case, for example, if the demand for the firm's final product is totally insensitive or inelastic to price changes, or if little or no change in the price of the final product occurs. Little change in the price of the final good is likely when the change in steel price is small or when the cost of steel constitutes a negligible portion of the costs of the final product.

This description of the traditional view of material substitution is to some extent an oversimplification. Economists have long known that production isoquants are not always the nice continuous convex curves pictured in *Figure 15.1* and in most introductory microeconomic textbooks. Technology may require that factor inputs be used in fixed proportions, especially in the short run when it can be difficult to modify or replace existing equipment on a production line. In such cases, a given output of final product requires one specific combination of factor inputs, and the isoquant curve collapses into a point, as illustrated in *Figure 15.2(a)*.

Alternatively, there may be two or three, mutually exclusive production processes that a firm may use to produce a given output, each requiring fixed, but different, factor proportions. Here the isoquant curve is replaced by two or three points.

Even where a continuous isoquant curve exists, it does not have to be convex. Aluminum and plastic are both used as sliding in the construction of houses in the United States. As plastic is increasingly substituted for aluminum, however, there is no reason to believe the amount of plastic required per pound of aluminum replaced has to increase. The isoquant curve in this situation is linear, as shown in *Figure 15.2(b)*, rather than convex, and firms typically operate at one end or the other.

The conventional theory of the firm also recognizes that the shape and location of isoquant curves is determined by existing technology, and that these curves may rotate or shift inward over time as a result of the creation and diffusion of new technology. Such shifts can also cause changes in the optimal mix of input factors including materials.

This traditional view of factor substitution provides the theoretical rationalization for the more common demand functions encountered in applied economic analyses of material demand, functions that typically assume demand depends on gross national product (GNP) or some other activity variable, the material's own price, and the prices of close substitutes. The activity variable affects the optimal level of final product output and hence the desired isoquant curve, while own and substitute prices determine the optimal mix of material inputs for producing the desired output.

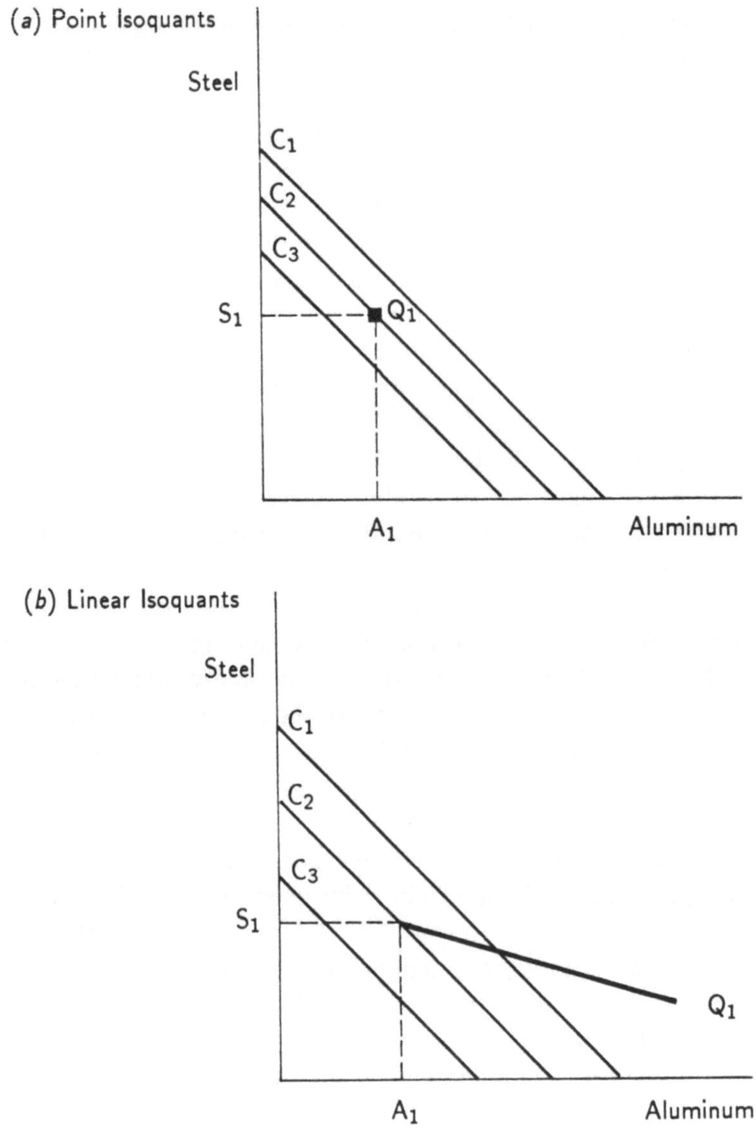

Figure 15.2. Material demand determined by point (a) and linear (b) isoquants and isocost curves.

15.3 The Alternative View

While the traditional theory of the firm recognizes that isoquant curves may possess discrete breaks, may be linear rather than convex, and may shift over time as technology changes, these possibilities are not emphasized or built explicitly into its analytical framework. They are raised more as afterthoughts, or as caveats and exceptions that in some instances may alter or undermine the theory's usefulness. The traditional theory views factor substitution, of which material substitution is a subcase, as a movement along a stable production isoquant made in response to changes in relative factor prices. The role of technological change is largely ignored, and in this sense the analysis is static.

The alternative view of material substitution, in contrast, emphasizes technological change and downplays material prices. In doing so, it does not reject the basic tenets of the theory of the firm, the existence of isoquant and isocost curves for example. Rather it postulates that in most cases the driving force behind material substitution is a shift in the isoquant curve, the result of new technology, rather than a shift in the isocost curve produced by changes in material prices.

New technology, it contends, is more important than changes in material prices for two reasons. First, in many applications material prices can vary over a broad range without producing any movement on the isoquant curve, and hence any change in the mix of materials. This is the case, for example, whenever the relevant isoquant is not a nice continuous convex curve, but instead is a point, a series of points, a straight line, or a series of linear segments, as *Figure 15.2* illustrates. Even with a continuous convex curve, changes in relative material prices may not produce any movement along the curve if production is currently taking place at one or the other ends. For example, solder, an alloy of lead and tin, is widely used in the automobile industry in smoothing over body joints, in filling dents, in manufacturing brass radiators, and in other applications. As Canavan (1983) has pointed out, the typical solder used for such purposes is about 98 percent lead and 2 percent tin, the latter being the minimum amount of tin required for the solder to retain needed physical properties. While there are no technical problems substituting in the other direction, that is, more tin for lead, it would require a tremendous change in the relative prices of these two metals, as tin is normally five to ten times more expensive per pound than lead.

Second, tremendous advances in material technology have taken place over the last few decades. New metal alloys, polymers, ceramics, composites have been created in the laboratory and introduced into the marketplace. New innovations have simultaneously enhanced the strength, corrosion resistance, and other physical properties of steel and the more traditional materials. As a result, design engineers in all material consuming industries – construction, automobile, aerospace, packaging, consumer durables, capital equipment – face a choice of materials that their predecessors a decade or two ago would greatly envy, and a choice that every year is richer and more extensive.

In contrasting the traditional and alternative views of material substitution, it is important to note that the two are not mutually exclusive. The real world is sufficiently complex and diverse that material substitution in some instances is driven entirely or largely by changes in material prices and in other instances by technological change. In still other situations, both may be important.

In addition, the influence of these two variables cannot always be cleanly separated. Changes in material prices causing a shift in the material mix being used may themselves be the result of new innovations in mining or metallurgy. Alternatively, the R&D efforts behind new resource-saving technology may originally have been prompted by shifting material prices.

Despite such complications, it is still worthwhile to inquire whether firms most often alter their material mix in response to a change in material prices or a change in technology. The answer to this question determines which of the two views of material substitution is the more realistic, and hence most useful. The section that follows investigates this question for material substitution in the US beverage container industry.

15.4 US Beverage Container Industry

The beverage industry produces beer and soft drinks, which it then ships in bulk containers, such as kegs, and in packaged containers, such as bottles and cans. In the United States, packaged containers are the more important, accounting for between 75 and 90 percent of the volume of both beer and soft drink shipments since 1950, and it is on material use in packaged containers that this section focuses.

Packaged containers can be further separated into glass bottles (both returnable and nonreturnable), plastic bottles, and metal cans. There are, in turn, four major types of metal cans – the three-piece tinplate can, the three-piece tin-free steel can, the two-piece aluminum can, and the two-piece tinplate can.

The three-piece tinplate can, first used for beer in 1935 and for soft drinks in 1953, was the original metal can. Composed of three pieces (a top, a bottom, and a cylindrical side), it was for a number of years made entirely of tinplate, or steel sheet covered with a thin layer of tin. In the 1960s, aluminum replaced tinplate in the top of this can and tin-free steel replaced tinplate in the bottom.

The three-piece tin-free steel can, introduced in the beer and soft drink industries in 1967 and 1969, is similar to the tinplate can. However, in place of tinplate it uses tin-free steel, a chromium-coated steel sheet. The top of this can is also made of aluminum.

The two-piece aluminum can is composed entirely of aluminum. The bottom and side are made from one piece; the top from another. First used for beer containers in 1958, the bottom and side were at that time formed by an impact extrusion process. This technology was never very successful, and five years later was replaced by the more effective drawn-and-ironed (D&I) process. In 1967 the two piece aluminum can was first used for soft drinks.

The two-piece tinplate can has a bottom and side made from one piece of tinplate, and shaped by the D&I process. Its top, the second piece, is made of aluminum. It was introduced into the beer and soft drink markets in 1971 and 1973.

These six different types of packaged containers and the materials they use – steel, tin, chrome, aluminum, glass, and plastic – are summarized in *Table 15.1*. Substitution among these materials has been examined in considerable detail up to 1977 by Demler (1980 and 1983). Drawing heavily on his work, this section looks next at the 1950–1977 period. It then examines more recent trends in material use in this industry.

15.4.1 The 1950–1977 period

The tons of steel, tin, chrome, aluminum, and glass consumed annually in the United States for every million barrels of beer and soft drinks shipped in packaged containers is shown in *Figures 15.3* and *15.4* for the years 1950–1977. The use of plastics is not shown. First introduced into the soft drink industry in 1976, its consumption was negligible over this period for soft drink containers, and zero for beer containers.

These figures highlight an important and striking aspects of material use in the US beverage container industry, its variability. Steel consumption

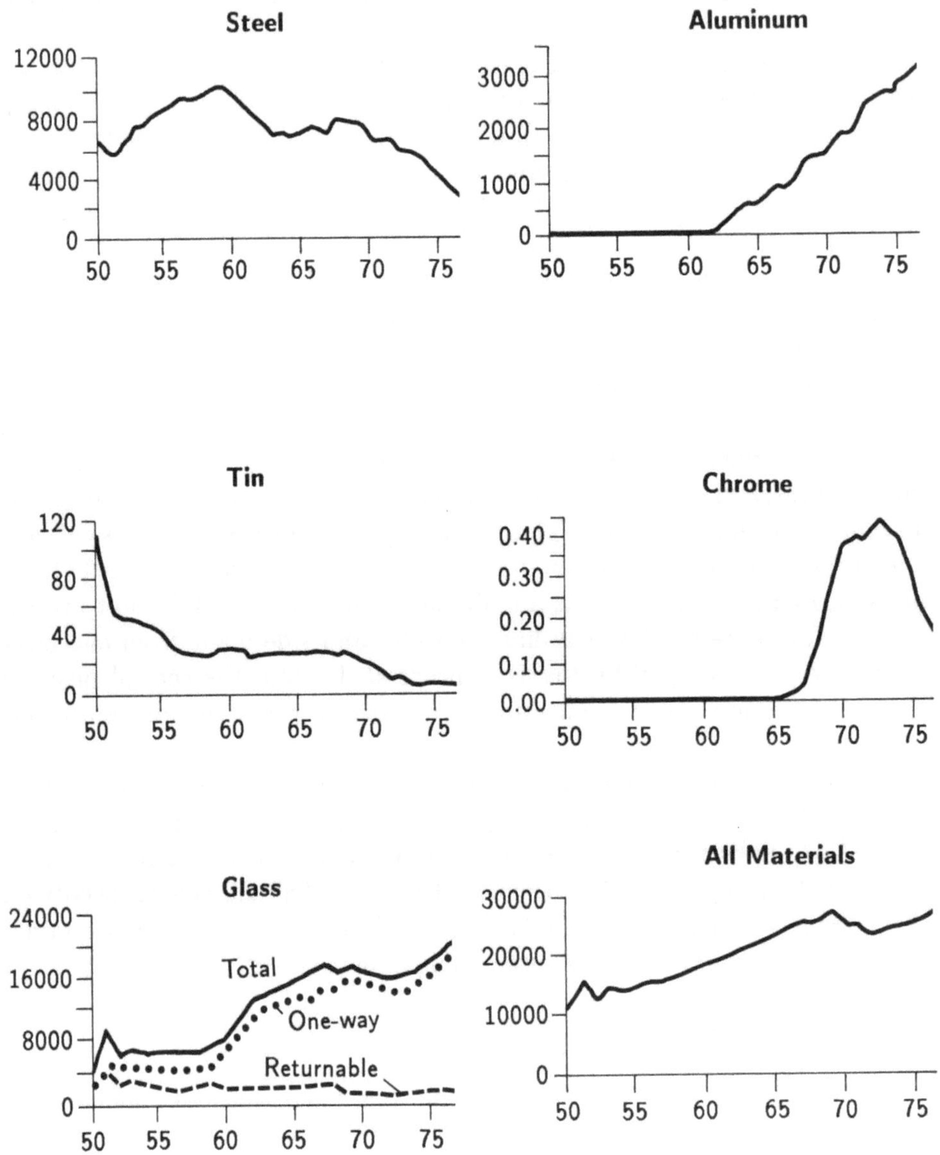

Figure 15.3. US material use in tons per million barrels of packaged beer shipments, 1950–1977. (Source: Demler, 1980, Tables A-1, A-16, A-17, and A-18, and the sources cited there.)

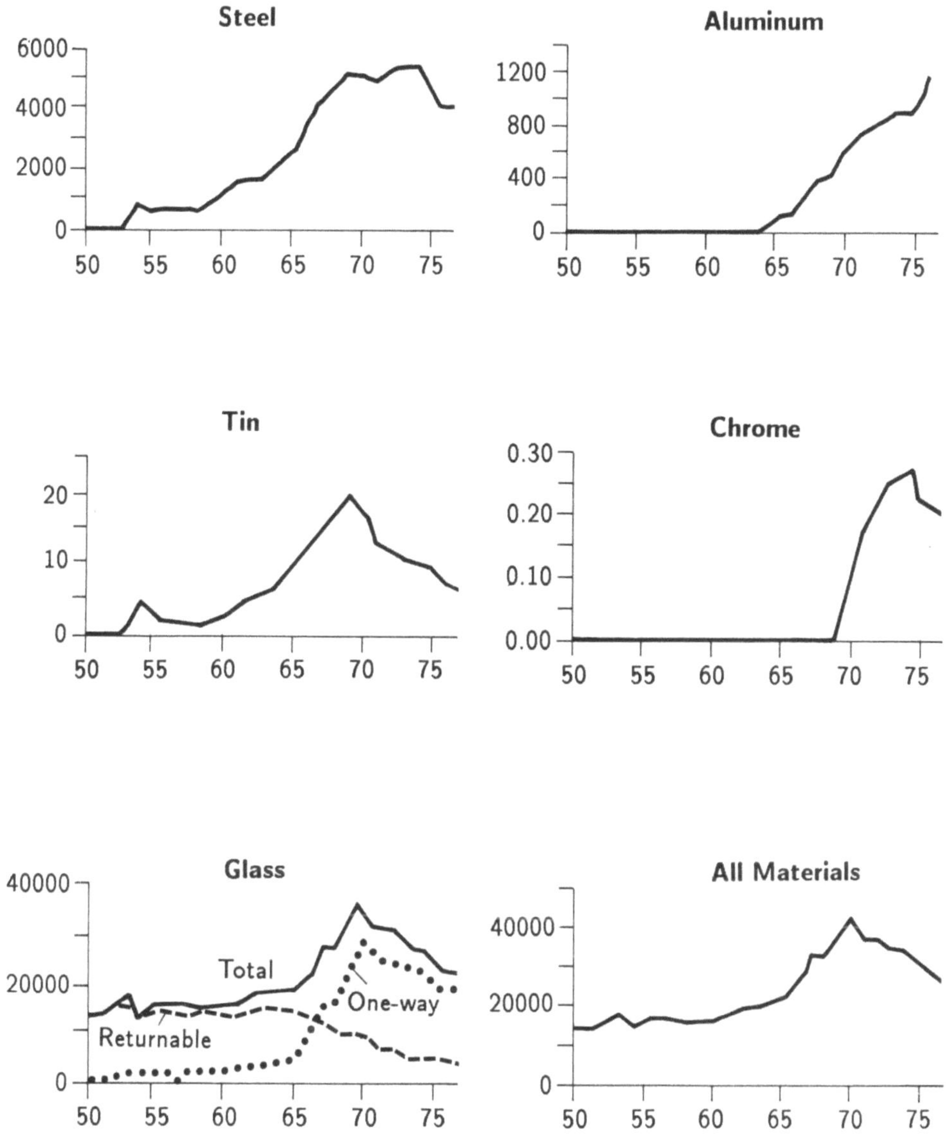

Figure 15.4. US material use in tons per million barrels of packaged soft drink shipments, 1950–1977. (Source: Demler, 1980, Tables A-2, A-16, A-17, and A-18, and the sources cited there.)

Table 15.1. Materials used in beverage containers.

	Material					
Container	Tin	Steel	Chrome	Aluminum	Glass	Plastic
Three-piece tinplate can	x	x	x (bottom end)	x (top end)		
Three-piece tin-free steel can		x	x	x (top end)		
Two-piece tinplate can	x	x		x (top end)		
Two-piece aluminum can				x		
Glass bottle					x	
Plastic bottle						x

Source: Demler (1980).

per million barrels of beer (*Figure 15.3*), for example, rises from 6,000 to 10,000 tons during the 1950s, and then enters a long period of decline. By 1977 it is less than 3,000 tons. The use of steel per million barrels of soft drinks (*Figure 15.4*), in contrast, starts at zero in 1950, and grows almost continuously until the mid-1970s.

Tin and chrome, each used with steel, follow their own individual patterns. The use of tin in beer containers is one of continuous decline, very rapid in the early 1950s, very gradual in the late 1950s and early 1960s, and then again more swiftly. For soft drinks, tin consumption rises until about 1970, and then declines. Chrome experiences a sharp but brief jump in usage both in beer and soft drink containers following the introduction of the tin free steel container in the late 1960s. By the early 1970s its use is already in decline.

The story for aluminum again is different. Following its first use for tops in three-piece tinplate cans in the early 1960s, aluminum consumption per million barrels of packaged shipments expands rapidly for both beer and soft drinks.

The use of glass, the final material examined, remains fairly constant during the 1950s for both beer and soft drinks, rises during most of the 1960s, and then declines somewhat toward the end of this decade. During the 1970s, this downward trend continues for soft drinks but is reversed for beer. For both beverages, there is a notable shift in favor of one-way glass bottles over the 1950–1977 period.

The total tonnage of all materials consumed per million barrels of packaged beverage shipments rises quite consistently over the entire period for beer, and rises for soft drinks until the 1970s. It is not surprising that these trends follow more or less closely those for glass. Glass bottles are heavy compared to metal cans, and glass consumption largely determines the combined weight of all the materials consumed.

Returning to the views of material substitution discussed in the preceding section, two explanations are possible for the substantial changes over time in the mix of materials used per million barrels of beer and soft drink shipments. The first is that rather dramatic shifts along stationary production isoquants have occurred in response to changes in material prices. The second is that technological change has caused the production isoquants themselves to shift inward over time.

Demler's analysis identifies two factors, or what he calls apparent determinants, principally responsible for changes in material use over time. The first is the mix of container types – the three-piece tinplate can, the returnable glass bottle, the aluminum can, and so on – and changes in that mix. The second is the material composition of individual container types, and shifts in that composition over time.

Trends in market share for individual container types are shown in *Figures 15.5* and *15.6* for beer and soft drinks. The beer market (*Figure 15.5*) appears to be particularly competitive and volatile. While the returnable bottle, the three-piece tinplate can, and the tin-free steel can all had their day as the industry leader, their fortunes by the 1970s were on the wane. The one-way bottle and the aluminum can, on the other hand, were expanding.

In the soft drink market (*Figure 15.6*), the returnable bottle managed to remain the leader over the entire 1950–1977 period. However, during the latter half of this era, it faced increasing competition from first the three-piece tinplate can, and then the returnable bottle and the aluminum can.

Since these changes in market share contribute significantly to the trends in material use portrayed in *Figures 15.3* and *15.4*, it is important to understand why they occurred. Here a variety of factors can be identified. Some consumers, for example, favor cans because they are lighter and easier to handle, and do not break. Others favor glass bottles because glass is inert and has no effect on taste. Deposit laws, now found in a number of states, encourage the use of (returnable) glass bottles. Among the cans, they favor aluminum, which is the easiest to recycle.

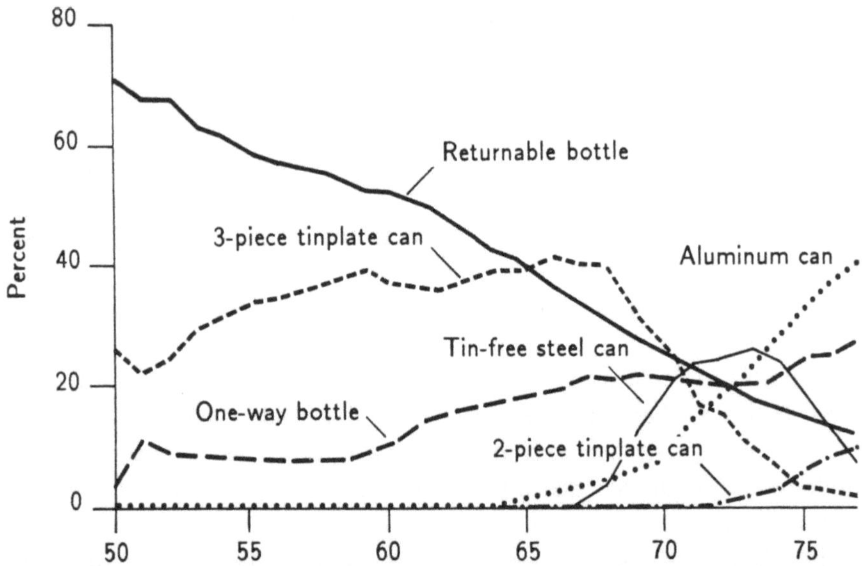

Figure 15.5. Percentage of packaged beer shipments by container type, 1950–1977. (Source: Demler, 1980, Table A-3 and the sources cited there.)

Perhaps most important, however, are the relative costs of different container types. Of concern are not just the costs of the glass, steel, tin, and the other materials used in making a containers, but the costs of producing, filling, and transporting containers as well.

As *Figure 15.7* shows, a close correlation does not exist between the market shares of various container types and the prices of their constituent materials. The sharp rise in the real price of tin in the early 1970s occurred after the peak in market share of the three-piece tinplate can, and is inconsistent with the rise in market share of the two-piece tinplate can that began just as the tin price began its ascent. The modest decline and then rise in the real price of glass is not very helpful in explaining why the market shares for the returnable and one-way glass bottle have moved in the opposite direction. While the aluminum can was rapidly penetrating both the beer and soft drink market, its real price changed very little.

While material prices influence the relative costs of different containers, they are not directly responsible for most of the shifts in market shares over the 1950–1977 period. These shifts are instead often identified with specific technological developments. Earlier, it was noted that the first aluminum can was made in 1958 using an impact extrusion process. However, it was not

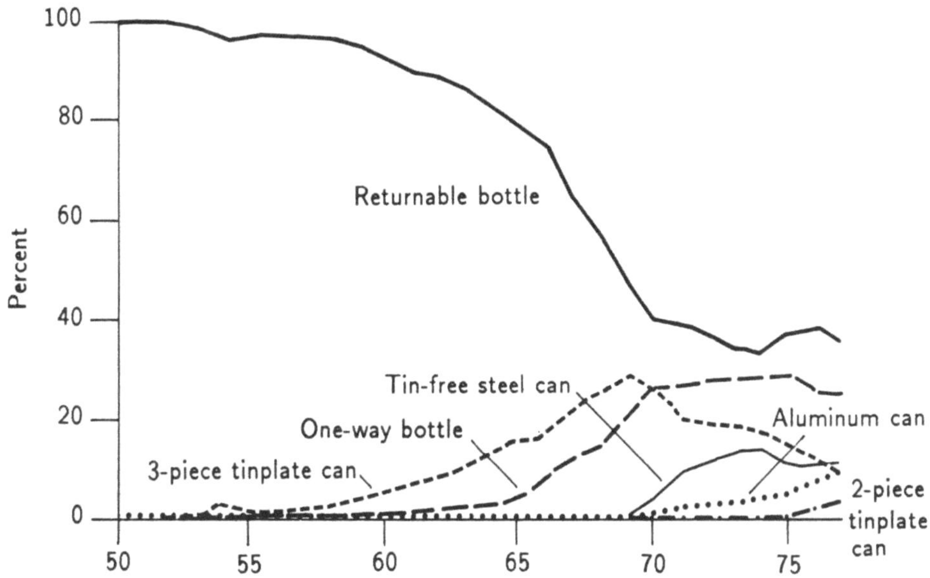

Figure 15.6. Percentage of packaged soft drink shipments by container types, 1950–1977. (Source: Demler, 1980, Table A-4 and the sources cited there.)

until the D&I process was introduced five years later that relative container costs shifted significantly in favor of the aluminum can, and its dramatic market penetration began. The steel industry responded by developing first the tin-free steel can and then the two-piece tinplate can. These innovations, coupled with the aluminum can, caused the demise of the three-piece tinplate can, which up to that time had been increasing its market share in both the beer and soft drink markets. The major innovations in the beverage container market, a number of which have produced new container types or enhanced the competitiveness of existing containers, have been identified by Demler and are shown in *Table 15.2*.

The second important factor affecting material use that Demler identifies is the change over time in the material composition of individual container types. This is most dramatically illustrated in the use of tin in tinplate cans, as shown in *Figure 15.8*. In 1950, some 2.77 pounds of tin were used to produce a thousand tinplate cans of the standard 12 ounce size. By 1977

Table 15.2. Major innovations in the beverage container industry.

Year of commercialization	Nature of innovation and sector of industry in which it occurred			
	Glass	Steel	Aluminum	Comment
Before 1935	Returnable bottle			First beverage container
Early in 1935		Three-piece hot-dipped tinplate can		Steel and canmaking industry penetrate the beverage market
Later in 1935	One-way glass bottle			Glass industry introduces a one-way container to compete with can
1935 to 1958	Glass container light-weighting			Competition to regain lost market share to metal can
1946 to mid-1950		Hot-dipped tinplate to differentially coated tinplate to 0.25 lb. electrolytic tinplate		World War II and Korean War and economics are the determining factors
1958			First aluminum beverage can	Manufactured by Hawaii Brewery and Coors brewery
1959	"Handy" glass containers			Glass industry's competitive container to metal can (weight efficiencies, strength, cost)
1961		Double-reduced steel		Steel and canmaking industry switch beverage can over to cost-efficient DR steel
1963			Reynolds Aluminum two-piece can	Aluminum industry enters beverage container market and increases market share
1965			Reynolds Aluminum "necked-in" can	Improves container & competes more effectively with DR steel

Table 15.2. Continued.

Year of commer-cialization	Nature of innovation and sector of industry in which it occurred			
	Glass	Steel	Aluminum	Comment
1967		Tin-free steel can		Steel and canmaking industries attempt to maintain beverage market
1969			Aluminum can from H19 alloy	Aluminum industry competes with TFS and increases market share
Late 1960–1976[a]				Plastics enter the beverage container market
1970	Plastic coatings			Glass industry maintains market share by threat from plastic containers
1971		Two-piece tinplate cans		Steel competes directly with aluminum can and attempts to regain lost market
1972–1974			H19 alloy more efficiently utilized	Aluminum industry in direct competition with steel
1974	Kerr-Heye glass			Press-and-blow technique, lighter and stronger glass container to compete with cans and plastics
1976		Miraform II	Miraform II	New bottom-profile can which both steel and aluminum use to compete with each other
1976[b]				After ban on AN resins, plastic industry attempts to gain market share with PET resins

[a]Plastics emerged in the form of AN-Resins.
[b]Plastics emerged in the form of PET containers.
Source: Demler (1983, pp. 34–35).

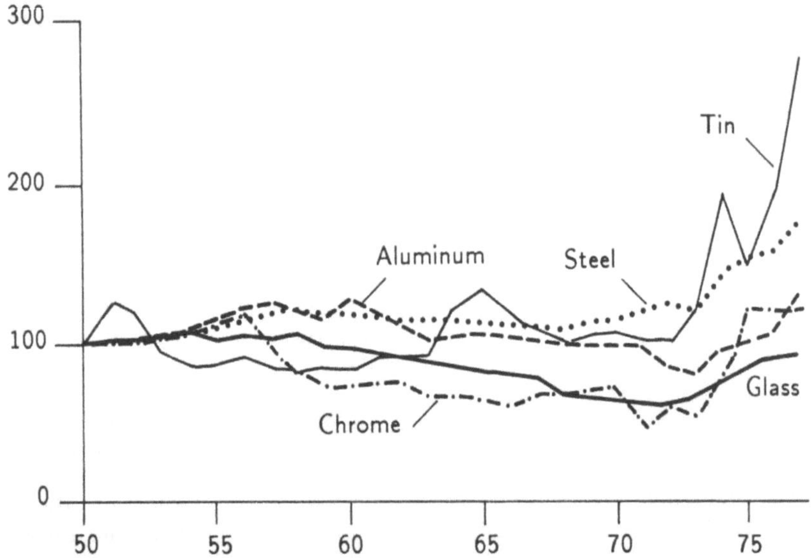

Note: The price of glass is derived from the prices of soda ash and silica sand, and assumes one part soda ash is used for every three parts of silica sand.

Figure 15.7. Trends in real prices for steel, chrome, tin, aluminum, and glass, 1950–1977 (1950 = 100). (Source: Demler, 1980, Tables A-19, A-20, and A-21.)

this figure has dropped to 0.16 pounds for the most metal efficient 12 ounce tinplate cans being produced, a drop of over 90 percent.

This decline was produced by a number of innovations. The substantial drop recorded in the early 1950s was largely the result of the widespread adoption of the electrolytic tinning process. Compared to the older, hot dipped process it replaced, electrolytic tinning provides a more uniform tin coating, which as a result can be much thinner.

In the 1960s two additional innovations – the introduction of the aluminum top and tin-free steel bottom on the three-piece tinplate can – further reduced the use of tin per tinplate can. Finally, during the 1970s research on the use of enamels which had been going on for years led to further reductions.

Declines in the weight of steel, aluminum, and chrome in the cans in which they are used, though less spectacular, also occurred, and again were largely the result of new technological developments. For example, the

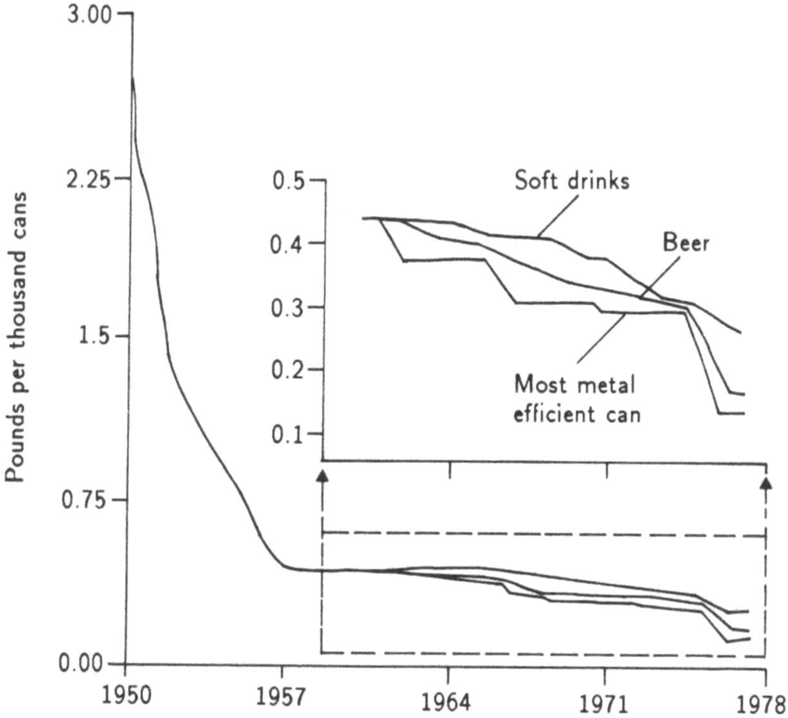

Figure 15.8. Tin content in pounds per thousand tinplate cans, 1950–1977. (Source: Demler, 1980, Figure IV-6.)

introduction of the more efficient H19 alloy substantially reduced the amount of aluminum needed per aluminum can in the late 1960s.

So technological change – the introduction and diffusion of new innovations – appears as the dominant force driving changes in both the mix of containers and the material composition of individual types of containers. These, in turn, as Demler shows, are the two principal determinants of trends in material use and substitution in the beer and soft drink container industries.

15.4.2 Recent trends

No study similar to Demler's has been published covering the years since 1977. As a result, estimates of the steel, tin, chrome, aluminum, glass, and plastic in packaged beverage containers comparable to those used to construct *Figures 15.3* and *15.4* are not readily available for recent years.

Still, from the work of Nappi (1986) on metal demand for the US packaging industry as a whole and other available information, it is clear that the important factors Demler identifies – the mix of containers and the material composition of container types – have continued to play a major role in shaping material usage in beverage containers.

Particularly dramatic shifts in market shares of individual types of containers have taken place in the soft drink industry. As *Figure 15.9* indicates, the market penetration of the aluminum can has continued unabated. In the early 1980s it became the most popular soft drink container. The plastic bottle whose market share was negligible in 1977 has also enjoyed very rapid growth. Used widely in the manufacture of large, two-liter containers, the plastic bottle by 1985 accounted for over 25 percent of all packaged soft drink shipments. The returnable bottle, the three-piece tinplate can, and the tin-free steel can were the big losers. The returnable bottle saw its share of the market shrink from 40 to 16 percent, while the three-piece tinplate can and the tin-free steel can, which together held nearly a fourth of the market in 1977, disappeared almost completely. These changes stimulated the substitution of aluminum and plastics for glass, steel, tin, and chrome.

In the beer industry, significant shifts in the mix of container types also occurred. The aluminum can increased its share of the packaged container market from 41 percent in 1977 to about 70 percent by 1985. The three-piece tinplate can, the tin-free steel can, and the two-piece tinplate can ceased to be used. The historical decline of the returnable bottle continued, while the rising share of the one-way bottle peaked in the early 1980s and then reversed itself. By 1985 these two glass containers held 30 percent of the beer market compared to 40 percent eight years earlier. The plastic bottle, despite its success with soft drinks, has yet to penetrate the beer market. So material substitution resulting from the changing mix of container usage has in recent years stimulated the consumption of aluminum at the expense of glass, steel, tin, and chrome.

The material composition of container types has also been changing over the last decade. A recent study of the beverage container industry by Chase Econometrics (1986, p. 4.41), for example, reports that the average gauge of the aluminum sheet used in the body of aluminum cans declined from 0.016 inches in the late 1970s to 0.0126 inches by 1986. The same study (p. 6.14) also notes that the PET (polyethylene terphthalate) resin used to produce a standard two-liter plastic soft drink bottle fell on the average from 60 to around 50 grams between the late 1970s and 1985. Such developments affect both the level and mix of materials used in beverage packaging. They also

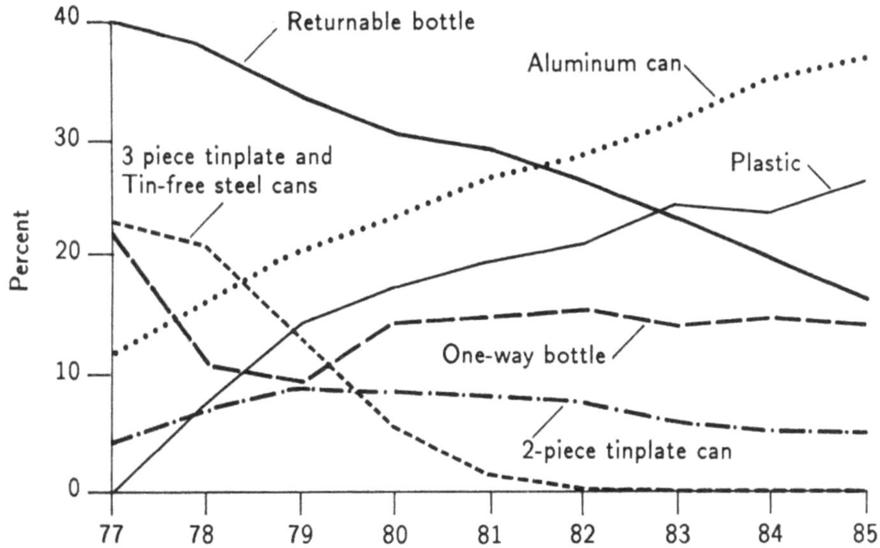

Figure 15.9. Percentage of packaged soft drink shipments by container type, 1977–1985. (Sources: Can Manufacturers Institute, 1986; National Soft Drinks Association, 1986; author's own estimates.)

affect the competitiveness of the different types of containers, and are in part responsible for the growing market shares of the aluminum can and the plastic bottle.

Looking behind these changes in the mix of container types and the material composition of individual containers in an attempt to identify their determinants, one again finds little evidence that changes in material prices are of great importance, except perhaps through the indirect influence they have on R&D and the subsequent course of new technology. As *Figure 15.10* shows the substitution of aluminum and plastics for steel, tin, chrome, and glass in recent years cannot be accounted for by a drop in their real prices, either in absolute or relative terms.

While the overall costs of various containers, which as noted earlier include the costs of manufacturing, filling, and transportation as well as the costs of materials, are significant, major shifts in these costs are typically the result of new technological developments. The first commercial use of the plastic bottle in the soft drink market, for example, was made possible by an innovation. Its subsequent rapid penetration of this market, in turn, hinged in part on improved technology that reduced resin requirements, and increased the number of plastic bottles filled per minute.

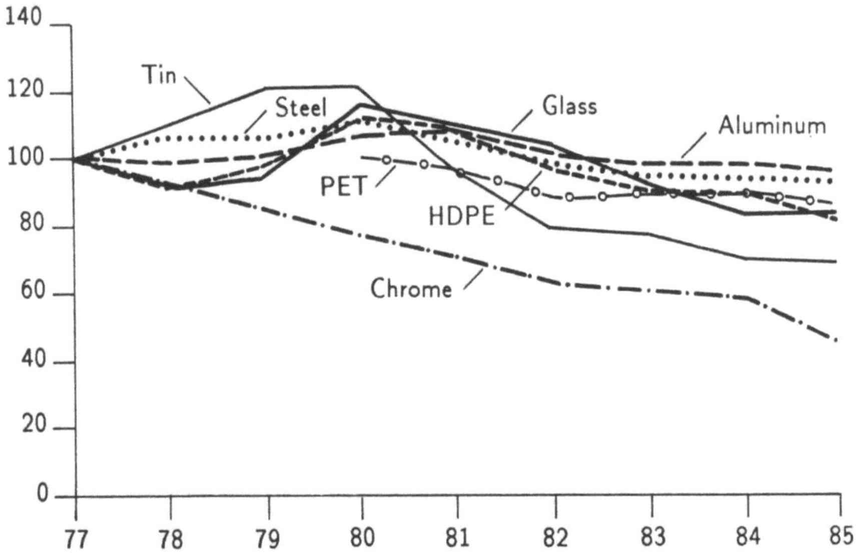

Notes: The price of glass is derived from the prices of soda ash and silica sand, and assumes one part soda ash is used for every three parts of silica sand.

Price shown for PET resins, used for sides of plastic beverage bottles, is for thermoplastic resins in general. Since data prior to 1980 are not available, for this series 1980 = 100.

Price shown for HDPE resins, used for the bottoms of plastic beverage bottles, is for high density polyethylene resins (blow molding) in general.

Figure 15.10. Trends in real prices of steel, chrome, tin, aluminum, glass, PET resins, and HDPE resins (1977 = 100). (Sources: *American Metal Market*, various issues; US Bureau of Mines, various issues; US Bureau of Labor Statistics, various issues.)

15.5 Implications

The experience of the US beverage container industry suggests that material substitution generally is the result of technological change and inward shifts in production isoquants, rather than changes in material prices and shifts along stationary isoquants. This, as noted earlier, does not mean that materials prices are unimportant. Some of the R&D efforts generating new innovations and producing shifts in isoquant curves presumably are motivated by material prices and the desire to cut materials costs. In such cases, however, material prices influence substitution indirectly by affecting the rate and direction of technological change. Material substitution, itself, is largely the product of technological change.

Beverage containers, of course, are but one of the many material markets, and one should be cautious in drawing general conclusions on the basis of one case study. Still, the strong support the fluctuating fortunes of glass, steel, tin, chrome, aluminum and more recently plastics in the beverage container market provide for the alternative view of material substitution, raises the question, does it matter? What are the important applications, if any, that follow from this finding?

One implication, discussed more fully elsewhere (Tilton, 1983, Chapter 1), is that our understanding of the relationship between material prices and demand, at least as conventionally portrayed in the theory of the firm and economics in general, is flawed. This relationship is described by the downward sloping demand curve, which is normally drawn as stable, continuous, and reversible. If, however, material prices primarily affect material demand indirectly by shaping the level and direction of R&D and the resulting flow of new technology, and if material substitution in turn is an important determinant of material demand, these assumptions are no longer likely to be valid. A rise in price over a limited range may have little or no effect on demand, but once beyond some threshold may stimulate new technology that shifts the demand curve downward. After the new technology has been introduced, it is unlikely the process can be reversed. A fall in price to its original level will not recapture the lost demand.

Perhaps of even greater importance, the traditional view of material substitution may impede our efforts to learn more about this important economic phenomenon. In examining material substitution, it encourages us to concentrate on changes in material prices and the factors responsible for these changes. It predicts we will find material substitution mostly in those uses experiencing significant changes in the prices of competing materials. The alternative view, in contrast, urges us to focus on new innovations and the important factors behind their generation and diffusion. It anticipates material substitution will be the most pervasive in those applications experiencing rapid technology change even if the prices of competing materials are relatively stable. So the two views suggest quite different research agendas for future investigations into the nature and causes of material substitution.

Acknowledgments

I am grateful to Carmine Nappi for his helpful comments on an earlier version of this study.

References

American Metal Market (weekly), Fairchild Publications, New York, NY, USA.

Can Manufacturers Institute, 1986, *Can Shipments Report 1985*, Washington, DC, USA.

Canavan, P., 1983, *The Determinants of Intensity-of-Use: A Case Study of Tin Solder End Uses*, Ph.D. dissertation, Pennsylvania State University, University Park, PA, USA.

Chase Econometrics, 1986, *Materials Competition in the Beverage Container Industry to 1995*, Bala Cynwyd, PA, USA.

Demler, F.R., 1980, *The Nature of Tin Substitution in the Beverage Container Industries*, Ph.D. dissertation, Pennsylvania State University, University Park, PA, USA.

Demler, F.R., 1983, Beverage containers, in J.E. Tilton, ed., *Material Substitution: Lessons from Tin-Using Industries*, Resources for the Future, Washington, DC, USA.

Nappi, C., 1986, Metals demand and the American container and packaging industry: 1970–1990, *Materials and Society* **10**(3):455–489.

National Soft Drink Association, 1986, *Statistical Profile 1986*, Washington, DC, USA.

Tilton, J.E., ed., 1983, *Material Substitution: Lessons from Tin-Using Industries*, Resources for the Future, Washington, DC, USA.

US Bureau of Labor Statistics (monthly), *Producer Price Indices for Commodity Groupings*, Government Printing Office, Washington, DC, USA.

US Bureau of Mines (annual), *Mineral Commodity Summaries*, Government Printing Office, Washington, DC, USA.

Chapter 16

Morphological Analysis, Diffusion, and Patterns of Technological Evolution: Ferrous Casting in France and the FRG

Dominique Foray and Arnulf Grübler

16.1 Introduction

The historiography of technical change has demonstrated that the process of technological diffusion is in itself also a developmental process. In other words, it is in its diffusion throughout the economy that a technology acquires its industrial and economic properties, transforms itself, and widens the initial market in which it was adopted. On the basis of these dynamic properties of the diffusion process, some authors have been hasty in inferring the theoretical impossibility of formal representation, since the objective of the diffusion is not the same at the beginning, in the middle, and at the end of the process. It appears to us, however, that the interest in a formal representation resides precisely in the possibility of periodizing the diffusion process, with the aid of criteria that can take into account the principal transformations of the technology under consideration. The diffusion process can thus be considered as a series of competitions at given times between a technology A, which is in the middle of a transformation, and other technologies (B, C, and D) with respect to those functions that A is successively able to assume. Generally these successive competitions will occur in ever larger markets as A progressively enlarges its initial functional characteristics. It is therefore possible to interpret the characteristics of the diffusion pattern of a given period on the basis of the manner in which competition developed throughout a previous period.

The first part of this chapter consists therefore in a complete and comprehensive morphological analysis (MA) of a set of (process) technologies for a particular industrial activity, in this case ferrous casting. Through the MA approach proposed, we will be able to define the criteria of the periodization of the diffusion process for the technology under consideration. More generally, we intend to show the importance and fruitfulness of an explicit and formal methodology in defining the technologies competing/diffusing in a particular market, which by its comprehensive nature, is not time-dependent or results simply from the aggregation level available in industry statistics.

In the second part we use the results of our MA of the technological trajectories in the casting industry to analyze their diffusion in two countries, France and the Federal Republic of Germany (FRG). We first describe the very different patterns of the technological trajectories in the two countries. We then continue to discuss the possible driving forces behind the *locking-out* of the gasifiable pattern process technology (GP process) in France and its diffusion in the FRG, followed by a quantification of the diffusion process based on standard diffusion methodology. This will be based on a simple Fisher–Pry (1971) type of technological substitution model. On the basis of the MA we describe the diffusion of the GP process as proceeding by successively filling two market niches: first, small batch-size production and later, following improvements in the technology, also mass production of ferrous castings. In the case of the FRG we point out the extreme importance of the early start of the diffusion process of the GP process technology inside a small initial market niche, which generates a process of accumulation of knowledge and learning (this was not the case in France) leading to the widening of the initial market niche.

The study of the diffusion trajectories (Section 16.3), which develop within a well-defined morphological space (Section 16.2), allows us to propose in the final section the historical pattern of evolution of casting technology. Thus our approach moves from a morphological arborescence to an evolutionary tree[1] with the help of the analysis of the diffusion and selection mechanisms for the technologies under consideration.

With respect to the results of this work, we can make one analytical and one methodological observation. First, this case study provides insights into the conditions for exit from a *lock-in* situation.[2] Second, the MA helps avoid misinterpretation and provides a clear theoretical rationale concerning the asymmetrical character and the discontinuities of the diffusion trajectory of the GP process.

Finally, it is our contention that the suggested three-step (morphological, standard diffusion, and evolutionary) analysis permits a better understanding of the historical pattern of evolution of a given technology.

16.2 Morphological Analysis of Technological Trajectories

In this section we propose a complete MA in order to construct the morphological space for the technological evolution of ferrous casting. The MA also permits us to define the relevant relations of rivalry between the technologies under consideration.

The Morphological Space of Casting Technology

MA is a technique for identifying, indexing, counting, and parameterizing a collection of all possible devices (processes) to achieve a specified functional capability. An MA is made up of the following steps: existence of a well-structured problem, identification of the parameters of the (technical, functional) characteristics, subdivision of each parameter into cases or *states* p^1k, p^2k, p^nk, and identification of the various combinations. In addition, we use the following definitions:

> **Morphological space** (p^jk) consists of a set of discrete points or *coordinates*, each corresponding to a particular combination of parameters. The space has as many dimensions as parameters.

Morphological distance between two points in the space is the number of parameters differing from one another in two configurations.
Morphological neighborhood is a subset of points, each of which is morphologically close to the other.
Technological breakthrough is achieved when a new configuration is obtained.

An MA starts with the construction of a morphological space for a particular set of technologies or products in order to understand comprehensively the whole environment into which they are embedded, and thus not to *miss* a technological route of possible future development. The morphological space is defined by any number of dimensions and subdivided into elementary spaces which show the *state* of the technology considered.

Firstly, the functional capabilities of the technology must be stated precisely. In this case, the problem consists of realizing ferrous metal products by a casting process (molding technology). Then in connection with this definition, four characteristic parameters are identified and subdivided:

P_1: The nature of the pattern (P_1^1: permanent, P_1^2: lost);
P_2: The nature of the mold cavity (P_2^1: hollow, P_2^2: full);
P_3: The stabilization force (P_3^1: chemical, P_3^2: physical);
P_4: The bonding method (P_4^1: simple, P_4^2: complex).

Finally, a hierarchy of these parameters is defined in order to take into account the compatibility constraints between the various states of the different parameters. For example, in our case, a permanent pattern (P_1^1) is not compatible with a full cavity mold (P_2^2) which in turn implies the use of a physical stabilization force (P_3^2). This hierarchical relation between the parameters ($P_1 > P_2 > P_3 > P_4$) leads to the morphological space of the molding processes, being represented as an *arborescent structure* (*Figure 16.1*), which gives a systematic representation of all possible alternatives to the casting problem.

In terms of graph theory, an *arborescent* structure is a tree with an original node [that is a point (a), where each other vertex can be attained by a part coming from (a)]. A graph which possesses an original node is *quasi-strongly connected* [for all pairs x, y, there exists a vertex $z(x, y)$ from which a path to x and a path to y begins].

The properties of a quasi-strongly connected graph will be used in the following to define the relevant relation of rivalry between technologies in the morphological space with which we are concerned.

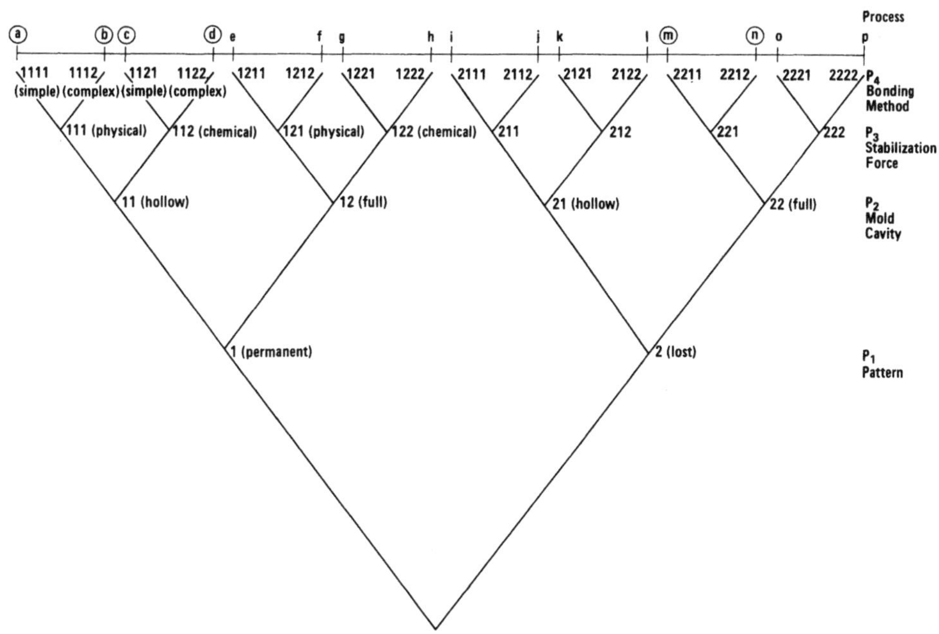

Figure 16.1. The morphological space of molding processes (with 4 parameters) and realizable (non self-contradictory) technological solutions suitable for mass production (*a, b, c, d, m, n*).

Let us now introduce some precisions:

- The 31 vertices of the tree do not represent the technical processes. These processes are located above the final branches of the graph. Thus, each process corresponds to a given combination of the states of the four parameters of the morphological space.
- The MA applied to molding technology results in 16 distinguishable combinations for four parameters (*a* to *p* in *Figure 16.1*), although some of them are self-contradictory: some states of one parameter are not compatible with some states of another parameter. Therefore, the combinations (*e, f, g, h*) are impossible, given the incompatibility between the permanent nature of the pattern and the full nature of the mold cavity. (*p*) is also a self-contradictory combination. When the impossible solutions are eliminated eleven solutions remain which must be considered.
- We are not yet capable of formulating any conclusion concerning the economic value of each combination, or their relative contribution to

the output (i.e., their market shares) of the sector. The goal of the MA is instead to provide a comprehensive definitional structure of the process technologies available and a taxonomy of their evolution. The second interest of the MA lies in the possibility of defining rigorously the competing technologies.

Morphological Neighborhood and Breakthrough: The Relation of Rivalry

The specification of rival technologies includes two notions:

- A notion of **substitutability**; two technologies that do not have the same basic function cannot be considered as being in competition. This basic function refers both to a dimensional criteria (for example, mass production) and to a qualitative criteria (for example, a given degree of complexity of products). According to this first constraint, we can conclude that five solutions (*i, j, k, l, o*) are inadequate for mass production and consequently not in competition. But, the solutions (*a, b, c, d, m, n*) are substitutable.
- A notion of **morphological distance** (MD); it is essential to define theoretically a technological change, either as an improvement of an existing technology, or as the emergence of a rival technology. We argue that competing technologies are separated by a given morphological distance which is estimated below. The MD will be calculated on graph G (*Figure 16.2*), from which the self-contradictory solutions are eliminated, as well as the solutions which are inadequate for mass production.

$G = (X, U)$, is the couple; constituted first by a set $X = (x_1, x_2, \ldots, x_n)$, and second by a family $U = (u_1, u_2, \ldots, u_m)$ of elements of the cartesian product $X \times X = [(x, y)/x \epsilon X, y \epsilon X]$.

This graph displays the properties of an arborescent structure as discussed above. In order to estimate the MD between two points in the space (i.e., the number of parameters differing from one another in two configurations), we use the notion of path.

A path of length $q > 0$ is a chain of a particular type: $\mu = (u_1, u_2, \ldots u_q)$, such as for each arc u_i (with $i < q$) the terminal extremity of u_i coincides with the initial extremity of u_{i+1}. The MD between two terminal vertices (two processes) is the length of the corresponding path μ, i.e., the number of arcs of the sequence:

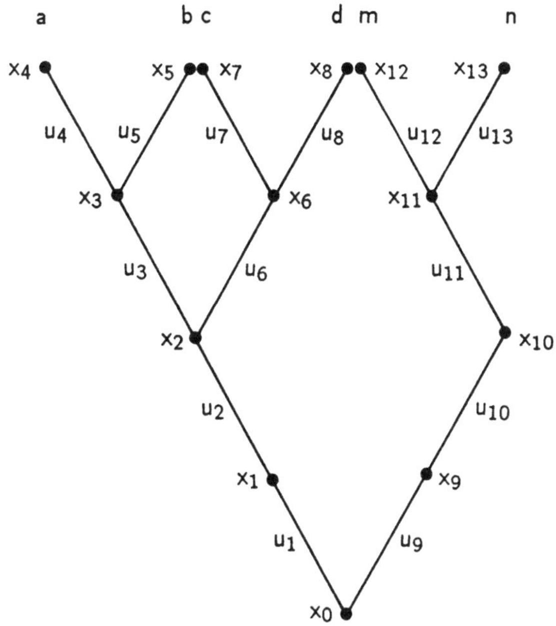

Figure 16.2. Representation of graph G defining process technologies for mass production of ferrous metal castings.

$$\begin{array}{lllll} MD(a,b) & = & \mu^2 & = & (u_4, u_5) \\ MD(a,c) & = & \mu^4 & = & (u_4, u_3, u_6, u_7) \\ MD(a,m) & = & \mu^8 & = & (u_4, u_3, u_2, u_1, u_9, u_{10}, u_{12}) \end{array} \quad (16.1)$$

On account of the hierarchical character of the graph, the estimation of the value of each arc should take into account a weighting coefficient reflecting its proximity to the original node.

We must then define a critical distance. Concurrent with this definition some technological changes occur inside a morphological neighborhood while others occur outside and can thus be defined as an emerging rival technology. According to the theory of the quasi-strongly connected graph, this critical distance is given by the radius of graph G.

The directed distance $d(x_i, x_j)$ is the length of the shortest path from x_i to x_j. The "associated number" of a vertex x_i is $e(x_i) = max\ d(x_i, x_j)$ with $x_j \epsilon X$ and $x_j \neq x_i$. The "center" is a vertex x_o with a minimum associated number. $e(x_o)$ is called the "radius" of graph G and is denoted as $p(G)$. In *Figure 16.2*, $p(G) = 4$. Thus, $(MD \leq 4)$ defines a morphological neighborhood and ($MD > 4$) defines a technological breakthrough (i.e., the emergence of rival technologies).

This morphological procedure results in the identification of two competing technologies: the sand molding (SM) process, corresponding to the combination of parameters (a, b, c, d) and the gasifiable pattern (GP) process, corresponding to the combination (m, n). Technological competition, which will generate a macrostructure in the industry, occurs therefore at the level of the parameter P_1 (permanent or lost pattern, *Figure 16.1*). Indeed, that is the level where the choice of firms can be analyzed in terms of continuity (i.e., technical change within a morphological neighborhood, for example from a to b) or of a morphological breakthrough (for example, changing from a to m). While technical change within a morphological neighborhood implies only a change of artifacts (incremental innovation), the incorporation of a rival technology (i.e., the commitment in another technological trajectory) implies both changes in artifacts and in the knowledge base.

Let us now discuss some of the aspects concerning the economics of technological competition for the case of molding technology.

Economics of Technological Competition

From an economic point of view, we attempt to characterize the technologies in competition (SM process versus GP process) at two complementary levels.

- *Technical complexity and simplification of the operating methods.* This first level refers to one of the characteristics of technical evolution (Foray, 1985): as technological processes become more complex, operating methods tend to become more simplified. The main steps of production used in both the SM and GP process are shown in *Figure 16.3*.

  Thus, the GP process enables an extreme simplification of the operating methods:

  > The GP process involves investing an injection molded foamed polystyrene pattern in a free flowing magnetizable molding material. Immediately prior to pouring, the molding material is rigidized by a powerful magnetic field. During casting, the polystyrene pattern volatilizes in the face of incoming metal stream which occupies the void left by the gasified pattern. Shortly after the casting has solidified, the magnetic flux is switched off and the flask containing the casting is taken to the knock-out station [Gupta and Toaz, 1978].

  But this simplification of operating methods is associated with increased technical complexity: a low level of complexity (SM process)

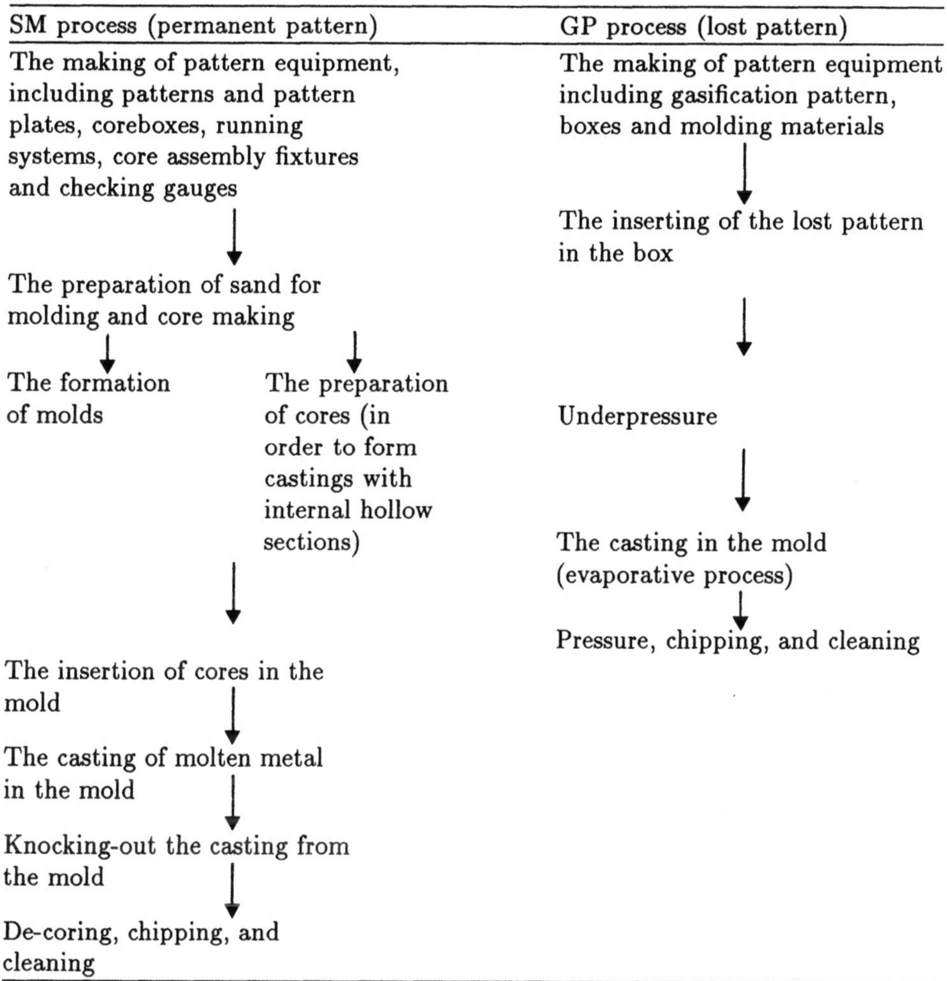

Figure 16.3. The main production operations used in the SM and GP molding processes.

corresponds to more complicated operating methods, while a high level of complexity (GP process) corresponds to more simplified operating methods. The history of the casting industry's technical progress clearly shows a process of increasing technical complexity and a corresponding simplification of operating methods.

- *Structure of costs and economies of scale.* The importance of learning in the finishing processes plus the relatively minor level of learning in the

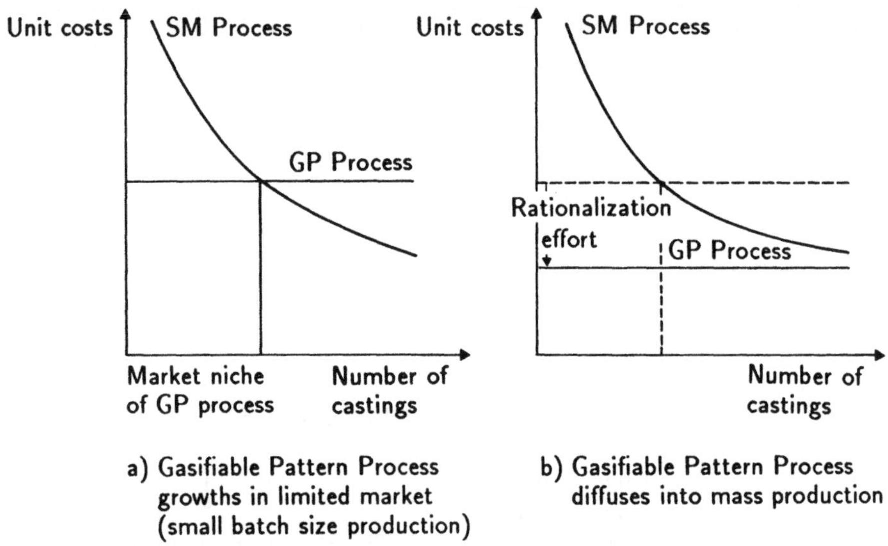

Figure 16.4. Evolution of the cost structure and two successive steps of market penetration for the gasifiable pattern (GP) process: (a) the GP process grows in a limited market (small batch-size production); (b) the GP process diffuses into mass production.

preparation and pouring processes, are features which affect the conditions for economies of scale in both SM and GP technologies. However, the problem as to whether pattern costs are included in the initial costs or not, represents a key-discriminatory feature between the competing technologies: in the case of SM processes one of the main economies to be achieved by increasing output of individual castings is the distribution of pattern costs. The higher the relative importance of pattern costs (the cost of a wooden pattern would be about 25% of the cost of a metal pattern) the more crucial is the search for mass production.

On the contrary the cost of a lost pattern cannot be included in the initial costs. Given that a lost pattern can be utilized for a unique casting, it is necessary to produce as many patterns as products. Therefore there is no direct relationship between the pattern cost per unit and the importance of the run, so that the decrease of the pattern cost per unit produced can be achieved only by the rationalization of the production of patterns. Until such rationalization efforts are effected, the GP process is thus inadequate for mass production [*Figure 16.4*(*a*)]. This flat pattern of the costs per number of castings explains both the limits of the GP process and its competitive advantage over the SM process for the production of small batch sizes: in this period the GP process diffused inside a small market niche only where it was in competition with the SM process for the unit production of very complex and large products. After the rationalization of the production of patterns [*Figure 16.4(b)*], the GP process also became economic for mass production: competition between the SM and GP processes becomes more and more important.

Thus, the evolution of the cost structures for the GP process implies a periodization of the diffusion process, the formal analysis of which is presented in the following section.

16.3 Diffusion Trajectories in France and the FRG

In *Figure 16.5*, which shows the output of the foundry industry in France and the FRG, two important features can be observed and documented. First, the evolution of the foundry industry follows a very similar path in terms of output volume both in France and in the FRG. A period of saturation and contracting markets followed the period of growth and expanding markets and in each case the turning point occurred in the early 1970s. Second, since 1960, the GP process started to diffuse in the FRG while in France it was *locked out*, and remained in a very minor market share position.

Figure 16.5 also shows that in the case of the FRG, the diffusion pattern of the GP process was not influenced by the contraction in the global market (i.e., decline in output volume) of the industry. Furthermore, the output figures of the GP process were apparently not affected by the strong fluctuations in the total market volume. On the other hand the evolution of the output of the SM process appears to follow closely the decreases in global output volumes and market fluctuations.

It is our contention that it is important to differentiate in diffusion research between two important situations with respect to the evolution of the market in which technologies compete. In the first place, when the market expands rapidly, diffusion takes place via differential growth rates, i.e., changing relative market shares are the result of one technology growing faster than another. This is in sharp contrast to the diffusion of a technology in a saturating, even declining market, as in our case. We maintain that under such market conditions, effective diffusion calls for a higher comparative

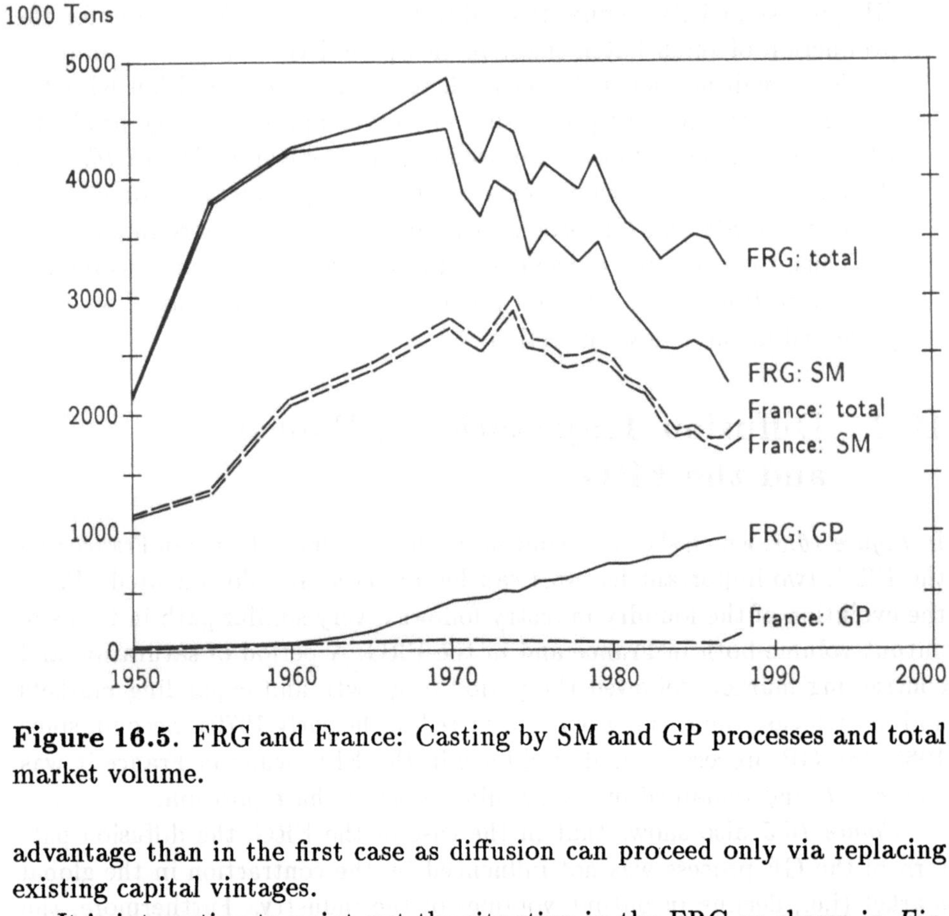

Figure 16.5. FRG and France: Casting by SM and GP processes and total market volume.

advantage than in the first case as diffusion can proceed only via replacing existing capital vintages.

It is interesting to point out the situation in the FRG as shown in *Figure 16.5*. Despite strong market fluctuations the output figures of the GP process evolve very regularly, i.e., they are not affected by short-term business cycle variations in market volume. Conversely, the SM process takes the full burden and acts as the *swing supplier*, i.e., in response to demand fluctuations.

It is our contention that the difference in behavior toward demand fluctuations is indicative of a high comparative advantage differential between the two processes in the FRG.

16.3.1 In search of specific factors of diffusion in France and the FRG

The preceding discussion of the morphological structure of the technological trajectories leads us to conclude that prior to 1970, the GP process could compete only for the casting of small batch sizes. In a second period, after a technological breakthrough involving the conditions of production of lost patterns, the GP process (which corresponded then to the combinations of parameters m and n) could effectively diffuse also in mass production and compete with the SM process. Thus, in order to explain the differences between the national patterns of diffusion, it is necessary to divide the adoption process of GP technology into two phases: the diffusion into the first market niche of complex, small-series production; and the subsequent diffusion into the mass production market.

The substitution curve, the parameters of which will be commented upon in Section 16.3.2, is illustrated in *Figure 16.6* and shows that a rapid substitution of the SM by the GP process in the first market niche of complex, small-series production took place in the FRG during the period 1960–1975. We must therefore explain the reasons for this rapid first diffusion period in FRG and then identify its influence on the diffusion trajectory of the second period.

Dynamics of Demand Structure and Profitability

The first driving force relates to the market niche for complex, small-series production. This highly specialized market expanded rapidly in the FRG in the early 1960s (this was not the case in France) and was (as discussed at the beginning of Section 16.2) an important factor in the rapid diffusion in the first phase. The documentation of this factor is, however, seriously hampered owing to the absence of relevant statistics prior to 1970. A second factor deals with the specific comparative (economic) performance of the GP process in the FRG during the first diffusion period. *Figure 16.7* depicts the sharp differences between the relative value-added for the two processes, in particular during the first phase of the diffusion of the GP process. The low level (factor 1.1) of the comparative advantage (value-added) in France could explain the disinterest by the French firms in the new process. Furthermore, the evolution of the relative value-added between the GP and SM processes (from 1.4 to 1.1) in the FRG between 1970 and 1987 correlates with our hypothesis that two phases exist in the diffusion of the GP process.

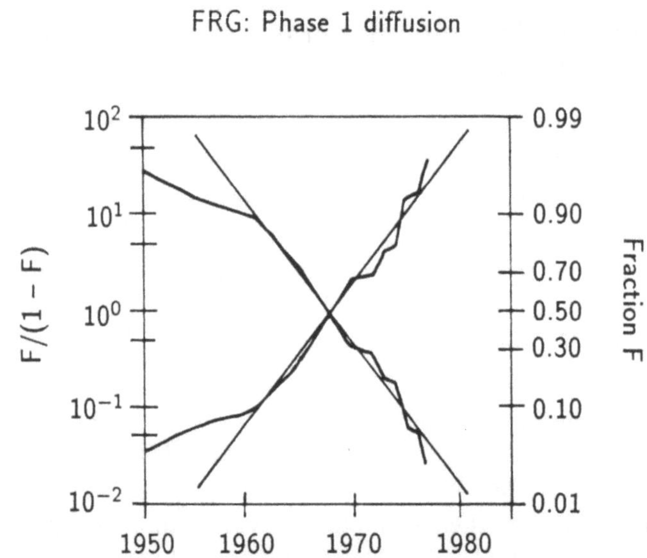

Figure 16.6. The substitution process of SM by GP technology in the first diffusion period (competition for small batch-size production only) in the FRG.

During the first phase, the market niche is made up of complex, small-series production and the comparative economic advantage of the GP process are correspondingly higher than during the second phase of diffusion where it approaches the value-added of mass production (i.e., a relative value-added ratio of 1).[3] Thus, the differential represents an initial explanation for the rapid diffusion of the GP process in its first market niche in the FRG. One question remains to be answered. How did the diffusion pattern in the FRG in the first period influence the outcome of competition in the second period?

Knowledge Accumulation and Learning During the First Period of Diffusion in the FRG

During the first period of diffusion the GP technology was rapidly adopted in the FRG, in spite of the fact that its adoption entailed a strong technological breakthrough for the innovative firms. The fundamental feature in this first diffusion phase is what occurred to some extent *underground*. The first diffusion phase generated a process of accumulation of knowledge, and included, via adequate institutional arrangements, the creation of a

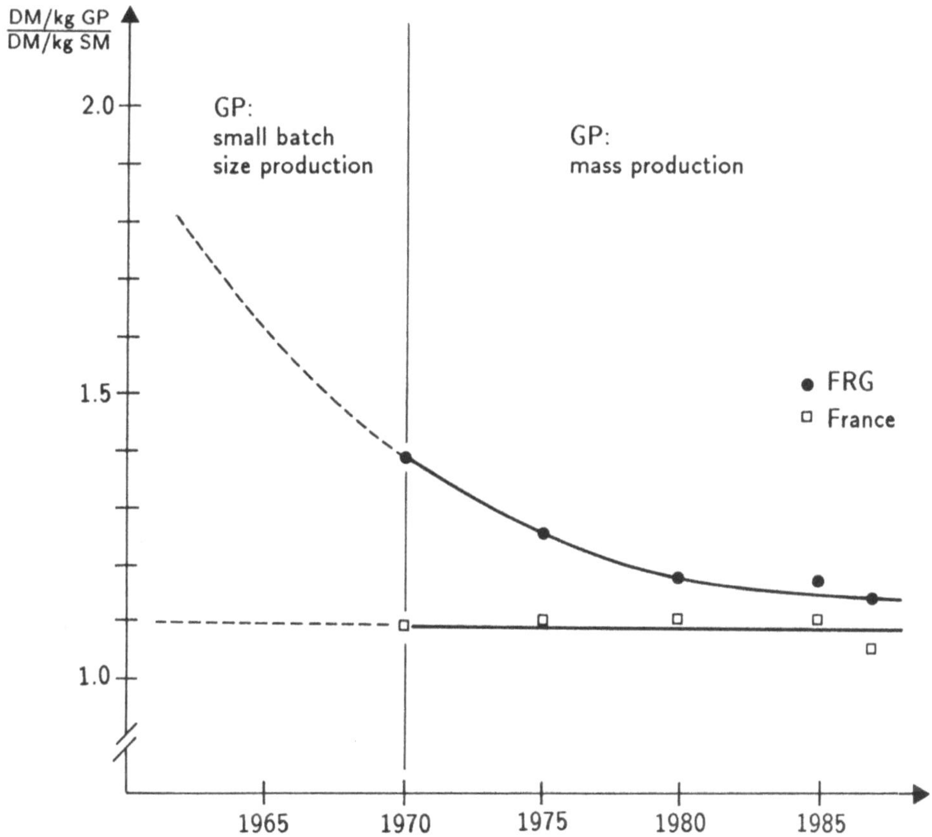

Figure 16.7. FRG and France: Relative value-added (profitability) between the GP and SM casting processes.

technological foundation in Ludwigshafen am Rhein, with strong participation by German firms (foundries and chemical enterprises). Research programs were oriented toward improvements in the use of polystyrene patterns to produce metal castings and the systematic generation of minor innovations, which were required for the industrialization of the GP process.

Thus, optimal pouring rate, adequate density of polystyrene, etc. were systematically investigated. After the seminal conception of the process (which can be interpreted as a jump in technical knowledge, i.e., a breakthrough, in our terminology), research programs were conducted in the FRG to solve the technological problems which continually occurred during the initial diffusion phase (Foray and Lebas, 1986). Thus, knowledge was accumulated during the first phase of diffusion also through an adequate institutional arrangement. More generally, this initial diffusion in a highly specialized market permitted the GP process to access, for the first time, those mechanisms (*cf.* [2]) related to increasing returns to adoption, learning by using, economies of scale in production, and informational increasing returns, while at the same time being *protected* by a high value-added differential.

Thus, the first phase of diffusion facilitated a learning process, resulting in the transformation of the technical process (from o to m or n, *Figure 16.1*), enlarging its initial functions, thus providing the basis for experimentation, incremental improvement innovations, increasing returns to adoption, etc., necessary for subsequent diffusion into the whole market niche.

The Dynamics of the Two Phases

As far as exiting from a *lock-in* situation is concerned, the German and French examples are quite instructive. The *lock-in* concept allows us to explain how a new and intrinsically superior technology may be impeded from supplanting an older technology. This is supported by the following quote:

> New inventions are typically very primitive at the time of their births. Their performance is usually poor, compared to existing (alternative) technologies as well as to their future performance [Rosenberg and Frischtak, 1983, p. 147].

Thus, when a new technology is introduced in its initial (and therefore primitive) form, it has virtually no chance of asserting itself, even if the old technology is *inherently inferior*. The latter has profited from its monopolistic period and entrenched itself materially (via technological interrelatedness) and intellectually (via *sui generis* evaluation norms) as the dominant productive paradigm. In this respect, our case study illustrates the crucial importance of an initial diffusion in a highly specialized market in order to overcome a technological *lock-in*. In this first period the new technology, *protected* by a high value-added differential, may improve within a *quasi in vitro* environment. Thus shielded, the new technology acquires industrial properties via the mechanisms related to increasing returns to adoption, gradually armoring itself for competition. Between 1950 and 1970, the GP process improved in a virtually underground fashion in the FRG; it was later able to enter the main competition arena under auspicious conditions. Having missed the first phase, France is now missing the second one also.

The diffusion of an innovation in a relatively minor market probably represents a unique tool for *preparing* the new technology for competition in its industry's major market. As Utterback (1987) suggests:

> Because performance will be initially unreliable and costs higher, a new technology will tend to start in a relatively small market niche where its unique performance advantages are critical – one ordinarily not occupied or of not great importance to the producers of the established product. Crude as it is, the new technology will gain ground by competing in these submarkets and its use will expand by means of its capture of a series of them.

The specificity of the national diffusion trajectories in France (*lock-out*) and the FRG (diffusion) is therefore based on the link between the two phases. According to Silverberg (Chapter 8): "A technology policy that does not take the interdependence of these two aspects into account will always be inherently flawed."

16.3.2 The formal analysis of the diffusion trajectories

A formal analysis of the diffusion trajectories of the GP process in the FRG through two successive market niches – small batch-size production prior to 1970 and mass production thereafter – is, however, seriously hampered by the absence of relevant disaggregated production statistics. For the diffusion trajectory within the first market niche for small batch-size production we assumed that a constant volume of complex castings was produced in small series in the FRG in the period prior to the mid-1970s in order to calculate the fractional market share of the GP process. For the second phase of diffusion we calculate the diffusion trajectory on the basis of the fractional share in total (tonnage and value) output. This is based on the conclusions of the morphological analysis, which has yielded that the GP process is also in effective competition for mass production in the post-1970 period.

Table 16.1 and *Figure 16.8* summarize the quantification of the diffusion trajectories in the case of the FRG, based on a simple Fisher–Pry type of technological substitution model. The properties and underlying assumptions of this now classical model will not be repeated here; details on the estimation algorithm used can be found in Grübler, Nakićenović, and Posch (1988).[4] In order to increase the analytical resolution of the formal description of the second phase of the diffusion (substitution) trajectory, we have used, in addition to output tonnage, output value (measured in current DM) by casting process in the period since 1970 (data source: Deutscher

Table 16.1. Phases in the diffusion of the GP process in the casting industry of the FRG: diffusion model[a] parameters.

	Fraction of GP in tonnage output	Fraction of GP in output value
Phase 1 (small batch-size market niche), period: 1960–1977	$\Delta t = 13.1$ (14.74) $t_o = 1967.8$ (14.74) n = 10 $R^2 = 0.965$	No data available
Phase 2 (total market including mass production), period: 1970–1987	$\Delta t = 52.4$ (45.76) $t_o = 1997.7$ (45.99) n = 18 $R^2 = 0.992$	$\Delta t = 61.58$ (17.70) $t_o = 1997.9$ (17.88) n = 18 $R^2 = 0.991$

[a]Δt: diffusion parameter, time in years to grow from 10% to 90% market share; t_o: inflection point (50% market share), time of maximum growth rate of market shares. Values in parentheses refer to t statistics of estimated diffusion model parameters.

Gießereiverband, 1975, 1980, and 1987). The estimated diffusion parameters are consistent between the two measures, with the diffusion rate of the GP process calculated on the basis of output value being around 17 percent slower than for output tonnage figures.

In keeping with the differential for the specific value-added (i.e., DM per kg of product) between the two process technologies discussed above, we note that the diffusion rate of the GP process into the first market niche of complex, small-series production is significantly faster (by a factor of 4) than in the second phase of diffusion, i.e., into the lower-value mass production market niche. This indicates that in addition to the higher specific value-added (as a proxy for its relative profitability) for the GP process technology (at least 1.4 in 1970, and most likely larger in the period before), other comparative economic advantages, such as lower production costs in small series, are influential factors which help explain the rapid diffusion of the GP process into the first market segment.

In *Figure 16.8* we show the diffusion (substitution) trajectories in the two successive market niches of the GP process. Particularly noticeable is the regular pattern of the second diffusion phase since 1970. In order to illustrate the decisive structural difference between the technological base in the casting industry between the FRG and France, we have compared the diffusion trajectory in the case of the FRG with the trajectory of the market share fraction of the GP process in France, which appears *locked*

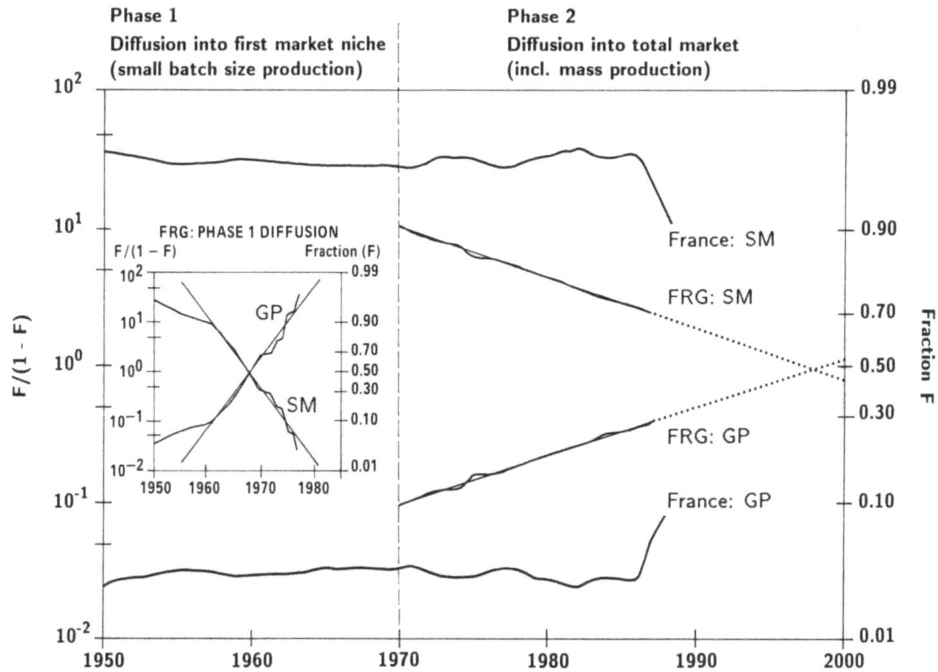

Figure 16.8. The two diffusion phases of GP casting technology in the FRG and its *lock-out* in France.

in at a constant market share fraction below the two percent level. Since 1986, however, this share has increased rather rapidly to the present level of below eight percent of total casting tonnage in France. This could be a first indication that the GP process might be at the beginning of a similar diffusion takeoff as was the case in the FRG some decades earlier.

16.4 Patterns of Evolution

The study of the diffusion trajectories, which develop into a well-defined morphological space allows us to reproduce finally the historical evolution of the casting technology. In relation to the MA (*Figure 16.1*) we are only interested now with the technical processes (located above the final branches in *Figure 16.1*), as industrial applications of the various possible combinations of parameters of the morphological space. However, the MA still remains the basis for the construction of an *evolutionary tree* by identifying two principal alternatives (i.e., the SM process versus the GP process route). All morphological combinations possible will, however, not be described. Only those that have actually evolved and diffused into the industry will be considered.

This last step of the analysis allows us to highlight some characteristics of technical progress: its cumulative character (i.e., evolution of trajectories defined on the basis of stable morphological combinations) on the one hand, and the localized character of learning processes on the other.

16.4.1 Construction of the graph

According to the result of the MA, two trajectories can be distinguished: SM processes (a, b, c, d) and GP processes (m, n). Both trajectories are based on the stability of the P_1 parameter (*Figure 16.1*) concerning the nature of the pattern. One trajectory describes the evolution of permanent pattern technology, and the other, the evolution of lost pattern technology. Apparitions of new morphological combinations are indicated by ramifications (b for SM and r for GP), while all other improvements, which do not create new morphological combinations, are incorporated simply by extending the existing branches of the trajectories. We make use of a data base consisting of 50 innovations in the foundry industry with a technical description and a historical dating of their introduction.

16.4.2 Describing the dynamics of technology

Figure 16.9 emphasizes three key features:

- Clustering of chemical based innovations between 1955 and 1975: lost molds predominantly or completely bonded by chemical means (development of the existing trajectory by changes at the level of the P_3 and P_4 parameters, see *Figure 16.1*).
- Emergence of a rival technology: the GP process (creation of a new trajectory by changes at the level of the P_1 parameter).
- Clustering of physical based innovations between 1970 and 1985: lost molds predominantly or completely bonded by physical means (development of the existing trajectories by changes at the level of the P_3 and P_4 parameters).

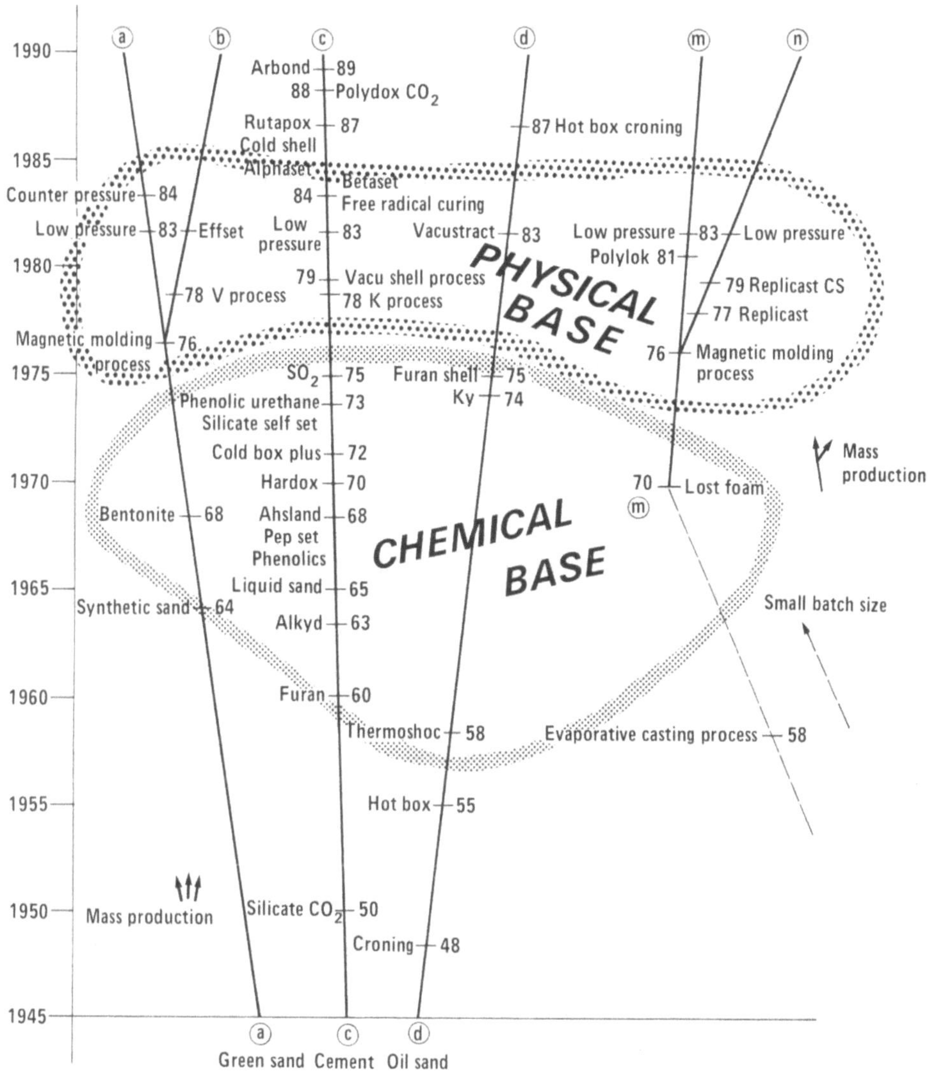

Figure 16.9. Trajectories of the molding processes and clusters of innovations (1945–1989).

The First Cluster of Innovations

The figure shows a first cluster of innovations during the period 1955–1975. This cluster was oriented toward the use of a chemical method for the stabilization of the mold. Originally the chemical methods were used by applying cement, CO_2 gas, oil sand, and shell molding (the croning process) (see bottom of *Figure 16.9*). Then improvements in the application of inorganic and organic binders determined a cluster of innovations (furan, alkyd, phenolics, pep set, bentonite, thermoshoc, etc.). According to the MA, these technological changes cannot be considered to be the emergence of a rival technology (all morphological distances are inferior to the radius of graph G). Since 1958, the GP process was used, but given its specific cost structure discussed above, it was devoted to small batch size and thus was not in competition with the mass production of castings.

The Emergence of a Rival Technology

In 1970, significant improvements concerning the GP process occurred. In particular rationalization in the production of lost patterns (pre-expansion and molding processes of expandable polystyrene) made this process adequate for mass production, so that the GP process (combination of parameters m) became substitutable for all existing SM processes (a, c, and d):

> The future of the gasifiable pattern process appears to be in large production runs using molded polystyrene patterns in unbonded sand. This is in contrast to its original use which was in the production of large short run castings [Bailey, 1982].

According to the MA, this technological change can be considered to be the emergence of a rival technology, given the substitutability of the processes and the morphological *distance* between the two competing processes (superior to the radius of graph G).

The Second Cluster of Innovations

The cluster of physical based innovations (the use of vacuum and magnetic fields) occurred after 1975, the year of the first industrial application of magnetic molding. The magnetic molding was introduced both for SM processes (magnetic molding, V process) and for GP processes. These technical changes were based on new morphological combinations (b and n) without altering the stability of both states of the P_1 parameter (i.e., the stability of both main trajectories).

16.4.3 The national patterns of evolution

Figure 16.10 shows the differences between the technological structures of France and the FRG. This figure is consistent with the results of our previous analysis concerning the diffusion trajectories in France and in the FRG. While the German pattern occupies the total area of the morphological space, the French structure leaves a large part uncovered, i.e., the GP trajectory is *locked out.*

16.5 Conclusion

Our case study was particularly appropriate in showing the advantage of a morphological analysis (MA) approach in technological diffusion analysis. Indeed, the MA of the structure of technological trajectories in the casting industry (*Figures 16.1* and *16.2*) avoids any misinterpretation concerning the asymmetrical character and the discontinuities of the diffusion trajectory of the GP process. On the basis of the morphological space of molding technologies, we can establish that the molding process under consideration (GP) cannot be thought of as a unique unaltered artifact throughout the period of diffusion. In fact, there are two diffusion trajectories corresponding to two combinations of parameters and therefore to two successive market niches. This breakdown into two periods allowed an exit from a *lock-in* situation by emphasizing the crucial nature of the first period of diffusion, where knowledge is accumulated and a process of learning within a *quasi in vitro* environment occurs, allowing a rival technology to develop capable of competing within the industry's entire market.

Notes

[1] Notions of arborescence and tree are used here in their specific meaning of graph theory.

[2] The theory of *lock-in* effects (Arthur, 1989) provides a clear understanding of the mechanisms (increasing returns to adoption) by which a technology may overcome its rivals and how it then generates its own defense mechanisms against – even inherently superior – technologies. The principal sources of the increasing returns to adoption are: learning by using, network externalities, economies of scale in production, informational increasing returns, technological interrelatedness, and the production of *ad hoc* evaluation norms. The last two sources allow us to explain the phenomena of maintaining mature technologies in the long term.

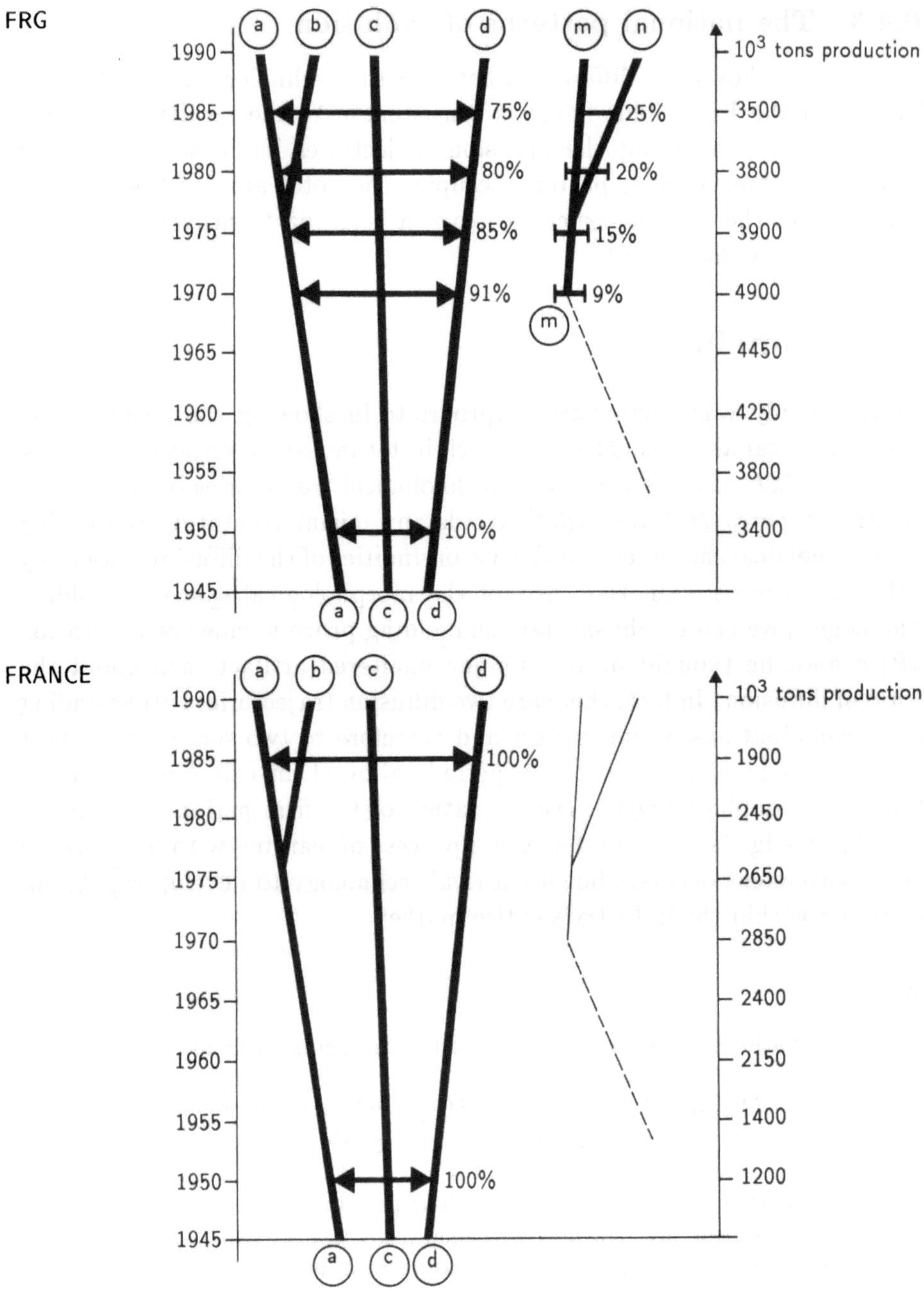

Figure 16.10. National patterns for the trajectories of the molding processes (mass production).

[3] Clearly, the nominal value-added differential illustrated in *Figure 16.7* should be presented in real terms. However, the estimation of real price deflators faces the difficulty that both the structure of the market and the product are changing (as demonstrated in the discussion above) and are consequently not reflected appropriately in the price index published by the industry.

[4] The use of the Fisher–Pry model to describe the diffusion of the GP process in two distinct periods is based on the argument that the theoretical structure of this model is appropriate for taking into account this mix between a phenomenon of continuity and a two-period analysis. However, the question of the use of other types of diffusion models (threshold/probit models) still remains open.

References

Arthur, B., 1989, Competing technologies, increasing returns and lock-in by historical events, *Economic Journal* **99**:116–131.

Bailey, M., 1982, Understanding the evaporative pattern casting processes, *Modern Casting* **4**:82–92.

Comité des Associations Européennes de Fonderie, *Die Europäische Gießereiindustrie im Jahre: 1984, 1975, 1986*, Köln, FRG.

Deutscher Gießereiverband, *Die Gießereiindustrie im Jahre: 1975, 1980, 1987*, Düsseldorf, FRG.

Fisher, J.C. and Pry, R.H., 1971, A simple substitution model of technological change, *Technological Forecasting & Social Change* **3**:75–88.

Foray, D., 1985, Innovation Majeure et Transformation des Structures Productives, *Revue Economique* **5**:1081–1113.

Foray, D. and Lebas, C., 1986, Diffusion de l'Innovation dans l'Industrie et Fonction de Recherche Technique, *Economie Appliquée* **3**:615–650.

Grübler, A., Nakićenović, N., and Posch, M., 1988, *Methods of Estimating S-Shaped Growth Functions: Algorithms and Computer Programs*, International Institute for Applied Systems Analysis, Laxenburg, Austria, mimeo.

Gupta, V. and Toaz, M., 1978, New molding techniques: A state of the art review, *AFS Transactions* **48**:1122–1134.

Rosenberg, N. and Frischtak, C., 1983, Long waves and economic growth: A critical appraisal, *The American Economic Review* **73**:146–151, Papers and Proceedings.

Syndicat Général des Fondeurs de France, 1988, *Production des Fonderies de France et d'Allemagne Occidentale de 1945 à 1985*, Paris, France.

Utterback, J., 1987, *Invasion of a Stable Business by Radical Innovation*, MIT Industrial Liaison Program, MIT, Cambridge, MA, USA.

Chapter 17

Competition and Complementarity in Diffusion: The Case of Octane

Robert U. Ayres and Ike Ezekoye

17.1 Introduction

The standard ontogenic (*life-cycle*) model of technological evolution can be characterized briefly as follows (Ayres, 1987): (1) a radical invention (**birth**) creates a new technology; (2) it is commercialized on the basis of performance and rapidly developed by a series of improvements and modifications (**infancy**); (3) it is successful enough in the marketplace to attract many variants and imitators who hope to exploit a growing market (**adolescence**); (4) the pace of technological change finally slows down enough to permit standardization and exploitation of economies of scale, and competition on the basis of price rather than performance (**maturity**); and finally a new and better technology supplants it (**senescence**).

The standard model involves substitutions in the adolescent and senescent stages. During the adolescent stage, the new and dynamic technology is gradually penetrating the markets of its predecessor. During the senescent stage it, in turn, is being displaced from its markets by its successor. The substitution of a new technology for an older one is often modeled as a deterministic process, following a simple mathematical formula such as a logistic function or a Gompertz curve (see, for example, Linstone and Sahal, 1976; Mahajan and Peterson, 1985; Mahajan and Wind, 1986).

However, complex social systems – including the system of innovation, adoption and diffusion of technology – are inherently nonlinear. As such, they must be expected to exhibit the characteristics of nonlinear dynamical systems. Among these characteristics is the occurrence of **non-equilibrium** steady-state behaviors (such as limit cycles and quasi-periodic motion) that temporarily emulate the behavior of simpler systems, but eventually depart from it (Crutchfield *et al.*, 1986). In short, social systems cannot be expected to always behave in accordance with any given simple model. Indeed, simple behavior, when it does occur, is likely to be an example of non-equilibrium steady state. Hence, from the standpoint of fundamental dynamical theory it seems likely that more can be learned by analyzing cases where the simple models fail than cases where they seem to work well (e.g., Fisher and Pry, 1971).

In particular, the simple deterministic substitution model that is normally assumed assumes that a substitution process, once it has proceeded past a certain threshold, inevitably proceeds to completion (unless it is interrupted by a further substitution). This implies the existence of an underlying self-reinforcing (*lock-in*) mechanism of some sort. Such mechanisms are intrinsically nonlinear in nature. A number of examples have been examined by Arthur (1983, 1988a, and 1988b). Obviously, the large number of cases where the substitution process has proceeded according to this script can be regarded as indirect evidence of the pervasiveness of self-reinforcing mechanisms. Yet, there are significant exceptions. Such a case is the subject of this chapter. We examine the technological evolution of fuels for spark-ignition internal combustion engines (e.g., automobile engines) since the beginning of the present century. The chapter concludes with a discussion of some possible explanations for the failure of "antiknock" additives to displace cracking as a means of raising gasoline octane, or conversely.

17.2 Historical Background

The automobile had no single inventor. It is usually traced to early models by Gottlieb Daimler and Wilhelm Maybach, and Karl Benz (*ca.* 1885). For the next twenty years and more, automobiles were essentially toys for the rich and adventurous. It was not until after 1905 (the year Ford Motor Co. was founded) that automobiles were technologically developed enough to be useful for simple transportation purposes. Even then, for many years, they remained expensive, unreliable and uncomfortable. However, by 1908 the dominant technological *trajectory* had been determined and the industry, led by Ford, began to standardize. The enormously successful *Model T* was introduced in 1908, which symbolically marks the end of the "childhood" phase of the auto industry and the beginning of **adolescence** and consolidation.

This chapter is not about autos, however, but about motor fuel. The relevance of the previous paragraph is simply that after 1908 demand for cars – and, consequently for gasoline – began to rise rapidly. It is important to note that in the earliest days automotive fuel was so-called *natural gasoline*, a medium volatility product of crude oil refining, consisting of fractions boiling in the range between 0–70° C and an octane of 72–75. But this light fraction averages only about 2.4% (by weight) of North American crude and no more than 4.7% of middle-Eastern crude.[1] To increase the output of motor fuel, early refineries blended natural gasoline with the next heavier fraction, *naphtha*, boiling in the range 70–140° C, but with less desirable combustion properties. The blend had an octane level of around 50. For North American crude oils the naphtha fraction averages 6.5% by weight (7.9% for middle-Eastern crude). Thus, while local details differed, petroleum refiners in the USA *ca.* 1910 could only utilize around 9% of their crude oil, by weight, directly for motor fuel.

At the time (1910), 9% of the crude oil was still adequate to supply the automotive demand, inasmuch as there were as yet relatively few vehicles on the roads. Indeed, the biggest market for petroleum products was still *illuminating oil* (kerosene), which constituted about 15.6% of the weight of the refinery product stream. However the heavier, lower-value fractions, *gas oil* (now known as heavy distillate, diesel oil, or heating oil), and *residual oil* together still constituted 75% of the refinery output. Gas oil, alone, accounted for about 60% of the product. There were already significant incentives to add value to the heavy fractions by somehow converting them into lighter fractions.

The breakup of Standard Oil Co. of New Jersey (NJ) in 1911 triggered a major innovation, the thermal cracking process. The chief inventor and innovator of the process was William Burton, a vice president of one of the spinoffs from Standard Oil, Standard Oil Co. of Indiana (now renamed Amoco). With its refineries on the shores of Lake Michigan, and its major market the rapidly growing Chicago metropolitan area, Indiana Standard was faced with an exceptionally rapidly growing market area, together with a rather limited access to crude oil.[2] A new technology promising to increase the fraction of crude oil that could be used for motor fuel was very welcome.

Burton's thermal cracking process – heating a batch of heavy gas oil in a closed tank or retort – effectively converted about 20% of the gas oil into a light fraction suitable for blending with natural gasoline and naphtha. This effectively doubled the output of motor gasoline from about 9% to around 21%, while simultaneously increasing its *research octane* (RON) rating from 50 to 55. The Burton process was first introduced in 1913–1914 and was enormously profitable to refiners. For this reason it was rapidly adopted by others (*Figure 17.1*). It also set off a great wave of competitive invention and innovation, since other oil companies did not like paying the high royalties demanded by Indiana Standard for what was, essentially a very simple invention. Burton and his colleagues began to improve their first crude batch process. Meanwhile, others entered the field with ideas for continuous thermal processes and (later) catalytic processes.

Table 17.1 summarizes the major innovations in refining after 1913 and *Figure 17.2* indicates the succession of substitutions in refinery technology in quantitative terms, as each technology replaced its predecessor and was, in turn, replaced. (Data for these exhibits has been taken primarily from Enos, 1962, and Lakhani, 1975). It is noteworthy that the substitutions displayed in *Figure 17.2* do seem to fit the standard ontogenic model reasonably well.

From the standpoint of the *octane* industry, Burton's radical innovation of 1913 marks the date of birth. But, what makes this case complicated (and interesting) is that there were two different – and noncomplementary – market interests and consequently two *driving forces* involved. The first, as suggested above, was the petroleum refineries' direct economic interest in increasing the output of high-value motor fuel per barrel of crude oil. Doubling the output of motor fuel per barrel from 9% to 20–21% meant, in effect, that less than half the amount of crude oil had to be discovered, pumped, shipped, and distilled to yield the same amount of salable product.

The second market interest – which created a demand for higher octane *per se* – was shared by the automobile users and manufacturers, but was

Table 17.1. Summary of major cracking technologies.

Name	Specific economic advantage over predecessor	Factors driving innovation
Burton batch thermal cracking process, Indiana Standard, 1913–1914	Increased octane to about 60 and motor gasoline yield per bbl of petroleum from about 9% to 21% or so.	Indiana Standard was created by the court ordered breakup of Jersey Standard; it was left with refining and distribution, but little crude supply. Demand in Chicago area was rising imperative to *stretch* each barrel.
Continuous *tube & tank* thermal cracking process (Clark, ESSO, 1922). Dubbs process (UOP, 1922). Cross process (1922).	Better suited to scale-up than batch process; increased octane to 72, mpg by 22% and output per unit of capital by 50%. Reduced process energy by 20%.	ESSO wanted to invent around Indiana Standard's processes and to invalidate other patents (e.g., Cross). Universal Oil Products (UOP) was created by a group of regional refiners to invent around Indiana Standard's patents because they were unable to license because they were in the same marketing area. UOP sued Indiana Standard to preempt.
Houdry fixed bed (batch) catalytic cracking process (Sun Oil, Socony-Vacuum, 1938)	Increased gasoline yield to 40% of crude, octane to 72. Cut process energy by 2/3.	Initial research in France was prompted by fears of shortages and lack of crude oil in Europe: Backing by Sun Oil Co. was due to a glut of heavy fuel oil and Sun's market niche with a single grade gasoline of higher octane than its competitors.
Continuous fluidized bed catalytic cracking process (ESSO *et al.*, Mobil, Houdry)	Better suited to scale-up than batch process; increased octane to 93–95.	Catalytic Research Associates was formed by Esso, with BP, Shell, Texaco, UOP, MW Kellogg and IG Farben to invent around the Houdry fixed bed process. Members of the syndicate could avoid royalties on the process. Mobil developed its own process for the same reason.

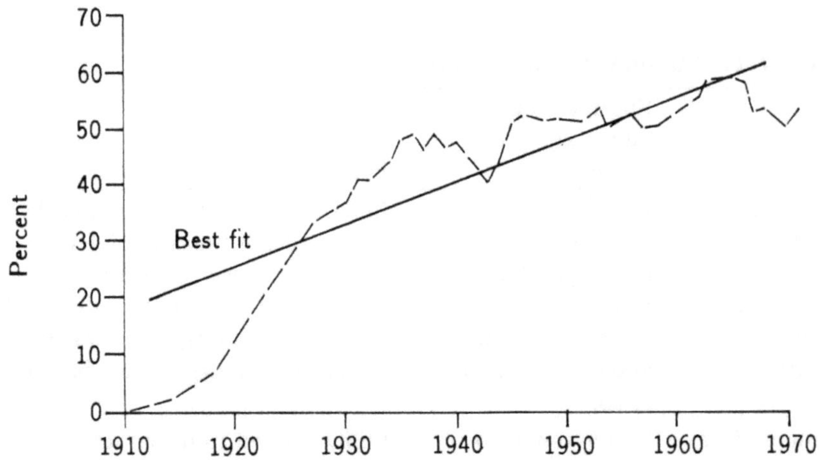

Figure 17.1. Cracking capacity as percent of total crude oil production capacity. (Source: Lakhani, 1975, p. 54.)

to some extent contrary to the interests of the petroleum companies. The conflict and its resolution are part of this story. The inherent characteristics of internal combustion engines are such that both output power and thermodynamic efficiency are functions of the compression ratio of the engine. Thus, high compression engines offer better performance for the car. The compression ratio is the ratio of the volume of combustion products after expansion (exhaust gases) to the volume of the fuel-air mixture at the point of ignition. Since the exhaust gases must be at atmospheric pressure, this is also a measure of the amount of compression in the engine.

For a spark-ignition engine – in contrast to a Diesel engine – the maximum compression is not limited by the geometry of the cylinder and crankshaft, or the tightness of the piston-rings, as might be expected, but by the tendency of the engine to *knock* or *ping*, which cuts power output sharply and can cause damage. Knocking means the octane level of the fuel being used is not high enough to operate at the design compression ratio. The attribute that permits higher compression is called the research octane number or RON, or simply *octane*. It varies considerably from fuel to fuel, depending on its chemical structure, oxygen content and other factors. In general, higher octane fuels permitted higher compression engines, which permitted better automotive performance as well as fuel economy. *Figure 17.3* shows

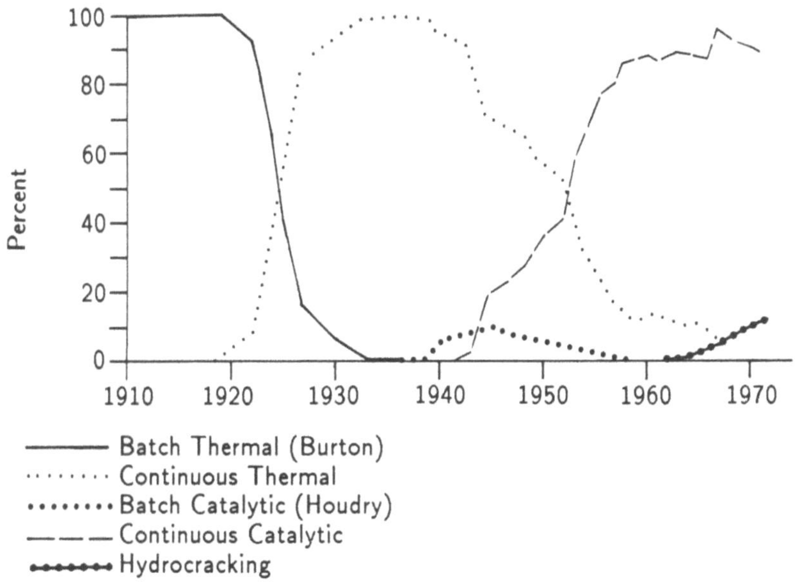

Figure 17.2. Petroleum cracking processes. (Sources: Data prior to 1958 from Ayres, 1987; data from 1958 on from Lakhani, 1975, p. 54.)

the historical progression of octanes from 1930 until 1970, while *Figure 17.4* shows the close parallel with increasing engine compression ratios.

Increasing fuel economy (due to increased octane levels) meant that gasoline sales in volume terms did not increase as fast as automobile usage. On the other hand, every increase in automotive performance attracted more first-time buyers of automobiles, and each additional vehicle in the fleet meant a guaranteed demand for gasoline throughout the life of the car. Thus, the petroleum industry had a somewhat contradictory interest in the *octane race*. On the one hand, as long as petroleum supplies were ample, better fuel economy was not in its direct economic interest. On the other hand, it did share the interest of the automobile manufacturers in attracting more and more people to buy cars, because the more cars people bought the more motor fuel the refiners could sell.

This conflict between short- and long-term interests on the part of the petroleum refiners had one direct implication, however. Given the possibility of increasing octane levels **independently** of changes in refining technology,

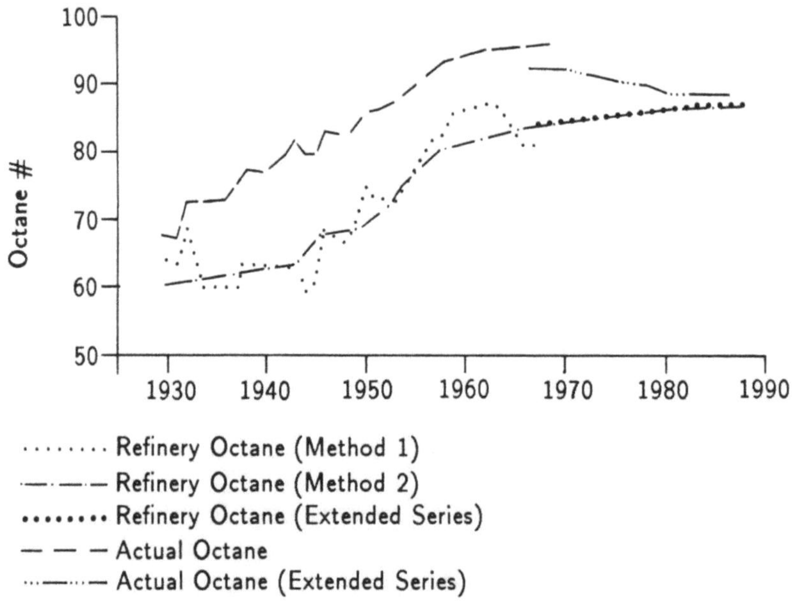

Figure 17.3. Average octane numbers.

vis-à-vis the possibility of increasing octane levels **in conjunction with** increasing the fraction of the crude oil that could be utilized for motor fuel, the latter was vastly preferable for the refiners. This preference explains much of the history of the octane race. Of course, the technology of increasing octane levels independently of refinery practice was introduced in the early 1920s. We discuss this next.

17.3 The Introduction of Tetraethyl Lead

The search for an antiknock additive for gasoline began in 1916, when engine compression ratios averaged only 4:1, yet *knocking* was a pervasive problem due to the low octane level of the motor fuels then available. At the time, however, the cause was not known. Charles Kettering's battery ignition system had been introduced only a few years earlier, and rival magneto ignition system manufacturers blamed it for knock. To counter this ploy (and find the real explanation, and a solution to the problem), Kettering and his colleagues Thomas Midgley, Thomas Boyd, and Carroll Hochwalt launched a research program at his Dayton Engineering Laboratories. It was subsidized

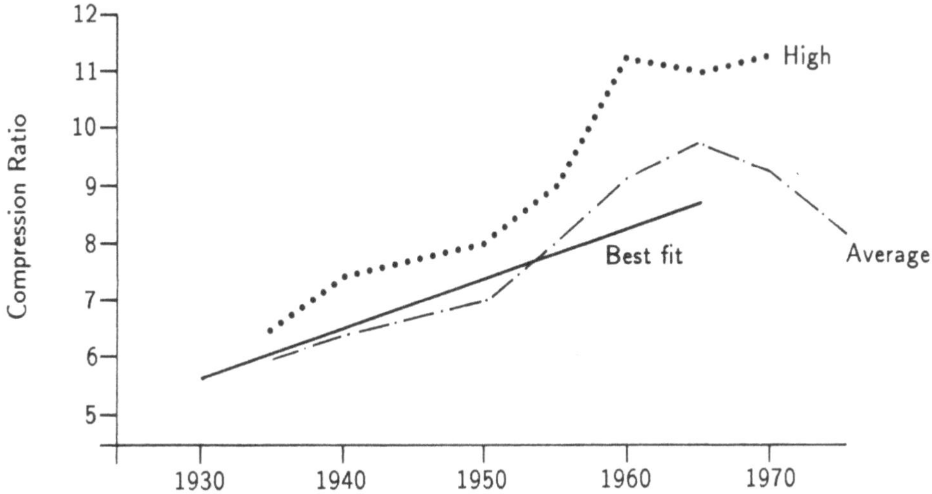

Figure 17.4. Automobile engine compression ratios.

by General Motors Corporation (GM), which later purchased Dayton Engineering Labs., (now known as DELCO Division) and made Kettering Vice President and chief scientist.

The first step was to test Midgley's theory that fuel volatility was the cause of the problem. (This had seemed plausible because increased demand for motor fuel had been met by increased blending of natural gasoline with less volatile naphtha). Volatility was ruled out by the end of 1916. Next, Kettering had a hunch that fuel color[3] might have an influence on knocking. This was quickly tested by adding various coloring agents to the fuel. The *color theory* was quickly discarded, but one of the chemical agents tested was iodine, which did have a measurably beneficial effect on knocking. For the next five years hundreds of compounds were tested, and some possible antiknock compounds were found, including aniline, selenium, and tellurium. They were all rejected for various reasons (such as odor). Finally, in December 1921, tetraethyl lead (TEL) was synthesized by Hochwalt. As an antiknock additive it has never been equalled, despite many millions of dollars of subsequent research by the German chemical cartel IG Farben.

For TEL to become a practical fuel additive, a manufacturing process was needed. This was developed by Charles Kraus, whose research was supported by Standard Oil Co., NJ. The GM patents on TEL and the Standard Oil Co., NJ manufacturing patents were consolidated by the formation of Ethyl Corporation in 1924, jointly owned by GM and Standard Oil Co., NJ.

Another problem that had to be overcome was the corrosion of spark plugs by lead oxide deposits. This was finally overcome by the addition of ethylene dichloride and ethylene dibromide in the additive. The latter, in turn, required a large and reliable source of bromine, which was finally achieved by the commercialization of the Dow process to extract bromine from seawater (1931).

From 1924 to 1930 the Ethyl Corporation was primarily involved in R&D, testing, advertising and marketing *premium* or *Ethyl* gasoline and building up its distribution network. Meanwhile, GM was actively promoting the higher performance cars that the new fuel made possible. Whether for this reason, or others, it was during this period that GM overtook Ford as the major US auto manufacturer.

Sales of TEL (in the form of *ethyl* fluid sold by Ethyl Corporation to refineries, and blended by the latter into commercial gasoline) took off. Motor fuel (gasoline) sales more than quadrupled from 1929 to 1967, with only a slight decline even in the worst year of the depression. Meanwhile, the average content of lead in grams per gallon of gasoline increased ten-fold and almost monotonically during the depression years (from 0.17 gm/gal. in 1929 to 1.75 gm/gal. in 1939) and reached an all-time peak of 4.71 gm/gal. in the wartime year of 1944. It hovered in the 3.5–3.9 range in the late 1960s before the first restrictions on TEL use – for environmental reasons – became effective. The average lead use, per gallon of gasoline used on highways, is shown in *Figure 17.5*.

17.4 Relative Contributions of Refining and TEL

In terms of the life-cycle model referred to briefly at the beginning of this chapter, one would expect the long-term competition between refinery technology and additives (notably TEL) to result in a clear superiority of one over the other, resulting in a well-defined displacement or substitution process. Before this hypothesis can be tested, however, we need a methodology for allocating the apparent *octane added* in each year (defined as octane per gallon above the base level of 50) among the various sources. From 1929 to 1970, roughly, the competition was strictly between refining and TEL. Since the environmental constraints on TEL have been gradually implemented, a new set of additives – basically alcohols – have appeared on the scene. These will be discussed later.

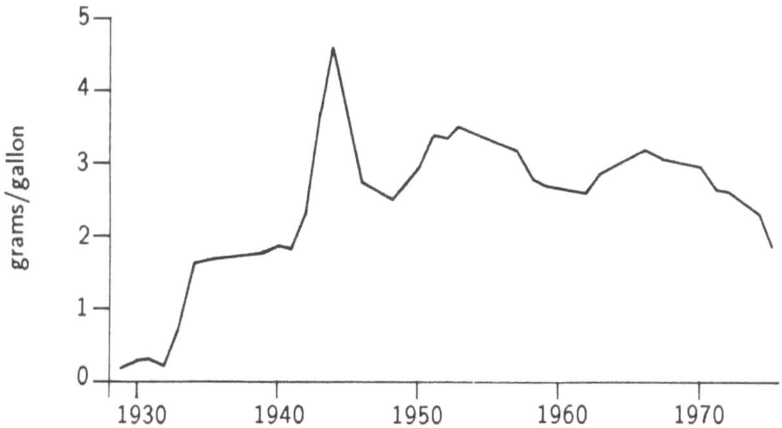

Figure 17.5. Lead use rate (grams/gallon). (Sources: Lead use: US Bureau of Mines, *Minerals Yearbook*, var. vols.; fuel consumption: US Federal Highway Administration, *Highway Statistics*, var. vols.)

There are two straightforward methodologies for estimating the octane-added, as defined above. Both start from the average octane level of fuel sold in a given year (see *Figure 17.3*). One approach is to use a *lead susceptibility* chart prepared by the Ethyl Corporation to determine the octane gain from a given amount of lead additive, based on the octane level of the *base fuel*, i.e., the gasoline as obtained from the refinery process alone. The chart in question is shown as *Figure 17.6*. It can be used to estimate the base fuel octane from the quantity of lead added (in grams per gallon). This method assumes, of course, that lead is added to *average* base fuel. In reality, high octane gasoline from some refineries has always been sold as *unleaded premium*, as long ago as the late 1930s.[4] This tends to lower the average octane level of the base fuel to which TEL *was* added, distorting the average picture somewhat.

On the other hand, the alternative approach – which can be termed "process accounting" – is to calculate the average octane of the base fuel from the fraction of gasoline produced by each refinery process in each year and the octane produced by that process. For purposes of this analysis we have assumed the octane levels indicated in *Table 17.1*, namely, Burton batch thermal cracking (55 RON), continuous thermal cracking (73 RON), Houdry batch catalytic (87 RON), continuous catalytic or fluidic (95 RON).

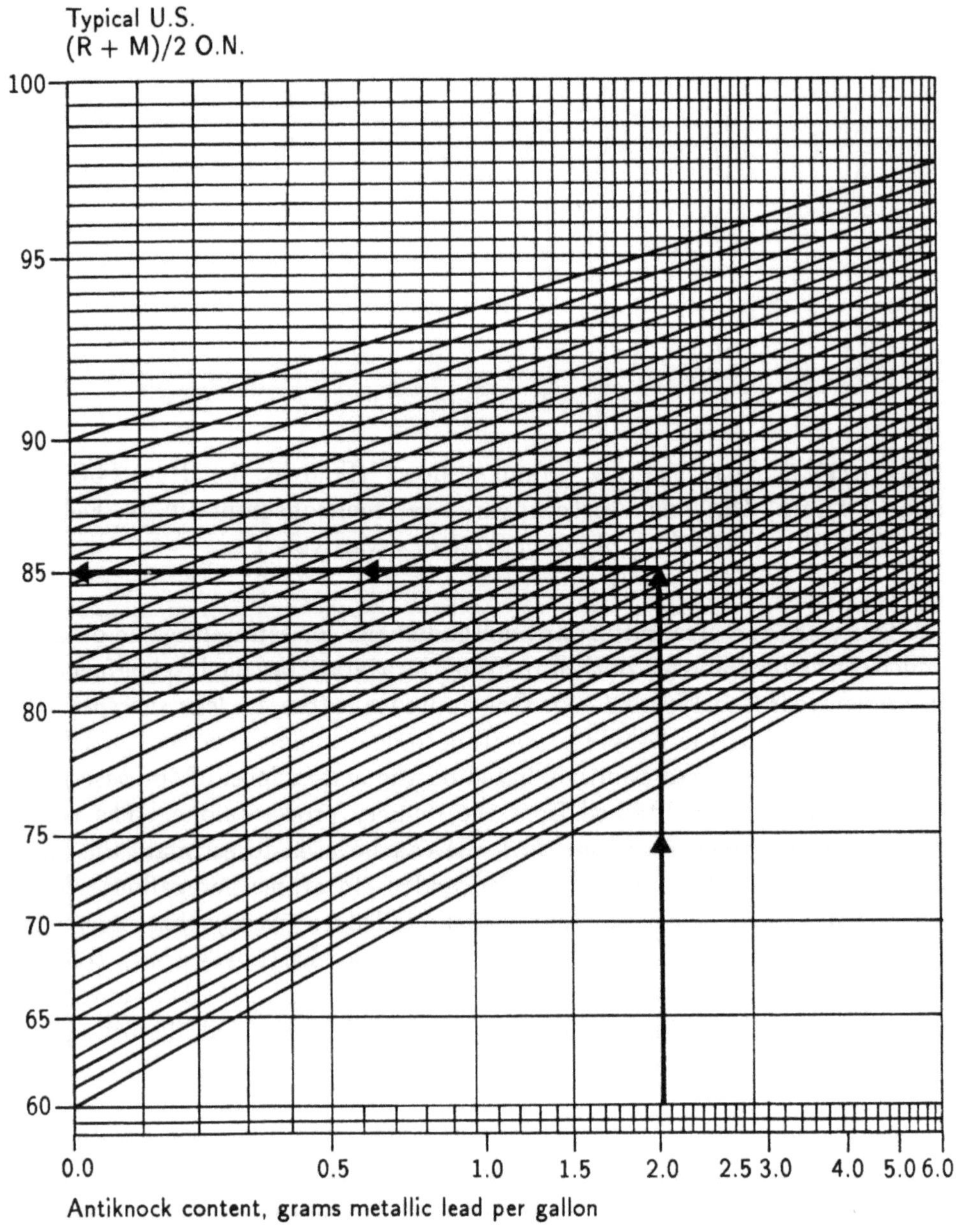

Figure 17.6. Ethyl Corporation lead alkyl antiknock susceptibility chart.

Here there are three difficulties. The first is the same as the one noted above, *viz.*, some high octane refinery products fuels have been sold directly as lead-free premium gasoline without added lead. The second problem is that the *average* octane number associated with each process is only approximate. Moreover, apart from the four main types of cracking process, refiners have had, since the 1930s, a variety of other octane-enhancing processes available, including hydrogenation, polymerization, alkylation, hydrogen reforming and catalytic reforming. In fact, each refinery is unique in its mix of processes and products. The third problem is that we do not have published data on production by process, but only on *capacity* by process. On the average, over a long period of time, the two probably track together roughly, but on a year-to-year basis there are likely to be significant variations as some types of capacity are more highly utilized than others.[5] Apart from wartime distortions, during the early years of penetration of a new process one might expect some *debugging* troubles to reduce capacity utilization; this is the pattern observed in other cases of new process introduction. By contrast, in the late stages of a displacement, a refiner might keep an old depreciated plant on-stream and available, but operating at a low level *just in case* of a sudden upsurge in demand. Thus, one would tend to expect capacity utilization levels for a new process technology to start at moderate levels, rising gradually due to *learning by doing* until fairly late in the life of that technology, before dropping to rather low levels immediately prior to being phased out.

For the several reasons given above, the two ways of estimating base fuel octane levels would not be expected to agree exactly. Of the two, the *lead susceptibility* method would appear to be more reliable. In fact, the agreement between the two methods is not remarkably close (*Table 17.2*). Using both methods of calculating refinery octane, the share of *added octane* attributable to refining technology versus that attributable to the addition of TEL is plotted in *Figure 17.7*. The results are very interesting, especially when the lead susceptibility chart is used to calculate base octane level. Starting in the late 1920s, the TEL share began to rise rapidly (except for the single relapse in 1932) to the 50% level, or more, which it held throughout the 1930s and even increased to a peak of 66% in the war year of 1944. Thereafter the TEL share began to drop, falling to 36% in 1950, with a slight pickup to 40% in 1953, followed by a further fall to a low point of 17% in 1963. Yet it rebounded once again to the 32% level in 1967.

Table 17.2. Refinery and actual octane plus additive share.

	Refinery octane[a]	Refinery octane[b]	Refinery octane[c]	Actual octane	Actual octane[c]	Additive share[a]	Additive share[b]	Additive share[c]
1930	64.0	60.50		67.5		0.200	0.400	
1931	63.5	60.88		67.0		0.206	0.360	
1932	69.0	61.00		72.0		0.136	0.500	
1933	63.5	61.25		72.0		0.386	0.489	
1934	60.0	61.50		72.5		0.556	0.489	
1935	60.0	61.75		72.5		0.556	0.478	
1936	60.0	61.88		73.0		0.565	0.483	
1937	60.0	62.00		75.0		0.600	0.520	
1938	63.0	62.13		77.0		0.519	0.551	
1939	63.0	62.38		77.0		0.519	0.541	
1940	62.0	62.63		77.0		0.556	0.532	
1941	63.0	62.75		78.2		0.539	0.548	
1942	63.0	63.00		79.2		0.555	0.555	
1943	63.0	63.38		81.5		0.587	0.575	
1944	60.0	65.75		79.6		0.662	0.468	
1945	61.0	66.98		79.5		0.627	0.424	
1946	68.0	67.75		82.7		0.450	0.457	
1948	66.5	68.25		82.5		0.492	0.438	
1949	69.0	69.00		83.8		0.438	0.438	
1950	74.0	70.25		85.88		0.331	0.436	
1951	73.0	71.00		85.95		0.360	0.416	
1952	73.0	71.75		86.75		0.374	0.408	
1953	73.0	74.25		87.50		0.387	0.353	
1957	82.0	79.75		92.20		0.242	0.295	
1958	83.0	80.38		93.45		0.241	0.301	
1959	85.0	80.82		94.00		0.205	0.300	
1962	87.0	82.13		94.90		0.176	0.284	
1963	87.0	82.50		95.10		0.180	0.279	
1966	81.0	83.75		95.65		0.321	0.261	
1967	81.0	83.50	84.1	95.83	92.25	0.324	0.269	0.193
1970			84.6		92.25			0.181
1975			85.4		90.63			0.129
1978			85.8		90.00			0.105
1979			86.0		89.60			0.091
1980			86.2		88.72			0.065
1981			86.3		88.60			0.060
1982			86.5		88.50			0.052
1983			86.6		88.55			0.051
1984			86.7		88.60			0.049
1985			86.7		88.50			0.047
1986			86.7		88.67			0.051
1987			86.8		88.61			0.047

[a]Method 1. [b]Method 2. [c]Extended series.

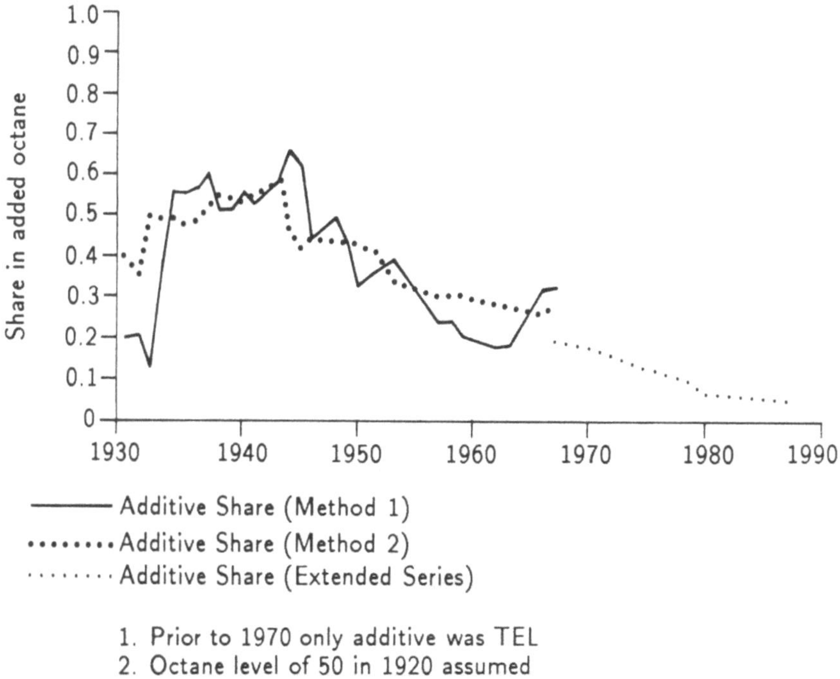

Figure 17.7. Contribution of additives to added octane.

17.5 Discussion and Conclusions

How can this behavior be explained at all, still less fitted into the conventional *substitution* picture? As noted earlier, one normally expects the *superior* technology to gradually displace the *inferior* one, following an S-curve or some similar path. In the present case, TEL became dominant rather soon after its introduction, but did not replace cracking, or even slow down its rate of adoption. In fact, since the 1940s the trend has been the other way. TEL has subsequently been displaced very largely by more advanced cracking and other refining technologies and new additives (*Figure 17.8*). This trend was well advanced even before the environmental regulations restricting the use of TEL.

On the other hand, neither of the alternatives has ever entirely displaced the other. Even as TEL was phased out, other octane-enhancing additives have begun to replace it (*Figure 17.8*, *Table 17.3*). If there are economies of scale or economies of adoption favoring *lock-in* to either approach, e.g.,

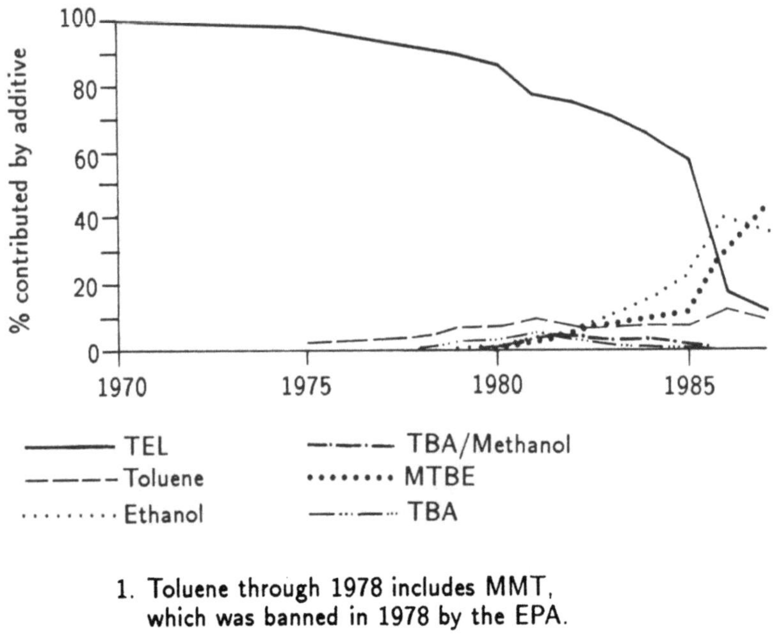

Figure 17.8. Percentage contribution of different additives.

Arthur (1988a), they are evidently compensated for by diseconomies (declining marginal returns), possibly associated with high-severity petroleum refining. No matter how sophisticated the refinery technology, it is apparently always economical (in the narrow sense) to gain additional octane by the addition of some TEL, or one of the alcohols. By the same token, there are also declining marginal returns to the use of TEL, or other additives, beyond a certain point. Thus the two technologies, while somewhat competitive, are also to some extent complementary.

Apart from the issue of complementarity, noted above, it is important also to observe that one of the two technologies, cracking, was evolving rapidly while the other remained static until regulation forced a change. In fact TEL is one of the few examples of a technology which essentially did not evolve at all after introduction. Its diffusion process was therefore *pure*, and not the more commonplace combination of technological change and diffusion together. Are there other cases like this one? Quite certainly there are, inasmuch as declining marginal returns and complementarity are not rare phenomena in economics.[6]

Table 17.3. Percentage contribution of different additives.

	TEL	Toluene	Ethanol	TBA	TBA plus Methanol[a]	MTBE
1967	100.0					
1970	100.0					
1975	97.3	2.7				
1978	91.4	3.8[b]		1.0		
1979	89.2	7.2	0.4	2.5	0.1	0.6
1980	86.3	6.8	1.2	2.5	0.8	2.5
1981	76.8	9.3	2.2	3.7	3.1	4.9
1982	74.9	6.9	6.0	2.5	4.0	5.7
1983	71.0	6.6	10.9	1.8	3.0	6.6
1984	65.1	7.2	15.5	1.0	3.3	7.9
1985	57.8	7.3	22.5	0.3	1.4	10.7
1986	17.6	12.0	41.2			29.2
1987	12.2	9.4	36.1			42.2

[a]For methanol, add (TBA plus methanol) to MTBE.
[b]Methyl manganese tricasbonyl (MMT) accounted for 3.8% (or 44% of additives used in unleaded gasoline in 1978). MMT was banned in 1978 by the EPA.

Notes

[1] In fact, for two early Pennsylvania refineries for which data is available – Pratt's and Downer's – the gasoline output was only 1.5% of the output stream (Williamson and Daum, 1959).

[2] Indiana Standard had oil wells in Indiana and Illinois, but the reserves were not large. The breakup of Jersey Standard left the parent company in possession of Humble Oil Co., with its large Texas crude oil reserves.

[3] Kettering was inspired by the red-green natural dyes in plants, such as the trailing arbutus, and an apparent relationship between leaf color and early blooming (Raymond, 1980).

[4] For instance, premium Sunoco "Blue" was made directly from the Houdry catalytic process; in the 1960s Amoco sold a premium lead-free gasoline of very high octane.

[5] During World War II this distorted the picture significantly, inasmuch as the demand for high octane *aviation gasoline* soared, soaking up virtually all of the refinery capacity for catalytic cracking. As a consequence, old thermal cracking plants were kept in service and the base octane level of fuels used by the civilian sector declined sharply. It was made up, in part, by extraordinarily high use of TEL, as shown in *Figure 17.5*.

[6] Another fairly obvious example is the complementarity between the basic oxygen furnace (BOF) and the electric arc furnace (EAF) in steel-making. The one converts pig iron from ore and scrap, but the other converts scrap only. The

balance between them depends on the scrap supply. On reflection, it must be clear that every coproduct relationship corresponds to some complementarity.

References

Arthur, W.B., 1983, *Competing Technologies and Lock-in by Historical Events: The Dynamics of Allocation Under Increasing Returns*, Center for Economic Policy Research, Paper 43, Stanford University, CA, USA.

Arthur, W.B., 1988a, Competing technologies: An overview, in G. Dosi *et al.*, eds., *Technical Change and Economic Theory*, Pinter Publishers, London and New York.

Arthur, W.B., 1988b, Self-reinforcing mechanisms in economics, in P.W. Anderson *et al.*, eds., *The Economy as an Evolving Complex System* (Proceedings of the Evolutionary Paths of the Global Economy Workshop, September 1987, Santa Fe, NM, USA.), Addison-Wesley, Reading, MA, USA.

Ayres, R.U., 1987, *The Industry-Technology Life Cycle: An Integrating Meta-Model?* RR-87-3, International Institute for Applied Systems Analysis, Laxenburg, Austria.

Crutchfield, J., Farmer, J.D., Packard, N.H., and Shaw, R.S., 1986, Chaos, *Scientific American* **255**:46–57, December.

Enos, J.L., 1962, *Petroleum Progress and Profits, A History of Process Innovation*, MIT Press, Cambridge, MA, USA.

Fisher, J.C. and Pry, R.H., 1971, A simple substitution model of technological change, *Technological Forecasting & Social Change* **3**(1):75–88.

Lakhani, H., 1975, Diffusion of environment-saving technological change: A petroleum refining case, *Technological Forecasting & Social Change* **7**:33–55.

Linstone, H. and Sahal, D., eds., 1976, *Technological Substitution*, Elsevier Science Publishers B.V., Amsterdam, The Netherlands.

Mahajan, V. and Peterson, R.A., 1985, *Models for Innovation Diffusion*, Sage Publications, Beverly Hills, CA, USA.

Mahajan, V. and Wind, Y., 1986, *Innovation Diffusion Models of New Product Acceptance*, Ballinger Publishing Co., Cambridge, MA, USA.

Raymond, L., 1980, Today's fuels and lubricants – and how they got that way, *Automotive Engineering*, pp. 27–32, October.

US Bureau of Mines, US Department of the Interior, *Minerals Yearbook*, various volumes, Government Printing Office, Washington, DC, USA.

US Federal Highway Administration, US Department of Transport, *Highway Statistics*, various volumes, Government Printing Office, Washington, DC, USA.

Williamson, H.F. and Daum, A.R., 1959, *The American Petroleum Industry*, Northwestern University Press, Evanston, IL, USA.

Chapter 18

Diffusion: Long-term Patterns and Discontinuities

Arnulf Grübler

18.1 Introduction

The characteristic S-shaped diffusion pattern and the resulting rates of diffusion are a macroaggregate of an underlying complexity of adoption causes. Diffusion is therefore not a unary process. Instead, diffusion phenomena are probably best conceptualized as proceeding through various stages of a *diffusion life cycle*. In each of these stages the process is characterized by different market niches, different determinants of diffusion, and different relationships to other diffusion processes – both of a competitive and interdependent nature. Diffusion processes should therefore be analyzed based on multivariate (i.e., considering an innovation diffusion case not in isolation) and multiattribute (i.e., using a number of measures to describe diffusion trajectories and developing comprehensive vectors of driving variables) approaches.

This multistage view of diffusion also raises the question of whether the term *diffusion* in any way captures the essence of most processes of technological or social/institutional change. Hardly any innovation diffuses into a vacuum. Along its growth trajectory an innovation interacts with existing techniques, depends on the development of a mediating framework for its effective absorption into the socio-technical system, and changes its technological, economic, and social characteristics. From such a multistage perspective, diffusion is probably best described as an "evolution resulting from a sequence of replacements" (Montroll, 1978), i.e., as a succession of substitutions along various specific (expanding) market niches.

If diffusion is defined as a sequence of replacements, one must analyze these processes comprehensively. It appears that much of the debate on the appropriate mathematical model(s) of diffusion, in particular the question of symmetrical versus asymmetrical diffusion models, may be the result of looking at an innovation from a unary (i.e., an innovation grows into a vacuum) or a binary (the market share of an innovation is analyzed *vis-à-vis* the remainder of the competing technologies) perspective. However, diffusion phenomena generally call for a multivariate approach, which has not yet found wide application in the various diffusion disciplines.

Diffusion and substitution phenomena can be observed along spatial and temporal hierarchies. They range from very short-term processes, such as the rapid spread (and disappearance) of fashion gadgets, to extremely long-term, pervasive transformations in the technological and social fabric as reflected in the growth of infrastructures or new forms of social and institutional organizations. Whereas shorter-term diffusion processes operate within a more or less equilibrium configuration, very long-term and pervasive diffusion processes are of an evolutionary, non-equilibrium type nature because they profoundly transform the very boundary conditions of the system within which they operate.

Although well established in geography, the importance of the spatial hierarchy of diffusion processes appears to have found only limited attention in the technological and marketing diffusion research disciplines. Yet consideration of the spatial hierarchy of diffusion processes could yield useful insight into the differences in diffusion rates and levels.

When analyzing the diffusion of large, pervasive systems, whose growth and interaction phases with other innovations span from several decades to centuries, one observes that these systems are not introduced nor grow and/or substitute for existing techniques and artifacts in a continuous way. An empirical investigation into the diffusion history of many processes of technological, economic, and social change in the USA reveals very strong nonlinearities, resulting from the discontinuous rate innovations are introduced and become absorbed into a socioeconomic system. The timing of these discontinuities are closely related to important historical turning points identified by researchers of the long-term *Wechsellagen* of economic development, pointing to a deeper causal relationship between the rate of acceptance (diffusion) and the rate of interaction (substitution) of pervasive social and technological innovations with successive phases and discontinuities of economic growth.

Despite the vast body of diffusion literature, our knowledge of the determinants of diffusion and substitution phenomena in different markets, products, and geographical spaces, as well as the differences and/or convergence of diffusion trajectories at national levels, is still fragmented and insufficient. A systematic collection of quantitative diffusion studies from different diffusion disciplines based on comparable measures and the development of a taxonomy appears necessary in order to advance the understanding of diffusion phenomena.

18.2 Phases of Diffusion

Economic, social, and technological development may be seen as being driven both by the introduction of fundamentally new solutions (basic innovations) and by incremental improvements in existing techniques and systems (product and process innovations). Although an overlap exists, it is important to distinguish between the two because it is the basic innovations that expand existing feasibility frontiers. In some cases they are pervasive, spilling over across many sectors, creating even entirely new industries and growth sectors. Over time, however, there is a gradual transition in the diffusion process from basic innovations that initially lead to the creation of new industries, to incremental improvements, and to product innovations as the diffusion process matures.

The early phase of the diffusion process is characterized by a highly volatile environment. Diffusion is primarily performance rather than price driven. The initial rate of adoption is usually slow, although monopoly opportunities arise for innovative entrepreneurs in a rapidly changing environment with a large diversity of competing designs that improve their technical and economic characteristics. As competition begins, one particular technological variant becomes dominant and standardization emerges in the new industry. Initial random fluctuations may result in the prevalence of a particular design that becomes perpetuated and amplified with increasing returns to adoption and the economic advantages of initial standardization, i.e., the *lock-in* effect of Arthur (1983). This *lock-in* effect usually occurs after an innovation has diffused into its first niche of application, which provides the experimental field for further technological improvements and cost reductions. In the case of pervasive systems, this first phase of diffusion is equally characterized by the development and emergence of a matching *technological style* (Perez, 1983), in the sense of an associated set of best technological and engineering practices (i.e., what has been referred to as a technological *trajectory* (Dosi, 1983; Nelson and Winter, 1982), as well as the emergence of a supportive social and institutional framework for pervasive diffusion (Freeman and Perez, 1988). Diffusion in this initial first phase can be rather fast, when adoption externalities, e.g., in the form of existing infrastructures, enable the rapid integration of a new technique or artifact within an existing technical or social context.

This first diffusion phase usually ends with the emergence of a dominant design and an accompanying increasing standardization. Emphasis shifts to incremental improvements and small cumulative innovations. Economies of scale and cost reductions along the learning curve sustain a rapid expansion phase of exponential growth rates, characteristic of the second phase in the diffusion of pervasive systems. Applications spill over to other sectors, generating additional positive feedback for sustained growth. The steady accumulation of benefits to both producers and consumers produces a vicious cycle (Brooks, 1986) of declining costs and expanding demand, which erects almost insurmountable barriers to the entry of new competitors (economic agents or potentially superior technologies or designs).

Nonetheless with gradual maturation many disadvantages become apparent. Cumulative and incremental improvements yield decreasing marginal returns and eventually market saturation sets in. Adverse social or environmental effects often increase nonlinearly with the scale of application of technology, so that the margin between social benefits and costs decreases rapidly, sometimes even producing net social costs. The social and institutional response to the attainment of limits of the dominant modes of development and growth can also be rather disruptive. Increasing awareness of social disbenefits and risks and the exhaustion of further incremental productivity and performance improvements constrain further diffusion. At the same time the saturation phase provides an opportunity window for a renewed period of experimentation and introduction of alternative technological and organizational solutions.

Some of the most important changes in socio-institutional frameworks and economic structure since the onset of the Industrial Revolution are indeed related to the pervasive adoption of new systems. Their adoption did not, however, always follow a smooth continuous classical diffusion trajectory nor can they be analyzed in isolation from other diffusion processes. For example, the diffusion of motor vehicles was contingent on the development of numerous other systems, such as paved roads, the internal-combustion engine, oil refining and motor fuels, sheet metals and high-quality steels, and electrical equipment among others.

The importance of distinguishing different diffusion phases is illustrated in *Figure 18.1*, which shows the growth of the automobile in the USA since the turn of the century. In the first phase the number of cars registered increased by a rate of approximately 30 percent annually. A major discontinuity by the end of the 1920s marks the transition to a second phase, where annual growth rates of approximately 5 percent are typical. The reason for the rapid growth of automobiles in the first diffusion phase and for the structural discontinuity by 1930 is also shown in *Figure 18.1*. The number of automobiles grew fast, first by replacing an existing form of individual road transportation: horses. This substitution process proceeded along a regular logistic market share fraction trajectory and was rather swift. The Δt of the process (i.e., the time to grow from a 10 to 90 percent share in the total number of road vehicles) is only 12 years. The fast diffusion (substitution) was not only the result of the comparative technical and economic advantage of automobiles over horses, but was also due to adoption externalities, i.e., the possibility of making use of existing infrastructures: surfaced roads had been already developed for horse carriages (Nakićenović, 1986). Another factor contributing to the rapid replacement of horses was the relatively short useful lifetime of the *rolling stock*, which in fact is very similar between automobiles and horses.[1]

Another example of a rapid first diffusion phase is the introduction of natural gas to most industrialized countries. This technology also used an already existing infrastructure grid to replace traditional (coal-based) city gas. Once the rapid first diffusion phase is completed, the second phase in diffusion begins. However, in this second phase markets, driving forces as well as diffusion speed and ultimate diffusion levels, are drastically different from the first phase of diffusion, which illustrates the importance of a differentiated analysis of the various phases in the diffusion of large, pervasive systems. Frequently this rapid growth in the first phase has been used to postulate a systematically decreasing diffusion coefficient and a resulting asymmetrical diffusion model describing the diffusion process over the entire

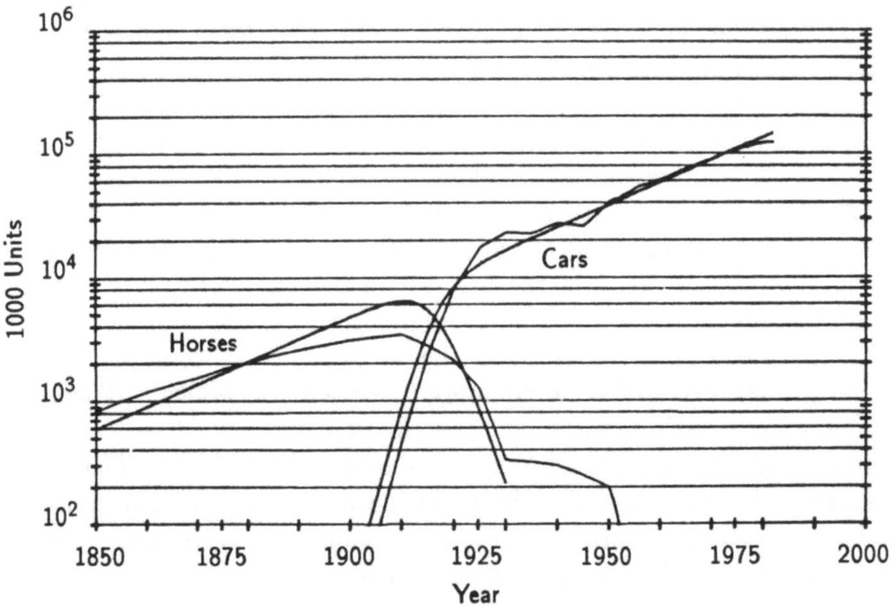

Figure 18.1. Number of draft animals and cars in the USA, data and model estimates from logistic substitution model. (Source: Nakićenović, 1986.)

life cycle. Such a continuous diffusion model application appears problematic considering the distinctly different driving forces and market potentials in the different phases of diffusion.

The general outline of the phases of the diffusion of large pervasive systems described above points to the fact that the diffusion process should not be considered as a single, univariate phenomenon, even if in mathematical terms the corresponding macro trajectory may be reduced to simple 3-to-4 parameter equations. Diffusion consists of successive phases; in each phase both the technological and economic characteristics of an innovation as well as the driving forces and the adoption environment for its diffusion constantly change.

A smooth S-shaped diffusion trajectory at the macro level is not necessarily a contradiction to a constantly changing adoption environment, diversity, and evolution of determinants of the diffusion process (such as prices, costs, performance, and learning within the industry) at the micro level, if the diffusion process is seen as portraying the features of a dynamic, self-organizational system, as demonstrated by recent diffusion models in economics (Silverberg *et al.*, 1988) and discussed in Chapter 8 by Silverberg.

Diffusion of an innovation of economic or social pervasiveness is a long process: diffusion time constants (Δts) range from decades to centuries. Even in the case of the diffusion of process innovations, diffusion time constants are considerable. In the summary of a study of international comparative technological diffusion (Nasbeth and Ray, 1974), Ray (1989) reports that "the time period to reach or to approach saturation is long – about three decades or even more". Thus, decades are required for the diffusion of significant innovations, and even longer time spans are needed to develop infrastructures or pervasive socio-technical systems.

18.3 From Measurements to Models

The problem of measures has been associated with diffusion studies in the technological area ever since the the first studies by Griliches (1957) and Mansfield (1961). In a stylized overview of the evolving perspectives on diffusion, one can summarize that diffusion was first viewed as a unary process (i.e., describing the diffusion of an innovation only by isolated measures such as the number of adopters). In a following step, studies analyzed diffusion as a binary process of interaction (substitution) with the remainder of processes or products available on the market. Finally, and what may be considered a more realistic representation of diffusion processes, studies analyzed diffusion as a multivariate phenomenon involving a number of competing alternatives with changing relative market shares over time.

Early measures of diffusion, such as the number of firms adopting a particular technique, have been criticized on grounds of the prevailing heterogeneity between economic agents as well as the different impacts of adoption decisions during various phases of the diffusion process. Capacity or output in the field of technological diffusion and first purchases in the case of consumer goods have been used later on as diffusion measures. Still it appears that relatively little systematic work has been carried out in the various diffusion research disciplines to analyze a diffusion process based on a systematic and comparable set of measures of the temporal and spatial spread of an innovation. Data availability, and not theoretical or methodological considerations, appears to govern the choice of particular measures used in diffusion studies. The most obvious criticism of this unary type of diffusion model is of course that hardly any innovation diffuses in a vacuum, i.e., that the diffusion of an innovation may be largely due to "cannibalizing" (to use a phrase from Chapter 6 by Vijay Mahajan) existing techniques or products.

In a next step, diffusion processes have been analyzed interpreting diffusion as a (binary) replacement process (Fisher and Pry, 1971). This binary diffusion model may be considered the predominant model in technological and marketing diffusion research. Despite its undeniable success in many case studies (e.g., Mansfield, 1968a and 1968b), there have been several critical observations of the model. The critique of the standard binary diffusion/substitution model (which is also valid for the original unary diffusion model, discussed above) primarily concentrates on the assumption of a static adoption environment. It is argued (e.g., Davies, 1979; Gold, 1981; Metcalfe, 1983) that in view of a constantly changing adoption environment and changing characteristics of an innovation, the diffusion path is likely to consist of an envelope of symmetrical (logistic) diffusion patterns, yielding as a macroaggregate an asymmetrical diffusion path with a decreasing diffusion coefficient.[2] Other arguments in favor of asymmetrical diffusion models considered the decreasing profitability of adoption at later stages of the diffusion process;[3] questioned the complete takeover assumption of the Fisher-Pry model (i.e., any diffusing innovation is assumed to result in complete market takeover), and finally perceived a *lack of flexibility* or goodness of fit of the standard model in view of empirical data.

In response to this perceived *lack of flexibility* alternative models have been proposed. These include the use of a modified exponential curve following Bass' (1969) distinction between external and internal influence driven diffusion, the Floyd curve, the Gompertz curve, or a cumulative normal or cumulative lognormal pattern. Other models have tried to develop more comprehensive formulations with additional parameters to accommodate a whole set of different S-shaped diffusion patterns. Examples of these models include the Sharif-Kabir (Sharif and Kabir, 1976) model, the NSRL model (Easingwood *et al.*, 1981) and the model proposed by Skiadas (1986) (for an overview see, for example, Hurter and Rubinstein, 1978; or Mahajan and Peterson, 1985).

However, the flexibility of these models (in terms of how they can describe a wide range of diffusion patterns) is achieved by paying a theoretical price. No theoretical rationale underlies various models to explain why a particular diffusion pattern should follow one of these asymmetrical models or what the (economic or behavioral) explanation of the additional model parameters (i.e., the various "delay coefficients" introduced into the models) might be. To some extent this line of research has thrown out the baby (behavioral/economic rationale of diffusion models) with the bathwater (the purported constraining conditions underlying the behavioral rationale of the logistic diffusion model).

A comment on the goodness of fit of flexible or asymmetrical diffusion models is appropriate. A prominent example of the application of asymmetrical diffusion models was the diffusion of the BOF (basic oxygen furnace) steel process (e.g., Skiadas, 1986). However, an innovation may not only replace existing techniques, but may be influenced (e.g., slowed down along the lines of an asymmetric diffusion pattern with saturation below 100 percent market share) also by the diffusion of newer competitors. The asymmetrical diffusion pattern in such cases is not an inherent characteristic of the diffusion of a particular technology, but simply the result of the diffusion trajectory of a newer competitor, which is omitted in the analysis under a binary view of diffusion.

This leads finally to a discussion of multivariate innovation diffusion models. From the above observations, one can conclude that the environment in which diffusion phenomena are embedded is only rarely so simple as to allow for a unary or binary view of the diffusion process. In reality structural changes in the technology base as well as in consumer products markets are characterized by a sequence of introduction, diffusion, and in turn replacement of a number of innovations, all of which compete simultaneously on the market. Thus diffusion phenomena are probably, in most cases, only adequately modeled when considering more than one or two competing innovations and their diffusion (or displacement) trajectories.

An extension of the standard Fisher-Pry model to a multiple substitution model was first proposed by Marchetti and Nakićenović (1979).[4] *Figure 18.2* illustrates such a multiple substitution pattern on the basis of the changements in the propulsion technology in the merchant marine of the USA. Despite the fact that the model is still too crude to represent all of the fine print and turbulences along the diffusion trajectories, two observations can be made. First, the diffusion trajectory of steam propulsion follows an asymmetrical diffusion pattern. Prior to 1890 steam propulsion replaces sails along a logistic substitution pattern. After that time, the diffusion rate slows down and levels off around the 90 percent market share level by the end of the 1930s. The reason for this can be easily attributed to the diffusion of a new propulsion system in the form of petroleum-based motor ships, replacing in turn coal-fired steam ships. The diffusion trajectory of steam ships can therefore only adequately be described by considering the whole

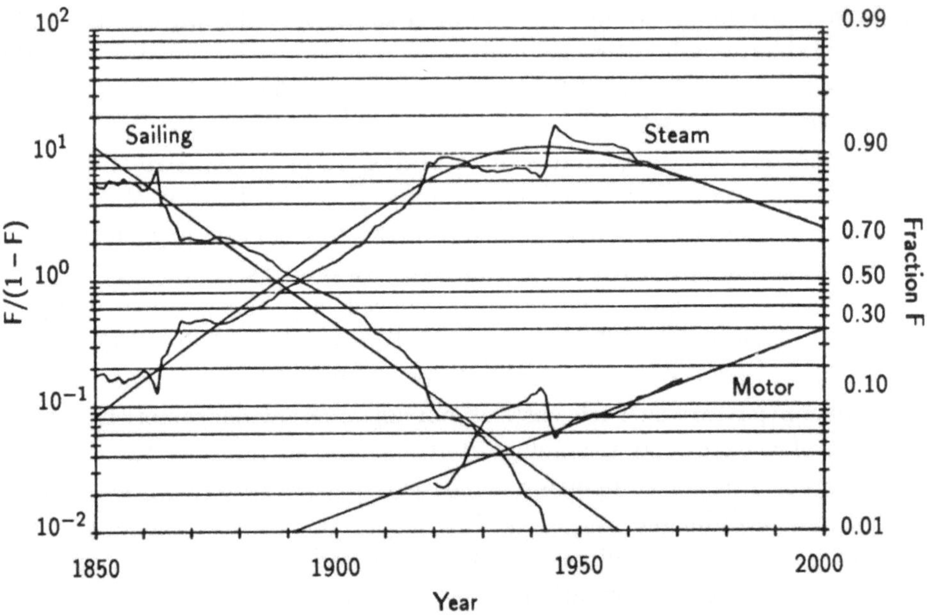

Figure 18.2. Successive replacements of the propulsion system in the merchant marine fleet of the USA, in fractional shares of gross tonnage registered, logit transformation. (Source: Nakićenović, 1988.)

technological environment into which it is embedded (i.e., the technologies an innovation replaces as well as newer technologies, the dynamics of which may in turn influence the diffusion trajectory of any particular technology).

The second observation deals with the displacement trajectory of sail ships. It has frequently been argued that the "sailing ship effect" (Ward, 1967), i.e., the improvement in the performance of an old technology (clippers in this case) when challenged by a new competitor, results in a noticeable effect of "striking back", i.e., retarding the diffusion trajectory of a new technology. On the basis of empirical evidence, doubts remain about whether this effect appears noticeable on the diffusion trajectory of steam ships and the retreating trajectory of sail ships. Thus, technological improvements in clipper technology have to be considered in relation to the parallel improvements in steam propulsion, as it is the relative technological performance of technologies, which among other factors, are driving forces of the replacement process.

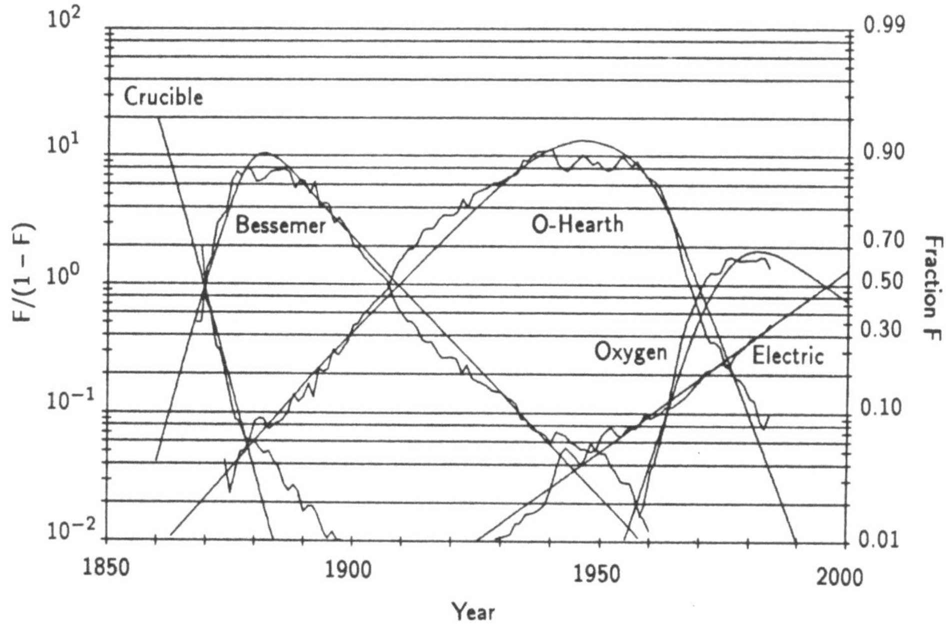

Figure 18.3. Successive replacements of process technologies for raw steel production in the USA, in fractional shares of raw steel tonnage produced, logit transformation. (Source: Updated from Nakićenović, 1987.)

Figure 18.3 presents an additional degree of complexity when considering multivariate technological substitution processes. Here as many as four different technologies (with decreasing and increasing market shares) compete simultaneously on the market (see the 1950s in *Figure 18.3*); and the diffusion trajectories of all processes show a high degree of diversity in their dynamics, i.e., diffusion coefficients, ranging in their Δts from less than two decades (replacement of the crucible process) to nearly seven decades (diffusion of electric arc steel).

Figure 18.3 provides empirical evidence of the argument that the (asymmetrical) diffusion trajectory of the basic oxygen process can be explained by the logistic diffusion pattern of electric arc steel, a continuation of which can be rationalized by the competitive advantages of mini-steel mills as well as by increasing tendencies for material (scrap) recycling. From this perspective, it is ironic that the diffusion trajectory of the basic oxygen process has figured so prominently as the underlying argument for suggesting alternative diffusion models to the standard logistic.

Figure 18.3 illustrates yet another characteristic feature of the *extinguishing process* (Poznanski, 1986), i.e., the displacement of a technology. Similar to the cases discussed above, where adoption externalities allow for a rapid first initial expansion to market shares below the ten percent threshold, one can observe an inverse phenomenon at the end of the diffusion life cycle of a technology. Old technologies appear – after a period of steady (logistic) decline in market shares – to retreat at a slower rate from the last few percent market shares, as a result of preserving competitive advantages in specialized market niches and taking advantage of production economics characterized by old, mostly written off, capital vintages.

Along the diffusion trajectories of the successive technological process generations in raw steel production shown in *Figure 18.3*, relative prices, costs, technical performance, etc., all change. Technological performance increases, and production costs and energy and material inputs decrease (Ayres, 1989) as a combination of both radical technological transformations (illustrated by the successive replacements of process technologies in raw steel production) and gradual improvements in the efficiency of existing process technologies. The combined effect, for instance, on the energy efficiency of raw steel production was certainly dramatic: specific energy requirements per ton of raw steel were in the order to 120 million BTU per (short) ton in the 1880s, compared with a present value of approximately 12, i.e., improvement by a factor of 10 over a period of 100 years (Grübler, 1990a).

In a similar way one can also illustrate the economic effects of the successive transformations in the composition of process technologies for raw steel production in the USA by looking at the long-term evolution of the real-term prices of iron and of steel rails in the USA. Two characteristic features of the long-term evolution of iron and steel products prices emerge from *Figure 18.4*. The first is the drastic decline of real-term prices in iron and steel products in the period 1870 to 1910 that goes along with the diffusion of the open-hearth process for raw steel production. A corollary of this observation is, that the rapid diffusion of the Bessemer process prior to 1870 was not driven by cost cutting, but rather by increased technological performance (quality of output, i.e., high-quality mass production of steel, and replacement of small-scale production of quality steel by the crucible process). *Figure 18.4* also illustrates the pronounced discontinuities in real-

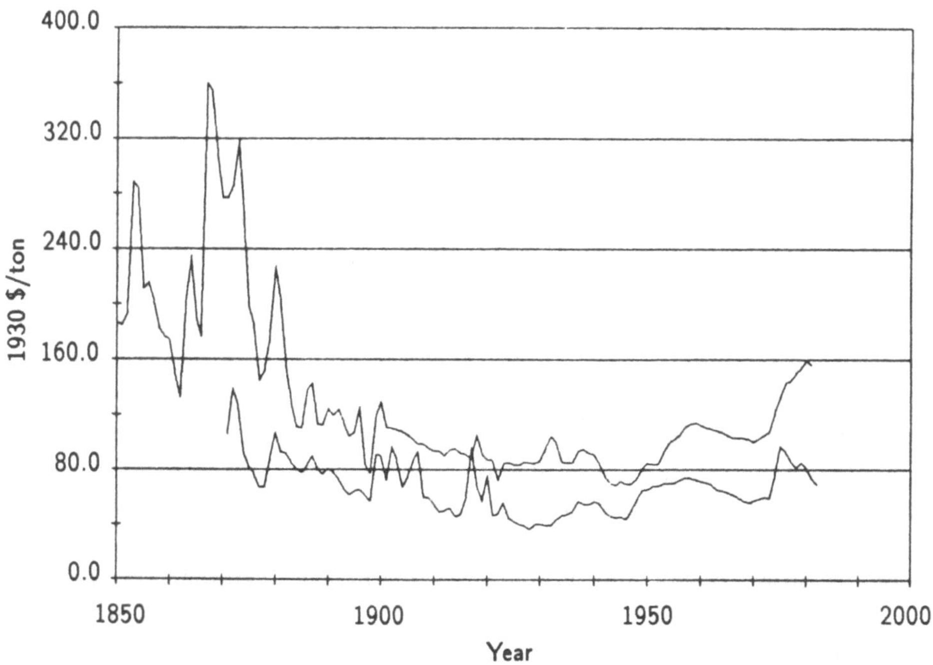

Figure 18.4. Real-term price of steel rails (top) and pig iron (bottom) in the USA from 1850 to 1980, in constant 1930 US$ per short ton. (Source: Adapted from Grübler, 1990a.)

term prices in the form of price *flares*, i.e., a succession of rapid inflationary and deflationary periods. These discontinuities in the real-term prices of steel products mirror discontinuities in the general wholesale and consumer price levels, used as an indicator by students of *Wechsellagen* of economic development ever since the time of Nikolai Kondratieff (1926).

The innovation diffusion cases discussed above demonstrate the complexity of diffusion phenomena: multiple interaction, sequence of successive replacements, importance of specific market niches in the introduction and phaseout of technologies, and a constantly changing adoption environment (in terms of relative technological performance, costs, and prices) appear to be inherent characteristics of diffusion and substitution processes. From such a perspective diffusion can hardly be reduced to single determining variables but emerges out of a complex vector of influencing factors. The importance of any individual driving variable such as relative costs or prices is not only different in the various phases of the diffusion life cycle of an innovation (i.e., relative costs appear to be of minor importance compared to technological performance in the initial diffusion phase) but also different in successive technological generations. The diversity and complexity (at the same time with a simple and consistent evolutionary structural change pattern at the macro level) suggest that the diffusion of large pervasive systems, which can span several decades in time, displays the features of dynamic self-organizational systems rather than operating in a classical equilibrium-type framework.

A valid point of critique on diffusion models has always been that the models tend to simplify the complex dynamics and transformations of both the market environment and the technological characteristics of innovations during the diffusion process. In other words, it is during its diffusion that a technology acquires its industrial and economic properties, transforms itself, and widens the initial market in which it is adopted. Some authors have consequently inferred the theoretical impossibility of formal representation, since the object of diffusion is not the same at the beginning, in the middle, and at the end of the process. It appears, however, that the usefulness of a formal representation resides in the ability to divide the diffusion process into periods, with the aid of criteria which can take into account the principal transformations of the technology under consideration. This, however, requires a methodological framework for a clear-cut definition of the technologies and their characteristics.

In this context it appears appropriate to comment on how one actually measures (defines) processes, products, etc., subject to diffusion studies. By which criteria does one consider *a priori* technological routes as competing or not competing with each other? What are the generic criteria and theoretical underpinnings of defining railways, continuous casting, or automatic teller machines as the object of the investigation of diffusion research without a comprehensive methodological analysis of the whole *space* in which technologies evolve?

The pragmatic answer that whatever leads consumers to purchase and industry to investment decisions, constitutes an appropriate unit for analysis of diffusion research, would appear rather unsatisfactory from its atheoretical nature. Furthermore, this view may lead to restrictions in the analysis, along the lines of unary or binary views of the diffusion phenomenon discussed above, i.e., missing the potential influence of other technologies or products on the diffusion trajectory of the particular innovation studied, or identifying possible technological routes that have been *locked out* from diffusion in the past. There clearly exists an asymmetry in diffusion research between approaches trying to identify and to define the object under investigation and the research efforts invested in the subsequent stages of research.

Obviously, when various process technologies differ in basic physical laws, or different generic functions provide a basis for dividing goods into product categories, the definition of the competing *species* may be clear-cut. This however is basically an *ex post* type of approach and does not represent a rigorous and systematic view of the whole environment, from which future competitors (ignored to date in the analysis) might emerge. Taking biology or anthropology as a guide, should we not attempt to define the object of diffusion research prior to analysis by some sort of *evolutionary tree*, spanning the whole diverse domain into which any particular case is embedded? Improvements and add-on innovations, which result in a particular technology becoming competitive within a given market segment could be represented as *branchings* along an evolutionary technological *tree* and allow for classification, taxonomy, and vigorous definition of technologies in competition. The task involved is admittedly complex, as not all technological settings provide a well-structured domain to carry out a theoretical analysis of the *technological space* and to deduce from such an analysis a definitional framework of competing technologies (as illustrated in Chapter 16). A formal analysis as a methodological framework for the definition of the object of diffusion research (see, e.g., Foray and Grübler, 1990) could, however, provide the possibility of ultimately developing a taxonomy and a classification system of technologies and their diffusion processes, which can be considered necessary to advance the theoretical foundations and practical uses of diffusion studies.

18.4 Hierarchy of Diffusion Processes

The diffusion of pervasive systems is charactererized by interrelatedness, interdependence, and cross-enhancing between a whole host of process, product as well as organizational and managerial innovations. This cross-enhancing results in the pervasive (economic and social) effects of whole *diffusion clusters* forming so-called *socio-technical paradigms* (Freeman, 1983). A mismatch in speed and frequency of diffusion or between technological innovations proper and the supportive social and organizational environment can result in disruptive effects or in the blockage of innovation diffusion, in case they are introduced outside appropriate "opportunity windows" (Soete and Perez, 1988) for their effective mediation into the social and economic context.

The diffusion of pervasive systems at the macro level involves therefore a host of related diffusion processes in a hierarchical structure, from the macro down to the micro level. One way to describe such a hierarchical structure of interrelated diffusion processes would be to develop a time scheme (much like a PERT graph) of all the different diffusion processes on which the growth of a pervasive system (a *socio-technical paradigm*) is contingent. However, no systematic research in this area has yet been performed. A simpler measure of a hierarchy of diffusion processes (without however considering their interdependence) would be to consider hierarchy in terms of the spatial or temporal patterns of diffusion.

The spatial hierarchy of diffusion processes is a well-recognized feature established in geography ever since the seminal contributions of Torsten Hägerstrand (1952 and 1967). *Figure 18.5* illustrates the spatial diffusion of railways in Europe (Godlund, 1952) and shows the two major characteristic features of spatial diffusion phenomena: the neighborhood effect and the hierarchy effect. Four spatial hierarchy levels can be identified in *Figure 18.5*. The spatial diffusion of railways originated in the UK (1826). Ten years later railway networks extended over much of England and reached a second spatial hierarchy level: Belgium, Lyon in France, and Bohemia in the Austro-Hungarian Empire. From these second hierarchy levels railways extended over much of central Europe by 1846, and by that time the third spatial hierarchy level was reached (Napoli in Italy and St. Petersburg in the USSR). The last (fourth) spatial hierarchy level was reached some 30 years later, when railway networks started to be constructed in Greece.

The railway network in all European countries was basically completed (i.e., achieved its maximum network size) by the end of the 1930s. Ever since, the railway network has undergone rationalization and decommissioning of links (the only exception being in the USSR). The completion of railway expansion throughout Europe by the 1930s, along with the later starting dates of railway construction in peripheral areas of Europe as shown in *Figure 18.5*, demonstrates a noticeable catch-up effect in the development of railway networks in Europe. Thus, while some 100 years passed in the UK between the start-up of railway construction and completion of the railway network, this development process proceeds much faster in the fringe countries of the European continent, such as Scandinavia, where this process took only about 50 years. Whereas the start-up dates of railway construction are thus spaced rather widely over time, there is stronger convergence in the ultimate

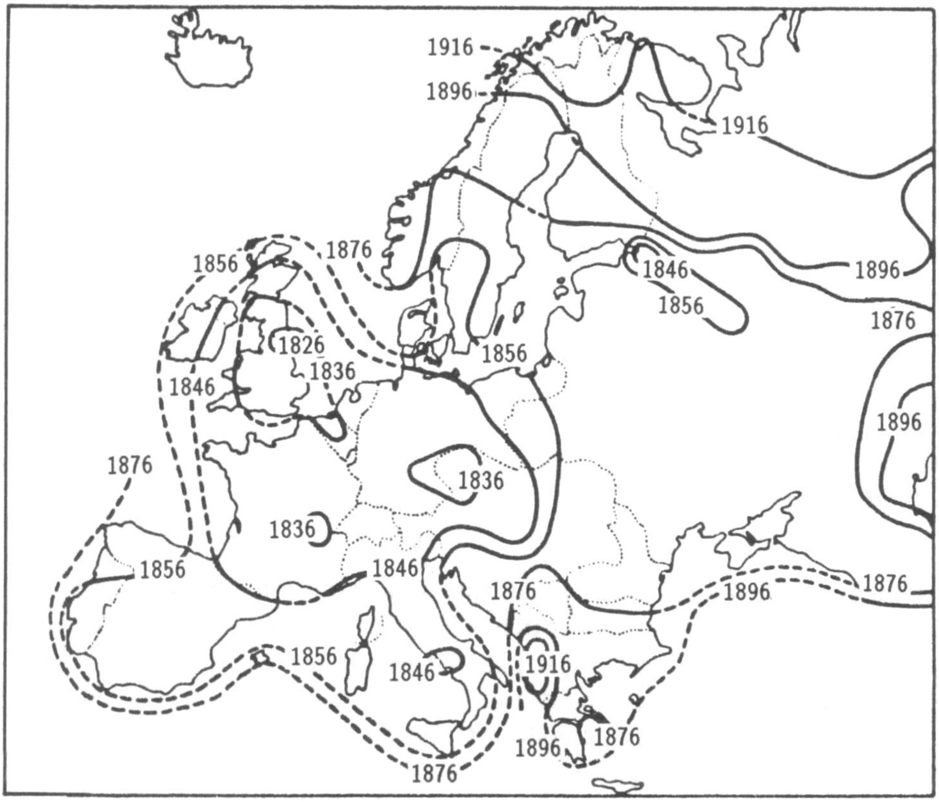

Figure 18.5. Spatial diffusion of railways in Europe showing spatial hierarchy and neighborhood effect, in isolines (10 year intervals) of spatial coverage by railway networks. (Source: Godlund, 1952.)

completion dates of this particular infrastructure expansion pulse, due to the catch-up effect.

Another observation on the spatial diffusion of railway networks is that the ultimate railway density between the core countries (i.e., the first hierarchical levels of the spatial diffusion of railways) and their hinterlands was in fact very different. The railway network density (either per capita or per unit country area) decreases with the distance to the innovation and the spatial hierarchy centers of diffusion (as shown in the next chapter by Nakićenović).

A second possibility of establishing a hierarchy of diffusion processes is to consider the temporal dimension, i.e., a hierarchy along the required diffusion

Table 18.1. Hierarchy of diffusion processes in the transport sector of the USA and the USSR, measured by their temporal diffusion parameters.

	USA		USSR	
	t_0	Δt	t_0	Δt
Total length of transport infrastructures	1950	80	1980	80
Growth of railways				
1830–1930	1858	54	1890	37
1930–1987	decline	decline	1949	44
Treated ties (USA)	1923	26		
Track electrification (USSR)			1965	27
Replacement of steam locomotives	1950	12	1960	13

time between introduction and saturation of the diffusion process. *Table 18.1* illustrates such a (temporal) hierarchy of diffusion processes in the transport sector of the USA and the USSR (Tzarist Russia prior to 1917). The time variables of diffusion and substitution processes are as follows (based on a logistic diffusion/substitution model): t_0 denotes the inflection point, i.e., the time of maximum growth (replacement) at the 50 percent diffusion (market share) level; Δt the time period in years that is needed to grow (substitute) from the 10 to the 90 percent diffusion (market share) level.

Table 18.1 shows that while the diffusion processes are shifted in time (time lag between t_0), the diffusion time constants appear to be of a similar order of magnitude between the two countries. Of course decisive differences remain. For instance, the railway network in the USA has been decreasing since the beginning of the 1930s (reduction by some 40 percent of the 1930 network), whereas the USSR has experienced a *second pulse* of railway construction since that time period, doubling the length of its railway network. However, also this "second wind" in railway construction appears close to saturation.

The similar duration in the diffusion of technologies and infrastructures between the two countries, with their distinct differences in history and market-clearing mechanisms, enables us to develop a rough hierarchical classification of diffusion processes, as reflected by their diffusion constants (Δt). At the macro level, when considering the growth of the total length[5] of transport infrastructures (i.e., the length of canals, railways, roads, and airways) as a diffusion process the time constants involved are very long indeed; some eight decades are required to go from 1 to 50 percent of the final network length. This process spans over successive generations of diffusion life cycles of individual transport infrastructures and across major social and economic transformations.

At the next hierarchical level, when considering the growth of individual transport infrastructures, we observe that the development of major infrastructures proceeds with a Δt of typically between four and five decades. The growth of the railway network in 19th century Tzarist Russia occurs some 30 years later than in the USA, but proceeds somewhat faster (i.e., catching up).

As a next hierarchical level in the temporal pattern of diffusion, we consider the case of technological change or upgrading within an already existing infrastructure grid, i.e., the substitution of treated wooden railway cross-ties in the USA (a substitution occurring in a contracting market, as the railway network of the USA has been decreasing since the 1930s) and the electrification of railway tracks in the USSR. Here diffusion time constants (Δt) of less than three decades are typical, i.e., significantly faster than in the case of new constructions of infrastructure grids.

Finally, at the last hierarchical level of the time constants of diffusion presented in *Table 18.1* above, we consider a replacement process within the capital vintage structure in the form of the rolling stock of railways. The replacement of the coal-fired steam locomotive by diesel traction (in the USA) and by diesel and electric traction locomotives (in the USSR) proceeds in both countries with a Δt of 12 to 13 years, i.e., a similar diffusion constant, which is even more noteworthy when considering the different adoption environments and economic driving forces prevailing between the two countries.

Proceeding further in the hierarchical structure of temporal diffusion processes to the micro level, one observes even faster diffusion Δts, in the order of a few years for consumer products, and one hierarchical level further down one can even observe Δts of a few months, e.g., in the diffusion of fashion items or novelty gadgets.

Thus we propose that diffusion processes can be characterized as operating within a temporal hierarchical structure, when considering the time constants (Δt) involved in diffusion. This hierarchy extends from the macro to the micro scale, throughout which one observes a self-similarity in the basic diffusion pattern, i.e., a fractal type of structure of diffusion processes all along their temporal (or spatial) hierarchical levels. Along the temporal hierarchy, diffusion processes appear to be faster the closer one gets to the micro level. Of course more empirical studies will have to be carried out before such a hypothesis can be firmly substantiated; however, a classification of diffusion processes along a hierarchical structure may indeed be a first step to develop a quantitative taxonomy of diffusion processes.

Two final observations on the temporal hierarchy of diffusion processes should be made. First, in view of the increasing time constants involved in diffusion when moving to the macro level (several decades), it is unlikely that equilibrium configurations can be expected at these higher hierarchy levels. Over such long time spans the technological and economic characteristics of technologies as well as the macroeconomic and social context into which they are embedded change drastically. This also means that the power of diffusion models in terms of establishing detailed causality and economic driving variables will dwindle the larger the system and the higher the level of aggregation.

Second, the importance of the diffusion phenomena, in that they ultimately transform the technological, economic, and social fabric into which they are embedded, also increases along the temporal diffusion hierarchy; the more important (pervasive) a particular diffusion process is in technological, social, and macroeconomic terms, the longer its diffusion constant Δt will be. From such a perspective, the diffusion rate of pervasive systems at the macro level appears to depend more on the *absorption capacity* and mediation rate of the socioeconomic system than being the inherent property of a particular innovation (like relative profitability in innovation diffusion at the micro level). The next section uses this hypothesis in the discussion of the average long-term diffusion rate derived from a large sample of diffusion histories in the USA.

18.5 The Discontinuous Nature of Diffusion Clusters

Interrelatedness, cross-enhancing, and a certain degree of clustering appear to be features in the diffusion of pervasive socio-technical systems. Within the area of influence of technological and social/organizational innovations on long-term economic growth, a major debate has been concerned with the discontinuous nature of technological change and its repercussions on *Wechsellagen* of economic development.

Ever since the work of Mensch (1975), this debate has been dominated by the discussion of the Schumpeterian hypothesis of the discontinuous rate of the *appearance* of innovations. Although evidence exists that the "bunching" of innovations in recessions and depressions is much weaker than originally maintained by Mensch a certain clustering of innovations can apparently still be identified at a statistically significant level (Kleinknecht, 1987). It is not the "bunching of innovations", however, that explains the whole sequence of discontinuities in economic growth, from slowdown to recession, recovery, and renewed growth. Innovations proper will only have a marginal impact on the economy as a whole. Only at a later stage, when being translated via diffusion into the creation of new infrastructures, industries, and product lines, would an innovation bunch provide for economic growth.

To some degree it is surprising that the debate has almost entirely concentrated on the *appearance* aspect of innovations (Mensch, 1975; Kleinknecht, 1987; Freeman *et al.*, 1982), whereas the *diffusion* aspect, i.e., how do innovations contribute to economic *growth* and – when saturating – to economic *slowdown*, has received relatively little attention. Although most researchers would agree on the theoretical rationale that economic slowdown and recession are saturation phenomena, extensive empirical evidence has not yet been assembled to demonstrate the coupled dynamics in the long-term pattern of introduction, diffusion, and saturation of a larger sample of innovations.

In the following the long-term history of innovations at a country level is described by considering their whole diffusion life cycle, i.e., from introduction to growth and eventual saturation. It can be shown that discontinuities in the long-term evolution result even without a rigid clustering in the appearance, growth, or saturation of innovations, although evidence indicates that the clustering effect is more focused toward the end of the diffusion life cycle (i.e., a kind of *season of saturations*) than during earlier phases.

In this context, innovations should not be defined only within a narrow technological context. Innovations consist of a host of interrelated technological, institutional, and organizational new ways to perform traditional or new tasks or to produce traditional or new products and services. The importance of such a set of interrelated, interdependent, technological, institutional, and organizational innovations as the major driving forces behind economic expansion periods was convincingly argued for and summarized under various headings, such as technological *trajectory* or *paradigm* or *socio-technical paradigm* (Perez, 1983; Dosi, 1983; Freeman and Perez, 1988). For our purposes, it suffices to stress the importance of the interdependencies between technological innovations, and the associated new forms of production organization, new institutions, and management methods, etc., required to realize fully the inherent growth potential(s) of a new industry.

The examples of innovation histories considered in the present context are therefore not only taken from the technological field. The empirical cases considered include the areas of energy, transport, manufacturing, agriculture, consumer durables, communication, military, and economic and also social diffusion and structural change processes such as the diffusion of literacy, reduction of infant mortality, and structural changes in employment. The diffusion histories of two samples of innovations in the USA since around 1800 are discussed below. The first sample consists of 117 diffusion cases analyzed at IIASA. This sample is augmented by additional cases found in the literature with a quantification of diffusion parameters. This brings the size of the sample to a total of 265 innovation cases.

Figure 18.6 shows the histogram of the diffusion rates as measured by their Δts for the two samples. As can be seen they range in duration from very short-term processes of only a few years to extremely long durations of up to 300 years. The mean value ranges between 40 and 60 years, with a standard deviation of about equal magnitude. It should be noted that the largest number of diffusion processes have Δts in the order of between 15 to 30 years, some of which we have given for illustrative purposes above (e.g. motor vehicles or steel production methods). In general, Δts appear to have a rank-size distribution.

The histogram gives one kind of summary about the distribution of diffusion processes: at any period of time, structural change can be decomposed into a large number of diffusion/substitution processes with a great variety in their Δts. Sometimes it is argued that the diffusion rate (i.e., Δt in our definition) should accelerate (systematically shorter Δts) over time, as a result of better communication and information channels. This hypothesis could not be substantiated on the basis of the two samples.[6] In fact the only significant acceleration tendency appears to occur *within* the diffusion of a given technological cluster. As a given cluster (e.g., the automobile industry) matures, it becomes increasingly characterized by an accelerating pattern of diffusion of incremental improvement innovations. *Figure 18.7* (Jutila and Jutila, 1986) illustrates this for the US automobile industry,[7] and shows the convergence tendency of the diffusion processes toward the end of of the diffusion life cycle (i.e., the *season of saturations*). Similar observations can also be made at the macro level when analyzing, for instance, the catch-up

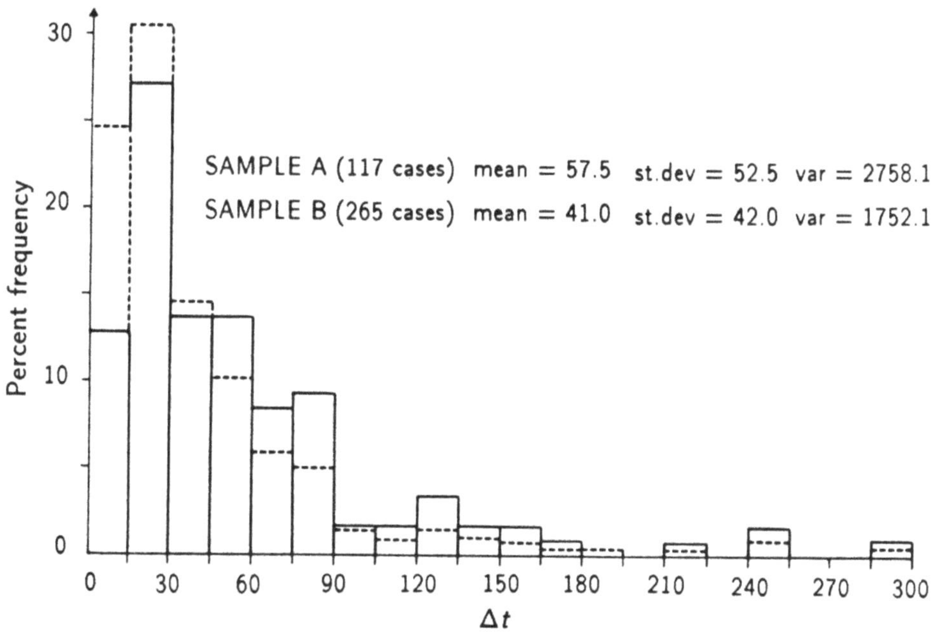

Figure 18.6. Histogram of diffusion constants Δt for two innovation diffusion samples in the USA, in frequency of Δts in percent. (Source: Grübler, 1990b.)

processes at an international level that result in a similar convergence in completion dates (see the next chapter by Nakićenović).

Another possible aggregate measure is the average diffusion rate over time for the whole socioeconomic system. For this measure we calculate the average diffusion rates of our innovation samples, i.e., the sum of the first derivatives of the diffusion/substitution trajectories[8] at any given point in time divided by the number of diffusion processes occurring at that moment. This indicator is the diffusion equivalent to the annual GNP growth rate. Our average diffusion rate measures the changing average rate of technical, economic, and social change at the country level: in our case the USA since 1800.

Figure 18.8 shows the average diffusion rate of 117 diffusion processes. It displays peaks and troughs, indicating that the process of change is not gradual and linear but is instead characterized by pronounced discontinuities. The general increase in the average rate of change is not indicative

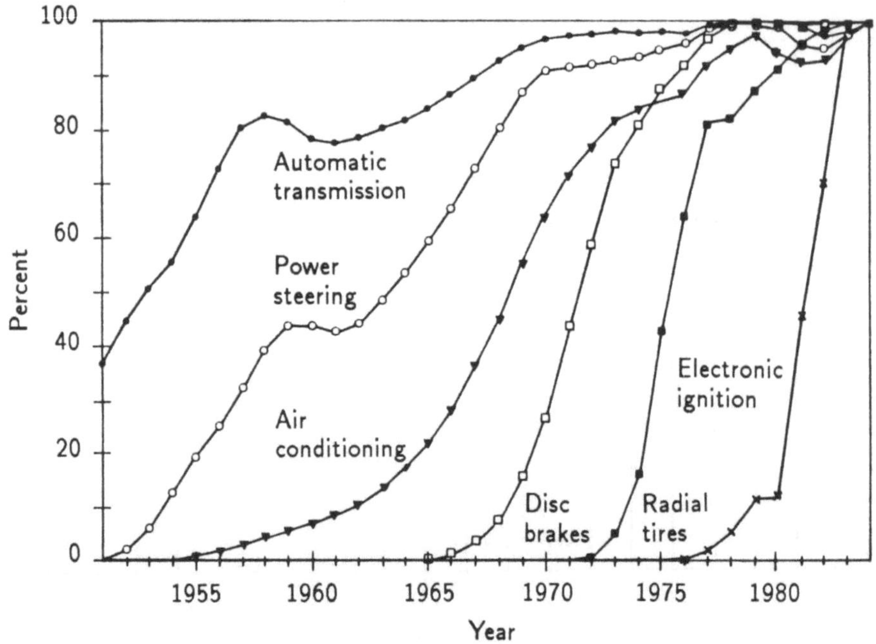

Figure 18.7. Diffusion of new technologies in the US car industry, percentage of cars manufactured. (Source: Jutila and Jutila, 1986.)

because, as one approaches the present, the number of shorter-term diffusion processes which can be documented increases. The increasing average rate of change is therefore likely a statistical artifact stemming from the bias in the sample. Although averages are used in the study, a higher number of overlapping short-term processes in one interval could result in a higher aggregate diffusion rate.

Figure 18.8 represents an aggregate rate of change over all diffusion processes, regardless of their social or economic importance. The question of how to weigh an extreme diversity in technical and social diffusion processes is indeed difficult. One could, for instance, resort to subjective weighting of different diffusion processes, e.g., by some kind of DELPHI method. Such an exercise was not attempted because of the inherently subjective criteria underlying such a relative weighting of processes of change, which span a period of about 200 years. Instead, it is assumed that the importance of any particular process of diffusion or change is directly related to the time constant of diffusion. Thus, the longer a process takes and the higher it is

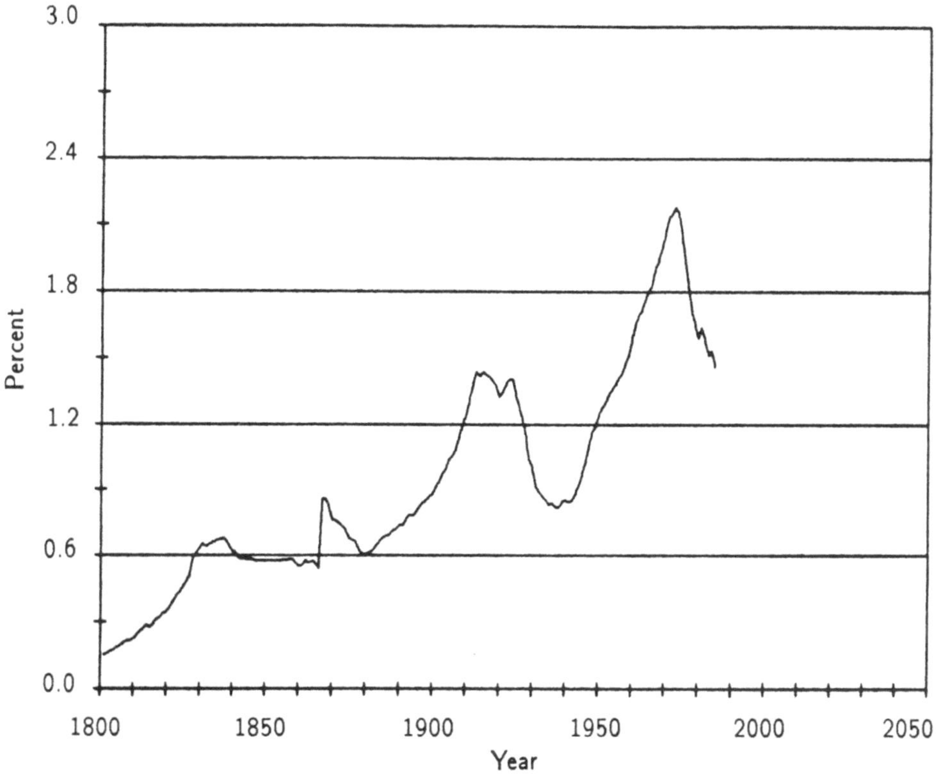

Figure 18.8. Average rate of technical, economic, and social change in the USA based on a sample of 117 diffusion processes, in percent/year. (Source: Grübler, 1990b.)

ranked in the temporal hierarchy of diffusion processes, the more pervasive its macro level effects are assumed to be.[9] Of course, this measure also remains debatable, but it is a plausible hypothesis. More important, however, is the fact that even this weighted average rate of socio-technical change still reveals pronounced long-term discontinuities as shown in *Figure 18.9*.

These long-term discontinuities in the rate of socio-technical change are the result of the complex coupled dynamics of the discontinuous rate at which innovations are first introduced and of their subsequent different rates of absorption (diffusion) in the socioeconomic system. Periods of accelerating technological and social diffusion rates indicate the emergence of a new socio-technical paradigm under which many (often interrelated) innovations diffuse into the economic and social environment contributing, via

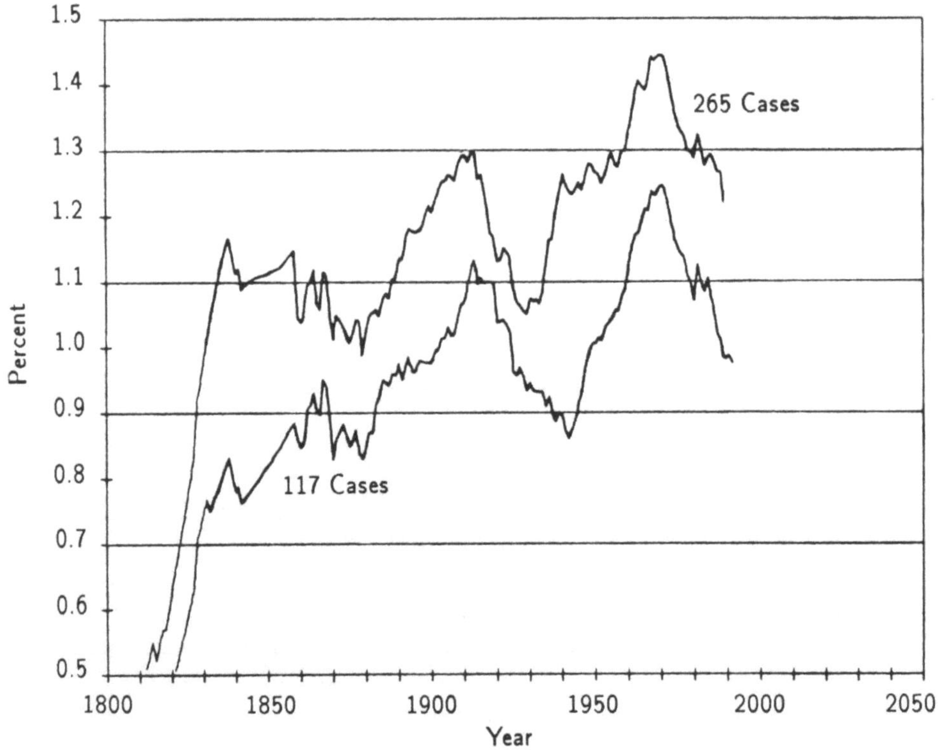

Figure 18.9. Weighted (by mean Δt) average rate of technical, economic, and social change in the USA based on samples of 117 (mean Δt 57.5 years) and 265 (mean Δt 41 years) diffusion processes, in percent/year. (Source: Grübler, 1990b.)

backward and forward linkages, to prolonged periods of economic growth. These periods are followed by periods where progressively more and more innovations enter their saturation phase of diffusion. Thus, each peak in the average rate of change in *Figure 18.9* characterizes the start of saturation of a corresponding cluster or family of diffusion processes. This *season of saturations* results in a significant decline in the average rate of technical and social change and, via market saturation and a decrease in investments, also to a slowdown in economic growth. The results also indicate the importance of new innovations, which initiate phase transition out of a "stalemate" position into a new period of accelerated rates of change. Presumably many innovations have emerged during the last decades that may turn out to be successful. If they were included they could perhaps lead to a trend reversal in the rate-of-change curve sometime after the mid-1990s – the time when these successful innovations, after a slow initial diffusion, would in turn enter into the steep exponential part of their diffusion life cycle.

The conclusion on the discontinuous nature of socioeconomic change is corroborated by analyzing the average diffusion rates of the second innovation sample comprising 265 innovation cases. Compared with the first sample, the greater preponderance of shorter-term diffusion and substitution processes after World War II results in a shorter mean Δt of the sample (i.e., of the weighting measure). Therefore, the weighted aggregate rate of change is higher in the second sample (*Figure 18.9*). However, this does not mean that this larger sample of diffusion processes yields higher rates of overall socio-technical change, but rather it indicates the emphasis diffusion research has placed on recent time periods and the resulting better documentation of also shorter-term diffusion processes. This is understandable both from the higher interest in the present rather than the past and from the data availability. Still, pronounced discontinuities remain, and the larger sample confirms the findings that the diffusion rate has been declining since 1970, indicating an increase in (market and diffusion) saturation phenomena ever since.

It should be noted that the turning points (discontinuities) in the diffusion rates of technological and social innovations coincide with the turning points of long-term *Wechsellagen* of economic growth identified by a number of long-wave researchers (van Duijn, 1983; Goldstein, 1988; Vasko, 1987). The resulting peaks (i.e., the maxima in the rate of socio-technical change and the onset of leveling-off and saturation phenomena) occurred in 1840, 1912, and 1970. Troughs (maxima of saturation periods and the slow beginning of a new phase of accelerated socio-technical change) occurred in 1820, 1875, and 1930. It is certainly not incidental that these troughs coincide with pronounced recessions, even depressions in the US economy.

The diffusion history of the large number of processes of technical, economic, and social change outlined above points to an essentially Schumpeterian view of long-term development. Major economic expansion periods appear driven by the widespread diffusion of a host of interrelated innovations, leading to new products, markets, industries, and infrastructures. These diffusion processes are sustained (in fact contingent) by mediating social and organizational diffusion processes. The growth (diffusion) of a dominant *socio-technical paradigm* cannot, however, be sustained indefinitely.

Market saturation, the dwindling improvement possibilities of existing process technologies, managerial, and organizational settings, and an increasing awareness of the negative externalities involved in the further intensification and extension of the dominant "development paradigm" pave the way to a *season of saturations.* During such periods opportunities arise for the introduction of new technological, organizational, and social solutions, some of which may have been latently in existence but were barred from market entry owing to the dominance of the previous "growth paradigm". Even when such innovations are introduced successfully, their penetration rates in the initial phase of their diffusion life cycle are rather slow and a matching new social and economic mediating context has still to emerge. This perpetuates the period of phase transition where the old is saturating and the new is still embryonic. It is only after such a period of transition, crisis, and mismatch that a new prolonged period of widespread diffusion of a new socio-technical *bandwagon*, and thus a period of prolonged economic growth, emerges.

Before precise linkages and causality mechanisms can be developed, such a conjecture clearly remains at the phenomenological level. A necessary step in this direction would be to establish linkages and interrelationships between and within socio-technical innovation *bandwagons* by a taxonomy of diffusion processes. Still, the picture that emerges from the approach presented here is that the overall development trajectory appears punctuated by phase transitions from an old saturating to a new but yet uncertain development path. As such, diffusion, and its discontinuities, may be an inherent feature of the evolutionary process that governs social behavior.

Notes

[1] The average active life of horses used for transportation purposes is from 12 to 14 years (recall here the Δt of their replacement in the order of 12 years) which at present is similar to the average useful life of an automobile: the replacement of the automobile fleet by cars equipped with catalytic converters also proceeds with a Δt of about 13 years (Nakićenović, 1986).

[2] The exponential diffusion coefficient of the standard logistic model should decrease over time, yielding an asymmetrical diffusion curve.

[3] This of course is a simplification of the assumption underlying the standard diffusion model, which considers *expected* profitability as the driving variable. At a later stage of the diffusion process, profitability declines along with the uncertainty associated with the outcome of an adoption decision; thus, the resulting vector on expected profitability may also stay constant over time.

[4] For the algorithmic and computer implementation of the model see Nakićenović (1979).

[5] One of the findings in the analysis of the growth of transport infrastructures in the two countries is that not only the development of individual systems like canals, railways or surfaced roads proceeds along a sequence of S-shaped (logistic) growth patterns (as shown in the next chapter), but also that the evolution of the aggregate of all these individual systems proceeds along an S-shaped growth trajectory. The growth pulses of individual transport infrastructures, which overlap in their various growth (even decline) phases, are thus consistent with a structured evolutionary path at the macro level of the whole transport system.

[6] In fact, such a hypothesis only appears warranted when the diffusion phenomenon is primarily reduced to a learning and communication process. It has been argued above that such a view might indeed represent an oversimplification in view of the different market niches, multiple competition environments and resulting changes in the driving forces over the diffusion life cycle.

[7] In the absence of quantitative diffusion model parameters, the cases shown in *Figure 18.7* are not part of the diffusion samples considered in the present context.

[8] Calculated from the parameters of a logistic model, i.e., Δt and t_0. In the case of asymmetrical diffusion/substitution patterns, growth rates are calculated based on piecewise linear trends in log *F/(1-F)* transformation. The contribution of Maximilian Posch in the development of the computational algorithm underlying the results is gratefully acknowledged.

[9] The weighting measure proposed links the *importance* of a particular diffusion process proportionally to its diffusion time constant Δt. Thus, we suggest that a one percent growth in the railway network of the USA (Δt of 55 years) is proportionally (55/12) more important than a one percent growth in the diffusion of diesel/electric locomotives (replacing steam locomotives) proceeding with a Δt of 12 years.

References

Arthur, W.B., 1983, *Competing Technologies and Lock-in by Historical Small Events: The Dynamics of Choice Under Increasing Returns*, Center for Economic Policy Research Paper 43, Stanford University, Stanford, CA, USA.

Ayres, R.U., 1989, *Technological Transformations and Long Waves*, RR-89-1, International Institute for Applied Systems Analysis, Laxenburg, Austria.

Bass, F.M., 1969, A new product growth model for consumer durables, *Management Science* **5**:215–227.

Brooks, H., 1986, The typology of surprises in technology, institutions, and development, in W.C. Clark and R.E. Munn, eds., *Sustainable Development of the Biosphere*, Cambridge University Press, Cambridge, UK.

Davies, S., 1979, *The Diffusion of Process Innovations*, Cambridge University Press, Cambridge, UK.

Dosi, G., 1983, Technological paradigms and technological trajectories. The determinants and directions of technical change and the transformation of the economy, in C. Freeman, ed., *Long Waves in the World Economy*, Butterworths, London, UK.

Easingwood, C.J., Mahajan, V., and Muller, E., 1981, A nonsymmetric responding logistic model for technological substitution, *Technological Forecasting & Social Change* **20**:199–213.

Fisher, J.C. and Pry, R.H., 1971, A simple substitution model of technological change, *Technological Forecasting & Social Change* **3**:75–88.

Foray, D. and Grübler, A., 1990, Morphological analysis, diffusion and lock-out of technologies: Ferrous casting in France and the FRG, *Research Policy* **19**(6):535–550.

Freeman, C., ed., 1983, *Long Waves in the World Economy*, Butterworths, London, UK.

Freeman, C., Clark, J., and Soete, L., 1982, *Unemployment and Technical Innovation: A Study of Long Waves and Economic Development*, Frances Pinter, London, UK.

Freeman, C. and Perez, C., 1988, Structural crises of adjustment, business cycles and investment behaviour, in G. Dosi *et al.*, eds., *Technical Change and Economic Theory*, Pinter Publishers, London and New York.

Godlund, S., 1952, *Ein Innovationsverlauf in Europa, dargestellt in einer vorläufigen Untersuchung über die Ausbreitung der Eisenbahninnovation*, Lund Studies in Geography, Series B. Human Geography No. 6., C.W.K. Gleerup Publishers, Lund, Sweden.

Gold, B., 1981, Technological diffusion in industry: Research needs and shortcomings, *Journal of Industrial Economics* **24**:247–269.

Goldstein, J.S., 1988, *Long Cycles: Prosperity and War in the Modern Age*, Yale University Press, New Haven and London.

Griliches, Z., 1957, Hybrid corn: An exploration in the economics of technical change, *Econometrica* **25**:501–522.

Grübler, A., 1990a, Technology diffusion in a long wave context: The case of the steel and coal industries, in T. Vasko, *et al.*, eds., *Life Cycles and Long Waves*, Lecture Notes in Economics and Mathematical Systems 340, Springer-Verlag, Berlin, Heidelberg, New York.

Grübler, A., 1990b, *The Rise and Fall of Infrastructures, Dynamics of Evolution and Technological Change in Transport*, Physica Verlag, Heidelberg, FRG.

Hägerstrand, T., 1952, *The Propagation of Innovation Waves*, Lund Studies in Geography, Series B, No. 4, Lund, Sweden.

Hägerstrand, T., 1967, *Innovation Diffusion as a Spatial Process*, University of Chicago Press, Chicago and London.

Hurter, A.P. and Rubinstein, A.H., 1978, Market penetration by new innovations: The technical literature, *Technological Forecasting & Social Change* **11**:197–221.

Jutila, S.T. and Jutila, J.M., 1986, *Diffusion of Innovation in American Automobile Industry*, Paper prepared for The Advanced Summer Institute in Regional Science, University of Umeå, Sweden, June 9–28, 1986.

Kleinknecht, A., 1987, *Innovation Patterns in Crisis and Prosperity, Schumpeter's Long Cycle Reconsidered*, Macmillan Press, London, UK.

Kondratieff, N.D., 1926, Die langen Wellen der Konjunktur, *Archiv für Sozialwissenschaft und Sozialpolitik* **56**:573–609 (German translation of original Russian article).

Mahajan, V. and Peterson, R.A., 1985, *Models for Innovation Diffusion*, Sage Publications, Beverly Hills and London.

Mansfield, E., 1961, Technical change and the rate of imitation, *Econometrica* **29**(4):741–766.

Mansfield, E., 1968a, *Industrial Research and Technological Innovation, An Econometric Analysis*, Norton & Co., New York, NY, USA.

Mansfield, E., 1968b, *The Economics of Technological Change*, Norton & Co., New York, NY, USA.

Marchetti, C. and Nakićenović, N., 1979, *The Dynamics of Energy Systems and the Logistic Substitution Model*, RR-79-13, International Institute for Applied Systems Analysis, Laxenburg, Austria.

Mensch, G., 1975, *Das technologische Patt, Innovationen überwinden die Depression*, Umschau Verlag, Frankfurt, FRG.

Metcalfe, J.S., 1983, Impulse and diffusion in the study of technical change, in C. Freeman, ed., *Long Waves in the World Economy*, Butterworths, London, UK.

Montroll, E.W., 1978, *Social Dynamics and the Quantifying of Social Forces*, Proceedings of the National Academy of Sciences, USA, **75**(10):4633–4637.

Nakićenović, N., 1979, *Software Package for the Logistic Substitution Model*, RR-79-12, International Institute for Applied Systems Analysis, Laxenburg, Austria.

Nakićenović, N., 1986, The automobile road to technological change, diffusion of the automobile as a process of technological substitution, *Technological Forecasting & Social Change* **29**:309–340.

Nakićenović, N., 1987, Technological substitution and long waves in the USA, in T. Vasko, ed., *The Long-Wave Debate*, Springer-Verlag, Berlin, Heidelberg, New York.

Nakićenović, N., 1988, Dynamics and replacement of U.S. transport infrastructures, in J.H. Ausubel and R. Herman, eds., *Cities and their Vital Systems, Infrastructure Past, Present and Future*, National Academy Press, Washington, DC, USA.

Nasbeth, L. and Ray, G.F., eds., 1974, *The Diffusion of New Industrial Processes: An International Study*, Cambridge University Press, Cambridge, UK.

Nelson, R.R. and Winter, S., 1982, *An Evolutionary Theory of Economic Change*, The Belknap Press of Harvard University Press, Cambridge, MA, USA.

Perez, C., 1983, Structural change and the assimilation of new technologies in the economic and social system, *Futures* **15**(4):357–375.

Poznanski, K., 1986, The extinguishing process: A case study of steel technologies in the world industry, *Technovation* **4**:297–316.

Ray, G.F., 1989, Full circle: The diffusion of technology, *Research Policy* **18**:1–18.

Sharif, M.N. and Kabir, C., 1976, A generalized model for forecasting technological substitution, *Technological Forecasting & Social Change* **8**:353–334.

Skiadas, C., 1986, Innovation diffusion models expressing asymmetry and/or positively or negatively influencing factors, *Technological Forecasting & Social Change* **30**:313–330.

Silverberg, G., Dosi, G., and Orsenigo, L., 1988, Innovation, diversity and diffusion: A self organizational model, *The Economic Journal* **98**:1032–1054.

Soete, L. and Perez, C., 1988, Catching up in technology: Entry barriers and windows of opportunity, in G. Dosi *et al.*, eds., *Technical Change and Economic Theory*, Pinter Publishers, London and New York.

van Duijn, J.J., 1983, *The Long Wave in Economic Life*, Allen and Unwin, London, UK.

Vasko, T., ed., 1987, *The Long-Wave Debate*, Springer-Verlag, Berlin, Heidelberg, New York.

Ward, W.H., 1967, The sailing ship effect, *Bulletin of the Institute of Physics and Physics Society* **18**:169.

Chapter 19

Diffusion of Pervasive Systems: A Case of Transport Infrastructures

Nebojša Nakićenović

19.1 Introduction

The objective of this chapter is to demonstrate that the diffusion of pervasive systems worldwide tends to cluster toward the saturation period. The advanced hypothesis is that the time span between the start of diffusion in leading and lagging countries decreases as the diffusion progresses. The process, therefore, appears much more focused internationally toward the saturation phase. However, this visible *catch-up effect* refers only to the relative diffusion rates and not to the absolute level of adoption. The leaders achieve higher diffusion levels and the followers lower ones, roughly in proportion to the lag in the introduction of a given innovation. This tendency will be illustrated on the basis of the evolution of transport infrastructures and systems in several countries.

There are three reasons for choosing transport infrastructures and transport systems. First, they are pervasive in the sense that they are an important aspect of the techno-economic development: they affect almost all facets of economic and private activities. Second, transport systems depend on numerous regional and national characteristics so that any congruence in their evolution must be related to the basic driving forces of economic and social development. The third reason for choosing this example is that historical data for a number of countries over relatively long periods are available.

19.2 Diffusion and Evolution of Technologies

Diffusion starts after the successful commercialization of a basic innovation leads to the creation of a new process, product, or service. Eventually, some innovations replace their predecessors, and some become pervasive in that they lead to a host of interrelated new activities in many sectors, creating whole new industries and thereby contributing to social and economic development. Electricity, automobiles, or computers are examples of such innovations that have changed many facets of everyday life.

During the early development phases of this process, the new industry is fluid with a high degree of diversity and experimentation. The initial emphasis is on improving technical performance without much regard for cost and price. Eventually one particular technological variant becomes dominant and is *locked in* leading to standardization in the new industry. This is usually a disruptive phase of development with a characteristic "shakeout" that only a few competitors survive. Prices decrease with increasing cost reductions owing to standardization, stronger competition, learning by doing, and many cumulative improvements resulting from incremental innovations. Economies of scale and cost reductions along the learning curve lead to advantages from which only a few competitors can internalize and benefit. This is also the development phase, when the pervasive nature of some technologies can be recognized. Some of them diffuse in many sectors.

As the technology and its applications mature, the possible disadvantages become evident. Improvements can cover an increasingly smaller domain of technical and managerial possibilities (Brooks, 1986). Saturation starts, and the problems associated with widespread and large-scale application become important. The social and institutional response is rather nonlinear and disruptive, and the awareness of social disadvantages and risks often increases rapidly making further adoption unacceptable. The perceived disadvantages of further applications can outpace incremental improvements as the technological and social potentials appear to be exhausted. Further adoption is virtually blocked. Just like the initial introduction and commercialization of a new technology or system, empirical evidence indicates that this saturation phase is also an international phenomenon. Innovations spread in *space* and *time*, but lag behind in the peripheral regions. However, the time interval between the leaders and the laggers tends to decrease as the diffusion progresses, so that the process appears much more focused toward the saturation phase. Thus, a visible *catch-up effect* in the diffusion of pervasive techno-economic systems throughout the world is evident. Saturation lags are smaller because diffusion processes in the peripheral or lagging regions accelerate over time (due to external learning, transfer of know-how and capital, etc.) proportionally to the introduction delay with respect to the diffusion duration in the leading regions. The intensity of diffusion is, however, lower in the lagging regions. In this sense, the diffusion processes can be seen internationally as Schumpeterian bandwagons or clusters of related families that emerge, with time lags in different parts of the world, to focus more strongly during the saturation phase.

During the periods when old systems saturate, new techno-economic paradigms emerge; the old development trajectory associated with the previous generation of pervasive technologies and institutional forms is not only challenged, but in time is also replaced with the new solutions. Dosi *et al.* (1988) have identified innovation diffusion and the resulting technological change as a fundamental force in shaping the pattern of transformation of the economy. They have shown that the dynamic adjustment mechanisms during these transformations have to do with both technical and institutional change, or the lack of it. With regard to technological change, it is both disruptive during the transition period (marked by fluctuations, frictions, and sometimes crises) and a source of order for the directions of change and dynamic adjustment processes, as new technologies diffuse through national and international economies. Thus, some of the most important changes in socio-institutional frameworks and economic structure are related to the pervasive adoption of new systems. Today, the emergence of new systems, such as the information and communication technologies, is often mentioned in this context because they may become pervasive throughout the economy; they will not diffuse in just a few selected sectors. Another way to formulate this phenomenon is to consider the interrelationships among different diffusion processes and their clustering in space and time.

Table 19.1. Clusters of pervasive technologies.

Period: 1750–1820	1800–1870	1850–1940	1920–2000	1980–2060
Dominant Systems:				
Water power, sails, turnpikes, iron castings, textiles	Coal, canals, iron, steam power, mechanical equipment	Railways, steam ships, heavy industry, steel, dyestuff, telegraph	Electric power, oil, cars, radio, TV, durables, petrochemicals	Gas, nuclear, aircraft, telecomm., information, photo-electronics
Emerging Systems:				
Mechanical equipment, coal, stationary steam, canals	Steel, city gas, indigo, telegraph, railways	Electricity, cars, trucks, Radio, roads, oil, telephone, petrochemicals	Nuclear power, computers, gas, telecommunication, aircraft	Biotech., artificial intelligence, space ind. & transport
Organizational Style:				
Manufacture	Industrial production	Standardization	Fordism-Taylorism	Quality control

Table 19.1 illustrates that this clustering effect reoccurred five times since the onset of the Industrial Revolution. It gives the dominant techno-economic systems for each epoch in the top row, and the list of emerging ones in the middle row. For example, the dominant cluster of the 1920–2000 period includes the growth of electric power, oil and petroleum, petrochemicals, cars, radio and TV, and consumer durables. During the same period we saw the emergence of new industries, which could become important in the future: nuclear power, computers, natural gas, telecommunication, and advanced aircraft. The last row lists the predominant organizational and management models (or "styles") during the respective periods. The list of clusters given in *Table 19.1* is of course not unique; it could be modified and extended. The timing is also not precise. However, it provides an illustrative account of interrelationships and clusters of pervasive technological systems (for a detailed analysis, see Freeman and Perez, 1988).

These changes in the techno-economic paradigms, as illustrated by the emergence and diffusion of interrelated clusters of pervasive technologies, show the history of development. For example, they show the history of economic restructuring, in terms of changes in employment and skills. They also show the history of transport systems. The following sections describe this evolutionary process of techno-economic and social change by illustrating the historical development of transport infrastructures and their clustering.

19.3 Transport Infrastructures

Transport and communication systems are the elements integrating human activities in space and time. As a rule, infrastructures used for the movement of people, goods, and information diffuse slowly and span many decades from their first introduction to obsolescence. Some are almost immortal, albeit usually providing different services from those originally intended. Old harbors are being converted into modern commercial and residential areas. A few of the obsolete canals were reused to build railways. Many ancient Roman roads are buried beneath modern highways.

In this sense many infrastructures serve as *rights-of-way* as actual transport systems replace each other in the eternal quest to increase quality of service, reliability, and speed, and make the movement of people, information, and tangible goods more efficient and convenient. While old infrastructures are frequently recycled, new systems appear that provide even greater possibilities, so that diversification and productivity usually increase in time. Thus, complexity increases resulting in numerous interlaced and overlapping niches occupied by competing modes of passenger travel, information channels, and transport of goods and services.

One clear trend emerges from the historical analysis – transport systems became ever faster and more productive. The first major improvement occurred with the age of canals. Canals represented a fundamental construction effort toward reducing natural barriers to connect coastal and inland waterways. At the same time, canals were a powerful motor of the industrial age: waterways allowed for new flows of goods, unprecedented exchanges between regions, specialization of labor, and access to more distant energy and raw material resources. The modern age of canals started about two centuries ago and lasted almost a hundred years. By the 20th century most national canal systems were in place, and many links already decommissioned. Eventually they yielded to the vicious competition from railroads.

The first railways were constructed in the 1830s and were able to extend the range, speed, and productivity offered by canals. More important, perhaps, was the capability to overcome even more imposing natural barriers. Bridges and tunnels were built for canals, but railways were capable of accommodating traffic and freight demands more directly – wherever demand existed it was almost always possible to build a railway line. In time, elaborate networks of railway systems were built in North America and Europe. Together with railways, a new era of coal, steam, steel, and telegraph began. The great railway era lasted until the 1920s.

Around the turn of the century the automobile was born and became the symbol of modern industrial development along with oil, petrochemicals, electricity, the telephone, and (Fordist) manufacturing. Paved roads reduced the time-space dimensions of modern societies. Speed and performance increased once again. The flexibility offered by an individual mode of transport became affordable for a wider social strata, and only recently have the disadvantages of the automobile become socially transparent although they have been known for a long time. This perception lag illustrates to what extent the automobile age was perceived as one of the preconditions for modern industrial development. This notion still appears to continue to prevail in many developing regions of the world.

Each successive mode of transport expanded into an infrastructure that was ten times larger than the previous one. For example, the length of transport infrastructures in the United States has increased by almost five orders of magnitude during the last two centuries. Each successive transport infrastructure was not only larger than the one it was replacing, but also faster. A hierarchy of space and time territories emerges. As Simon (1988) noted, most of the natural and man-made systems are often hierarchical, i.e., they have boxes-within-boxes architecture; transport systems portray a similar structure. A man walking or using waterways can cover a mean circle of a few kilometers diameter in one hour, the size of a village or small town. A person traveling by rail or horse could travel more than a dozen kilometers in the same period a hundred years ago. The automobile and rapid rail systems offer a larger range – up to 100 km – and can, thus, effectively connect cities; air travel extends the radius to almost 1,000 km. As connected territory increases so does travel, tangible goods transport, and information flow increase in a unit of time.

The sequences of development of canals, railways, and roads appear in *Figure 19.1* as regular diffusion processes when their size is plotted as a percentage of the saturation level. In addition, the figure shows the development of telegraphs and oil and gas pipelines. In time the complexity and diversification of transport and communication systems increased. While ancient roads and later canals were used for all transport and communication services, telegraphs evolved along with the railways as a specialized communication system. Oil pipelines represent a dedicated transport system

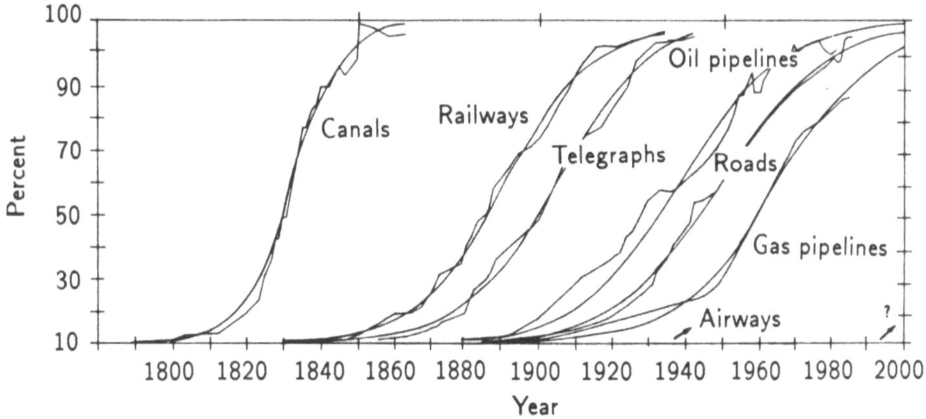

Figure 19.1. Diffusion of canals, railroads, telegraphs, roads, and oil and gas pipelines in the USA, in percent of saturation length.

related to the development of roads, the automobile, and the oil industry. The six S-shaped growth pulses can be subdivided into three groups: canals followed by railways and telegraphs, followed by oil pipelines, roads, and gas pipelines.

Figure 19.2 reproduces the diffusion of canals, railways, and roads on a logarithmic scale as the ratio of the growth level reached in a given year, divided by the amount of growth left to the saturation level.[1] The diffusion of telegraphs and oil and gas pipelines is omitted from *Figure 19.2* to enhance the clarity of presentation. This is a convenient way of presenting S-shaped diffusion processes as straight lines. The growth pulses of canals, railways, and surfaced roads are separated in time by 55 years when measuring the distance between the respective inflection points. The saturation and the onset of decline of all five infrastructures coincides with the beginning of prolonged recessions (in the 1870s, 1930s, and 1980s) of the last three long waves.[2] The development of canals, relative to the achieved saturation level, was much quicker than the expansion of railways and roads. The time constant of growth, Δt, is about 30 years for canals, 54 years for railroads, and 56 years for surfaced roads.[3] Thus, the life cycles of infrastructures are indeed very long, spanning periods of a century from introduction to saturation. The duration of senescence can be even longer.

The most vital of the structures, however, are permanent. They may provide different services than originally envisaged. More than a century after the canal era, the remaining inland waterways are used for leisure

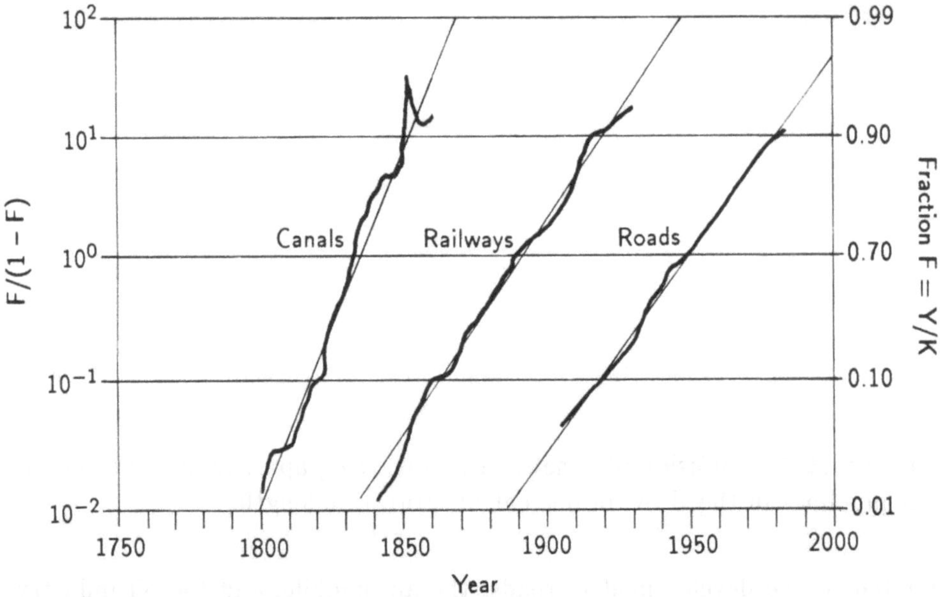

Figure 19.2. Logistic growth of infrastructures in the USA: length achieved in a given year divided by growth left to saturation level.

activities, transport of low-value goods, and irrigation. There are also more sails today than in the heyday of ocean clippers, but most of them are on pleasure boats.

19.4 Diffusion Clustering

The great canal expansion was initiated in the United Kingdom between 1750 and 1850. By the 1850s the canal construction saturated with a total network of about 5,630 km including 60 km of water tunnels and a large number of locks and aqueducts. The successful English experience in canal development, and especially the *canal mania* of the end of the 18th century, became the development model on the Continent and in the United States, albeit with a considerable time lag. For example, the growth of canals in France and the United States started much later and occurred almost in parallel and ended in practically synchronous saturation toward the 1860s with a total network length of about 4,825 km and about 7,010 km, respectively. Thus, there is a pronounced catch-up period in canal construction in France and the United States compared with England. Despite a considerable time lag in the introduction of canals, growth proceeded much faster, and as a result the expansion of canal networks reached saturation in all three countries (and most other industrialized countries) within a relatively short period of about two decades.

Canal construction did not come to a complete standstill, even after the networks had reached their maximum size and proceeded to decline. For example, substantial canal network extensions occurred in Germany and Russia after the 1860s.[4] However, these new constructions could not compensate for the overall network decline after saturation. The canal business was eroded to such an extent by competition from the railways, that many important canals were decommissioned and the lengths of the networks declined.

In general, the diffusion of canals was swifter in the countries that were the "late starters" leading to congruence and a *season of saturations*. The onset of decline under the competitive pressure of railways was an international phenomenon apparently affecting both the early innovators and the followers at about the same time. The consequences of these developments, however, are different. The leading countries developed canal networks over longer periods, and consequently the systems were larger and more pervasive than in the countries that exhibited a pronounced catch-up effect. In those countries the saturation was initiated before the canals became a vital and important component of overall economic development. They became isolated links and never achieved the great economic importance as in the leading countries.

The development of railways was initiated in the United Kingdom during the 1830s, and within two decades the first major railways were built in other countries. *Figures 19.3(a)* and *19.3(b)* document both the expansion and decline phases of the railway network in six industrialized countries, illustrating to what extent the development of railroads has converged internationally. *Figure 19.3(a)* shows the diffusion of railroads in the United States, the United Kingdom, and Germany (Federal Republic of Germany after World War II) and *Figure 19.3(b)* in the Austro-Hungarian Empire (Austria after World War I) and France, and the two growth pulses of the railways in Russia and the Soviet Union. While the railway construction processes differ both in slope and duration, there is a high degree of congruence in the ultimate saturation of railway networks in the industrialized countries during the 1920s.

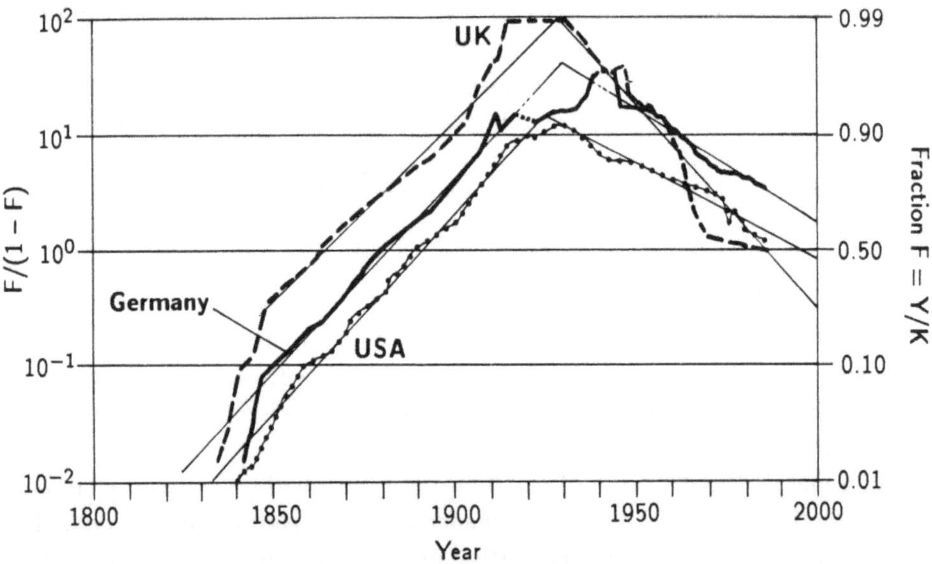

Figure 19.3(a). Growth and decline of railway networks in the USA, UK, and Germany/FRG. (Source: Grübler, 1990.)

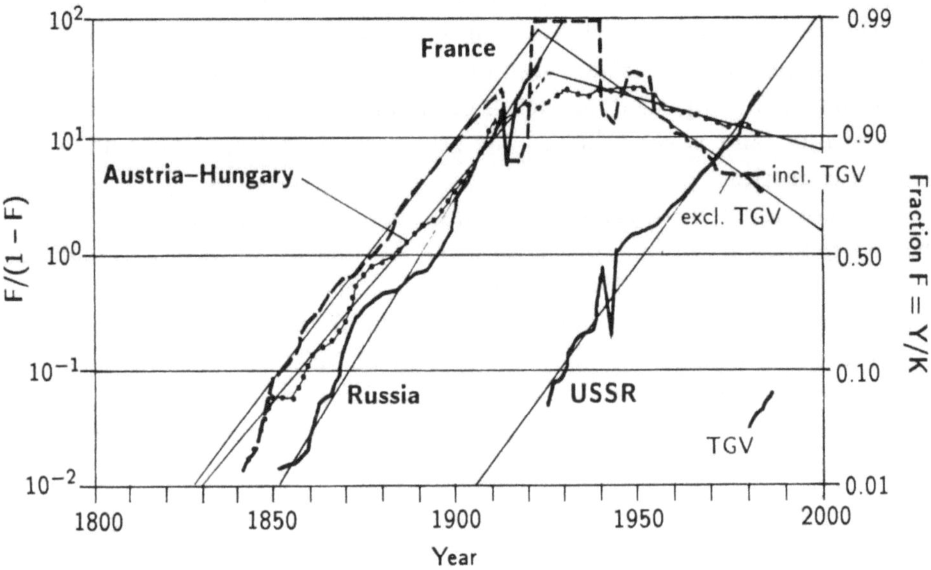

Figure 19.3(b). Growth and decline of railway networks in Austria, France, and Russia/USSR. (Source: Grübler, 1990.)

These six examples clearly show that the development of a particular techno-economic trajectory can follow similar paths even in countries with fundamentally different social and economic relations, different technological bases, and certainly different initial conditions. In this sense we can speak of international bandwagons in the diffusion of pervasive techno-economic systems. The most interesting cases are France and the Soviet Union; they show departures from the development pattern of the other countries.

In France there are two unusual features worth noting. The first is that the turbulence during the saturation phase is very large compared with other countries; the second feature is the introduction of the TGV (*train à grande vitesse*) during the 1970s. Without the additional infrastructure dedicated to the TGV, the length of the French railway system continues to decline along the historical path, while the inclusion of the TGV links could indicate the beginning of a trend reversal. Thus, one could speculate whether the introduction of rapid rail transport systems does not, in fact, represent the beginning of a new transport infrastructure. To document this possibility, the growth of TGV lines is plotted in the lower right corner of the graph.

The development path of railroads in Tzarist Russia is almost identical to the patterns observed in the other five countries, until the onset of saturation in the 1920s. This period also coincides with the October Revolution. The reconstruction period is the possible reason for the further expansion of railroads in the Soviet Union. Thus, two consecutive expansion pulses of the railway network (which are analogous to the two phases of canal construction mentioned in [4]) are evident. After saturation of the first pulse, the second followed a similar trajectory with a slightly longer duration, and is now entering its own saturation phase.

The diffusion cluster of railway networks that reached saturation by the 1930s accounts for almost 70 percent of all railroad lines constructed to date. This implies that the development trajectory based on the extensive construction of railroads was not repeated by latecomers, i.e., by those countries not participating in the expansion pulse. In fact, the global railway network reached a length of about 1.3 million kilometers by the 1930s and has remained at that length ever since. Decommissioning of lines in the core countries was compensated by new construction in the peripheral regions. In these core countries railroads reached completion within a very brief period starting in the 1930s and ending during the 1940s. The railway bandwagon focused as it evolved, converging toward the saturation period. Thus, the focusing increased as the diffusion cluster matured. In other words, the early introduction of railroads occurred with great lags between the early and late adopters. However, the latecomers appeared to achieve faster diffusion rates than the original innovators. A pronounced catch-up effect is convergent toward the saturation. This focusing of the expansion pulse of the railway era is noteworthy considering that the whole development process lasted about 100 years in the leading countries and only a few decades in the late-starters.

Some evidence indicates, however, that the absolute level of adoption is much lower for the laggers toward the end of the cluster. In other words, the ultimately achieved railroad density is in general higher the earlier the railroads are introduced; leaders achieve the highest diffusion levels. *Figure 19.4* shows this phenomenon. The density of the railway networks is measured as a ratio of the peak of national railway length over land area. *Figure 19.4* includes both the countries where the networks were completed by the 1940s and those that built railways later and thus did not belong to the railway bandwagon. For each country two different empirical measures are used to assess the starting date of railway construction: first, the year when the network reached one percent of its ultimate maximum length, and second, the date when the first railway line of national importance was constructed.

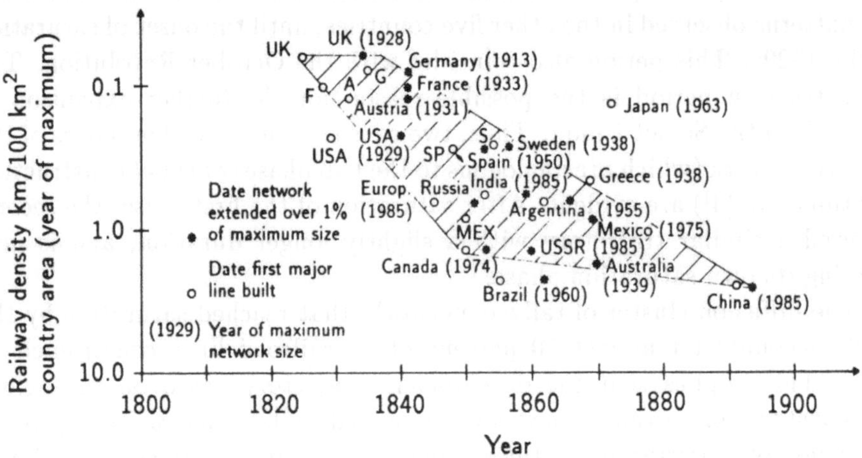

Figure 19.4. World railway network density as a function of introduction date. (Source: Grübler, 1990.)

Figure 19.4 shows that the railway densities can be grouped within a declining *density envelope* as a function of the introduction date. This result shows that the diffusion rate accelerates with the introduction lag, but that the intensity of the final adoption level decreases at the same time. Furthermore, the intensity falls further for those countries that introduced railways after the completion of the expansion pulse in the core countries. This identifies an *opportunity window* for the diffusion of pervasive systems like railways, which were the dominant transport system during the second half of the last century.

Roads and automobiles are the most pervasive transport system of the present. Roads existed long before the automobile. For example, the French road network of the 1890s had a total length of 500,000 km compared with 800,000 km in 1985; in the United Kingdom roads extended over 283,000 km in 1920 compared with 349,000 km in 1986; and in the United States the road length was 5 million km in 1921 compared with 6.2 million km in 1985. Thus, the total mileage of all roads increased only slightly in most of the industrialized countries during this century as the road transport systems evolved. However the quality of roads improved dramatically during this period, and in time surfaced and paved roads emerged as a new transport infrastructure replacing the old, dusty road system. Unfortunately, this process is not well documented in most countries, so that very few comparable statistics are available concerning the road system development.

Figure 19.5 shows that the diffusion of the surfaced road networks in the United States and the Soviet Union occurred at about the same rate (Δt of 64 and 66 years, respectively). However, the development of roads in the Soviet Union lags by about 30 years compared with the United States. In contrast to the diffusion of canals and railroads that have saturated in the past, the diffusion of surfaced roads in the Soviet Union is still far from the potential (estimated) saturation level. The current length of about 812,000 km represents about 56 percent of the estimated saturation level (K of about 1.4 million km). In comparison, the current length of paved roads in the United States is about 5.6 million km with an estimated saturation level of about 6 million km. Owing to limited data describing the international development of the modern road systems, the diffusion rates and adoption levels of the automobile will now be analyzed.

During the last century horse-driven vehicles were the predominant form of road travel. In the United States the number of road horses peaked at more than 3 million in the 1920s, but declined rapidly thereafter. The horseless carriage was introduced toward the end of the last century. The spread of the automobile was very swift until the 1930s when a structural change occurred in the growth path followed by lower expansion rates (see the previous chapter by Grübler).

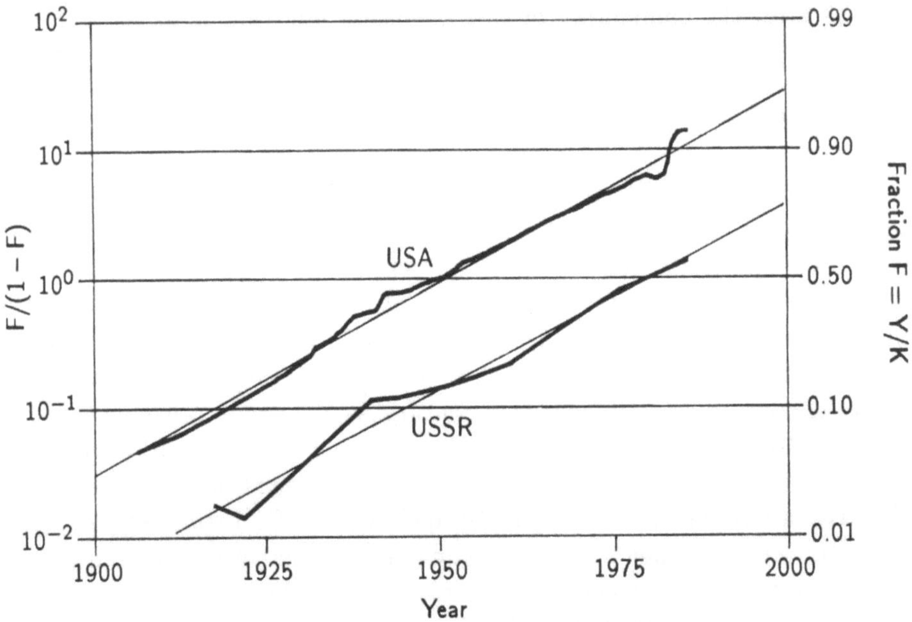

Figure 19.5. Diffusion of surfaced roads in the USA and the USSR. Note that there is a factor 4 difference between the estimated K in the two countries. (Sources: Grübler, 1990; Nakićenović, 1988b.)

By 1930 horses had virtually disappeared from the major roads in most industrialized countries, indicating that motor vehicles filled the niche previously occupied by horses. Thus, the replacement process lasted altogether about 30 years. Additional evidence is seen in the fundamental transformation of the vehicles themselves after the 1930s and the numerous innovations in production methods and vehicle design that provided for higher performance, more comfort, and lower price. Several changes in other sectors also made the automobile more attractive and accessible to the public. Examples include innovations in the steel industry (higher-quality alloys and wider sheet metal), the petrochemical industry (high-quality rubber and catalytic cracking, see Chapter 17 by Ayres and Ezekoye), and a host of other institutional changes, which eventually even led to automobile *compatible* settlement patterns. Incidentally, automobile use today has led to numerous environmental problems, but in a historical perspective the replacement of horses by cars alleviated one of the grave environmental problems of the cities of the last century, namely, horse manure in the streets.

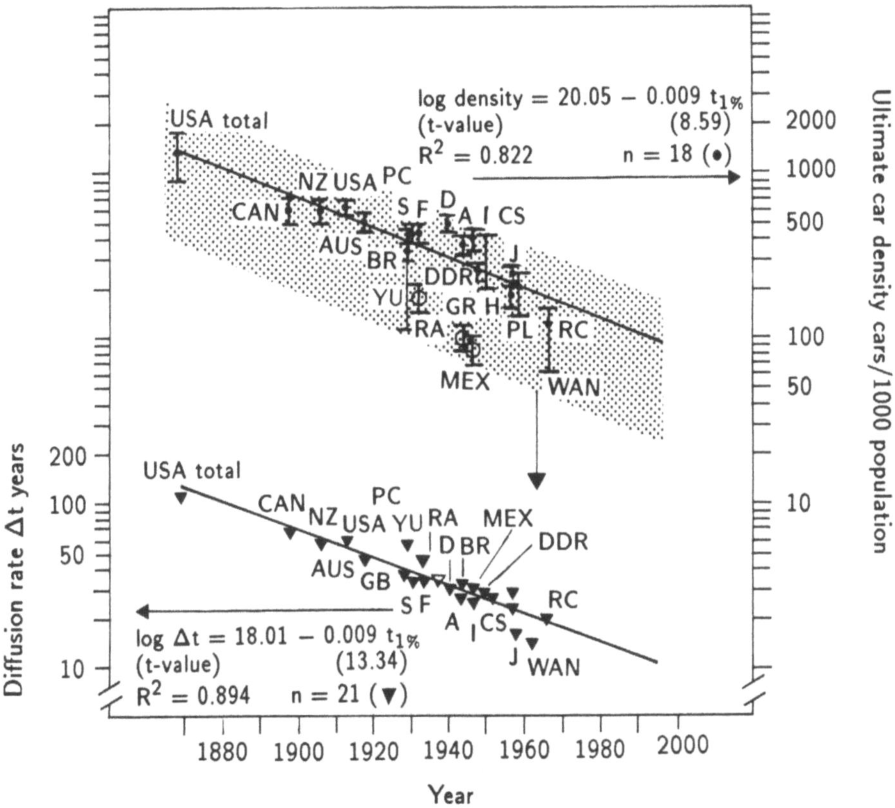

Figure 19.6. Diffusion rates and ownership densities of passenger cars as a function of introduction date. (Source: Grübler, 1990.)

The adoption level of the automobile in different countries follows the same pattern as the observed saturation levels of railroads. Of all industrialized countries, the diffusion rate was the slowest in the United States and Canada, faster in most of the European countries, and fastest in Japan. This tendency is also confirmed at the global level; the adoption of the automobile was even quicker in the developing countries, albeit that it has only recently started and has led, at least for the time being, to relatively low levels of car ownership.

Figure 19.6 reports these results by giving the estimated diffusion rate (Δt, given on the lower curve and labeled on the left vertical axis) as a function of the beginning of the innovation diffusion given on the horizontal axis. For example, the diffusion of the automobile started in the 1920s in

Australia, with a diffusion time constant (Δt) of about 50 years. The shaded band on the top of the figure gives the estimated automobile adoption levels (plotted on the right vertical axis in number of cars per 1,000 capita with the associated statistical uncertainty bands), again as a function of automobile introduction dates. Thus, a pronounced *acceleration* of the diffusion rates can be observed that is proportional to the time lag in the innovation adoption and also that the ultimate adoption levels *decrease* with the lag in the innovation adoption.

This result is consistent with spatial patterns of innovation diffusion. Originating from the innovation centers that reach the highest adoption levels, the innovation generates additional gravity centers in space; however, the adoption levels remain lower in the peripheral regions compared with innovation centers. Our result confirms this finding at an international level in both time and space. Early innovators achieve higher adoption levels and have the longest diffusion or *learning* times. Peripheral regions catch-up so that the diffusion process focuses toward the saturation period, but the adoption level in the periphery remains much lower than in the leading centers. The data confirm this acceleration tendency, which leads to lower adoption levels and explains as much as 89.4 percent of the variance (R^2 adjusted for the degrees of freedom). This acceleration tendency in the global diffusion process indicates that a country that just started adopting private automobiles now would have a diffusion rate (Δt) of less than a decade. Such a rapid diffusion of car ownership appears quite infeasible from a practical point of view. Should it actually occur, it certainly will not result in a significant growth of both absolute and relative car adoption levels.

19.5 Substitution Waves

The evolution of transport infrastructures can also be seen as a substitution process. Instead of analyzing their development as a sequence of individual diffusion processes (as the succession of growth pulses shown in *Figure 19.2*), they can be viewed as systems that replace each other in time. *Figure 19.7* reproduces the growth in the length of the transport infrastructures in the United States (from *Figure 19.2*) as a substitution process[5] and shows for comparison the equivalent substitution process in the Soviet Union. It shows the successive substitution of the four transport infrastructures in the two countries: canals, railroads, surfaced roads, and airways. The shares of each infrastructure in the total length are plotted as the ratio of the

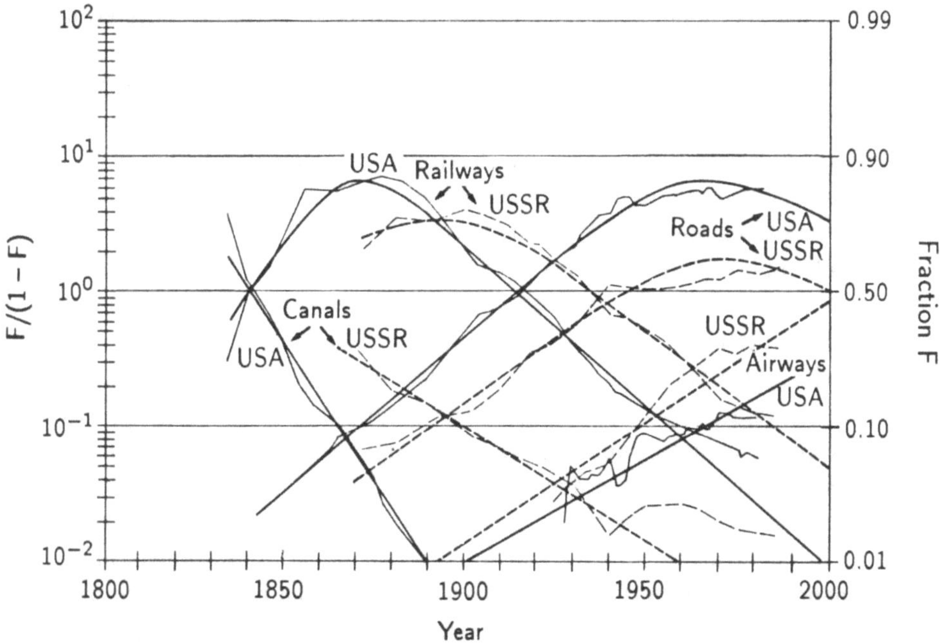

Figure 19.7. Substitution of transport infrastructures in the USA and the USSR, fractional shares in length. (Sources: Grübler, 1990; Nakićenović, 1988b.)

share of one infrastructure divided by the sum total shares of all others on a semi-logarithmic scale. This particular representation shows the relative importance of competing infrastructures and the dynamics of the evolution process since 1800. In any given period, there is a clear market dominance (i.e., more than a 50 percent share) and simultaneous spread of transport activities over two or three different systems. Thus, while competing infrastructures are all simultaneously used, their mix changes over time leading to the expansion of the area connected in a unit of time. Projecting this competitive process into the future leads to the increasing importance of airways and their possible market dominance after the 2050s, notwithstanding the likelihood of a new competitor emerging in the coming decades, such as magnetic levitation trains (maglevs) and advanced forms of air transport (e.g., hypersonic aircraft) for transcontinental distances. Although air transport and the automobiles are still expanding modes of passenger travel, railroads are now in a phase of postsaturation in most industrialized countries. Their once dominant position in intercity passenger traffic has been eroded. Two symbols of this decline are the discontinuation of the transcontinental railway service in the United States, and the burgeoning deficits of most national railway systems in other countries. Similar evolutionary changes in the development of transport infrastructures also occurred in the European countries. Thus, the international diffusion bandwagons of transport infrastructures can be also described as an evolutionary substitution process, revealing the changing structure of transport systems.

Even the planned economies have undergone a similar evolutionary path. The pattern of temporal changes in the Soviet Union (and Russia before the revolution), like in the United States, is marked by a high degree of regularity and a quest for higher speed and productivity. The phase transitions, however, are lagged by a few decades when compared with the United States. For example, the dominance of railways lasts until the 1940s while in the United States it ended two decades earlier. The decline of canals occurred much later as well, while the growth of national airways follows the same path as in the United States. During the last decades, the development of the transport infrastructures in the two countries has been converging. Also the rate of relative growth in the importance of road infrastructure and their saturation appear synchronized in the two countries. Thus, there is an increasing congruence and similarity in the structural and functional evolution of the transport system in the two countries. To a large extent this is also due to the fact that both countries have relatively low population densities and vast territories that modern transport systems must bridge in a matter of hours.

This irreversible process of evolutionary change (the *survival of the fittest* in an ever-changing environment) is nevertheless only a proxy for the real dynamics in the development of transportation systems, which should be measured in some common performance unit. Because transport systems provide a whole range of services, such a common descriptor is difficult to define. Furthermore, most transport services require a mix of different transport modes. On average this mix apparently changes in time.

Two obvious choices for an appropriate indicator are ton- or person-kilometers per year. These units distinguish between freight and passenger services, although they are usually offered simultaneously, but they do not distinguish between short and long distances. Thus, there is no obvious shortcut and it appears necessary to analyze passenger and freight transport separately for both short and long distances. Fortunately, the available data make it possible to reconstruct the dynamics of these substitution processes for at least some countries. As these results have been reported elsewhere (see, for example, Nakićenović, 1988a and 1988b; Grübler, 1990) we will describe briefly the modal split for long-distance passenger transport in the United States and the Soviet Union, and then show the evolution of freight transport in the Federal Republic of Germany. These examples are representative of the developments in most industrialized countries and illustrate the convergence in the development of transport systems. They show more explicitly the linkages in the development of pervasive systems between the transport infrastructures, vehicles, and other related technologies and the overall pattern of economic and social restructuring.

Figure 19.8 shows the substitution of different transportation modes in intercity (long-distance) passenger travel in the United States and the Soviet Union. By excluding urban and metropolitan transport, the competition for intercity passenger traffic is reduced to four major transport modes in the United States: railways, buses, cars, and airways. The major competitors over this period in the Soviet Union were boats, railways, buses, and airways; the automobile never gained any significant importance. Today, most of intercity travel is by car and plane in the United States and by rail, bus, and plane in the Soviet Union. Comparison of the two countries shows that in the past the intercity passenger transport development portrayed a phase-shift in the two countries. Rail and bus are virtually extinct in the United States, while in the Soviet Union they are still the dominant means of transport.

The substitution dynamics indicate that by the end of the century airways may become the dominant form of intercity travel in the United States and the Soviet Union. The evolutionary paths appear to be convergent. In both countries road transport has currently the largest share, albeit by automobile in one case and by bus in the other. What is important is that the average choice of different modes of passenger travel changes consistently in both countries and favors faster and more productive systems.

Zahavi (1979) has shown that traveling is optimized under the constraints of individual time budgets and family income. On average roughly 15 percent of disposable family income is allocated to travel so as to *maximize distance*. Szalai (1972) analyzed time budgets of the inhabitants of 12 cities throughout the world including industrialized and developing countries. His work indicates that the time spent for traveling (work and non-work) is roughly 1.5 hours a day. In a given situation, each individual will make different choices, but on average the modal split will change as income increases, despite the assumed invariance of travel-time budgets. Incomes have increased and the cost of travel has decreased in real terms, leading to an increase in the volume of travel (passenger-kilometers) and the range in a given unit of time.

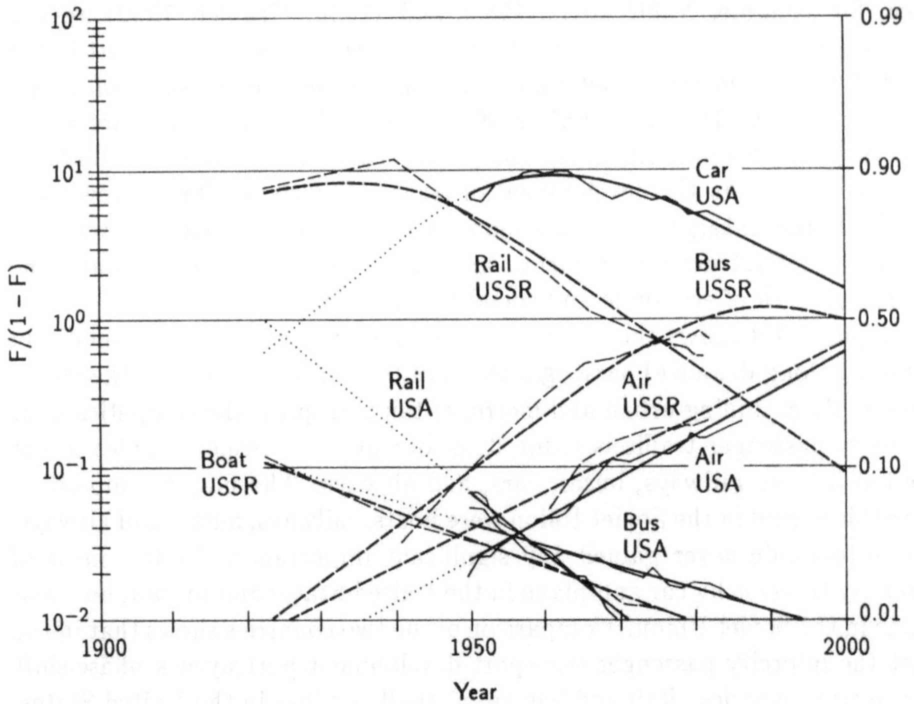

Figure 19.8. Substitution of intercity passenger transport modes in the USA and the USSR, fractional shares in inter-city passenger-km. (Sources: Grübler, 1990; Nakićenović, 1988b.)

Over time, larger shares of travel are thus allocated to faster modes. The slower modes recede to service fewer people over shorter distances (fewer passenger-kilometers) and low-value segments of freight transport. Their shares in passenger transport decline, resulting in the rationalization of the respective infrastructures. The least productive links are decommissioned, and the infrastructure declines. The diffusion of more productive and faster means of transport leads to increased range and connects even larger areas into one single complex. During the last two centuries villages merged into towns, towns into cities, and cities into metropolitan areas. Large urban corridors have evolved throughout the world, some approaching a hundred million inhabitants.

While the possibility of constant travel time allocation would tend to increase the demand for faster modes of passenger transport subject to monetary budgets, the changes toward faster modes of freight transport are caused by different driving forces. Transport of tangible goods imposes more stringent requirements with increasing value-density (value per unit weight or volume). Long transport times require larger inventories and correspondingly higher capital expenditure. Furthermore, the risk of loss or damage also tends to increase with the exposure time. All told, more valuable goods are transported by faster transport modes. With the introduction of flexible manufacturing, larger worldwide transport of perishable goods and fashionable garments, the air-road systems are increasing their competitiveness. The increasing value density of tangible goods and the decreasing materials intensiveness tend to generate higher value added on most efficient routes.

Figure 19.9 shows a snapshot in time of imports of manufactured goods into the Federal Republic of Germany as a function of value-density. The figure illustrates that the most valuable goods are transported by more efficient and faster means of transport. Basic materials such as coal, gravel, scrap, and ores in the value range of a few DM per kilogram are mostly transported by sea, canal, and rail. But even in this lowest segment, trucks are more competitive and are the dominant transport mode with value of up to about 100 DM per kilogram. Most of the manufactured goods such as automobiles and machine tools fall into the range below this threshold. Electronics, computer, and precision instruments are usually shipped by air. The highest value manufactured goods that are exclusively transported by air are aerospace products and aircraft themselves.

Recently, General Motors started a 5,000 kilometers *production line* connected by airfreight. Cadillac Allanté car bodies are assembled at Pininfarina in Turin, Italy, and transported by Lufthansa and Alitalia B-747 cargo aircraft to Detroit for final assembly of engine, power train, and electronics. This airlift assembly line is apparently cost-efficient considering all the direct and indirect costs of potential damage risks and the production inventories that would be locked in ocean freighters for weeks. On a more speculative note, the further dematerialization of manufactured goods and an increase in value through software and information content would tend to increase the share of air transport in freight. Collocation of production facilities and services close to airports may proceed in a similar way as industrial activities condensed along previous transport infrastructures: railways and later roads.

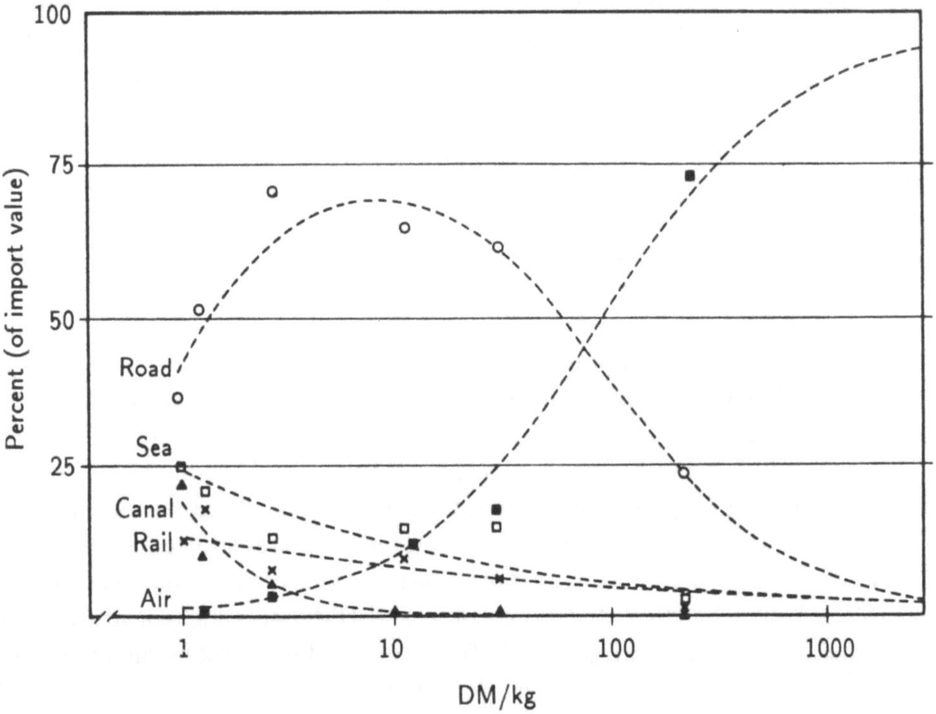

Figure 19.9. Substitution of transport modes in the import value of manufactured goods versus value-density (DM/kg) in the Federal Republic of Germany. (Source: Grübler, 1990.)

19.6 Clusters and Families

The substitution of older transport infrastructures for new ones has shown strong linkages in the evolution of pervasive systems. These examples show a degree of synchronization in the diffusion of transport systems that converge toward the onset of the saturation phase. They reflect strong linkages in the evolution of techno-economic systems denoting major techno-economic paradigm shifts. Countries in which the innovation is introduced with a lag behind the leading regions tend to catch-up but saturate at lower adoption levels. A comparison of the evolution of transport systems in the United States and the Soviet Union indicates that the catch-up effect occurs also over longer periods: from one generation of transport system to another. While the modal structure was out of phase between the two countries only

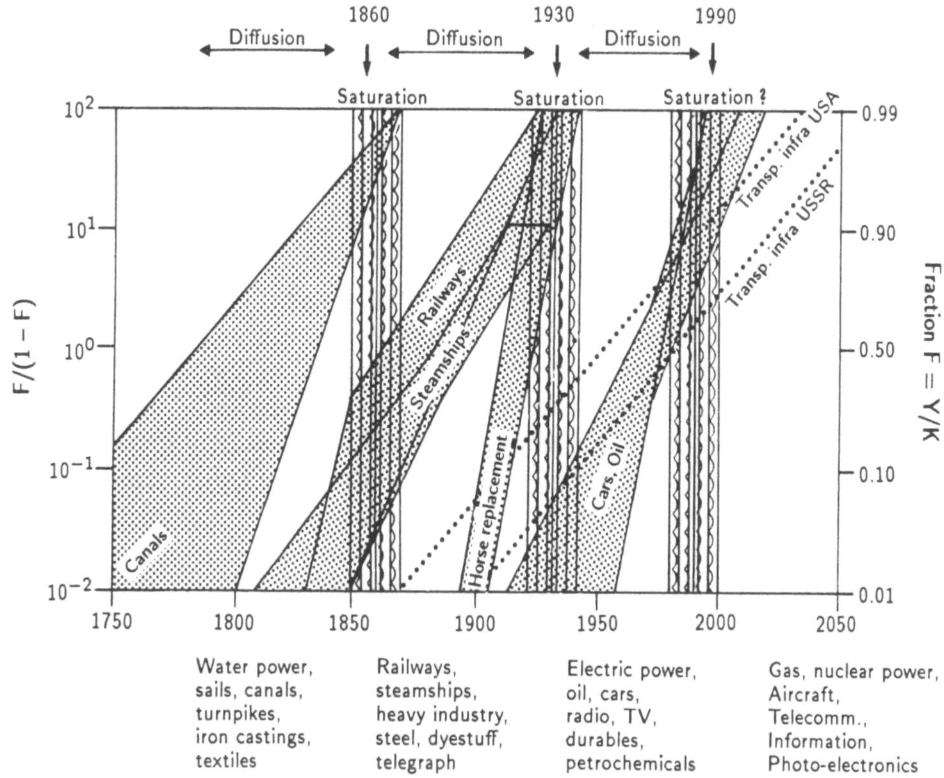

Figure 19.10. Three clusters of interrelated diffusion processes at the world level.

a few decades ago, there is convergence in the recent stagnation of road development and vigorous growth of air travel.

These phenomena can be documented internationally and not in just a few selected countries or sectors. *Figure 19.10* summarizes our findings from the illustrative examples given above and other case studies and shows the international diffusion bandwagons of canals, railways, and automobiles. The various diffusion processes that define the bands are listed below each cluster. As mentioned earlier, the focusing is not very pronounced at the time when the innovations are introduced, and there is a high degree of overlap among the various bandwagons due to differences in the time constants (Δt) among individual diffusion processes. Nevertheless, three rather clear clusters can be distinguished. The first saturates about 1860, the second about 1930, and the third is centered a few years away in 1995. Each cluster is converging toward the saturation period. The focusing increases as the diffusion cluster matures. In other words, the introduction of innovations is associated with great lags between the early and late adopters. However, the latecomers appear to achieve faster diffusion rates than the original innovators. Thus, a pronounced catch-up effect is convergent toward saturation.

We have also observed, however, that the extent of the absolute diffusion level is much lower for the latecomers toward the end of the cluster. The saturation density of railroad networks of a particular country is lower if the diffusion rate is higher. In other words, the ultimately achieved railroad density is, in general, higher the earlier the railroads are introduced; leaders achieve the highest diffusion levels. We have observed the same phenomenon in the spread of motorization in different countries, the early innovators such as the United States having the highest per capita density with the achieved density decreasing in proportion to the lag in automobile introduction.

In this sense the saturation phase of each cluster represents a barrier to further diffusion at the international level. Very few diffusion processes can *tunnel* through this barrier. If it is true that this marks the beginning of paradigm shifts, it is not surprising that the further diffusion of systems associated with the old techno-economic development trajectory is blocked and thus makes way for the new. It is the disruptive crisis of the old that provides the fertile ground for new systems to develop; it is a process of creative destruction (Schumpeter, 1939).

Our working hypothesis is that the diffusion processes within each cluster constitute families of interrelated systems that enhance each other and promote the pervasiveness of the three successive development trajectories. Classification or a taxonomy of each innovation that belongs to a given cluster and which makes it viable, would probably reveal a hierarchical system with one successful diffusion having a positive feedback and catalytic effect on the development of many others within the whole cluster. If this is the case, then the clustering is not coincidental as Schumpeter (1939) pointed out. It is likely that it is the interdependence of individual diffusion processes that in time focuses each cluster.

19.7 Conclusions

A certain degree of regularity, synchronization, and recurrence can be noted within each family of interrelated technologies, although the clustering or bundling is not very rigid. There is stronger evidence for the congruence in the saturation of diffusion processes rather than in their emergence, so that the focusing increases as the systems mature. In other words, the time span between the beginning of the diffusion of new pervasive systems, i.e., between the leaders and the laggers, tends to decrease as the diffusion progresses; the process appears much more focused toward the saturation phase. This visible catch-up effect refers only to the relative diffusion rates and not to the absolute levels of diffusion. The leaders usually achieve higher diffusion levels; the levels of the followers are lower roughly in proportion to the lag in the introduction of a given innovation.

The results, however, cannot be conclusive before explicit linkages are established between the individual diffusion processes that constitute each cluster of techno-economic development. A taxonomy and establishment of hierarchies within each cluster could be a worthwhile route toward determining the relationships and driving forces behind the clustering effect observed in the samples of innovation diffusions. This could perhaps answer some of the questions that emerge from the analysis. For example, why are the clusters more focused toward the saturation phase? Alternatively, under which conditions could unbundling occur and perhaps lead to a more even pace of structural change? The overall development path is punctuated by crises that emerge in the transition from an old saturating cluster to a new one. The creative destruction of the old systems is characteristic for evolutionary processes in biology. The extinction of old ecosystems is followed by expansion of new ones. Techno-economic development paths portray apparently similar features of evolutionary change.

Many new techno-economic systems and innovations made over the last few decades have not diffused sufficiently to make an impact at the aggregate level, while the older systems, based on the old paradigms, are approaching or have already reached market saturation. Depending on whether and how societies adopt the systems, we may witness new clusters of diffusion in the future that would have fundamental impacts on employment and competitiveness, on the structure of the economy, and consequently on the transport systems. Since their adoption and mediation by society will not be homogeneous and will affect countries and regions differentially, the process can be expected to be disruptive.

Notes

[1] Historical data and the fitted logistic curve are transformed as $Y/(K - Y)$, where Y denotes the actual infrastructure length in a given year and K the estimated saturation level. The data and the estimated logistic trend are plotted as fractional shares of the saturation level, $F = Y/K$, which simplifies the transformation to $F/(1 - F)$, the level of relative growth achieved divided by the remaining potential. Transformed in this way, the data appear to be on a straight line, which is the estimated logistic function. Without this transformation, the data and the trend curve would portray the same S-shaped growth as shown for the three transport infrastructures in *Figure 19.1*: canals, railways, and surfaced roads.

[2] The evolutionary path of successive replacements of traditional by new paradigms as seen in the diffusion of technologies and institutions, economic restructuring, and transformations in social relations is captured in the Schumpeterian notion of the long waves in economic development, i.e., the seesawlike pattern of the Kondratieff (1926) pulses experienced in the market economies during the last two centuries. Freeman (1963), Marchetti (1985), and many others have subsequently extended both the conceptual and empirical description of long waves, diffusion, invention, and innovation processes, albeit from different methodological and theoretical perspectives. For a comprehensive review of this research area see van Duijn (1983), Vasko (1987), or Goldstein (1988).

[3] The duration of the diffusion process is conveniently measured as the time that it takes to grow from 1 percent to 50 percent of the saturation level. We call this measure Δt. In this example Δt equals 30 years. Due to the symmetry of the logistic function, the same time is required for an increase from 50 to 99 percent of the saturation level. An alternative definition of Δt is the time required to grow from 10 percent to the 90 percent level. In this case the value of Δt would be slightly different from the other definition, but for all practical applications both definitions can be used interchangeably.

[4] The canal growth pulse in Russia was slower than in other countries and continued beyond 1870, but it also saturated toward the end of the century with a total length of about 900 km. After the Revolution an ambitious new canal development program was implemented leading to the second pulse in canal growth extending the size of the network close to 21,000 km (it will be shown below that railroads also developed through two growth pulses). Despite this enormous effort, canals currently represent less than two percent of all transport infrastructures in the Soviet Union (as will be shown below).

[5] The fractional shares, F, are not plotted directly but as the linear transformation of the logistic curve, i.e., $F/(1\text{-}F)$, as the ratio of the market share taken by one infrastructure over the sum of the market shares of all other competing systems. This form of presentation reveals the logistic substitution path as an almost linear secular trend with small annual perturbations. Thus, the

presence of some linear trends in *Figure 19.7* indicates where the fractional substitution of transport infrastructures follows a logistic curve. In dealing with more than two competing technologies, we must generalize the Fisher and Pry (1971) model, since in such cases logistic substitution cannot be preserved in all phases of the substitution process. Every competitor undergoes three distinct substitution phases: growth, saturation, and decline. This is illustrated by the substitution path of railways (and later roads), which curve through a maximum from increasing to declining market shares (see *Figure 19.7*). In the model of the substitution process, we assume that only one competitor is in the saturation phase at any given time, that declining technologies fade away steadily at logistic rates, and that new competitors enter the market and grow at logistic rates. As a result, the saturating technology is left with the residual market shares (i.e., the difference between one and the sum of fractional market shares of all other competitors) and is forced to follow a nonlogistic path that joins its period of growth to its subsequent period of decline. After the current, saturating competitor has reached a logistic rate of decline, the next competitor enters its saturation phase and the process is repeated until all but the most recent competitor are in decline. A more comprehensive description of the model and assumptions is given in (Nakićenović, 1979).

References

Brooks, H., 1986, The typology of surprises in technology, institutions, and development, in W.C. Clark and R.E. Munn, eds., *Sustainable Development of the Biosphere*, Cambridge University Press, London, UK.

Dosi, G., Freeman, C., Nelson, R., Silverberg, G., and Soete, L., eds., 1988, *Technical Change and Economic Theory*, Pinter Publishers, London and New York.

Fisher, J.C. and Pry, R.H., 1971, A simple substitution model of technological change, *Technological Forecasting & Social Change* **3**:75–78.

Freeman, C., ed., 1983, *Long Waves in the World Economy*, Butterworths, London, UK.

Freeman, C. and Perez, C., 1988, Structural Crises of Adjustment, Business Cycles and Investment Behaviour, in G. Dosi *et al.*, eds., *Technical Change and Economic Theory*, Pinter Publishers, London and New York.

Goldstein, J.S., 1988, *Long Cycles: Prosperity and War in the Modern Age*, Yale University Press, New Haven and London.

Grübler, A., 1990, *The Rise and Fall of Infrastructures, Dynamics of Evolution and Technological Change in Transport*, Physica Verlag, Heidelberg, FRG.

Kondratieff, N.D., 1926, Die langen Wellen der Konjunktur, *Archiv für Sozialwissenschaft und Sozialpolitik* **56**:573–609 (German translation of original Russian article).

Marchetti, C., 1985, Swings, cycles and the global economy, *New Scientist* **1454**: 12–15.

Nakićenović, N., 1979, *Software Package for the Logistic Substitution Model*, RR-79-13, International Institute for Applied Systems Analysis, Laxenburg, Austria.

Nakićenović, N., 1988a, *Dynamics of Change and Long Waves*, WP-88-74, International Institute for Applied Systems Analysis, Laxenburg, Austria.

Nakićenović, N., 1988b, Dynamics and replacements of US transport infrastructures, in J.H. Ausubel and R. Herman, eds., *Cities and their Vital Systems, Infrastructure Past, Present and Future*, National Academy Press, Washington, DC, USA.

Schumpeter, J.A., 1939, *Business Cycles, A Theoretical, Historical and Statistical Analysis of the Capitalist Process*, Volume I and II, McGraw-Hill, New York, USA.

Simon H., 1988, *Prediction and Prescription in System Modeling*, Paper presented at IIASA: Perspectives and Future, International Institute for Applied Systems Analysis, Laxenburg, Austria, 14–15 June.

Szalai, A., 1972, *The Use of Time, Daily Activities of Urban and Suburban Populations in Twelve Countries*, Mouton, The Hague and Paris.

van Duijn, J., 1983, *The Long Wave in Economic Life*, Allen and Unwin, London, UK.

Vasko, T., ed., 1987, *The Long-Wave Debate*, Springer-Verlag, Berlin, Heidelberg, New York.

Zahavi, Y., 1979, *The "UMOT" Project*, Report DOT-RSPA-DPB-20-79-3, prepared for the US Department of Transportation and the Ministry of Transport, FRG, NTIS, Springfield, VA, USA.

Chapter 20

Determining the Service Life Cycle of Computers

Theodore Modis and Alain Debecker

20.1 Introduction

Nowadays most people in business are familiar with the application of logistic growth to the sale of products even if they ignore its mathematical formulation. They all use a bell-shaped curve to refer to a product's life cycle. The corresponding S-curve, for a product's cumulative sales, is also something with which they feel comfortable.

Business people are, however, less in tune with the mortality of their products. Hardware maintenance contracts for computers closely follow systems sales. At the beginning the number of contracts increases at the same rate as sales, but instead of the familiar S-curve, contracts reach a peak and then slowly start declining as machines become old and obsolete. In *Figure 20.1* we see the familiar forms: the sales life cycle, the corresponding S-curve of cumulative sales, and the less familiar service life cycle. Notice that the last graph peaks only after the product has completed its sales life cycle. Qualitatively the business world is aware of the service life cycle. They refer to it as the product's *end-of-life* curve. Quantitatively, however, they fall

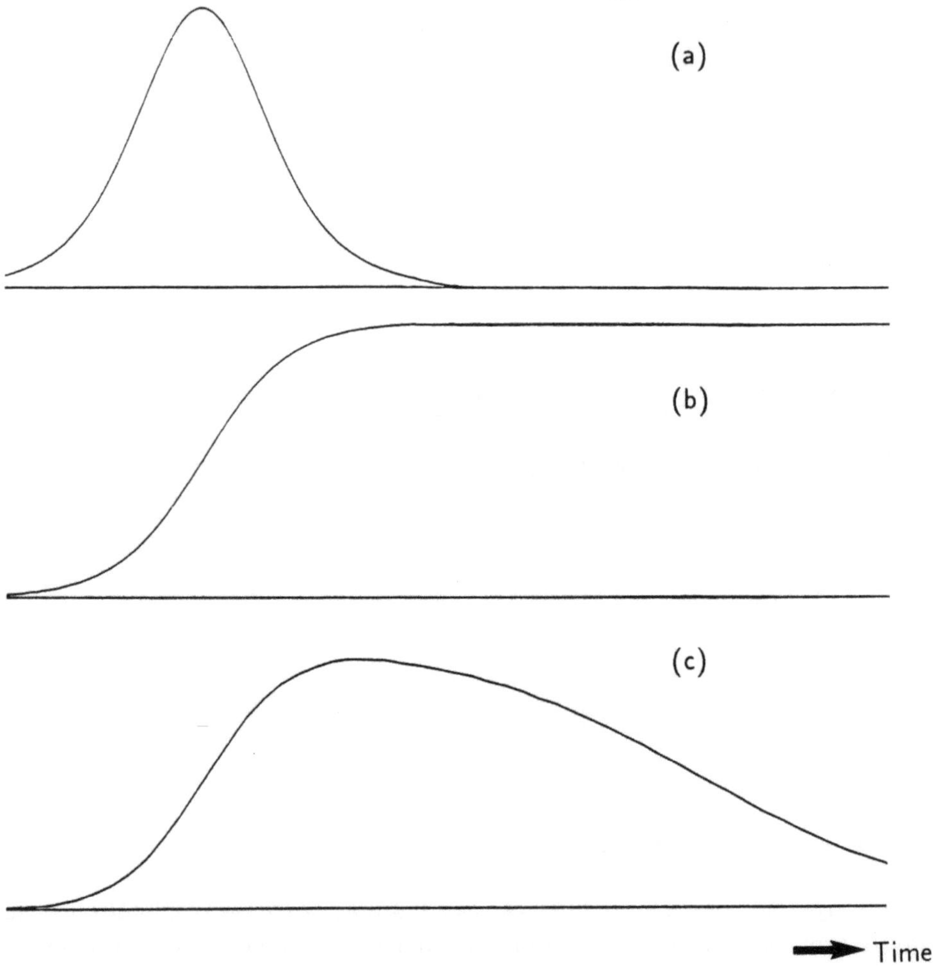

Figure 20.1. (a) Life cycle of sales, (b) Cumulative number of sales, (c) Number of active service contracts at a given time.

short of being able to predict when the product will peak and at what rate revenue will decline afterwards.

The life cycles of products are becoming shorter. A specific computer model used to sell for 4–5 years in the 1970s; they now only sell for 1–2 years. Service life cycles have been known to last for 12–14 years. By how much are they going to decrease in the future?

A recent article in the *Harvard Business Review* addresses this question (Potts, 1988). Graphs and figures liven up the discussion, but the treatment remains qualitative and rather pictorial. It corroborates the intuition but does not help with quantitative forecasts. Determining the service life cycle is interesting for other reasons besides the ever-increasing importance of service revenue. It provides knowledge of the installed base, its actual and future size, which is crucial in marketing strategies for add-ons (follow-up sales in terms of accessories, upgrades, etc.), software, and a variety of services. This is why we propose a quantitative approach for determining a product's *life span* as opposed to its *sales life cycle* only.

20.2 Logistic Diffusion

Logistic growth formulated in the Volterra-Lotka equations accurately describes situations where a niche is being filled under natural competition, be it an ecological or a market niche. The situation becomes more complicated when one is interested in tracking down the survivors of a generation over time. The total number of units sold of a particular product increases with time reaching its ceiling at the end of the product's life cycle. The number of products *in use*, however, never reaches the same ceiling as there is a certain mortality among the products sold. Out of a hundred units sold in month one, 98 may be still "alive" a year later but only 50 in five years time. This erosion comes from aging and obsolescence, and sets in right from the beginning. While populations grow happily along their S-shaped birth curve, they are at any point in time subject to a certain mortality.

Mathematically the case calls for a convolution function. The folding of logistic growth with a mortality function. What shape should the latter have? Sales follow a logistic growth, namely the rate of sales over time, $P(t)$, proceeds along a bell-shaped curve of the form:

$$P(t) = \frac{M}{(1 + e^{-\alpha(t-t_o)})(1 + e^{\alpha(t-t_o)})} \tag{20.1}$$

where M, α and t_o are constants. To introduce mortality we begin with the simplest possible assumption of a constant percentage decay rate, in other words an exponential decay:

$$\frac{1}{R(u)} \; \frac{dR(u)}{du} = -k$$

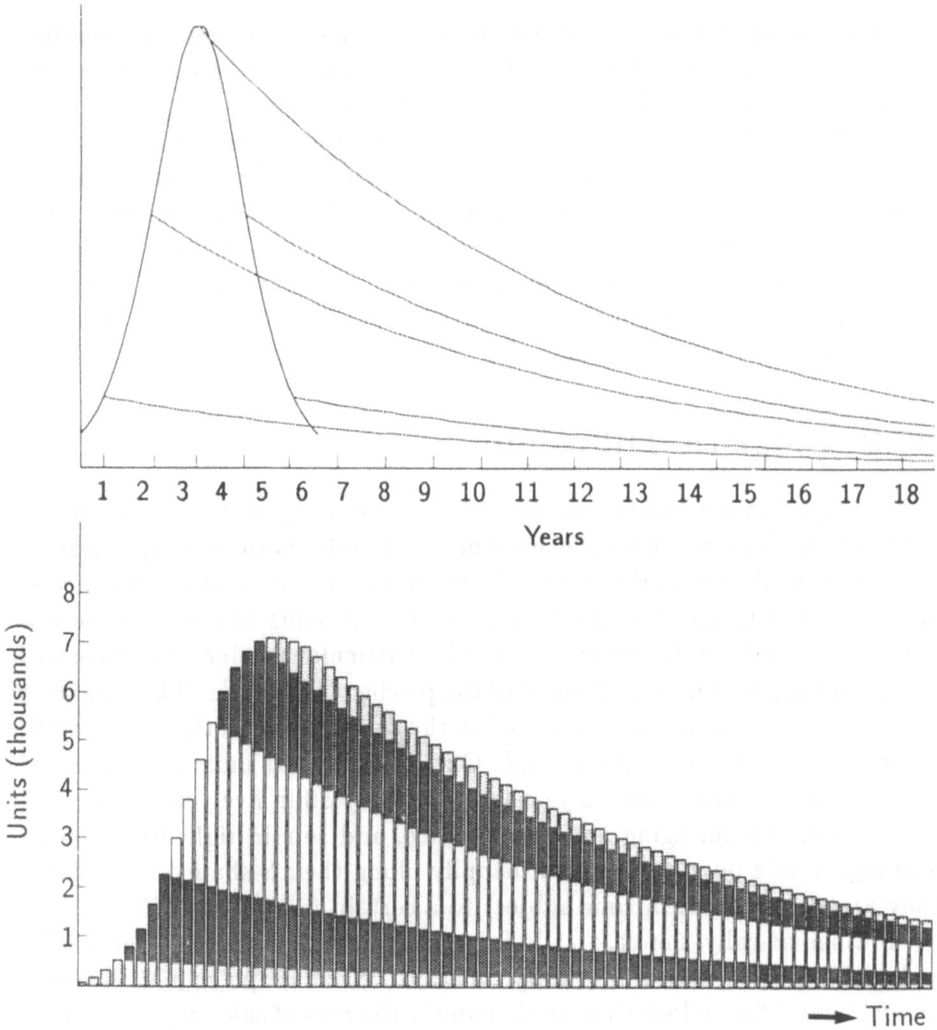

Figure 20.2. Above: life cycle outlining sales of service contracts and exponential mortality rates over a five-year period. Below: the integrated number of contracts with finer time bins (trimesters). The overall envelope represents total survivors at a given time.

where $R(u)$ is the remaining contracts of age u, and k is a constant. The convolution function is given below and graphically it is shown in *Figure 20.2*.

$$
\begin{aligned}
Q(t) &= \int_0^\infty P(t-u)R(u)du \\
&= \int_0^\infty \frac{Me^{-ku}}{(1+e^{\alpha(t-u-t_o)})(1+e^{-\alpha(t-u-t_o)})}du
\end{aligned} \tag{20.2}
$$

When we first applied this formulation to real data we obtained good results, particularly for old products for which the sales life cycle had finished years ago and service contracts were already beyond their peak and declining. However, two difficulties soon emerged. One was related to relatively young products where the rate of decay was so low that it implied they would be around for a very long time. This fact contradicts the increasingly rapid cycling of products witnessed in the computer industry. The second difficulty was more revealing. When tracking decay rates in a number of areas we found that they were not constant over time. These observations prompted us to raise the level of sophistication for mortality, i.e., to try a second degree function for the decay rate:

$$
\frac{1}{R(u)} \frac{dR(u)}{du} = a\ R(u) + b
$$

where a and b are constants. Setting $N = -b/a$ we can rewrite this expression as

$$
\frac{dR}{du} = -a\ R(N-R) \tag{20.3}
$$

which unveils the logistic nature of the assumed form. Furthermore, we can eliminate one constant by setting $N = 1.0$ since the mortality ceiling is 100%.

We have now arrived at a logistic mortality, however, it was not entirely circumstantial. An older study on the appearance and survival of supertankers had revealed to us an exemplary S-shaped mortality for that "species". In addition, Marchetti has investigated human mortality (Marchetti, 1990) and has also established logistic laws for the process. However, what is important is that the logistic mortality better fits the data on computer service contracts we were trying to describe. The new convolution function described below involves five parameters, three from the logistic growth of contract "births", equation (20.1), and two from the logistic mortality, equation (20.3), (u_o is an integration constant). This is graphically depicted in *Figure 20.3*.

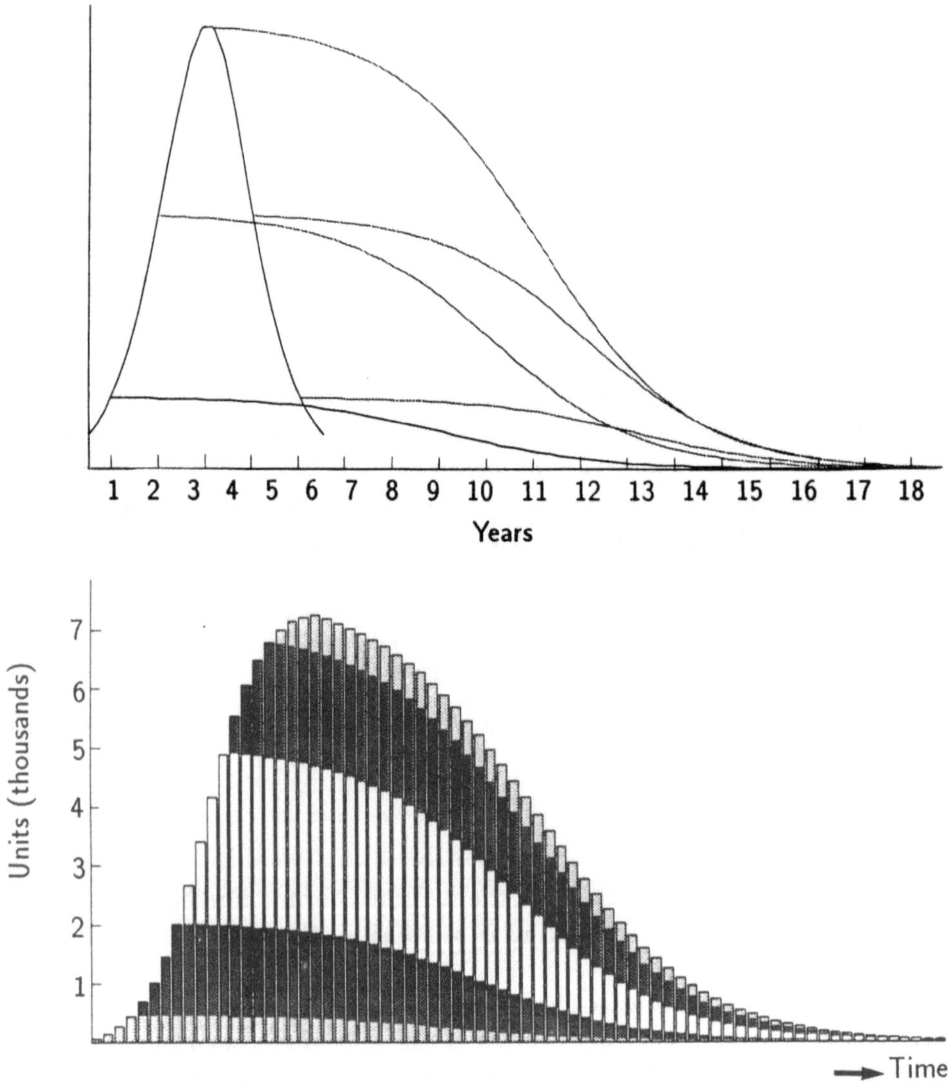

Figure 20.3. Above: life cycle outlining sales of service contracts and logistic mortality rates over a five-year period. Below: the integrated number of contracts with finer time bins (trimesters). The overall envelope represents total survivors at a given time.

$$\begin{aligned} Q(t) &= \int_0^\infty P(t-u)\ R(u)\ du \\ &= \int_0^\infty \frac{M}{(1+e^{\alpha(t-u-t_o)})(1+e^{-\alpha(t-u-t_o)})(1+e^{a(u-u_o)})} \end{aligned} \quad (20.4)$$

20.3 Applications

Using equation (20.4) in forecasting or business planning implies that the five parameters have to be determined. To do this experimentally one needs at least five measurements – monthly or quarterly data points on active contracts. In order to limit the uncertainties involved in logistic fits one must have *many* data points. Consequently parameter determination becomes a fitting procedure where the five unknowns are to be determined from many more than five data points. The procedure we used is one of a X^2 minimization. The X^2 is formed as follows:

$$X^2 = \sum_i \frac{(D_i - Q_i)^2}{w_i} \quad (20.5)$$

where D is the data array, Q from equation (20.4), w a series of weights to be supplied, and i the index of time bins.

A function minimization program was used to search for the values of the five parameters. The weights were usually taken to be uniform, but in some cases they were adjusted through business knowledge. An example is shown in *Figure 20.4*. For this older computer model, even though we are missing the early data, we have enough points (monthly data for six years) to make a reliable determination. In fact, by ignoring the most recent year, the last 12 points from the fit have a negligible effect on the parameter values. Repeating the operation, i.e., dropping the last 24 points, still replicates closely the parameter values originally found. In conclusion, we can say that as long as the historical data go far enough to hint at the decline beyond the peak, the parameter determination shows a remarkable stability.

We did, however, encounter two unexpected difficulties.

20.3.1 Young products

The first difficulty involved young products, i.e., products for which the historical data have not yet reached the peak. In these cases a 4- or 5-parameter fit, involving equations (20.2) or (20.4) respectively, produced *a*

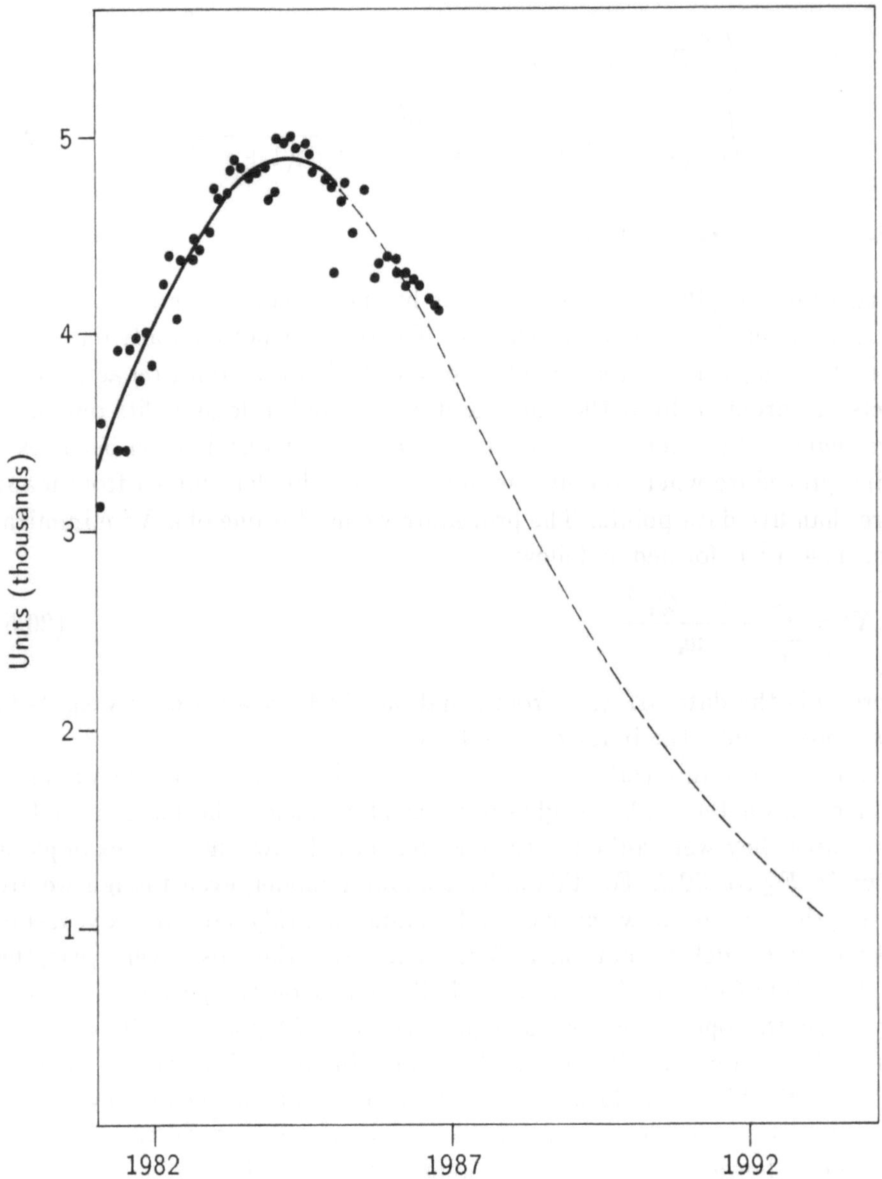

Figure 20.4. Monthly data and fit for the hardware service contracts of an old computer model. Each point represents the number of active contracts at the time. The fit (solid line) is based on the data up to the end of 1985, two years before the end of the historical window. The dotted line is the extrapolation of the convolution function determined from the fit.

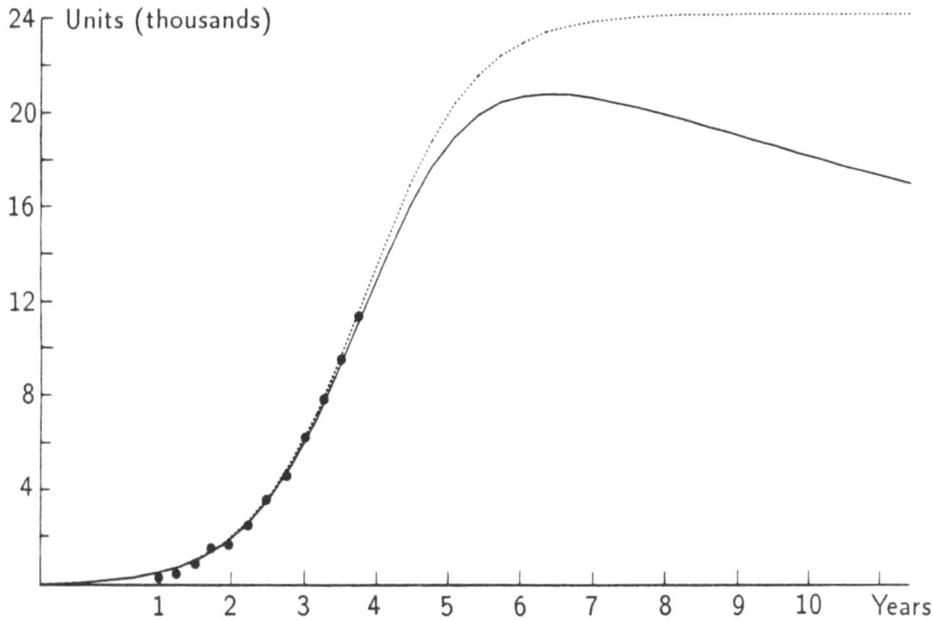

Figure 20.5. Trimesterly data and fit for the active contracts of a recent computer model. The fitting program yielded a null mortality (dotted line). The solid line suggests the realistic evolution taking mortality into account.

null mortality most of the time. The best fit found by the program would be a simple logistic growth with zero mortality. This was seen in terms of the fact that the difference between contracts sold and contracts active was very small in the early days of a new product. The program could not determine mortality parameters if a simple 3-parameter growth logistic could describe the data just as well (see *Figure 20.5*). Although this is a comforting feedback on the robustness and stability of the parameter determination method, it precluded the determination of contract mortality for young products.

At the same time it inspires a resolution to the problem through factorization. We adopted a two-step procedure. First we fit the contract sales data, i.e., the appearance of new contracts, to a simple 3-parameter logistic. Once we determine the ceiling this way, we fit the active-contract data to the convolution function (20.2) or (20.4), but this time with the maximum M fixed at the value already determined from the sales fit. Clearly the procedure demands knowledge of the "birth" of contracts as well as the contracts active at a given time.

20.3.2 Speeding up the computation

The second difficulty concerned the computer time required. Working for a computer company one is often spoilt because there are usually unlimited computing resources. Nevertheless, human nature is such that *having* does not quench *greed.* When we realized that the minimization of a X^2 involving the convolution of two logistic functions, with five parameters needing to be determined, required more than a few minutes of real time computing, we found the situation unacceptable. Rather than searching for programming tricks to reduce the computing time we adopted a mathematical trick. The calculation of the exponentials involved in the logistic expression takes a long time to calculate. Logistic growth, however, can also be obtained through a simple hyperbola and the recursive relation (Meade, 1985):

$$x_{n+1} = \frac{x_n}{a\ x_n + b} \tag{20.6}$$

where a and b are constants and the third constant is the starting value x_o.

The new approach then is to construct an array, element by element, using relation (20.6) and a similar one for mortality with two parameters only (the starting value is taken close to 100%, representing mortality at age zero). The five parameters are then determined by using the minimization program to best match the constructed array to the data, the active contract array is shown at the bottom of *Figure 20.6.* Whenever the data are available in this way, in the full diagonal matrix form, a further factorization becomes possible and mortality can be determined independently from each line (across), while sales keep growing along their life cycle (down).

20.3.3 Phasing out as a generation

It is common knowledge that the usefulness of a generation of computer models decreases with time not because of aging, e.g., material fatigue resulting in frequent breakdowns, but due to obsolescence. Computer models do not drop out of use individually like used cars. They phase out together as their technological generation becomes outdated.

This phenomenon is confirmed here in a formal way. From *Figure 20.3* we see that the number of models in operation declines logistically from the day of sales. The end-of-life point, taken as the 1% of sales that remains at a certain date, is reached at approximately the same date for all models. People who bought the very first models will keep them for up to 16 years. Those who bought the last few models sold say six year later, will only

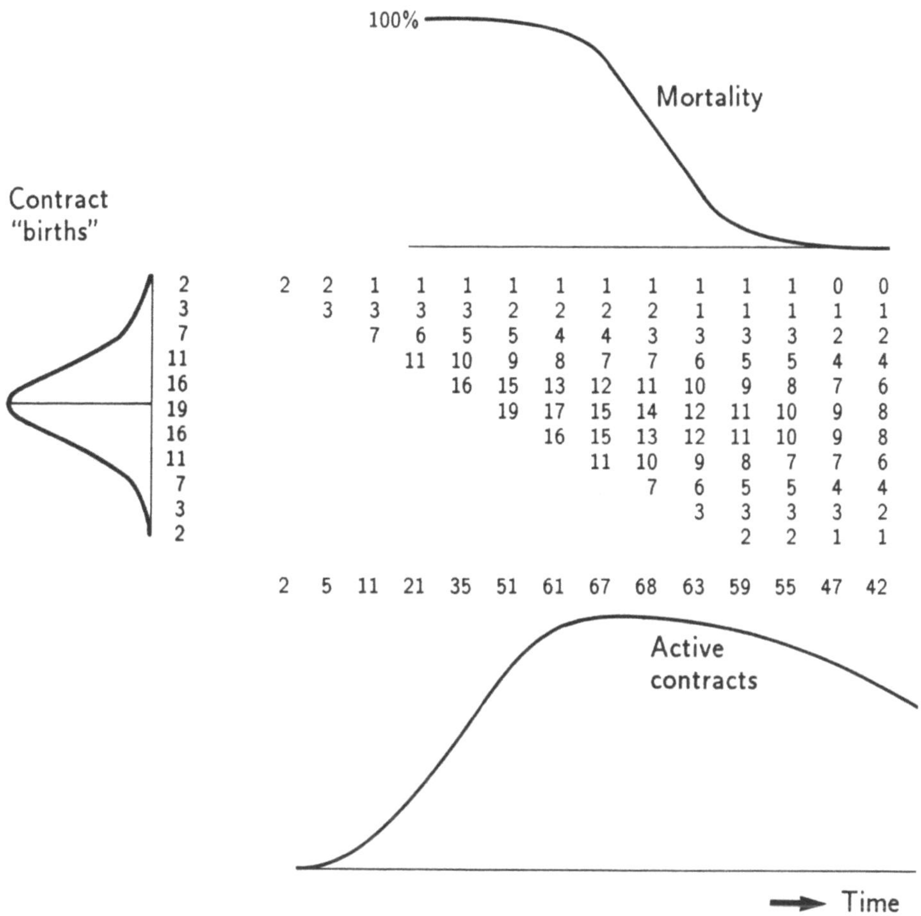

Figure 20.6. The full matrix of contracts for a hypothetical model. The diagonal elements represent initial contract sales. Projected vertically on the left they give rise to the life cycle. Each horizontal array displays mortality over time. The bottom line represents active contracts as a function of time.

keep them ten years. This generation of computers had a sales life-cycle of six years and a service life-cycle of 16. Perhaps unfairly, the models fabricated last, even though perfected and more reliable are endowed with shorter lifetimes.

20.4 Conclusion

Unlike cumulative product sales, service contracts display a mortality over time. Computer service contracts are best described and forecast through the use of a convolution function of two logistics: one representing the growth of contract sales (three parameters) and the other representing contract mortality (two parameters).

For young products it is recommended to factorize the process into two fits, births of contracts first and active contracts afterwards. To keep computing time within convenient limits, one should use a recursive hyperbolic relationship instead of the analytic function involving exponentials.

Computer models in operation phase out independently of the date they were sold. As their technological generation becomes obsolete, early sales and late sales all drop out of use at the same time.

Beyond hardware service revenue forecasts, the approach offered here helps determine the strategically important installed base which is the market for add-ons, software, and a multitude of other services.

References

Meade, 1985, Forecasting Using Growth Curves – An Adaptive Approach, *Journal of the Operational Research Society* **36**(12):1103–1115.

Marchetti, C., 1990, Longevity – An operational definition, private communication.

Potts, G.W., 1988, Exploit Your Product's Service Life Cycle, *Harvard Business Review* **66**(5):32–36 (September–October).

Chapter 21

Diffusion of Process Technology in Dutch Banking[1]

Paul Diederen, René Kemp, Joan Muysken, and Rombout de Wit

21.1 Introduction

Structural changes have been occurring in Western economies since the 1970s. It has gradually become apparent that the service sector can no longer absorb the labor force which cannot be employed in the manufacturing or government sectors. In reaction to these structural changes, it was generally felt that the flexibility of the economy should be improved: economic activity should adapt much quicker to changes in its environment.

The diffusion of new information technologies plays an important role in this process of structural change. Since these technologies facilitate decentralization, they have a profound impact on the division of labor, within and between organizations, and on patterns of employment, e.g., part-time and self-employment.

Banking is a sub-sector of the service industry, in which the diffusion of applications of information technology is particularly prominent. Hence it can be considered as an appropriate case study when analyzing the developments mentioned above. This chapter deals with the diffusion of information technology in the Dutch banking sector. It begins with an overview of technological development in banking. A model is then introduced to analyze the diffusion of technology. Finally our experiences in estimating this model for applications of information technologies in handling bank accounts are reported.

21.2 Dutch Banking: An Overview

The banking industry is a conglomerate of many different financial services firms, most of which carry out several functions. According to the International Standard Industrial Classification (ISIC), the Dutch banking industry can be divided into four branches. The first branch consists of the Central Bank and general commercial banks (wholesale, retail, trading, and investment banks); the second branch comprises cooperatively organized (agricultural) banks, postal giro services, and savings banks; the third consists of other credit and financial institutions like building societies and brokers; finally, the fourth contains complementary financial firms like commission-agents in bonds and stocks and financial administrative firms.

The banking industry has four functions. Firstly, we mention its intermediating function in the payment system. Secondly, *assets management* is important, by which banks are directing the composition on the *assets side* of the balance sheet. Activities rated among this function are participations and financing (loans, credits, mortgages, and the like). The third function consists of activities directed towards the acquisition of financial means like savings and (demand and time) deposits, classified as *liabilities management*, indicating its effects on the *liabilities side* of the balance sheet. Finally, banks perform a number of other financial services like issuing shares and stock-jobbing, on the one hand, and services originally not part of the banking profession, like acting as an agent on behalf of insurance companies and travel-agencies, on the other hand.

The analysis in this article will mainly focus upon the first two banking branches, i.e., general commercial banks, cooperatively organized banks, postal giro services, and savings banks. In the Netherlands, these branches are currently dominated by six large banking organizations, accounting for between 80% and 90% of the banking industry's economic output and more than 90% of its employment. The sort of technological change considered in this chapter mainly concerns innovations influencing the first function mentioned, the intermediary function in the payment system. The six largest Dutch banks take care of the payment system in the Netherlands.[2]

21.2.1 Main developments

Until the early sixties the banking world was relatively quiet. Traditionally, the different banks were strongly specialized, and their activities were restricted to their own territory. The former postal giro services took care of the mass payment traffic of wage earners and consumers. The trading banks (general commercial banks) financed loans and credits for trade and manufacturing and accepted money deposits and savings from the same economic sectors as well as from the wealthy. Cooperatively organized agricultural banks financed activities and acquired savings in the agricultural sector, while laborers had their accounts at a savings bank or with the postal giro service.

Since the mid-sixties, however, the situation has changed dramatically (see Peekel and Veluwenkamp, 1984). The prosperous economic growth of the whole economy led to a great demand for loans and credits. This resulted, on the one hand, in a series of mergers between the banks and, on the other hand, in the penetration of each other's markets. Wholesale banks started to operate on the retail market with the ultimate objective of acquiring savings to finance industrial loans and credits. Conversely, cooperatively organized banks with huge savings balances entered the wholesale market to sell loans and credits. Thus, besides the process of concentration, a process of branch *blurring* started and competition increased.

While the number of independent banks decreased dramatically, the number of offices in the country increased until the early eighties. Since then the reverse trend can be seen, i.e., a decrease in the number of offices. The total number of offices grew steadily from 7,520 in 1971 to more than 8,600 in 1981, and then it decreased to 8,232 in 1986. In 1986 the number of inhabitants per bank office was estimated to be 2,616 or 1,765 if post offices are included. International comparable figures are 2,310 in the United States and 1,524 in France.

21.2.2 The economic growth of banking

Table 21.1 gives an indication of the growth of banking compared with the market sector as a whole. It gives an overview of gross value-added in 1980 prices for the period 1971–1986.

Table 21.1. Economic growth of banking and of the market sector at large (amounts in Dutch guilders $\times 10^9$).

	Banking industry			Whole market sector		
Year	Value added	Index number	Annual rate	Value added	Index number	Annual rate
1971	6	100		226	100	
1976	9	155	9.1%	268	119	3.5%
1981	12	206	5.9%	287	127	1.4%
1986	14	237	2.9%	307	136	1.4%
1971/86			5.9%			2.1%

Sources: CBS (1971–1987) and CPB (1971–1987).

The figures indicate that the banking industry, compared to the whole market sector, flourished during that time period. Average annual growth rates in each five-year period were well above the comparable growth rates of the market sector as a whole, with an average annual growth rate over the last fifteen years of 5.9% versus 2.1% for the whole market sector.

As mentioned, the banking sector carries out different functions, the payment function, the *assets* and the *liabilities* management, and other financial services. In *Table 21.2* we present some indicators referring to the first three functions. The indicators do not represent (any kind of) value added but the aggregated balance figures at the end of the year. Therefore, although they do not represent the development of economic output, they give us a rough idea of how total production volumes of different banking activities have changed.

The average annual growth rate of payment transactions by more than 25% is striking. The highest growth rates, however, occurred in the early seventies (an average annual rate of 75%), while the growth rates during the last ten years varied between 5% and 6%. The banking industry has only been capable of processing these huge volumes because it changed to an automated payment system.

The other two functions have been developing more or less parallel to the banking industry as a whole, growing at about 6% per annum. Retail functions both on the *assets* side (consumer credit and mortgages) and on

Table 21.2. Development of different banking functions (index numbers 1971 = 100; average annual growth rates 1971–1986).

Function	Banking	Payment	Assets Management				Liabilities Management	
Year	Value added	Transactions	Loans	Mortgages	Cons. credit	Debt. balance	Savings balances	Near money
1976	155	1680	167	205	195	211	107	225
1981	206	2194	166	165	179	301	155	217
1986	237	2945	223	201	177	248	160	306
1971/86	5.9%	25.3%	5.5%	4.8%	3.9%	6.2%	3.2%	7.7%

Source: CBS (1971–1987).

the *liabilities* side (savings balances) have experienced lower growth rates than wholesale activities such as loans and debtors balances on the *assets* side and near money on the *liabilities* side.

21.3 Technological Developments[3]

Technological developments in the banking industry began in the system of payment. It is generally accepted that a distinction is made between countries that might be characterized as having a *cheque* payment system and others having a *giro* payment system. *Table 21.3* shows clearly that the Netherlands has a *giro* payment system.

Table 21.3. Composition of payment instruments in 1983, in percent.

	Transactions		Amounts	
	Cheques/ creditcard	Giro transfers	Cheques/ creditcard	Giro transfers
France	85	15	10	90
Netherlands	23	77	1	99
United Kingdom	71	29	10	90
United States	97	3	28	72
Germany, F.R.	11	89	16	84

Source: Bank for International Settlements (1985).

The banking industry's path of technological development, its technological trajectory (this term was coined by Nelson and Winter, 1982), differs according to the prevailing payment system. Technological developments in countries like the Netherlands with a *giro* payment system started with the automation of, firstly, records of the accounts of clients in the sixties and, secondly, of *giro* transfers. This was accomplished by investments in big mainframe computer systems in connection with the development and introduction of new *regular* payment instruments like giro salary accounts and automated debits and credits. In *cheque* countries, however, one was less able to develop comparable automated instruments for periodical payments like salaries, mortgages payments and monthly rents. Consequently, payers were forced to continue writing cheques periodically. Technological developments in countries like the USA, the UK and France were mainly directed at the automation of the labor intensive processing of cheques.

The automation of payment transactions began after World War II, when the economic revival induced a strong growth in payments. Mechanical bookkeeping machines were introduced in the fifties. The first mainframe computer (1962/1963), combined with the punched card (1960), made it possible to automate financial mutations in the account records of the administrations at the central offices of the banks. This process of *central automation* was reinforced by the establishment of a so-called automated clearinghouse. The *decentralization of computer and communication technology* started with the automation of local offices and banks and was intensified by the establishment of information networks. Consequently, the following phases of technological development can be distinguished:

(1) Central automation
   (a) at the head offices of individual banks (1960–1970);
   (b) of clearinghouses and external integration with business clients (1970–1980);
(2) Decentralized automation
   (a) of the back offices and front offices of local branches (1980–1990);
   (b) of information networks as the merger of computer and communication technologies (1990–...).

Both stages of central automation took place at only a few *production* centers of commercial banks and clearinghouses. Its applications concern the centralized registration of the loans, savings, and securities accounts, but above all the central administration of current and salary accounts as well as the payment transactions involving debiting and crediting these accounts. The effect of this process of technological development was mainly a strong increase in the labor productivity associated with the central processing of the payment transactions.

We now witness, in the two decentralized automation phases, as opposed to the period of central automation, a process of technological development in which offices of certain local banks could be qualified as early adopters and others as laggards with respect to the adoption of computers and communication technology. In other words, the *process of diffusion* of technologies through the branch network of banks is paramount. Moreover, the effects on employment will not be restricted to a small group of production workers at clearinghouses and production centers at the head offices, but will apply to the majority of bank employees, about two-thirds of which are working at local offices throughout the country.

21.3.1 Local banks and offices

The technological development at branches and local banks can be divided into four stages; the first two interact with phases 1(a) and 1(b) of central automation and the last two are a further differentiation of phase 2(a) as distinguished above.

1a Manual production process, making use of mechanical bookkeeping machin(
(1960–1970);
1b *Optical character recognition* (OCR) equipment (1970–1980);
2a (i) Mini computer system with back-office terminals (1980–1985);
2a (ii) Counter terminals in the front office (1980–1990).

During phase 1 of central automation, the technological developments at local banks and offices were restricted to the use of mechanical bookkeeping machines [stage (a)] until the early seventies. From the beginning of the seventies onwards, OCR equipment [stage (b)] came into use. The equipment was used for typing the payment orders of clients on *counting slips*, which could be read optically by the mainframe computers at the central head offices of the banks. Most machines were equipped with controlling functions correcting simple mistakes and producing a cleaner *input* at the mainframes. Up to this stage there were no major differences regarding technological developments at the local offices between the organizations involved.

The start of computer and communication systems at local banks and offices [stage 2a(i)] can be located around the middle to late seventies. This led to a productivity increase in back-office work which had little consequences for employment due to an increase in output. The next stage of technological development at local banks was the installation of terminals at the bank counters [stage 2a(ii)] and so-called quick-cash terminals. These counter terminals are "smarter", having more in-built functions. The possible effects on employment and the organization of work are more pronounced. In particular, the work of cashiers and counter clerks is now automated.

The adoption of more advanced information-technology-based production techniques has consequences for commercial policy: before the introduction of off-line counter terminals the commercial policy could be qualified as product-oriented, implying that employees were specialized in certain products, for instance, cash, insurances, consumer credits or business loans. With the introduction of counter terminals in the eighties, commercial policy changed into some type of *integrated client management.* This kind of management implied that clients were, in principle, served the whole range of banking products by a *personal banker* who was assisted by some counter clerks. This meant on average that the commercial employees were expected to have a higher education and more skills, but at the same time that they also had to fulfill routine activities and tasks in which they were not so well trained.

Banks are currently at the stage of designing, establishing, and implementing information networks [stage 2(b)]. An information network originates when the management at offices and local banks can retrieve data (about accounts and characteristics of clients as well as external financial data) from the central or local computer systems, use the data with the aim of gathering information for management purposes, and possibly send the newly processed information back to the central computer systems or to clients. The actual applications of the networks are expected to mature in the nineties.

21.3.2 Banking products and processes

In banking, it is not so easy to distinguish between process innovation, on the one hand, and product innovation, on the other. Usually, the two develop hand in hand. For instance, the process of central automation at the head offices of banks and clearinghouses could not have been realized without a simultaneous introduction of new products like the salary and current accounts, automated debits and credits, and giro-cards inviting payment. The fast increase in the number of machine-readable payment instructions at the automated clearinghouse (from 5.8 million in 1970 to 278 million in 1985) illustrates the growth of the bank's new payment products. Diffusion of new process technology is reflected in the gradual introduction of new product innovations.

Asset management is relatively unaffected by information technology so far. The contracting of loans still predominantly consists of face-to-face negotiations and an analysis of the client's financial data. The use of personal computers for financial analysis, the recording of the client's financial data, the word processing of standard financial contracts as well as the associated administration, have only just started. Applications in the field of mortgages are a little bit more advanced in the sense that it is possible to produce standard offers; however, it is felt that the applications do not go much further than a qualitative support of the negotiations with the client. In the case of insurance policies and travel arrangements, some use is made of on-line communication when processing the contract for a final insurance policy or the booking of a journey.

Finally, some remarks should be made regarding the relatively new electronic banking products like cash dispensers, point-of-sale and electronic banking terminals. These products were mainly developed in the United States. In Europe, *cheque* countries like the United Kingdom and France soon followed. The developments in *giro* countries like the Netherlands and the Federal Republic of Germany (FRG) started much later because originally no action against unsecured cheques was needed. Nowadays high labor costs of teller transactions and pressures from the retailers and petrol companies accelerate the diffusion of cash dispensers.

21.4 Modeling Technology Diffusion

In the previous section the process of technological change has been analyzed in a qualitative way. In this section an analytical framework will be presented which should enable us to analyze this process more quantitatively. To this end a diffusion model has been developed, which describes changes in production volume, manufactured by different techniques, as a result of adopting new techniques and scrapping existing ones. The model generates a life-cycle pattern in the course of time for each technique. Since the inputs required for each technique are known, the model also generates the changes in employment volume, work requirements and occupational structure.

The model presented in subsequent sections tries to express the idea that learning and adjustment processes are the most important ingredients guiding the process of technology adoption.[4] To learn and to adjust established routines takes time and money. Firms weigh these sacrifices against expected benefits from new technology. As benefits grow, risk decreases and learning proceeds, and the diffusion of new production techniques progresses. First a brief frame of reference will be presented, expressing the angle from which we try to look at technological change. Then the model will be outlined, specifying what variables are considered to be relevant in the case of the banking industry.

21.4.1 The diffusion of innovations and the role of information

Two stages in the process of technological change can be distinguished. First of all, there is the decision to invest in new technology. Second, there is the process of learning to operate a new technique. Common to both is the fact that complex information processing is involved. Information processing plays a key role in the determination of the speed of technological progress. In the first stage of technical change information about investments in alternative techniques has to be gathered. This information has to be structured and valued in order to estimate the costs and benefits of switching to new production techniques. In the second stage new techniques have to be learned and new skills have to be mastered.

These processes of learning, gathering, structuring, using and mastering information are at the core of technological change. We are flooded with enormous amounts of various types and forms of information. Also, information comes from sources of varying reliability and importance. However, learning is a costly activity, since human capacity to interpret and process all this information is limited. Thus, to be of practical use, information is usually structured in some hierarchical order.

To introduce a hierarchy in available information, we assume that the more information ties in with present practice, the more attention it receives and the better we are able to value the information. New facts are easier to interpret, the more they correspond to what is already known. Furthermore we suppose that decision makers are not equally susceptible to information from different sources. The closer the source, the more influential the information will be. Moves by direct competitors, operating in the same submarket, in the same geographical area, using similar techniques, are watched most closely. Consequently, perception of alternatives and ensuing decisions are conditional upon one's position in a competitive environment.

Relevant information can be of three types, as indicated in *Table 21.4*. First, we distinguish global information: general facts and knowledge about the sector structure and its working conditions. Then there is firm-specific

Table 21.4. Categories of information.

Information	Examples
Global	Product-, inputs- and labor market development; technological trends; industry structure and macro-economic development; government policy.
Firm specific:	
External	Competitors marketing strategies (product differentiation, specialization, pricing, geographical spread, promotion); firm structure; production techniques.
Internal	Organizational and financial structure; technical features; market position; firm specific knowledge and values; entrepreneurship.

external information, coming from the environment, its impact depends on the position of the firm in relation to its competitors. Finally there is firm-specific internal information, depending on the conditions within the firm itself.

The core problem of dealing with information in real-life situations is precisely how to combine data of qualitatively different dimensions systematically, such that conclusions lead to unequivocal decisions. Decision algorithms are not only intricately complex, but also tend to be rather situation specific. It is impossible to observe decision making on the basis of all relevant information directly in a model and to describe quantitatively the evaluation of multidimensional information leading to a decision. Models can only approximate the weight or impact of information by looking at a limited number of key variables. We use (1) production volumes, (2) so-called technical distances between techniques, and (3) disparities in efficiency.

The *production volumes* in the present period and the distribution of production over techniques are the direct reflection of the outcome of strategic behavior in the past. Thus there is a lot of information implicit in production volume figures. The behavior of the system in the next period is approximated by relating the actual values of production volumes to their potential values. The ratio of the two is a determinant of the speed of change.

Suppose techniques can be ranked according to increasing efficiency from 1 to n. *Figure 21.1* pictures production volumes corresponding to the available techniques. The actual amount produced by means of technique i is indicated by the starred column (**). Under conditions of a constant total production in time, and assuming an efficient use of techniques, the potential maximum amount that can be produced with technique i is this column plus

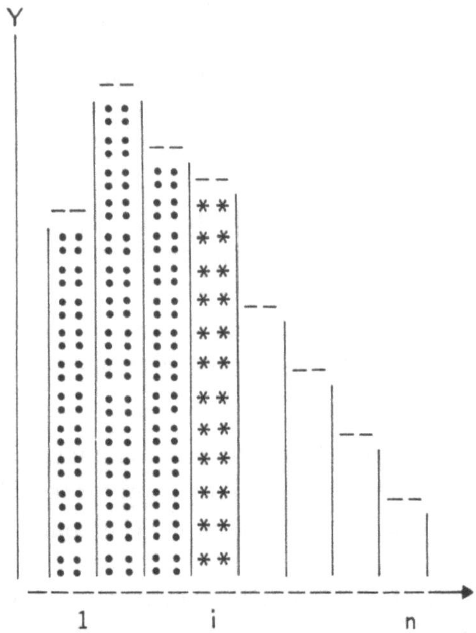

Figure 21.1. Production volumes by techniques 1 to n.

the amount produced less efficiently at present, represented by the speckled columns (::). This corresponds to the total production volumes of techniques 1 to i. The ratio between these actual and potential volumes is used in the expression we call the *competitive force* of technique i. Similarly we define the *competitive pressure* on technique i, using the ratio of the actual production volume and the maximum effectively competing volume in terms of production costs. The latter is represented by the starred plus open columns in *Figure 21.1* and corresponds to the total production volumes of techniques i to n.

We characterize production techniques by means of vectors of input coefficients. This enables us to define a distance function in the vector space of techniques. This so-called *technical distance* between two techniques is assumed to represent the extent to which these techniques resemble each other. When a firm scraps an old technique and adopts a new one, this distance represents the technical barrier that has to be overcome. The concept tries to embody a host of information of a different character, concerning

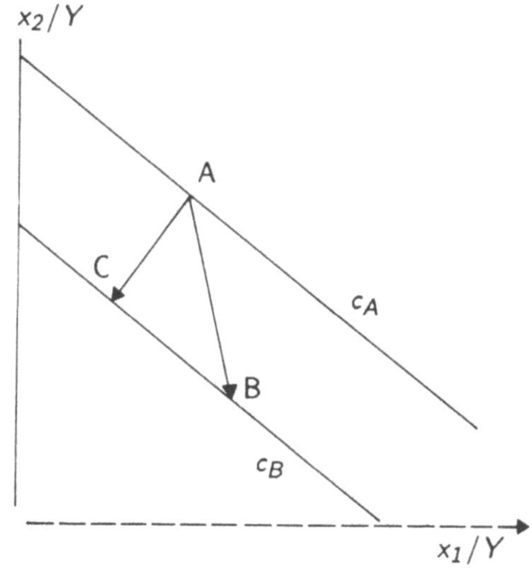

Figure 21.2. Costs and benefits when switching from techniques A to B.

adjustment problems, friction costs, learning efforts, and attitude towards risk.

In the model, technical distance is balanced with the *efficiency disparity* variable, which captures the gains from switching techniques. The gains are expressed by computing the cost difference of producing one unit of the product by the old and the new technique.

If a technique uses only two factors of production, x_1 and x_2, the situation can be pictured in a diagram as in *Figure 21.2*. Suppose a firm operates a technique at A. A new technique is available at B. The lines c_A and c_B depend on factor prices and connect all input combinations with equal costs. If a firm switches from the old to the new technique, it has to cross barrier $A - B$. The benefits of this move are proportional to the distance between the cost lines, distance $A-C$. The ratio of benefits $A-C$ and costs $A-B$ will determine the extent to which the firm is prepared to switch. When factor prices change, this affects the numerator $A - C$ but not the denominator $A - B$ in this ratio.

21.4.2 The diffusion model

The elements introduced above are now combined in a diffusion model. Given the inputs, a range of valuable techniques and the development of demand, the model can describe the changes in the use of a production technique in the production of a banking product. This change is the result of two components. The first element is the increase in the use of a technique, because the technique is adopted by firms that formerly operated using less efficient techniques. The second component is the decrease, caused by the fact that former users of the technique scrap it and switch to a better one. The model is defined by:

$$\text{Change} = \text{Increase} - \text{Decrease}. \tag{21.1}$$

Firms that decide to switch techniques do this by balancing expected benefits against estimated costs. In appraising benefits and costs, they take into account the production techniques used by competitors as a source of information on their market position and relative efficiency. When it considers the virtues of a new technique, the firm must determine the following:

(1) The riskiness of adopting a new technique.
(2) The extent of its present use, relative to its potential use: competitive force.
(3) The efficiency benefits.
(4) The barrier to be overcome when the firm adopts the new technique.

These are represented in the model as follows:

(1) The risk of adopting a technique i is assumed to be due to a lack of information and experience and is assumed to vary with the extent of its use in the last period: $Y_{i_{t-1}}$, where t is the time index.
(2) The extent of its present use, relative to its potential use is the ratio of $Y_{i_{t-1}}$ and $\sum_{k=1}^{i} Y_{k_{t-1}}$, where summation is over the less efficient techniques. We use this ratio in an expression symbolizing the attractiveness of a technique, called the competitive force – CF[5] (see also *Figure 21.1*).
(3) The efficiency benefits – EB – is supposed to vary with the cost difference per unit of output.[6] This corresponds, in the two dimensional case pictured in *Figure 21.2*, to the distance $A - C$.
(4) The barrier to be overcome – TD – is also expressed in terms of input coefficients.[7] This corresponds to the technical distance $A - B$ in *Figure 21.2*.

Thus we can conclude that:

$$\text{Increase} = a_1 * Y_{i_{t-1}} * \frac{EB}{TD} * CF \tag{21.2}$$

where the parameter a_1 is a scale factor with respect to the speed of diffusion resulting from competitive force.

Analogously, we specify the decrease in the use of a technique. Whereas the impulse to increase the production volume made by using a technique i is related to the use of less efficient techniques, the decrease is related to characteristics and volumes of more efficient techniques:

$$\text{Decrease} = a_2 * Y_{i_{t-1}} * \frac{EB'}{TD'} * CP \tag{21.3}$$

Here summation is over the production volumes of better techniques; when defining competitive pressure, CP, the technical distance, TD', and efficiency discrepancy, EB', are between technique i and the average better alternative, where the parameter a_2 is a scale factor with respect to the speed of diffusion resulting from competitive pressure.

Pulling all elements together, the change in the production volume made by using technique i in period t is described by:

$$Y_{i_t} - Y_{i_{t-1}} = a_1 * Y_{i_{t-1}} * \frac{EB}{TD} * CF - a_2 * Y_{i_{t-1}} * \frac{EB'}{TD'} * CP \tag{21.4}$$

The model is a combination of two expressions that closely resemble logistic curves. Together they result in a life-cycle path for every distinct technique.

Given input coefficients per technique, we are able to compute technical distances, as can be seen from the definition of TD above. Given wages and factor prices, we can determine the efficiency gap EB between techniques. New techniques are expected to be cheaper than older ones, if not from the outset, then at least after a process of learning by doing and after capacity utilization, which tends to be low immediately after the introduction of new equipment, has reached normal operating levels. Production volumes are used to compute competitive force and pressure CF and CP, respectively. This being done, the parameters a_1 and a_2, influencing the general reaction speed to innovations, can be estimated.[8]

21.5 Estimating Diffusion of Technology

The model presented in the preceding section has been estimated, using data from one large Dutch banking organization, which consists of more than 900 relatively independent local banks. The local banks decide when to adopt new technology. Out of this population of almost 1,000 banks, a representative sample of 100 was included in the data set.

The model has been used to describe changes between 1979 and 1987 in the technique of handling accounts. This product mainly involves the transfers of payments from one account, business or private, to another and the withdrawal and depositing of cash. For account handling four techniques are to be distinguished during this period: (1) conventional OCR equipment, (2) back-office automation, (3) front-office automation, (4) cash dispensers/automatic teller machines.

Both OCR equipment and back-office terminals are used for data input activities (cheque and payment orders). Counter terminals and cash dispensers affect data processing (which is done automatically), but primarily deal with teller transactions. Counter terminals came into use in the early eighties, and rapid growth started after 1983.

21.5.1 Data

Given the inputs, a characterization of the four available techniques and the development of demand, the diffusion model outlined above enables us to describe the diffusion of techniques in handling accounts in the branch offices of our case study bank. Local branch offices appear to have different attitudes toward the introduction of new technology. There are early adopters, cautious imitators, slow responders, and conservative laggards. Before elaborating on this, we will first briefly describe the main techniques used in the model, together with the relevant factors of production.

In *Table 21.5* the different inputs are presented. We subdivide labor into five categories, and we divide capital into two parts: one that does embody technological change (equipment) and one that does not (housing). Finally, we distinguish two other production inputs: mail, which also includes the intermediary services of the central office, and other factors such as energy, advertising, and cleaning. With respect to labor requirements, it should be noted that these are not measured by numbers of employees or man-hours of an employee, but by *job content*, by hours of activity of a certain type.

The production factors can briefly be described as follows:

- Commercial work: negotiations with clients over loans, selling of insurance and travel services and administrative work like the conclusion and continuation of contracts and handling claims.
- Counter work: all counter activities, such as dealing with deposits and withdrawals and the opening and closing of accounts.

Table 21.5. Factors of production in banking.

Labor	Capital	Other
Commercial work	Buildings	Mail
Counter work	Inventory including computer equipment	Other
Data input work		
Administrative work		
Managerial work		

- Data input work: consisting of data input for payment transfers and control (checking balances of accounts).
- Administrative work: other administrative activities including accounting, catering services, and small repairs.
- Managerial work.
- Central computer processing and mailing (ccpm): processing of transfers, storing accounts and savings information, and mailing at the central office.
- Inventory including computer equipment: inventory and computer and communication systems at the local branch office.
- Buildings.
- Remaining inputs: energy, advertising, and cleaning.

Vectors describing a technique will have nine elements, corresponding to the production factors in *Table 21.5*. Furthermore prices of the various capital services, mailing and maintenance, as well as wages for five different types of labor, are needed. Finally, we need the production volumes made every year by the distinct techniques.

For the 1973–1987 period, cost figures for inputs, employment figures, and production volumes of products were collected from the 100 banks sample. Price indices for inputs stem from local offices (like labor costs per man year) and from other sources (price indices on computer equipment coming from the Central Bureau of Statistics in the Netherlands). Production volumes for accounts are based on the number of payment transfers, teller transactions, and the number of payment accounts, both personal and business.[9]

In every local branch office during every time period, a technology for payment transfers is identified through detailed data on automation from individual offices. We also know production volumes and inputs per branch office. This makes it possible to determine aggregate production and input coefficients for each technique in use. Production volumes for techniques are computed by aggregating the production volumes of local offices which used the same kind of technology. Input coefficients for accounts are calculated by aggregating the costs of inputs of offices using the same technology and deflating these costs by their price index, and dividing this by their (aggregated) production volume. For the entire estimation period input coefficients are computed for each technique, not only showing differences in the amount of an input per unit of output between techniques, but also within a single technique over time.

The aggregate input coefficients are presented in *Figure 21.3(a)*, for the 1975–1987 period.[10] Accounts experienced substantial labor savings. Automation is reflected (directly) through input coefficients for computer equipment and inventory. *Figure 21.3* clearly shows how input coefficients for computer equipment increased.

Looking at the decomposition of labor input per unit of output for accounts [*Figure 21.3(b)*], we learn that technological change decreased the input of counter work, data input work, and administrative work. Although labor input per unit of output decreased, employment increased in this period due to the growth of production volumes (5.3% in the whole 1975–1987 period).

Comparing disaggregated input coefficients for different techniques of account handling shows that input coefficients for computer equipment were higher for every consecutive technique. On average, each new technique leads to less labor input, mainly as a result of less data input work and less administrative work. Labor savings for counter work do not appear in the data. Implementation of counter terminals does not lead to large labor savings, at least initially. Also in the case of cash dispensers, input coefficients do not reflect the labor savings on counter work.

21.5.2 Estimation results

Measures for competitive forces and competitive pressures of techniques have been computed using the production volumes of techniques from the previous period. Efficiency gaps and technical distances between techniques have been calculated for several years and the diffusion model has been estimated using these figures.

The inclusion of these variables, *EB* and *TD*, respectively, does not improve the performance of the model. It turned out that sometimes new techniques are more expensive than old ones. Measurements of technical

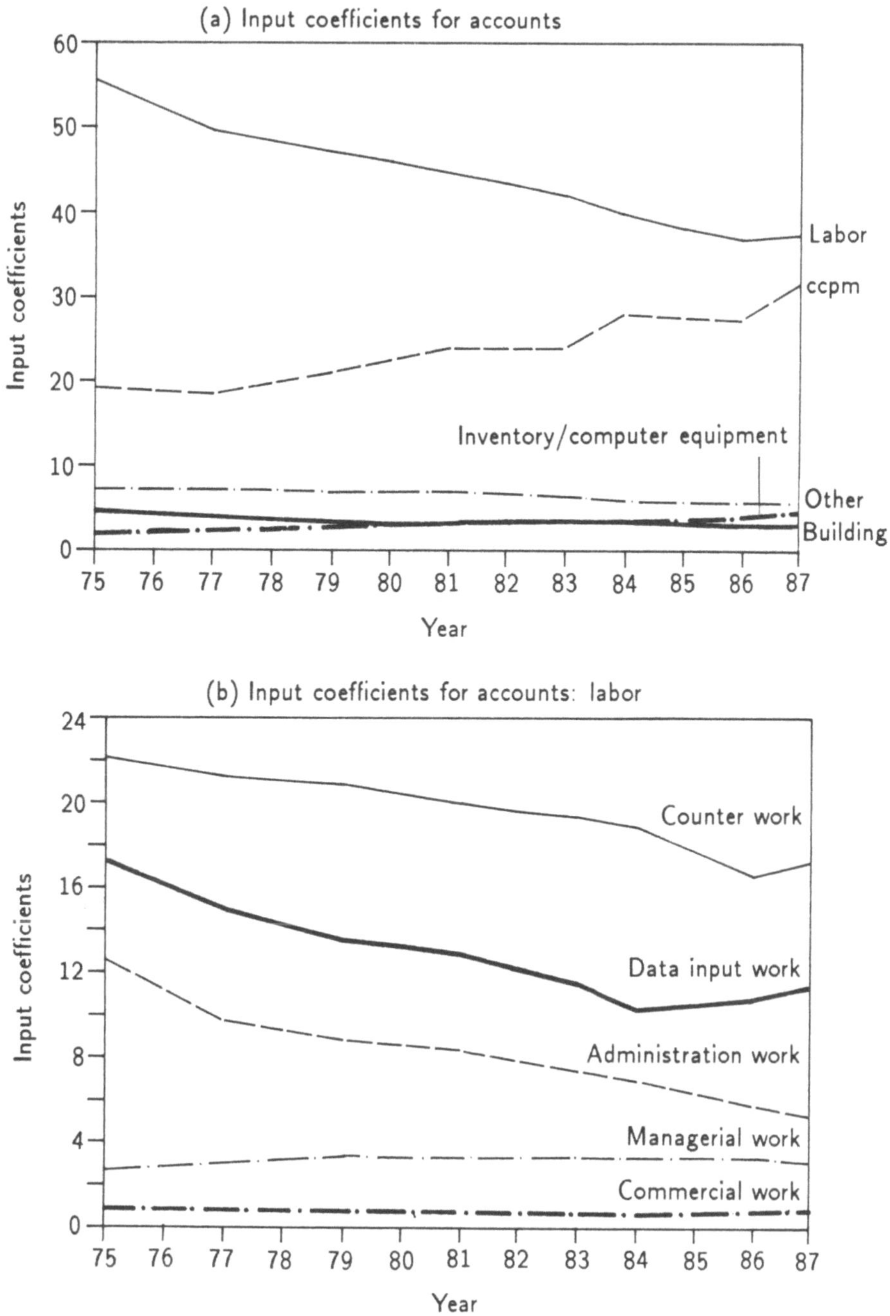

Figure 21.3. Input coefficients for accounts.

distances, although of the right sign, fluctuated strongly over time, and therefore did not provide accurate information about technical barriers. This can be caused by several factors:

- Measurement problems: accurate determination of a vector of technical coefficients per production technique is nearly impossible due to factors like economies of scale, joint production, and special circumstances; accurate calculation of costs of production is equally difficult, due to problems of attribution.
- Aggregation and definition of the product: the product made with the new technique is not really identical to the product made by using the old technique. Mostly there is some type of improvement, e.g., in the flexibility or speed with which the service is delivered.
- The current difference in production costs of the new and the old technique might not be a good approximation for the expected benefits of adopting a new technique. A new technique can be adopted because it is expected to be cheaper in the future, despite the fact that it might be more expensive in the introduction phase. Due to learning by doing and economies to scale, the price of a new technique may fall considerably after it is first introduced.
- The efficiency gain may not be one of the main arguments to adopt a new technique. The firm is merely afraid to fall behind in the level of technology or is motivated to keep up with its competitors.
- At the moment of introduction the new technique does not operate at full capacity. Investments in a new technique often lead to excess capacity, which in the short term disproportionally raises the costs of the new technique.

The diffusion process of technology was best explained by competitive force and competitive pressure as the only variables. Model results for this simple version of the diffusion model will be given. The estimations are produced by minimizing least squares (under the restriction of a_1 and a_2 being the same for all techniques, but not a_1 and a_2 being equal). Market expansion is reflected by the estimates for a_1 and a_2, 1.45 and 0.74, respectively, a_1 being almost twice as high as a_2.[11]

In *Figure 21.4*(*a*) real production volumes of techniques for accounts in the 1978–1987 period are presented, together with the production volumes for techniques as explained by the diffusion model [*Figure 21.4*(*b*)].

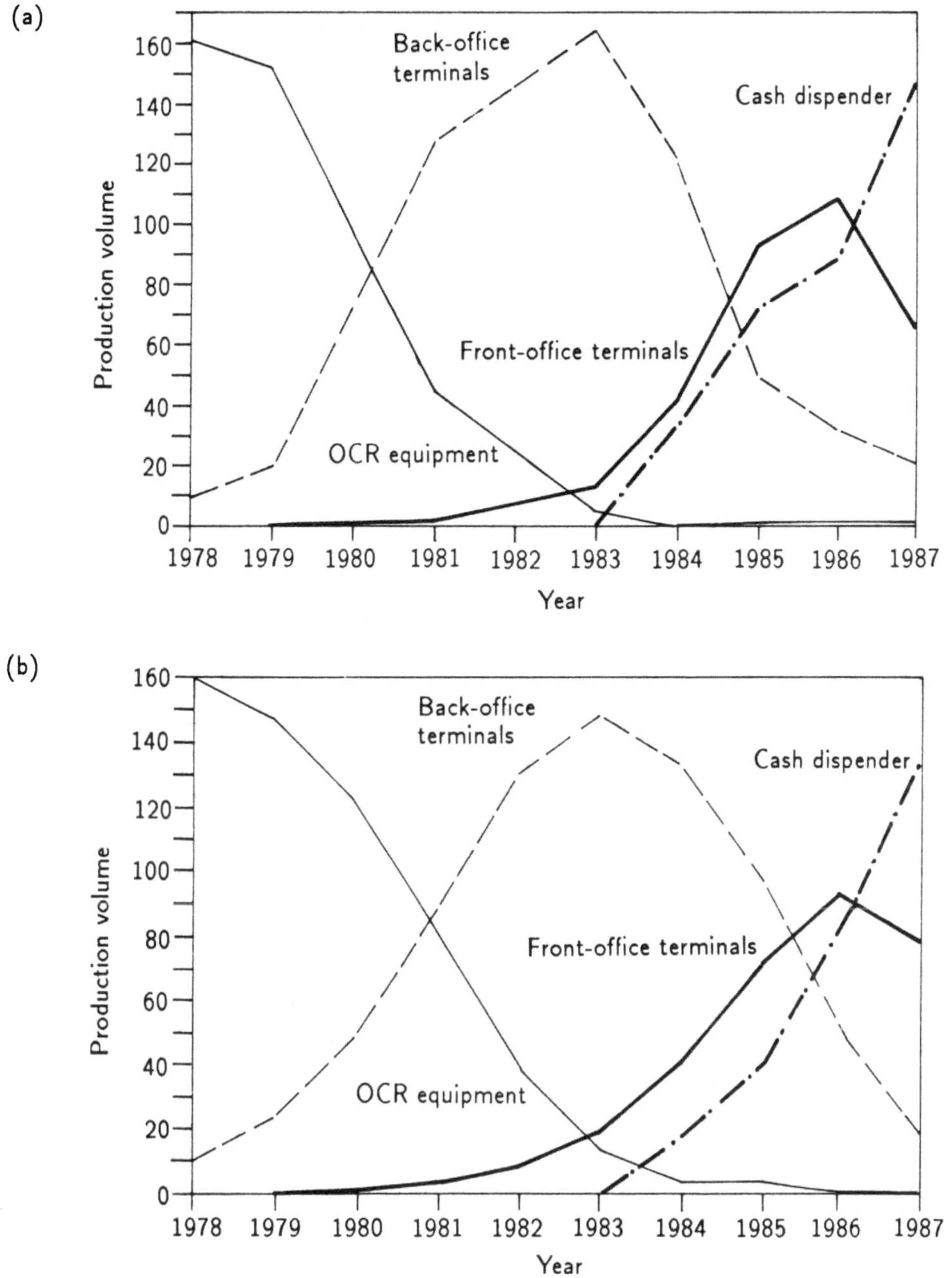

Figure 21.4. The diffusion of techniques for handling accounts: (a) data and (b) model results.

Figure 21.4 shows how conventional OCR equipment is replaced by back-office computer equipment. Back-office terminals, first adopted in 1978, experienced a fast growth within five years. In 1983 nearly all banks possessed back-office computer systems. In the years following 1983, front-office terminals and cash dispensers were implemented. Like back-office terminals, front-office terminals grew rapidly over a five-year period. Cash dispensers in local branch offices experienced an even faster growth. It is likely that this fast growth will continue in the coming years.

Changes in the choice of technology, among other factors, lead to changes in factor demand. Since each technology is characterized by input coefficients, reflecting the amount of input per unit of output, the consequences of technology diffusion on factor demands can be computed. However, in order to calculate overall changes in factor demands of the banking industry, the above estimation exercise has to be performed for every banking product.

21.6 Summary and Concluding Remarks

In this chapter we gave an overview of the application of information technology in Dutch banking and presented a diffusion model to explain several of the developments in more detail. To this end, we employed data of a large Dutch banking organization. The diffusion model analyzes the diffusion of techniques in local branch offices. The model is based on the notion that technological change involves a learning process under uncertain conditions. Therefore the availability of a more profitable technique does not imply that it will be used immediately: different techniques will be employed at each moment to produce a certain product. This is a consequence of phenomena such as the limited capacity of individuals and organizations to process multi-type information, the balancing of current expenditure against uncertain future benefits, and the dependence of firm decisions on perceived reference groups. These phenomena are captured in the model by the notions of technological distance, efficiency barrier, competitive power, and competitive pressure. The notions lead to two expressions that closely resemble logistic curves, one for the growth in the use of each technique and one for its decline. Together they result in a life-cycle path for every distinct technique.

In the model a technique employed for a certain product is characterized by its input coefficients, i.e., the amount of inputs necessary to produce one unit of output. Since this requires very detailed information, we confined the implementation of the model to one large bank over the period 1979–1987, for which we obtained data on about 100 local branch offices.

For our case study bank the impact of technological change is analyzed at the local branch office level (where most people employed in banking are situated). Technological change at the central office, predominantly through the use of large mainframe computers and information technologies in trading rooms, is exogenous in the model analysis. From our case study bank detailed cost figures, data on production volumes, and information on investments in automation were available for each local branch office, making a more profound analysis possible.

Although all activities at local branch offices are influenced by technological change to some degree, model results are only reported for the aggregated product that is most strongly affected by technological change: account handling. Accounts include personal accounts and business accounts. Relevant activities are teller transactions (deposits and withdrawals) and the processing of transfers (data input and control).

With respect to the techniques used, for accounts four techniques are distinguished during this period: (1) conventional, (2) back-office automation, (3) front-office automation, (4) automatic teller machines. From data on the contributions of the different technologies to output growth, it is clear that new technologies in banking were not installed at once. Back-office automation was installed in the late seventies by our case study bank over a five-year period. Front-office automation began in the mid-eighties and also took five years. Cash dispensers were implemented just after front-office automation was installed but experienced a more rapid growth.

In addition to the four different kinds of non-labor inputs, five kinds of labor inputs are distinguished: commercial work, counter work, data input work, administrative work, and managerial work. These labor inputs relate to activities (job content) instead of occupations.

Technological change has had an important impact on factor input with respect to some banking activities. In the case of accounts there were substantial labor savings on data input work (including the checking of the balances), administrative work, and counter work. Computer equipment in the back office led to savings on the processing of transfers that involve data input work, and the checking of the balances of accounts; personal computers and mainframe computers, storing information, led to savings on administrative work.

On a broader level, technical change has induced changes in commercial policy in banks. There has been a move from a product-oriented approach to an approach called *integrated client management*, where customers are served the whole range of banking products by a single *personal banker*.

The gradual adoption of new techniques, and the development of the input coefficients of a single technique over time, reflects the notion that technological change involves learning and adjustments. Benefits that are related to the new technologies are not captured immediately. Employees must be retrained and must adjust to new technology: it takes time to learn to work more productively.

A necessary extension of the model is to allow for the fact that techniques diffuse which are temporarily inefficient, at least in their initial stages of introduction. In this context the relationship between technological change and new products is also important. Another important extension would be to determine more explicitly what elements in the evolution of specific firms are favorable to the introduction of new techniques. The skill structure of the work force is an obvious candidate. Finally, the change in input coefficients over time of existing techniques should also be explained. This change can be considerable. These possible elaborations point at limitations of the model in its present form. However, the possibility to identify these limitations illustrates that a model is a very useful way to analyze causal relationships. Moreover, in spite of these deficiencies, the model fits the data well. Also one should realize that by using the model some causal relationships between variables could be firmly established.

Notes

[1] This article is based on a report entitled Technological Change, Employment, and Skill Formation in Dutch Banking, University of Limburg, Maastricht, April 1989. This report was part of the Dutch contribution to the OECD/CERI-project *Technological Change and Human Resources: the Service Sector.*

[2] The six largest Dutch banks are: ABN–AMRO [Algemene Bank Nederland (General Netherlands Bank) and Amsterdam Rotterdam Bank], RABO (Coöperatieve Raiffeisen-Boerenleenbank; Cooperative Raiffeisen-Agricultural Bank), NMB–Postbank [Nederlandsche Middenstands Bank (Netherlands Retailers Bank) and the former postal giro services and postal savings bank], CLBN (Crédit Lyonnais Bank Nederland; Lyon Credit Bank Netherlands), NCB (Nederlandse Crediet Bank; Netherlands Credit Bank) and NSBB (Nederlandse Spaar Bank Bond; Netherlands Savings Bank Association).

[3] This section is based on de Wit (1988).

[4] The theoretical background has been exposed in Diederen *et al.* (1988). It draws on ideas put forward by Kornai (1971) and by Nelson and Winter (1982) and subsequent authors. See also Iwai (1984) and Day (1987).

[5] $CF = 1 - \frac{Y_{i_{t-1}}}{\sum_{k=1}^{i} Y_{k_{t-1}}}$.

[6] $EB = \sum_{k=1}^{m} w_k(x_{i_k} - x_{j_k})$, i being the new and j being the average old technique, and summation being over all input coefficients x_k.

[7] $TD = \sqrt{\sum_{k=1}^{m} (x_{i_k} - x_{j_k})^2}$.

[8] Due to disembodied technological change and learning by doing, techniques can change to some extent over time. This process can influence their relative costs and technical distances. Therefore, we measure technical coefficients of one technique at several points in time. We expect the size of a_1 and a_2 for different products to reflect the toughness of competition in the distinct product markets. The a_1 will be slightly larger than the a_2 in expanding markets.

[9] Resulting production volumes for accounts and savings are composed of these components by weighing their importance by unit costs: for accounts, the number of payment transfers, teller transactions (deposits and withdrawals), and the payment accounts (to include the actual establishing of an account) are aggregated according to the ratio of their unit costs, i.e., 1:4:60.

[10] The input coefficients, depicted on the vertical axis, give the amount of input per unit of output, as computed, using price indices for inputs and composed production volumes for products.

[11] It is hard to draw conclusions with respect to the competitiveness of the market (reflected by the absolute value of the estimates) because the values for a_1 and a_2 cannot be compared with the values in other sectors.

References

Bank for International Settlements, 1985, *Payment Systems in Eleven Developed Countries*, Basel, Switzerland.

CBS, 1971–1987, *Annual Statistical Yearbooks*, Staatsuitgeverij, The Hague, The Netherlands.

CPB, 1971–1987, *Centraal Economisch Plan*, Staatsuitgeverij, The Hague, The Netherlands.

Day, R.H., 1987, The general theory of disequilibrium economics and of economic evolution, in D. Batten *et al.*, eds., *Economic Evolution and Structural Adjustment*, Springer-Verlag, Berlin, Heidelberg, New York.

Diederen, P.J.M., Kemp, R.P.M., Muysken, J., Palm, F.C., Bartels, C.P.A., and Webbink, A.H., 1988, *Diffusion of Technology: Effects on Employment and Occupational Structure, Part 2*, Maastricht/Oudemolen, The Netherlands.

Iwai, K., 1984, Schumpeterian dynamics: An evolutionary model of innovation and imitation, *Journal of Economic Behaviour and Organization* **5**: 159–190.

Kornai, J., 1971, *Anti-equilibrium, on Economic Systems Theory and the Tasks of Research*, North Holland Publishing Company, Amsterdam, The Netherlands.

Nelson, R.R. and S.G. Winter, 1982, *An Evolutionary Theory of Economic Change*, The Belknap Press of Harvard University Press, Cambridge, MA, USA.

Peekel, M. and J.W. Veluwenkamp, 1984, *Het girale betalingsverkeer in Nederland*, PCGD/RPS, Amsterdam, The Netherlands.

Wit, G.R. de, 1988, Automatisering in het bankwezen, *Economisch Statistische Berichten* **73**(3642):124–129.

Additional References

Day, R.H., 1984, Disequilibrium economic dynamics, a post-Schumpeterian contribution, *Journal of Economic Behaviour and Organization* **5**:57–76.

Franke, R.H., 1987, Technological revolution and productivity decline: computer introduction in the financial industry, *Technological Forecasting and Social Change* **31**:143–157.

Freeman, C. and Soete, L., 1985, *Information technology and employment, an assessment*, University of Sussex, UK.

Iwai, K., 1984, Schumpeterian dynamics, Part 2: technological progress, firm growth and economic selection, *Journal of Economic Behaviour and Organization* **5**:321–351.

NIBE, *Bankenboekje 1972–1987*, NIBE, Amsterdam, The Netherlands.

Osterman, P., 1986, The impact of computers on the employment of clerks and management, *Industrial and Labor Relations Review* **39**(2):175–186.

Petit, P., 1983, *Information technology and economic perspectives, automation of services: the case of the banking sector*, OECD, Paris, France.

Revell, J.R.S., 1983, *Banking and electronic funds transfers*, OECD, Paris, France.

Wit, G.R. de, 1987, Technologische ontwikkelingen in het Nederlandse bankwezen: de internationale context, *Tijdschrift voor Politieke Ekonomie* **10**(2):27–48.

Wit, G.R. de, 1990, Technoloigcal development at automated clearing houses in the Netherlands, Research Memorandum, University of Limburg, Maastricht, The Netherlands (forthcoming).

Chapter 22

The Chain Saw in Swedish Forestry[1]

Jonny Hjelm

22.1 Introduction

The kind of technology and the kind of machines and tools that are used in the labor process are always chosen by one or several individuals. The people who select the technology and those who use it are usually not identical, but, as we shall see, sometimes they are. The choices are never made in a social vacuum. The social structure imposes constraints on who can make the decisions as well as on how the decisions are made.

Technical change has rarely been studied from the perspective of the technology users. When, however, technical change has been considered from such a perspective, the following two questions have been asked:

- How does technical change affect the content of the working process and the demands on labor skills?[2]
- How do workers adjust to technical change?[3]

In this kind of research the employers and affiliated technical groups are looked upon as the only decision makers, the workers being ascribed a passive role.

This study describes how a group of workers actively introduced a new technique to their own labor process. The intention of the research is to analyze a technical change, where the workers' own views and actions are of central importance. Great emphasis, however, is also put on the economic and social structure in which they worked. With such an approach it is hoped to shed new light on the dynamics of technical change and, in turn, to provide a sound basis for discussions of future technical developments. For a more detailed presentation of the benefits of research from such a starting point, see Edquist and Edqvist (1980).

More specifically, this study is concerned with how the Swedish forest workers, during the 1950s gradually replaced the manual one-man crosscut saw with the motor-driven saw. The introduction of the chain saw in forestry was only the first step in the rapid and accelerating mechanization of the Swedish forestry during the postwar period. The point of departure for this study is the following: *It was the forest workers themselves who financed and bought the chain saw*. It is unusual that wage earners participate in technical change in this way and in this example, because of its marked profile, we can discern some important features of technological change that might be difficult to observe in many other cases.

22.2 Technical Development in Swedish Forestry

Swedish forestry (*Figure 22.1*) delivers the raw material, i.e., the timber, to the forest industry where it is then processed. Timber production in forests is usually divided into primary and secondary production. It is in the secondary production stage, where the trees are sawed down, that we find the forest workers.

Before mechanization the forest worker felled the trees, crosscut them into suitable pieces, removed the branches and bark, and either alone or with the help of a horse or some technical device, transported the timber to the nearest river. This work was done during the winter. The most important reason for this was that the transport of timber required frozen ground covered with snow. Afterward, usually in spring but sometimes even in summer, the timber was floated down the rivers to the forest industry on the coast.

During the cutting operation, the axe was originally the universal tool in all phases of the work. During the second part of the nineteenth century the crosscut saw made its entry successively. It replaced the axe both in felling

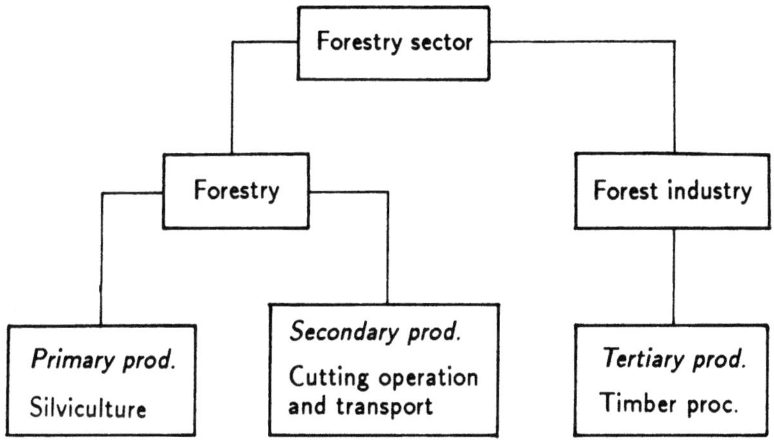

Figure 22.1. Structure of the forestry sector.

and in crosscutting operations. The crosscut saw had a handle on each end and was used by two men. During the first decades of the twentieth century, it was replaced by the one-man crosscut saw which, as the name implies, had only one handle. The same decades also witnessed the appearance of the buck saw. It was mainly used for the felling of weaker timber dimensions. A new tool was also used for removing the bark from the stock; the so-called barking spade replaced the axe here. As of this time, the forest worker used the axe only to remove branches from the trees.

The cutting operation was organized by a contractor until the 1940s, usually a farmer with a horse. He competed with other farmers to fell the timber and transport it to a drivable water course at the lowest price. A contract was drawn-up between the forest contractor and the buyer or owner of the wood to settle the financial terms, as well as the rules stipulating how the cutting operation should be done. The rules where rather detailed, and breach of contract could lead to prosecution. For example, it could be stipulated that the timber had to be delivered by a certain date and that the stumps were not allowed to be too high, which would mean loss of timber. In order to meet the stipulations of the contract, contractors employed and paid as many forest workers as were needed. Often the workers were relatives or neighbors of the contractors. Workers used their own tools; tools which they either bought for a relatively small amount of money in the village shop or made themselves. As a result, there were considerable variations in tool design and efficiency.

The contract system meant that the real employers (the wood owners and buyers; for example, the large private forestry companies and the state company Domänverket) were not directly involved in the organization of the cutting operation. As long as the tools and the methods of work did not violate the rules of the contract, the forest worker could freely choose how to work, what kind of tools to use, and also what time of day he wanted to work. The individual contractor was in charge of the cutting operation and had to ensure that it was done as stipulated.

During the 1930s and 1940s conditions in the Swedish forestry came under attack. Some people questioned whether it really was a task of the workers to decide upon tools and methods of work. Should not, they asked, forest work be analyzed in a scientific and systematic way? After such an analysis one could perhaps find more effective ways of organizing and carrying out cutting operations.

It is clear that many who discussed these views were influenced by the ideas of *the Scientific Management movement*, a movement that had its real breakthrough in Sweden in the 1920s and 1930s (see, for example, Bergren, 1981). The first concrete result of this new movement occurred quite rapidly. During the 1930s, Domänverket and the larger forestry companies in Norrland established two research institutes which immediately began a systematic study of the different logging operations, for example, the cutting operation. At the beginning, the main purpose was to provide the employers with useful information for wage negotiations. However, very soon the institutes began to study the efficiency of the tools and tried to find the best way of using them.

Machines for motorized felling and cutting that were already in existence at that time were also studied (see Staaf, 1983; Leijonhuvfud, 1953). Attempts to mechanize the heavy, and time demanding, manual felling and crosscutting operations of forest work had already been made during the last part of the nineteenth century and the first decades of this century, e.g., in the USA and the Nordic countries. There were various sporadic experiments in which people, steam engines, and combustion engines were used as energy sources. However, these attempts were not successful (Lidberg and Burénius, 1961).

The first chain saw that was of practical use was constructed during the 1920s. It was the so-called two-man chain saw. It was not only the first somewhat functional chain saw, it was also the first that both looked like and was technically similar to the chain saw which made its breakthrough in the middle of this century.

The two-man chain saw was not used in Sweden very much. During the 1940s it became clear that a functional one-man chain saw at a reasonable price was needed, i.e., a price which would make it possible for the worker to buy his own. Hitherto, and particularly in the case of the two-man chain saw, the high price had made such purchase impossible. The two-man chain saws were mainly owned by the forest companies.

In 1948, six hundred one-man chain saws were sold in Sweden. The number increased slowly but steadily until the middle of the 1950s when sales figures suddenly increased. Over the following ten years, about 25,000–30,000 chain saws were sold per year (see *Figure 22.2*). Sales advertisements promised a great deal.

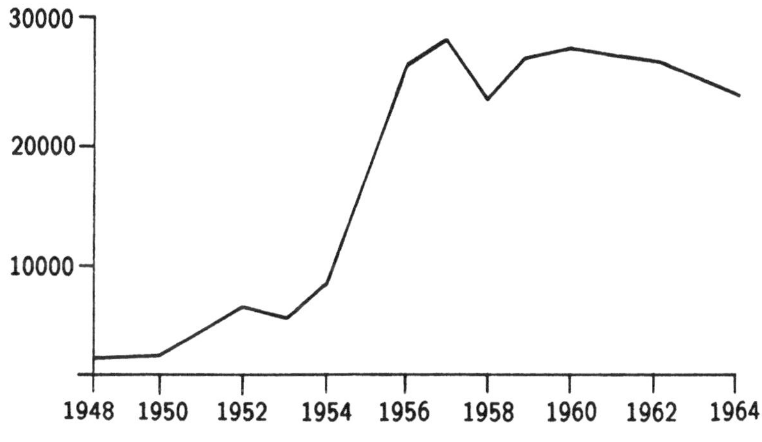

Figure 22.2. Sales of one-man chain saws in Sweden during the period 1948–1964. (Source: Helgeson *et al.*, 1979.)

An advertisement from the mid-1950s stated: "Now it is coming – the new one-man chain saw which makes the heavy work in the forest an easy game". But it was not only salesmen who described the chain saw in such lyrical words. Similar opinions were formulated both by top managers involved in large-scale forests and by groups within the Swedish Forestry Workers Union (SSAF). To summarize, the one-man chain saw was expected to reduce the physical burden of forest work and to give the forest worker a higher income; but it was also anticipated that it would lead to an increase in well-being and give the forest worker an increased social status.

Between the introduction of the first chain saws and those that where used in the middle of the 1960s, there was a period of technical improvement (from now on the term *chain saw* is used instead of the older and longer term *one-man chain saw*). The weight of the saw was reduced from about 15–20 kilograms to 7–10 kilograms. At the same time the construction was made more reliable and effective. Until 1972 the forestry workers financed and bought their own tools, including their own chain saws. But at about this time the large logging machines became common. Later, during the 1970s and 1980s, they were to supersede the forestry worker with his chain saw. These new machines are usually handled by one operator who fells, crosscuts, and piles up the timber for further transport.

Toward the end of the 1970s, about 70 percent of all the cutting was done by the logging machines. This percentage has remained the same during the 1980s. More than half of these expensive machines are today owned privately by forest workers – a fact that has been subject to a lot of attention and critical discussion.

The critics point at several aspects, but in general they argue that the current system is similar to the 1930s contractor system. They argue, for example, that the economic risks are too heavy for the individual with a long and stressful workday. Several studies have shown that the working conditions of the machine operator have caused physical and psycho-social problems (Lidén, 1989).

A similar development – the machine replacing human and animal energy sources – can be seen in other parts of the forest industry during the postwar period. The horse has been replaced by the tractor and river floating by lorry transport. Another important change is the great effort and resources that the forest industry has spent on silviculture. The development after the postwar period can be illustrated in *Figure 22.3* for productivity development.

22.3 Interviews

In traditional studies of the technical development of forestry, analyses and descriptions of the transition from the one-man crosscut saw to the chain saw are rare (Stridsberg and Mattson, 1980;[4]). Usually, it is only mentioned that the chain saw was introduced during the 1950s and that the new tool improved productivity. It is characteristic for this type of research that

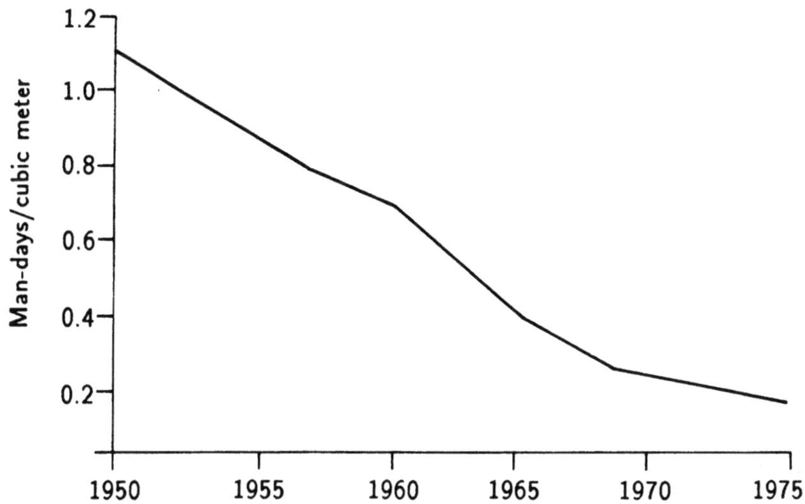

Figure 22.3. Productivity development measured in man-days/cubic meter. (Source: Andersson and Hultman, 1980.)

forestry's technical development is described with only costs and productivity in mind. Why the forest workers themselves bought a chain saw that was about twenty times more expensive than the one-man crosscut saw is a question that has not been given a satisfactory answer. When a reason is given, it is that the forest worker wanted better pay and less burdensome work.

Against this background, i.e., how the changing process has been dealt with and the problems of finding an answer in the forestry history books, it was decided that the best and only way to get a proper answer is to ask those who where part of the changing processes. Therefore fifty elderly forest workers, who had worked in forests in the county of Västerbotten in northern Sweden (*Figure 22.4*) during the 1950s when the chain saws superseded the one-man crosscut saws, were interviewed.

All of the people interviewed had worked the greater part of their active life in the forest, and they had not only experienced the introduction of the chain saw but also the rapid technical changes during the last decades. Moreover, it could be presumed that the majority had lived the greater part of their life in an environment which was influenced not only by forestry

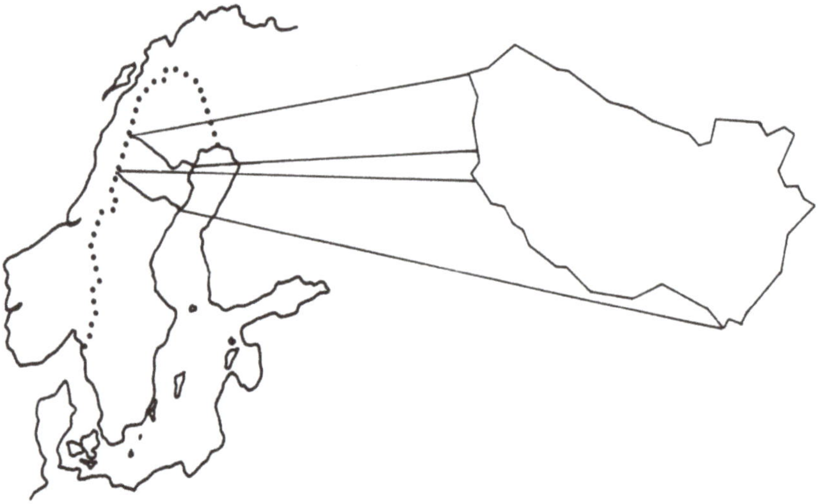

Figure 22.4. Map of Sweden and research area – Västerbotten County.

development, but also by the transition from small-scale farming to large-scale mechanized production.

The first part of the interview was spent in asking the forest workers about their life history, since this was of more interest than the tools themselves. During the first part of their lives they had one foot in a small farmer's society and the other foot among the wage earners of the growing industrial society. However, when forestry and farming were rationalized, the base for small farming was slowly pulled away from under their feet. Some stayed in the countryside and became wage earners on an annual basis and some moved to industrial work in more densely populated areas.[5] One facet of this development is the technical development of forestry; in particular, the introduction of the chain saw. This meant that forest workers themselves took part in the process which in the end undermined their traditional small farming way of living. This meant that the research was not only concerned with technical development, but that, in fact, it was also probing into the forest workers' collective history.[6]

22.3.1 The pioneers[7]

The first chain saws used by those interviewed were the two-man chain saws; they were heavy, clumsy, and not very reliable. Most of the two-man chain saws that they knew about were owned by the companies. Usually, both one- and two-man chain saws were demonstrated by the employers. Sometimes a few saws were available on loan at no or very little cost to the forest workers. However, neither those who hired nor those who bought their own one-man chain saw during the early 1950s were satisfied (nobody interviewed bought a two-man chain saw). Under normal circumstances the forest workers still preferred to use the one-man crosscut saw. But in felling and crosscutting, especially in very thick and hard-to-saw forests, the chain saws might be used. For this type of forest work the chain saws could both lessen the physical strain and improve the income.

The forest workers who bought or hired these chain saws were the *pioneers*. They had some, but very little knowledge of the chain saw before they used it for the first time. Until the forest workers could see with their own eyes the chain saw being used, they were not sure what type of machine it was. The fact that a motor replaced muscle power was impressive, and it fostered great hopes. It was through the pioneers that the most important information and knowledge about the chain saw was spread at this stage. The predominately negative experiences among the pioneers during the first part of the 1950s made most forest workers uninterested or somewhat skeptical about the chain saw, believing that it could never be used as an effective tool.

22.3.2 The breakthrough

In the middle of the 1950s, the reputation of the chain saws slowly improved and more forest workers began to think about buying them. One reason for this change was persuasive advertising, but even more important was the fact that the chain saws had been significantly improved. These improvements changed the forest workers negative opinion into a more positive one. Those interviewed who, during the breakthrough, bought and began to use the chain saw, were clearly more pleased with their new tool than the pioneers had been. Under certain working conditions, (e.g., felling where snow is not deep and with a majority of thick older timber free from knots) a chain saw was considerably superior.

However, not all forest workers had such positive experiences. There were many who were disappointed with the chain saw since their high expectations were not fulfilled. Their experiences differed mainly because of (1) the labor process, which involved heavy physical work and also required continuously changing methods, and (2) the piecework wage system.

The cutting operations were physically trying, and in addition, required a great deal of planning. It was important to be able to adopt quickly to the changing working conditions. The extent of the difficulties and how the corresponding problems were solved, varied among the forest workers. How well they managed to handle their problems was reflected in the size of the wage packet.

One of the main reasons behind the piecework wage system was that it would compensate for the differences between cutting areas. Every cutting area was evaluated by means of several criteria, e.g., the number of knots in a tree or the length and form of the tree. The cutting areas were divided into five *classified groups.*[8] How successful the forest workers were in their local wage negotiating procedures depended on many factors. In general, the majority of the interviewees were of the opinion that during the second half of the 1940s as well as in most of the 1950s they had a strong negotiating position because of the scarcity of workers.

The two main reasons for buying a chain saw were that the forest workers hoped to lessen the physical work load and to improve their wages. Their hopes were fulfilled to a limited extent. The chain saw first replaced the one-man crosscut saw for such work as felling and crosscutting; two tasks that had been physically hard. But as this work was about only half of the total work, and as the chain saw was heavy to carry around in the cutting area, the workers were just as tired as before after a day's work in the forest. The other important feature was the piecework wage system. Under this type of wage system every possible work- and time-saving device or method was used to increase output. In other words, the time saved by the chain saw was not used for rest, but to increase the individual forest workers production volume.

Wage expectations were fulfilled to a greater extent by those who had bought a well-functioning chain saw and had the opportunity to use it in the type of forest where motorized felling was effective. Most respondents, however, said that their wages did not change in any specific direction after they had begun to use the chain saw. But such a comparison is difficult to make because of the piecework nature of wages since it has to relate to a number of factors which vary from work place to work place. The consequence was that the forest workers had to evaluate the purchase of a chain saw with rather extreme measures. The purchase was from an economic point of view either *good*, *bad* or *so-so.* It was almost impossible to be more exact. These difficulties in evaluating the economic consequences provided hope for improvement.

With the introduction of the chain saw the psychological work load increased. The main reason was the uncertainty about the reliability of the chain saw. For example, would it start in the morning without a problem and function the whole working day? If anything went wrong, which was not unusual, the irritation was great; every involuntary break in work put the forest worker under pressure. These problems decreased during the second part of the 1950s when the forest worker's own knowledge increased and when the retailers of the chain saws, as well as smaller motor repair shops, started to give quick and effective service.

22.4 Forestry and Farming

In this particular case, because the forest workers used and cared for their own tools, their attitude toward using more effective tools and methods of work was taken into consideration. Many of those in favor of a rationalization doubted that forest workers were positive to such changes. Sometimes it was explicitly stated, but usually it was only insinuated that an alarming number of workers were traditionalists and preferred to work in the same way and with the same kinds of tools that their forefathers had used (see, for example, Hultmark, 1946 and 1957).

However, the forest workers' background in small farming did not, as the rationalization spokesmen feared, slow up or hinder the technical development of forestry. Instead, this background (together with the wage system) made them take an active part in the transformation. At the same time, it lay the foundation for their disappointment with mechanization in the long term. The manual forest work did not become physically less burdensome or a more well-paid occupation.

The close relationship between forestry and farming is easily seen in one particular area: *the ownership of tools*. It was natural that every farmer owned the tools necessary for forest work. Furthermore, the tools for forest work were also used for providing the household with wood for heating. When the more expensive chain saws appeared on the market, and became a realistic alternative to the one-man crosscut saw, all interviewees took it for granted that they themselves had to bear the costs for the chain saw. The idea that their employers would do this was alien to them. Three people who were actually given their first chain saw thought it was a mistake or a deed of goodwill on the part of their employers. It was not expected to happen again.

The forest worker received information about the chain saw in more or less the same way as information about farming tools was received. Neighbors, relatives, and work-mates were the most important information channels. The decision process was carefully thought-out and made by the individual forest worker himself. Every forest worker tried, as far as possible, to make a rational judgment and take as many factors as possible into account.

22.5 Individual Actions and Their Consequences

The basic attitude among the forest workers to the purchase of a chain saw, and to its use in the cutting operation, was not to any appreciable extent influenced by traditionalism or disinterest in innovations. From the very beginning, attitudes were positive but cautiously pragmatic. Of course, the workers themselves were satisfied if they could find a tool which lessened the extent of manual labor while improving wages. The forest workers purchased the chain saw as soon as they realized that it was a more functional tool than the one-man crosscut saw.

Every forest worker who bought a chain saw used all the accessible information channels. After having thoroughly considered the different alternatives, a decision was taken. It is quite clear that the forest worker conformed to small farmer behavior. The interviews testify, for example, that the workers never considered using their trade union to obtain information about the chain saw. Also, it was never questioned what standpoint they, as a collective, should have toward the new tool. Relations with the Swedish Forestry Workers Union (SSAF) were clear-cut. The union's task was to negotiate the best possible wage agreement, not to influence production technology. Neither the union meetings nor the union magazine was used as a forum to exchange information about the chain saw.

Later on, at the beginning of the 1960s, when almost all forest workers had obtained a chain saw, unemployment slowly started to increase. Since the 1940s, unemployment had been almost nonexistent. At the same time the piecework wage system slowly began to change. According to those interviewed, these changes often meant a decrease in wages per unit produced. The benefit of increased production, with the more effective chain saws, could not and would not go solely to the forest workers. Their employers and the forest owners were also to have a share. The forest workers responded by increasing their work tempo. In this way salaries remained at a high level for the most efficient workers. For others, the older and those who were less productive, salaries decreased. The increase in work intensity implied an increase in accidents and injuries which meant that the level of job satisfaction deteriorated[9] (see also Nohrstedt, 1975).

The forest workers slowly began to realize that their individual space of action was limited. A proper decision and a rational choice may have completely different consequences when the whole collective of forest workers takes the same course of action. Philosophers have labelled such mechanisms as *counterfinality*. A rational action with a definite goal produces the opposite outcome (see, for example, Elster, 1978).

The lesson the forest workers learned, which guided their actions when new machines were later introduced into forestry, was the importance of acting as a collective. Equally important, however, was the insight that good working conditions and positive wage developments are dependent not only upon technology, but upon the social order in which the technology is embedded. The wage system increased the tempo and could turn all technical improvements into something that simply increased the rate of accidents. This was the reason why all fifty interviewees were critical of what had happened within forestry during the last decades. The time-rate system, which the forest workers fought for during the 1960s and were still striking for in 1975, was established between 1975 and 1980. However, this has by now been slowly undermined by the decline in wage levels and by the introduction of the so-called *bonus system*, a type of group piecework wage system. The machine operators have a specific entrepreneurial system which, like the piecework wage system, makes them work at the highest possible rate with as long a workday as possible, often well into the night.

The technical development in forestry has not only changed the labor process and the working conditions; the repercussions in the social milieu have also been significant. When forest workers changed from being seasonally employed to being employed on a yearly basis, the already ongoing closure of farms and rural depopulation increased. The restructuring of both the farming and the forestry industries, and the accompanying decrease in the demand for workers changed the social milieu. The countryside of Västerbotten, which in the 1950s was full of life, is today very sparsely populated. Deserted houses and abandoned farms and schools meet those who travel through the countryside. That this state of affairs would be the result of the entry of chain saws and tractors into forestry, and of the corresponding technical development in farming, was something that very few forestry workers realized during the 1950s.

22.6 Conclusion

Studies of technical change within capitalist production are seldom carried out from the workers' perspective. Usually, the worker is regarded as the passive receiver of a new technique. An attempt has been made here to show that this outlook is too limited. Workers can participate in technical developments in a variety of ways, even though the case of forestry may be a very special case. The forest workers both introduced and used a new technology. This summarizes the first aim of this chapter.

The second purpose was to underline the importance of analyzing the economic and social structure within which the actors work. The introduction of the chain saw can only be understood if the forest workers' small farming milieu is taken into account, an environment which both formed and constrained the actors. While they were actively engaging in the technical development of forestry, they were undermining their position as small farmers. The small farmer who seasonally worked in the forest was transformed into a full-time employed forest worker. Those who participated in this process gradually abandoned their individualist attitude to new technology and replaced it with a more collective outlook on forestry. They also developed a more complex view of the effects of technology on society, and they became fully aware of the complex relationship that existed between the individual and collective rationality in technical change.

Notes

[1] This chapter is a summary of a book that will be published in 1991 by Arkiv, Lund, Sweden. The preliminary title is *Skogsarbetaren och motorsågen. En studie i teknik- och arbetsprocessförändring* and will contain a complete list of references.

[2] Perhaps the most well-known researcher is Harry Braverman, *Labor and Monopoly Capital*, Monthly Review Press, New York, 1974. See also discussions of the so-called Braverman-debate in, for example, Paul Thompson, 1983, *The Nature of Work*, Macmillan Publishers Ltd., London, UK.

[3] This was the main question for many researchers in the industrialized countries during the early postwar decades. For a more exact account of this research in Sweden, see Chapter 2 in Torsten Björkman and Karin Lundkvist, 1982, *From Max to Pia*, Arkiv, Lund, Sweden (in Swedish).

[4] For a more historical aspect of the technical development in forestry, see, for example, Sven Embertsén, 1975, *SCA Secondary Production in the Area of Kramfors 1911–1965*, Lund, Sweden (in Swedish).

[5] This is a transformation which Dan Bäcklund has analyzed in *On the Outskirts of Industrial Society*, Umeå, Sweden, 1988 (in Swedish).
[6] For a discussion about collective biography – prosopography – see Lawrence Stone, 1981, Chapter 2 in *The Past and the Present*, Cambridge University Press, London, UK. An earlier representative of this way of writing of labor history was Edvard Bull, 1958, *Arbeidermiljö under det industrielle gjenombrudd*, University Press, Oslo, Norway.
[7] Where nothing else is indicated the sections on *The Pioneers* and *The Breakthrough* are built on the interviews.
[8] The basis for the piecework wage system was presented in 1945 at a Conference arranged by SDA. See Lectures at an SDA – an organization for work and time studies – conference on forestry rationalization issues), *SDA Meddelande*, Nr. 21–29, 1945 (in Swedish).
[9] This was something that many interviewees emphasized.

References

Andersson, B. and Hultman, S.-G., 1980, *The Value of the Forest and the Role of Forest Industry*, Swedish Academy of Agriculture, Uppsala, Sweden (in Swedish).

Bergren, C., 1981, Did Taylorism ever generate Swedish industry? *Arkiv för studier i arbetarrörelsens historia*, No. 19–20:31–50, Lund, Sweden (in Swedish).

Edquist, C. and Edqvist, O., 1980, *Social Carriers of Technology*, Zenith, Lund, Sweden (in Swedish).

Elster, J., 1978, *Logic and Society: Contradictions and Possible Worlds*, John Wiley & Sons Inc., New York, NY, USA.

Helgeson, B., Johansson, J., and Lindberg, U., 1979, Mendner in the Forest, *Nordiskt Forum* **23** (in Swedish).

Hjelm, J., 1991, *The forest worker and the chain saw. A study in a technical and labor process change*, Arkiv, Lund, Sweden (in Swedish).

Hultmark, N., 1946, Tools in forestry models and diffusion in Northern Sweden, in *Norrlands Skogsvårdsförbunds Tidskrift*, pp. 36–71 (in Swedish).

Hultmark, N., 1952, Labor problems in forestry, in *Svenska Skogsvårdsföreningens Tidskrift*, Årgång **50**(2):137–167 (in Swedish).

Leijonhuvfud, A.C., 1953, Skogsarbetarnas yrkesutbildning, *Svenska Skogsvårdsföreningens Tidskrift*, Årgång 51:97–112.

Lidberg, B. and Burénius, Å., 1961, *Chain Sawing*, Research Foundation SDA, Stockholm, Sweden (in Swedish).

Lidén, E., 1989, *The Forest Machine Owner's Work Situation*, No. 178 from a series of reports from Institutionen för skogsteknik, Sveriges Lantbruksuniversitet (in Swedish).

Nohrstedt, S.-A., 1975, The background to the lumberjacks' strike 1975, *Arkiv för studier i arbetarrörelsens historia* **11–12**:3–39 (in Swedish).

Staaf, A., 1983, *A History of Forestry Technology*, 251/1983, Sveriges Lantbruks-universitet, Institutionen för skogsteknik, Skogshögskolan Garpenberg, mimeo (in Swedish).

Stridsberg, E. and Mattson, L., 1980, *A History of the Forests*, LT Publishing House, Stockholm, Sweden (in Swedish).

Chapter 23

The Spatial Spread of the Aids Epidemic in Ohio: Data Analyses via Expanded Regressions

Emilio Casetti and C. Cindy Fan

23.1 Introduction

The spatial realities of the spread of AIDS (Acquired Immune Deficiency Syndrome) have not attracted the attention commensurate with their importance. A few spatial geographical studies of the epidemic have begun to appear (Dutt *et al.*, 1987; Jager and Ruitenberg, 1988; Wood, 1988; Gardner *et al.*, 1989; Gould, 1989; Shannon and Pyle, 1989). However, most studies concerning the *spread* of this epidemic focus upon homosexuals and intravenous drug users, and are often concerned with determining to what extent the members of these groups modified their patterns of behavior out of concern for AIDS (Curran *et al.*, 1985 and 1988; Gong and Ruknick, 1986; Kulstad, 1986). These studies clearly imply that the spatial spread of AIDS will differ across geographical spaces and environments, but these implications are not translated into systematic investigations of which patterns occur where and when; of which *types* of spatial spread of AIDS exist; of which relations exist between such types of dynamics and the spatial geographical environments to which they are associated.

It is useful to focus on the theoretical and practical significance of investigations concerned with the spatial dimensions of the AIDS epidemic. These investigations have an important role within the context of the war against AIDS, are critically useful to planners and policymakers, and are important components of a body of knowledge for which a critical need exists.

Education and prevention are the strongest weapons against AIDS available at present (Johnson and Adler, 1987). AIDS education and the *danger awareness* it produces can bring about behavior modifications that will slow down the diffusion of AIDS within the groups mostly at risk today, and into the segments of the heterosexual population in which a multiplicity of sexual partners is more prevalent (teenagers and singles).

It is well known that efforts to educate people about AIDS and to induce requisite behavior modifications encounter a resistance rooted in values and mores that is very difficult to overcome. Only when opinion leaders and the public at large have a perception of impending danger are these resistances more easily overcome, and effective, widely publicized, educational efforts become possible. Understanding the realities and scope of the spatial spread of the AIDS epidemic can induce a perception of impending danger in communities that AIDS has not yet touched to any significant degree. The awareness of spatial spread can do much to convince people and communities that AIDS is not somebody else's problem and to realize that it is very much their problem also. However, consciousness-raising actions about the spatial spread of the AIDS epidemic will have the greatest impact when they are based on solid scientific research.

Community leaders, planners, and policymakers have a critical need for knowledge concerning the spatial spread of AIDS. They are likely to be confronted by spatially and geographically disaggregated realities and need spatially disaggregated forecasts. They need to know which types of spatial spread of AIDS are likely to unfold in the regions for which they are responsible. They require this knowledge to anticipate needs, to plan, to target scarce resources to where they can do the most good, and to request from appropriate governmental agencies a support commensurate to forecasted needs. A knowledge base that does not address the spatially disaggregated dimensions of the spread of the AIDS epidemic does not provide the type of information required to allocate resources, to plan, and to inform.

The empirical analysis of the spatial spread of AIDS constitutes an important component of the body of knowledge concerning this epidemic. Let us place the pertinent issues into perspective. The compartment models using differential equations or stochastic processes formalizing the transitions from "susceptibles" to "infectives" to "removals" (Bailey, 1957 and 1975; Bartlett, 1960) have a critical role in the modeling of epidemics including AIDS. The parameters in these models define modalities of any epidemics, and their estimation is very important to understanding the specifics of the AIDS epidemic and to design policies to cope with it. However, the estimation of the parameters of the spatial versions of these models has proved especially difficult because very different mixes of parameter values can account for the same data. Certainly, the spatial variants of the classical epidemiological models constitute one direction of research that is very much worth pursuing, but this is not likely to produce in the short-run spatially disaggregated portraits and projections of the spread of AIDS (Gould, 1989).

The patterns of spatial spread of the AIDS epidemic constitute the realities that the classical epidemiological models and their spatial extensions are designed to explain. In fact, the contributions that identify these patterns and test hypotheses concerning their occurrence and change define the realities that the classical epidemiology models need to explain. The complementarity between the two classes of models is obvious, and is the counterpart of similar complementarities in other research fields. For instance, in economics, structural models based on economic theory are employed side by side with forecasting models that emphasize the temporal patterns and well-established empirical relations between variables, as in the case, for instance, of the *leading economic indicators.*

23.2 Research Approach

The research approach employed in the analyses presented in this chapter involves adding spatial dimension(s) to models formalizing the temporal dynamics of the AIDS epidemic, or adding a temporal dimension to models of the spatial distribution of AIDS cases.

The analyses implement the expansion method techniques and paradigm (Casetti, 1972, 1982, and 1986), and are aimed at investigating the drift of models relating the number of AIDS cases to time with respect to spatial geographical variables, and the temporal drift of models relating the number of AIDS cases to spatial geographical variables. From the exposition that follows it will be apparent that these twin investigations correspond to the primal and dual formulations of the same *expansion* formalisms.

The expansion methodology is a technique and a research philosophy or *paradigm*. As a technique, it can be used for the orderly introduction of complexities into simpler mathematical formulations, for modeling parametric variation, and for investigating whether substantively meaningful models *drift* across substantively meaningful *contexts*. As a paradigm, the expansion method suggests asking questions concerning the empirical occurrence and the theoretical bases of parameter drift, while at the same time providing orderly and easily implementable routines to answer these questions.

The expansion method involves:

- Specifying an *initial model* in which some or all of the parameters are in letter form.
- Specifying *expansion equations* that redefine some or all of these letter parameters into functions of substantively relevant variables or random variables.
- Generating a *terminal model* by substitution of the expanded parameters for their counterparts in the initial model.

The terminal model encompasses into the same structure the initial relationship, and a specification of its potential drift across a substantively relevant *context*. When the terminal model is estimated and tests of statistical hypotheses are carried out upon its coefficients, conclusions concerning the occurrence of contextual drift may be reached, and mappable or graphable mathematical portraits of this drift can be obtained. Reviews of applications of the expansion method are contained in Casetti (1986) and Casetti and Jones (1987).

In the paragraphs that follow, the research approach employed in this chapter is demonstrated focusing upon the variation (*drift*) in the temporal growth of AIDS in response to population densities, which is a convenient descriptor of *geographical environmental contextual* differentiation. In fact, population density constitutes an especially useful proxy for many indices of spatial differentiation at the county level of resolution. Low population densities are generally an attribute of rural environments. However, rural population densities tend to be higher with proximity to urban centers, the more so the larger and the more closely spaced these centers are. Furthermore, theoretical and empirical research in the mathematical land-use tradition has shown that urban population densities tend to decline with distance from city centers, and tend to be higher in larger urban centers and agglomerations. This suggests that population density can position a county on a continuum with remote rural environments at one end, and the core portions of large urban agglomerations at the other. Here, we propose to model and investigate the *drift* that occurs when a polynomial relating AIDS counts to time holds with different parameter values depending upon the population density of the county considered.

The implementation of the expansion methodology begins with the definition of an initial model capable of expressing the change in the number of AIDS cases over time. The following initial model was selected

$$A = \alpha_0 + \alpha_1 t + \alpha_2 t^2 + \alpha_3 t^3 + \epsilon, \tag{23.1}$$

where A stands for the natural logarithm of the cumulated number of AIDS cases, t denotes time, and ϵ is an error term. A was specified as a third-degree polynomial in time in order to allow for the possibility that the growth rates of AIDS increase at first, and then decline.

We wish to investigate the spatial geographical unfolding of the AIDS epidemic in Ohio. This research objective can be operationalized as a search for the parametric drift of equation (23.1) with respect to variable(s) specifying significant dimensions of geographical differentiation at the county level of resolution. Specifically, the coefficients of equation (23.1) were expanded into quadratics in D, where D stands for population density. A quadratic was selected to allow for rates of growth of AIDS higher in urban environments, and possibly comparatively lower in larger metropolitan agglomerations in which the awareness and fear of AIDS has induced substantial changes in behavior.

The expansion equations used are:

$$\alpha_0 = \gamma_{00} + \gamma_{01} D + \gamma_{02} D^2 \tag{23.2}$$

$$\alpha_1 = \gamma_{10} + \gamma_{11} D + \gamma_{12} D^2 \tag{23.3}$$

$$\alpha_2 = \gamma_{20} + \gamma_{21} D + \gamma_{22} D^2 \tag{23.4}$$

$$\alpha_3 = \gamma_{30} + \gamma_{31} D + \gamma_{32} D^2. \tag{23.5}$$

By replacing the right-hand sides of equations (23.2) through (23.5) with the corresponding parameters in equation (23.1), the following terminal model is obtained

$$\begin{aligned} A = {} & \gamma_{00} + \gamma_{01} D + \gamma_{02} D^2 + \\ & \gamma_{10} t + \gamma_{11} t D + \gamma_{12} t D^2 + \\ & \gamma_{20} t^2 + \gamma_{21} t^2 D + \gamma_{22} t^2 D^2 + \\ & \gamma_{30} t^3 + \gamma_{31} t^3 D + \gamma_{32} t^3 D^2 + \epsilon. \end{aligned} \tag{23.6}$$

Equation (23.6) incorporates both the temporal trend specification contained in the initial model and the spatial drift specification represented by the expansion equations. Hypotheses concerning the occurrence of geographical drift can be easily tested using (23.6). To this effect the parameters of equation (23.6) are estimated from empirical data, and then tested for significance. For instance, we can test the null hypothesis that the population parameters of the terms in which the drift variable D appears are zero. If the hypothesis is rejected, the occurrence of drift may be taken as proven, and the estimated terminal model and related estimated expansion equations provide a statistically significant mathematical portrait of how, in Ohio, the unfolding of the AIDS epidemic varies depending upon the population densities of the counties involved. However, before describing the results of our empirical analyses, the *dual* of the mathematical formulation represented by equations (23.1) through (23.6) needs to be focused upon.

Whenever a terminal model has been generated from a linear initial model by linear expansion equations, there is a second linear initial model and associated linear expansion equation(s) that will yield the same terminal model (Casetti, 1986). If a terminal model is given, as soon as a linear initial model and linear expansion equations capable of producing it are defined, a second linear initial model and associated linear expansion equations that will produce the same terminal model become defined. The two initial models and associated expansion equations may be called, respectively, *primal* and *dual.* Which expansion is primal is arbitrary, but, when an expansion is defined as primal, the second one becomes the dual of the first one.

The intrinsic duality of the linear expansions is illustrated by the fact that the same terminal model (23.6) can be arrived at from an initial model relating the logarithm of AIDS counts, A, to population densities, D, and by expanding its coefficients in terms of the t variable. To show this, assume the dual initial model

$$A = \beta_0 + \beta_1 D + \beta_2 D^2 + \epsilon \tag{23.7}$$

and the dual expansion equations

$$\beta_0 = \gamma_{00} + \gamma_{10}t + \gamma_{20}t^2 + \gamma_{30}t^3 \tag{23.8}$$

$$\beta_1 = \gamma_{01} + \gamma_{11}t + \gamma_{21}t^2 + \gamma_{31}t^3 \tag{23.9}$$

$$\beta_2 = \gamma_{02} + \gamma_{12}t + \gamma_{22}t^2 + \gamma_{32}t^3. \tag{23.10}$$

By replacing the right hand sides of (23.8) through (23.10) for the corresponding parameters in (23.7) terminal model (23.6) is again obtained. Suppose we estimate the parameters of the terminal model (23.6) from empirical data, and replace the numerical parameter estimates obtained for the corresponding letter parameters in the primal and dual expansion equations. The "estimated expansion equations" thus produced are mathematical portraits of the empirically observed drift of the relationship between AIDS counts and time represented by the primal initial model, and of the empirically observed temporal drift of the dual initial model that expresses AIDS counts as a function of population densities. Also, the estimated primal expansion equations will yield an estimated relationship between AIDS counts and time at any arbitrary level of population density. Similarly, the estimated dual expansion equations can produce an estimated relationship between AIDS counts and population densities at any arbitrary point in time.

Equations (23.1) through (23.10) relate the logarithm of AIDS counts to t and D. However, in this research, the focus is upon the *dynamics* of the AIDS epidemics, which is best expressed by the growth rates of the AIDS cases. Consequently, while equation (23.6) was estimated using empirical data, the interpretation of the results obtained was based on a *derived* formulation obtained by taking the time derivatives of equations (23.1), (23.6), (23.8), (23.9), and (23.10). The population densities, D, were assumed to be invariant over the time horizon considered.

23.3 Population Density Expansions

The empirical analyses were based on quarterly county-level AIDS counts for the 88 Ohio counties for the time horizon spanning from 1982 to the fourth quarter of 1987. The population densities were calculated using 1985 populations and areas. The D variable is scaled in thousands of people per square mile. The following stepwise regression estimate of equation (23.6) was obtained:

$$\begin{aligned} A = & -0.2486 + 0.0315tD^2 + 0.0107t^2 + 0.0766t^2D - 0.0028t^3D^2 \\ & \quad (-2.32) \quad (2.32) \quad (3.57) \quad (15.34) \quad (-7.94) \qquad (23.11) \\ R^2 = & \ 0.80. \end{aligned}$$

t-values are in parentheses under the regression coefficients. Equation (23.11) was used to parametrize the *derived* system, thus producing the following:

$$A' = a_1 + a_2(2t) + a_3(3t^2) \qquad (23.12)$$

$$a_1 = 0.0315D^2 \quad (23.13)$$

$$a_2 = 0.0107 + 0.0766D \quad (23.14)$$

$$a_3 = -0.0028D^2 \quad (23.15)$$

$$A' = b_0 + b_1D + b_2D^2 \quad (23.16)$$

$$b_0 = 0.0107(2t) \quad (23.17)$$

$$b_1 = 0.0766(2t) \quad (23.18)$$

$$b_2 = 0.0315 - 0.0028(3t^2), \quad (23.19)$$

where A' denotes percentage rate of change of AIDS counts.

Equations (23.13), (23.14), and (23.15) are the "estimated expansion equations" of the primal derived system. They portray the drift of the relationship between percentage rate of growth of the AIDS counts and time. This means that a time trend in growth rates can be specified for any arbitrary value of D. Plots of these trends corresponding to values of D of 0.1, 0.4, and 1.6 are shown in *Figure 23.1*. These A, B, and C trends in the Figure correspond, respectively, to population densities of 100, 400, and 1,600 people per square mile. The D variable is dimensioned in thousands of people per square mile. To evaluate these densities, consider that the D values for the Ohio counties range from 0.028 to 3.167 with an average of 0.271.

Clearly, at the low densities represented by the A trend, the growth of the AIDS counts increased throughout the time span considered, but moderately so. At intermediate densities (the B curve) it did increase more. At the higher densities represented by the C curve it increased even more but in 1986 it started to decline. No evidence of such decline is apparent for the lower densities. The percentage growth rates in AIDS counts in the vertical axis of the figure are best interpreted in terms of "doubling times". Consider that A' values of 0.2; 0.4; 0.6; 0.8; 1.0; and 1.2 percent will bring about a doubling of a county's AIDS cases, respectively, in 3.4; 2.31; 1.15; 0.85; 0.69; and 0.58 years.

The estimated expansion equations of the dual derived system, namely, equations (23.17) through (23.19), are mathematical portraits of the temporal drift of the relationship (23.16) between AIDS growth and population densities. *Figure 23.2* shows these relationships for the years 1983 (A curve), 1985 (B curve), and 1987 (C curve). Notice the increasingly strong tendency

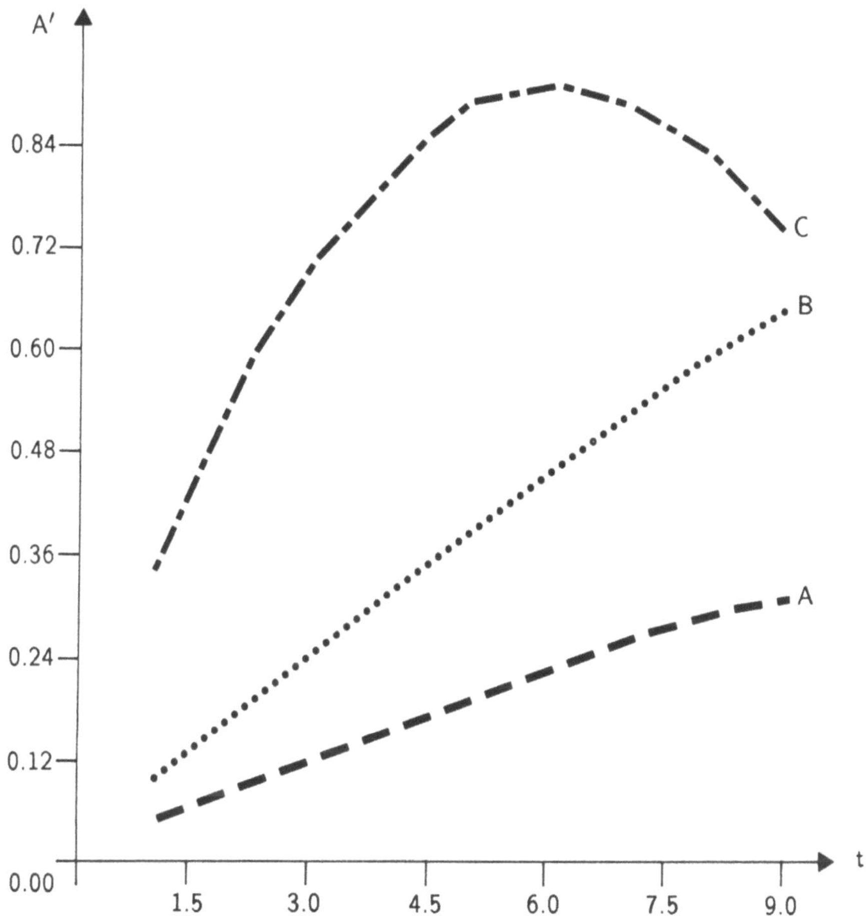

Figure 23.1. Estimated relations between percentage rates of growth of AIDS counts (A', vertical axis) and time (t, horizontal axis; $t = 0$ in 1980). The 'A', 'B', and 'C' curves are estimated trends, respectively, corresponding to levels of population densities of 100, 400, and 1600 people per square mile.

for the AIDS growth to become lower at higher population densities in more recent years, while no such tendency existed in 1983.

Figures 23.1 and *23.2* point toward the notion that different mixes of the mechanisms propagating AIDS are at work in different geographical environments. The AIDS scare of more recent years had the effect of differentially slowing down the diffusion of AIDS among homosexuals and to a lesser degree among drug users, which is reflected by the AIDS dynamics in higher

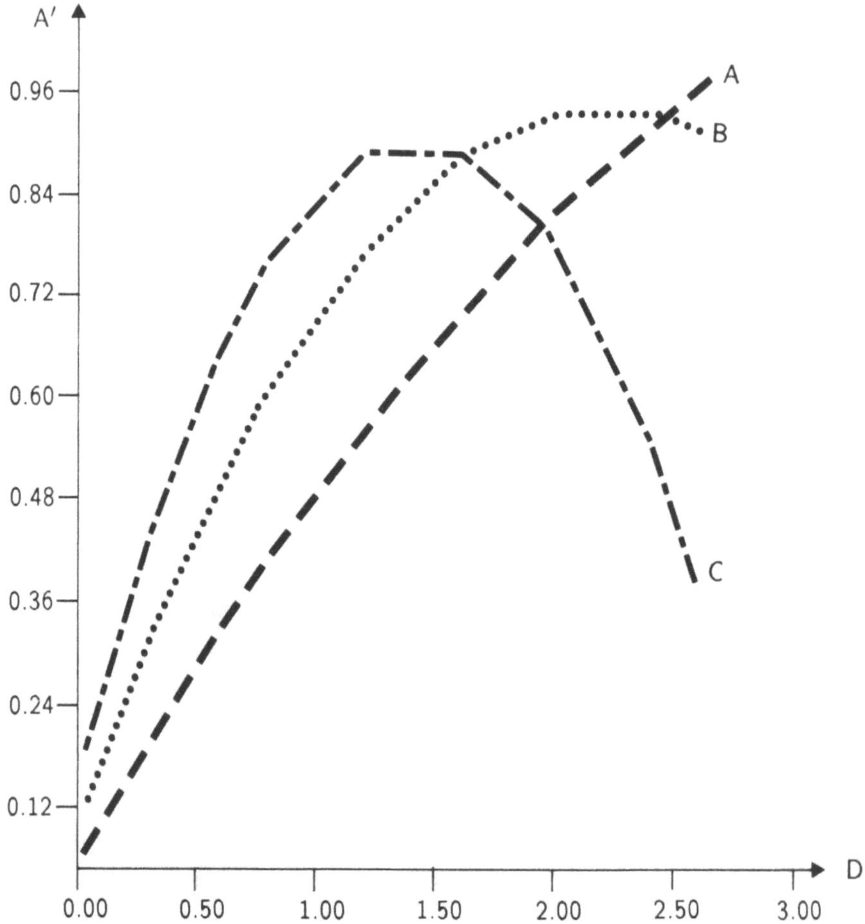

Figure 23.2. Estimated relations between percentage rates of growth of AIDS counts (A', vertical axis) and population densities (D, horizontal axis, in thousands of people per square mile). The 'A', 'B', and 'C' curves are estimated relationships evaluated respectively for 1983, 1985, and 1987.

population densities counties. At the same time other mechanisms more diffuse, more broadly based, and less controllable, which are operating in the background, render the AIDS epidemic ubiquitous and pervasive. Perhaps, these other mechanisms tend to have a greater relative importance in counties with lower population densities, and, in perspective, they may be having an increasingly important role over time.

23.4 Spatial Polynomial Expansions

The analyses discussed in the preceding section involve investigating the temporal dynamics of the AIDS epidemic in terms of its *drift* in response to differences in geographic environments. In the example, the *initial model* is a polynomial in time, and the *expansion equations* relate deterministically the polynomial's parameters to population density, which is a good proxy for a number of variables differentiating geographic environments. Portraits of the spatial drift of temporal dynamics of AIDS, and of the temporal drift of relationships between AIDS and environmental characteristics, constitute the results of an analysis implementing this approach.

The second type of expansions reported in this chapter is in terms of *spatial polynomials*. The spatial polynomials considered here are trend surfaces and two-dimensional Fourier polynomials, which are widely used in the spatial disciplines such as geography, geology, and regional science for the mathematical analysis of spatial variation. They both involve relating geographically distributed phenomena to polynomials in the spatial coordinates of the observations.

Of the two types of spatial polynomials, the trend surfaces (Agterberg, 1984; Chorley and Haggett, 1965; Tobler, 1969; Yeates, 1969) have been used more often and have proved to be very useful to separate gentler trends at the regional level of resolution. On the other hand, the two-dimensional Fourier polynomials (Casetti, 1966; James, 1966; Rayner, 1971; Tobler, 1969) are better suited to depict spatial patterns characterized by multiple maxima and minima. In particular, they can better represent geographical landscapes with urban centers of varying size arranged into the patterns suggested by central place theory.

In this study, the parameters of the same initial model (23.6) employed in our earlier analysis were redefined into $f(x, y)$ functions of the spatial coordinates x and y. These functions were specified as trend surfaces in the first analysis, and as two-dimensional Fourier polynomials in the second.

The trend surface expansions and the Fourier polynomial expansions extend the spatial polynomial analyses to the expansion method environment. While trend surfaces and Fourier polynomials model and portray the spatial variation of a variable, trend surface expansions and Fourier polynomial expansions are aimed at investigating the spatial variation of a model. Here, they are used to analyze the spatial variation of the dynamics of AIDS counts specified as a polynomial in time. The terminal model obtained by trend surface expansions is best suited to investigate broad *regional* trends in the spatial variation of the AIDS dynamic. Instead, its counterpart obtained by Fourier polynomial expansions will filter out and display the spatial variation of the AIDS dynamics at a finer level of resolution. In fact, the results of the empirical analyses reported later in this chapter suggest that the two expansions reveal two distinct and complementary aspects of the spatial temporal manifestation of the AIDS epidemic.

As noted earlier in this chapter, terminal models can be generated by primal or dual initial models and expansion equations. With regard to the trend surface expansions and the Fourier polynomial expansions, the primal formulations involve expanding the parameters of an initial model, relating AIDS counts to time, into functions of the spatial coordinates x and y. However, the same terminal models can be obtained by expanding dual initial models relating AIDS counts to their spatial coordinates, into polynomials in time.

Upon estimation, the terminal models generated by spatial polynomial expansions can be used to portray the temporal trend in AIDS counts at any given point in the study area (primal formulation), or to portray the spatial distribution of the AIDS epidemic at any given point in time (dual formulation). A few simple manipulations can extract from these, temporal trends or spatial distributions in the growth rates of AIDS. The results from our analyses are presented in the form of maps of the spatial distribution of AIDS growth, which correspond to realizations of the dual initial models at selected values of the dual expansion variable, that is time.

The regressions on which the maps are based yielded an R^2 of 0.25 with 6 variables in the case of the trend surface expansions and an R^2 of 0.72 with 20 variables in the Fourier polynomial expansions.

Figure 23.3 shows the county outlines and the location of the major urban centers and interstate highways in Ohio. All the shaded polygons refer to counties included in Metropolitan Statistical Areas (MSAs) with boundaries as defined on 30 June 1983 (US Bureau of Census, 1988). The darkest shading identifies counties that contain one (or more) central cities of the MSAs in the state. They are usually referred to as "central city counties". For example, Cuyahoga county in the northeast contains Cleveland, which is the central city of the Cleveland MSA. The names shown in *Figure 23.3* are names of the central cities rather than of counties, since the former are more directly linked to the location of major urban centers and of the MSAs. The lighter shading represents other counties that are included in the state's MSAs. They are usually called "suburban counties". The central

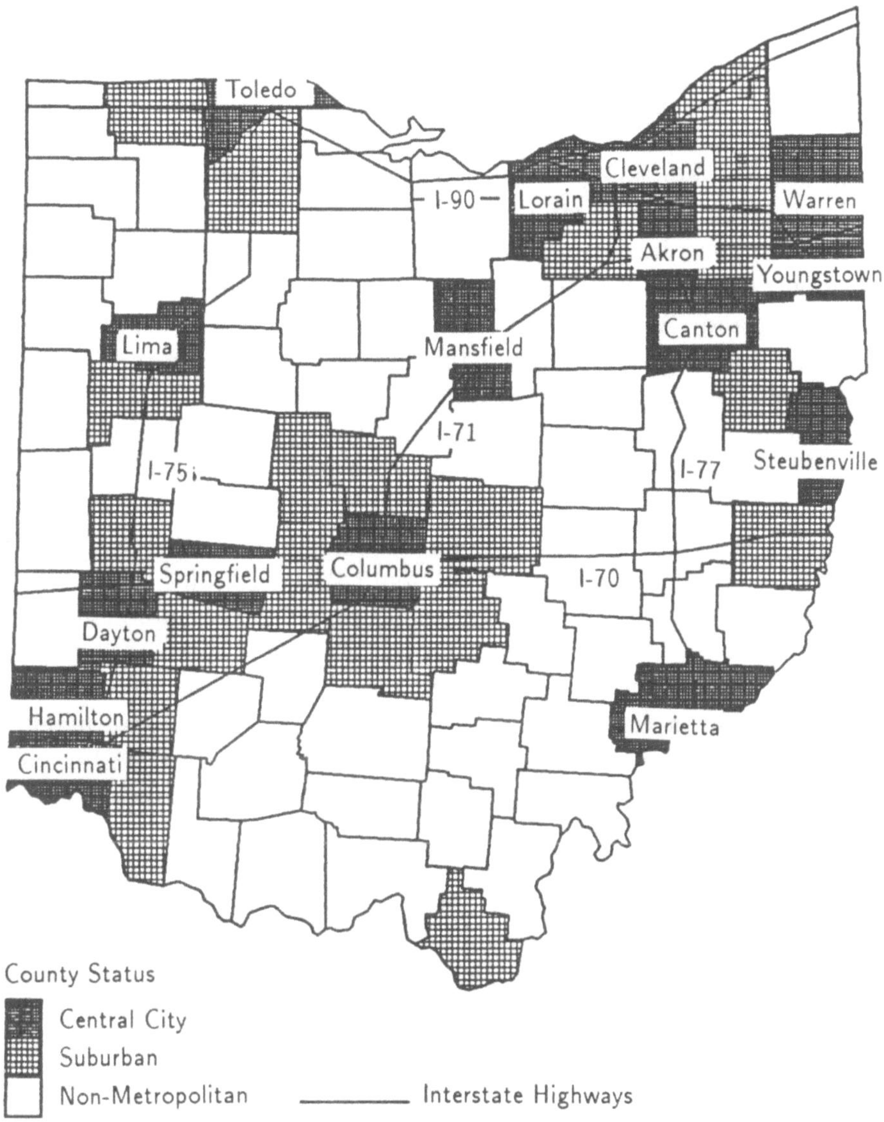

Figure 23.3. Metropolitan and non-metropolitan areas in Ohio.

city and suburban counties together form the metropolitan areas, whereas the unshaded counties are the non-metropolitan areas of the state.

In Ohio, the largest urban centers are found along a diagonal running from northeast to southwest. Among them, Cleveland, Columbus, and Cincinnati developed earlier and are also the most populated. In the northeast and southwest, there are conglomerations of urban centers referred to as Consolidated Metropolitan Statistical Areas (CMSAs). These are the Cleveland-Akron-Lorain CMSA and the Cincinnati-Hamilton CMSA. Other metropolitan areas include Toledo and Lima in the northwest, and smaller urban centers in the Appalachian areas in the south and east.

Maps (a) and (b) in *Figure 23.4* display the estimated growth rates in AIDS cases, for the years 1984 and 1987, based on the trend surface expansion analysis. The patterns in both maps are dominated by the Cleveland-Akron-Lorain CMSA in the northeast, by the Cincinnati-Hamilton CMSA in the southwest, and by the northeast-southwest alignment. These patterns do indeed reflect the fact that the AIDS epidemic spread earlier and faster (1) in the largest urban centers, (2) where several major urban centers are relatively close to one another, and (3) along major transportation routes. All of these increase the opportunities for human interactions, and heighten the chance for AIDS transmission. Clearly, the trend surface expansions bring forth the dominant aspects in the spatial spread of the AIDS epidemic in Ohio at the regional level of resolution. This spread manifests itself in a widening of higher growth areas around the two CMSAs, and in the tendency for the two areas to grow closer along I-71, which is the major routeway linking Cleveland, Columbus, and Cincinnati.

The estimates of growth rates of AIDS for 1984 and 1987 based on the Fourier polynomial expansions are portrayed in Maps (c) and (d) in *Figure 23.4*. The growth rates intervals controlling the shading in these maps are the same as those used in Maps (a) and (b), to facilitate the comparison between the results of the two analyses. Maps (c) and (d) show highest growth rates for most of the metropolitan counties. Also, unlike the growth estimates produced by the trend surface expansions, the ones displayed in Maps (c) and (d) capture the growth in AIDS counts in urban centers away from the northeast-southwest corridor, and, more specifically, in the smaller metropolitan areas in the southeast and northwest. Clusters of counties with high growth rates are also found along the I-90 highway running from east to west along the lake shore and along the two interstates from north to south, I-77 and I-75. Maps (c) and (d) suggest that most of the spatial diffusion of AIDS between 1984 and 1987 runs along these arteries. In fact the two

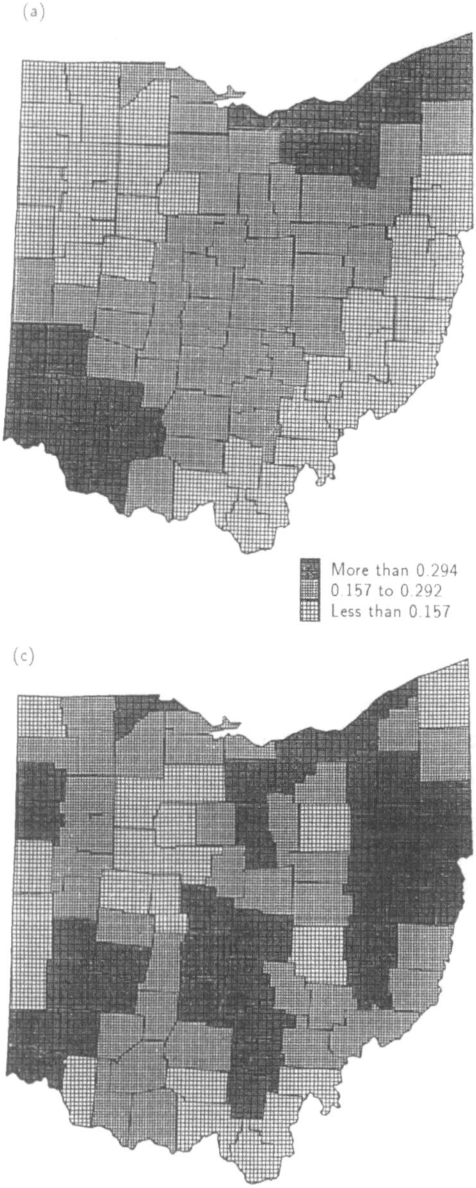

Figure 23.4. The diffusion of AIDS in Ohio: estimated growth rate; (a) 1984 and (b) 1987 (based on trend surface expansions); (c) 1984 and (d) 1987 (based on Fourier polynomial expansions).

maps show that except for some counties south of Columbus, the interstices not serviced by interstate highways generally display the lowest growth in AIDS counts.

The differences between Maps (a) and (b) on one hand, and Maps (c) and (d) on the other, indicate that the Fourier polynomial expansions have a greater capability to resolve the spatial drift in the temporal dynamics of the AIDS epidemic at the county level. However, the trend surface and the Fourier polynomial expansions can filter distinct aspects of spatial variation. The trend surface expansions give a broad and more general picture of dominant regional trends. In contrast, the Fourier polynomial expansions yield generalized but fine-grain portraits at a considerably higher level of resolution.

23.5 Conclusion

In this chapter, the expansion method's research philosophy is applied to investigating the spatial spread of the AIDS epidemic in Ohio. To this effect, the spread of the epidemic was conceptualized as the parametric drift of mathematical relationships capable of representing the temporal dynamics of the cumulative count of AIDS cases. One approach to formalizing this spatial drift involved redefining the relationship's parameters into functions of a variable, population density, selected for its ability to differentiate geographical environments. In terms of the second approach, the spatial drift of the AIDS dynamics was investigated by expanding its parameters into functions of spatial coordinates.

The results of our analyses indicate that in Ohio the AIDS epidemic grew earlier and faster in larger urban centers and along major transportation routes. Our results also suggest a trend toward a slower growth of the epidemic in the core areas of the major metropolitan agglomerations.

Acknowledgments

We wish to express our appreciation and thanks to Dr. A. Herzog of the Ohio Department of Health for his encouragement, and for providing the data on which the analyses in this chapter are based. Comments and criticisms by Drs. L.A. Brown, S.A. Foster, W. Gorr, P. Gould, and A. Herzog are gratefully acknowledged.

References

Agterberg, F.P., 1984, Trend surface analysis, in E.L. Gaile and C.J. Willmott, eds., *Spatial Statistics and Models*, D. Reidel Publishing Co., Dordrecht, The Netherlands.

Bailey, N.T.J., 1957, *The Mathematical Theory of Epidemics*, Charles Griffin & Co. Ltd., London, UK.

Bailey, N.T.J., 1975, *The Mathematical Theory of Infectious Diseases*, Charles Griffin & Co. Ltd., London, UK.

Bartlett, M.S., 1960, *Stochastic Population Methods in Ecology and Epidemiology*, Methuen & Co. Ltd., London, UK.

Casetti, E., 1966, Analysis of spatial association by trigonometric polynomials, *Canadian Geographer* **10**:199–204.

Casetti, E., 1972, Generating models by the expansion method: Applications to geographic research, *Geographical Analysis* **4**:81–91.

Casetti, E., 1982, Mathematical modelling and the expansion method, in R.B. Mandal, ed., *Statistics for Geographers and Social Scientists*, Concept Publishing Company, New Delhi, India.

Casetti, E., 1986, The dual expansion method: An application for evaluating the effects of population growth on development, *EEE Transactions on Systems, Man, and Cybernetics* SMC-16, No.1:29–39, January–February.

Casetti, E. and Jones, J.P. III, 1987, Spatial applications of the expansion method paradigm, in C. Dufornaud and D. Dudycha, eds., *Quantitative Analysis in Geography*, University of Waterloo, Waterloo, Ontario, Canada.

Chorley, R. and Haggett, P., 1965, Trend-surface mapping in geographical research, *Transactions of the Institute of British Geographers* **37**:47–67.

Curran, J.W., Morgan, W.M., Hardy, A.M., Jaffe, H.W, Darrow, W.W., and Dowdle W.R., 1985, The epidemiology of Aids: Current status and future prospects, *Science* **229**:1352–1367.

Curran, J.W., Jaffe, H.W., Hardy, A.M., Morgan, W.M., Selik, R.M., and Dondero, T.J., 1988, Epidemiology of HIV infection and AIDS in the United States, *Science* **239**:610–616.

Dutt, A.K., Monroe, C.B, and Dutta, H.M., 1987, Geographical patterns of AIDS in the United States, *The Geographical Review* **39**:456–471.

Gardner, L.I., Brundage, J.F., Burke, D.S., McNeil, J.G., Visintine, R., and Miller, R.N., 1989, Spatial diffusion of the human immunodeficiency virus infection epidemic in the United States, 1985–1987, *Annals of the Association of American Geographers* **79**(1):25–43.

Gong, V. and Rudnick, N., eds., 1986, *AIDS: Facts and Issues*, Rutgers University Press, Brunswick, OH, USA.

Gould, P., 1989, Geographic dimensions of the AIDS epidemic, *Professional Geographer* **41**(1):71–78.

Jager, J.C. and Ruitenberg, E.J., eds., 1988, *Statistical Analysis and Mathematical Modelling of AIDS*, Oxford University Press, Oxford, New York, Tokyo.

Poznanski, K., 1986, The extinguishing process: A case study of steel technologies in the world industry, *Technovation* **4**:297–316.

Ray, G.F., 1989, Full circle: The diffusion of technology, *Research Policy* **18**:1–18.

Sharif, M.N. and Kabir, C., 1976, A generalized model for forecasting technological substitution, *Technological Forecasting & Social Change* **8**:353–364.

Skiadas, C., 1986, Innovation diffusion models expressing asymmetry and/or positively or negatively influencing factors, *Technological Forecasting & Social Change* **30**:313–330.

Silverberg, G., Dosi, G., and Orsenigo, L., 1988, Innovation, diversity and diffusion: A self organizational model, *The Economic Journal* **98**:1032–1054.

Soete, L. and Perez, C., 1988, Catching up in technology: Entry barriers and windows of opportunity, in G. Dosi *et al.*, eds., *Technical Change and Economic Theory*, Pinter Publishers, London and New York.

van Duijn, J.J., 1983, *The Long Wave in Economic Life*, Allen and Unwin, London, UK.

Vasko, T., ed., 1987, *The Long-Wave Debate*, Springer-Verlag, Berlin, Heidelberg, New York.

Ward, W.H., 1967, The sailing ship effect, *Bulletin of the Institute of Physics and Physics Society* **18**:169.

Chapter 24

Branching out into the Universe

Cesare Marchetti

Work on diffusion models started at IIASA in 1974 when we were searching for a solution to the problem of putting some internal logic into the dynamics of energy markets (Marchetti, 1975). The problem was formally solved by using multiple competition dressed in logistics which are the simplest coder for the simplest diffusion process, the epidemic one.

The immensely complex phenomenon of using energy in various forms with all the interfacing of economics, technologies, and politics, and over a period of more than 100 years, showed up as a crystalline substitution, i.e., a multiple diffusion process (*Figure 24.1*). The result was a philosophical shock to me, because it robbed the process of its contemporaneity. A single set of equations, each with an input of only two parameters, was capable of describing the whole process even 100 years or more ago. In addition the description did not require the notion of money or other economic paraphernalia. Everything could be reduced to the timing and speed of the introduction of each new competitor. All the rest was a consequence, even 100 years later. Apart from the isolation resulting from the effects of daily affairs, the system showed an incredible long-term self-consistency, in spite of the continuously changing technical, economic, and political substrate. The

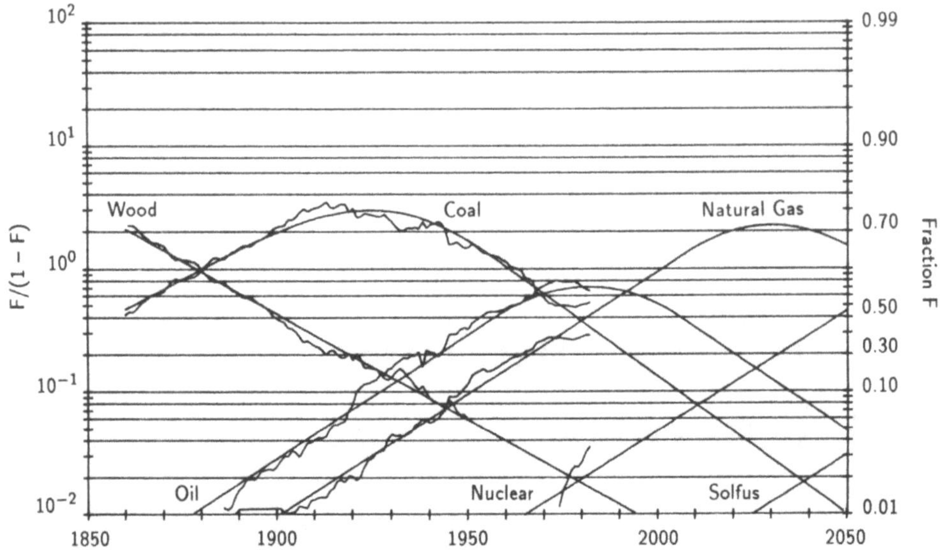

Figure 24.1. World primary energy substitution, in fractional shares of primary energy consumption by source. (Source: Marchetti and Nakićenović, 1979.)

precision of the match discouraged any thoughts that it was a pure coincidence. The next step was in fact the initial branching out into a whole set of cases of primary energy substitution, from single countries to electricity production, to single industries.

Altogether we collected about 300 cases (Marchetti and Nakićenović, 1979), showing that the multiple diffusion process was an unsinkable descriptor of the dynamics of the real world in the energy area. The first branching out from energy proper started here, where we were using proxies, like the number of diesel versus steam locomotives to pattern the substitution of oil versus coal for railway transport, or the amount of coal extracted using various technologies, to pattern the competition between these technologies. Everything worked very well, and the suspicion was that the methodology could be of very broad significance.

Our ancestors operated in the thirties under the spur of the work of Volterra in Italy and Lotka in the USA. Both gentlemen were dealing with biological systems, but there was an overspill of interest for people working in the area of economics and sociology. A number of papers appeared, some with penetrating insights into the possibility of using logistic analysis to map social processes. I still wonder what inhibited an explosion of research at that time. Certainly multiple competition or diffusion was not under control or even in the minds of these precursors. Nor was the strong effect of Kondratieff cycles in single penetration logistics. Nor the frantic hunt around the asymptote for certain economic systems.

My exploration was, however, prudent, choosing contiguous subjects. Having a loose consulting contract with FIAT, I was asked to peep into the car system. The original stimulus was to see if technological innovations or promotional campaigns had an effect on sales and on the ownership of cars. The first case that I studied was Italy where I looked at cars in circulation, a statistic more easily available than the one on sales (see, for instance, Marchetti, 1983). I was again philosophically shocked. The diffusion runs with an astonishing stability and precision, insensitive to any initiative coming from the industry or sudden changes in economics and politics on the outside. The first and second oil shock did not produce the tiniest dent, not to speak of promotional campaigns into which the car industry sinks fortunes.

Sponsors and colleagues were equally upset. The faith in pretty girls had deep emotional roots. *Obviously* all the *mise en scène* had the purpose of beating competition, and not really to increase the ownership of cars. That was the explanation of my results. So I went to see the evolution in time of market shares between the main car producers in the USA and Japan. The results confirm that *obviously* all the fuss was not about beating the competitor, but to keep him at bay.

My personal interpretation, based on hundreds of evident examples, is that people talk and the system does its business. The French have a delicious fable written down by Lafontaine, *La Mouche Cochère*, where a fly by actively flying up and down helps a pair of oxen to pull a heavy cart. And it is also very tired and self-satisfied at the end. A great deal of the talk of *decision makers* seems to fit this description very well. I wonder how many ponderous decisions had to be taken to keep the curves of *Figure 24.1* in such good shape.

The philosophical breakthrough in the application of these diffusion equations came when a friend of mine, president of the Italian Marketing Association, asked me to prepare an innovative paper for a conference on innovation which he had organized in Turin in 1979 (Marchetti, 1980). I had on my desk

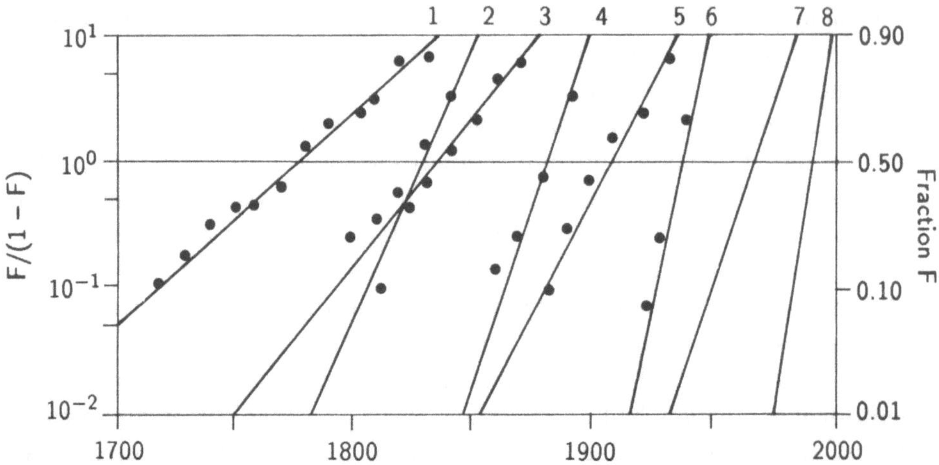

Figure 24.2. Invention and innovation waves, the secular set: The first three waves of the series are historical. We live in the fourth. Odd numbers indicate invention waves and even ones innovation waves (number 8 indicates the fourth innovation wave). (Source: Marchetti, 1980.)

the Mensch book on innovation, "*Das Technologische Patt*", with statistics on basic inventions and innovations dating back to 1800 (Mensch, 1975). The quality of these statistics comes from a homogeneous definition for the *dates* when inventions and innovations appeared.

In order to make these statistics treatable with our diffusion equations, I then made the hypothesis that an innovation can be considered as an object, a product, like a car, generated in response to market demand. The set of innovations would progressively fill the market, like cars. The thread was thin but it held. Not only innovations, but also invention waves fit the model very well. This was the first hint of the fractal nature of the social and economic system we shall talk about later on.

The analysis of the inventions and innovations waves *à la* Mensch revealed internal regularities that led me to find some arithmetic rules to calculate the ones that followed. The result of this exercise is reported in *Figure 24.2*. That the system regulates so precisely such fancy action like inventing and innovating was really an unexpected result.

But also entrepreneurship is equally regulated. For instance, an analysis of the opening of firms producing cars and of firms quitting the market follow perfect (cumulative) logistics. Nitety-nine percent of the firms did quit, and that gives a glimpse of what could be defined as the *Holocaust of the enterpreneurs.*

The chart of *Figure 24.2* reveals also another detail in the behavior of the system: that the centerpoints of the innovation waves are spaced at 55 year intervals, the distance of one Kondratieff long cycle. It puts the mechanisms into a Schumpeterian vision of markets progressively saturated leading to recession. Innovation and entrepreneurship reopen the game into a new spurt of activity. A completely different approach was taken by Stewart (1982) of Nutevco who measured the deviations of primary energy consumption and electricity consumption from the secular trends. Also his analysis shows in crisp form the pulsations of the global activity. With a periodicity of 55 years.

Faith and daring come with success and at this point all inhibitions were removed. After all experimenting was cheap, and failures could find their way to the paper basket. Only a short sampling can be taken out of the 3000 odd examples which Nakićenović, Grübler, and I have collected. I must honestly say that being beyond my career I was always the most daring. I did not have much to lose and have no professional peers to keep me in line. Much stimulation comes when critics start gnawing at the borders to check the consistency of the stuff. One of the first observations was that the model only operates for Western countries during the last century or so. It was true that all our examples fell into this time slot, but the obvious reason was the availability of sound statistics.

But a number of things are well recorded from a deeper past, so I analyzed the diffusion wave of the construction of gothic cathedrals. Their cumulative number in terms of *first stones* is reported in *Figure 24.3*. The regularity of the process, over such a long period of time, with economic revolutions, wars and plagues, is really astonishing. In my view it is a milestone in revealing the deep regulatory feedbacks of the social system. This self-regulation is pervasive. In a completely different field we can observe, e.g., the perfect resilience of the air transport system growth to the oil price shocks of 1974 and 1979. The immunity was acquired through a complex and painful *internal* reorganization of the air companies. Going to centralized economies does not change the situation either, as much work done by Grübler (see, for instance, Chapter 19) shows.

At this point the concept started taking shape, i.e., the diffusion process is really at the cultural level and *paradigms* are generated and selected somewhere, and then move into the heads of people to become finally *action.* Hägerstrand's field measures of the mechanisms of this diffusion through very

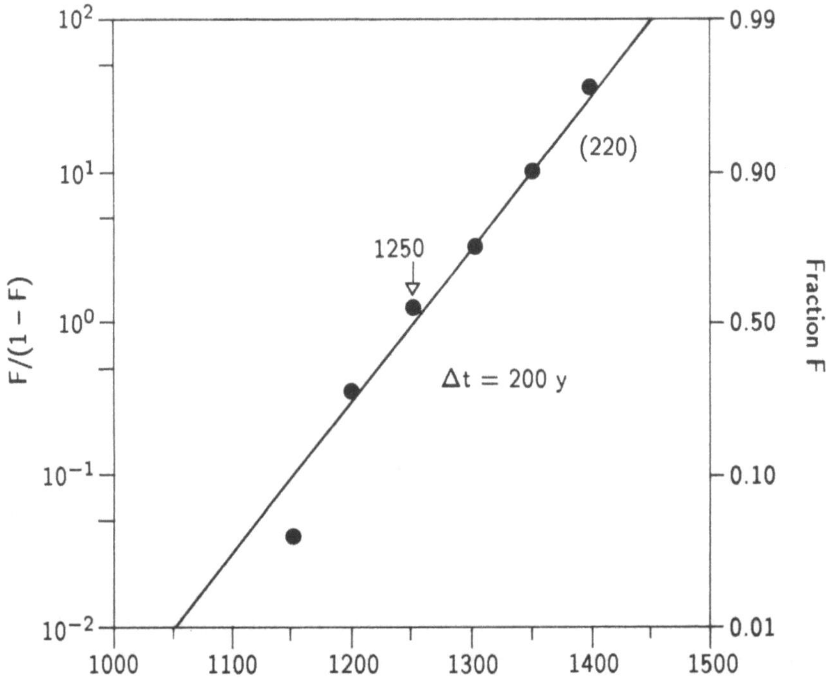

Figure 24.3. Building gothic cathedrals (an incredible burden to the populations of the Middle Ages) analyzed as a diffusion process. The chart gives the cumulative number of ***first stones*** for what will in time become a cathedral. The dates are well documented. (Source: Marchetti, 1986.)

stable social networks made by sets of about 100 people, gives a hint about the causes of the stability of the process.

These friendly gangs, as they emerge from anthropological studies, already existed in the neolithic age. Fast travel and telecommunication gives them larger territories without changing the structure and timing of their interactions very much. This communication by internal lines neatly explains why the information pouring by the media does not seem to dent the rates of a diffusion process, even over a span of 100 years.

Speaking of media, a couple of years ago I did some research for the CEE on nuclear energy and the social system (Marchetti, 1988). Measuring the opinion of people is a hard task because people rarely know the deeper meaning of their actions, and sometimes lie about this. But the media act on strict feedback from their customers, and try desperately to match what the customers want to hear. So I started measuring the media in order to check public opinion. The hypothesis behind the use of the media as a proxy for public opinion may be right or wrong, but the result I want to show is that the coverage (cumulative number of articles or TV spots) follows exactly the scheme of a diffusion process. *Figure 24.4* reports the coverage of nuclear power by the media in the USA. Dailies and TV are much the same as they fish in the same pond of public opinion. The periodical press has a slightly higher time constant.

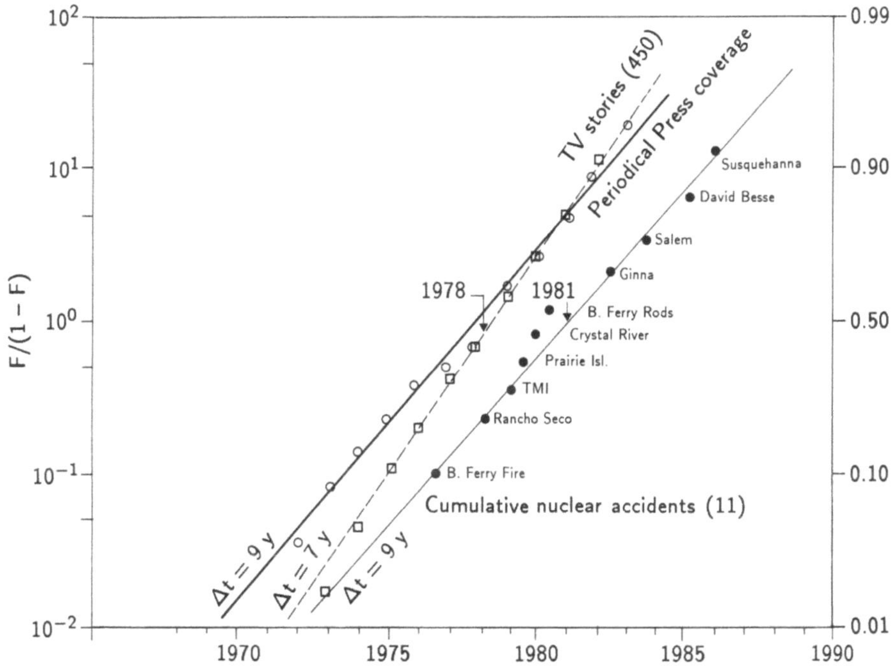

Figure 24.4. Media coverage of nuclear energy in the United States (measuring cumulative number of evening news reports in US TV networks and coverage index by US periodicals on nuclear power plants; data from Mazur, 1984). Also shown are the cumulative number of major US nuclear accidents. Observed is a parallel of the accident wave with the press wave *as if* opinion intensity preceded accident probability by a couple of years. The Three Mile Island accident, perceived as the worst, happened when coverage by the periodical press was at a maximum (F = 50%). (Source: Marchetti, 1988.)

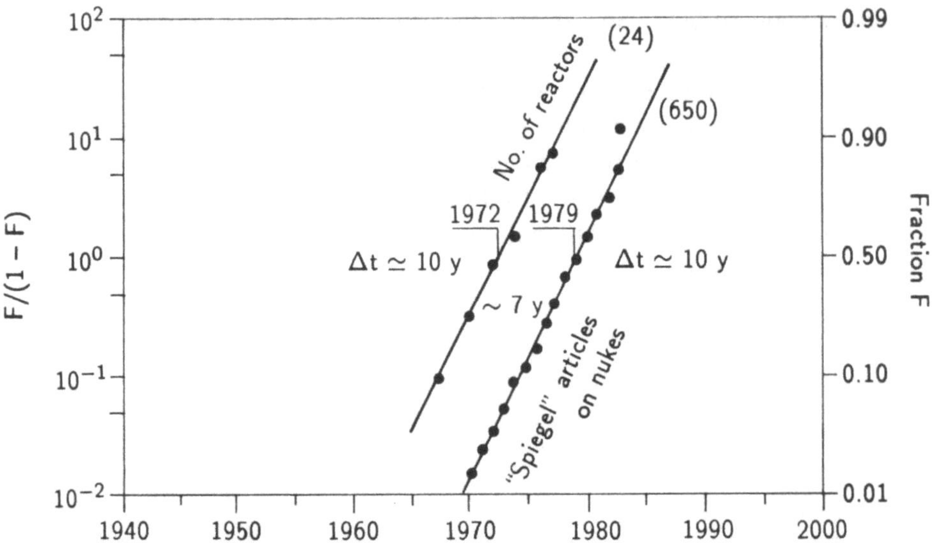

Figure 24.5. Media coverage on nuclear energy as represented by *Der Spiegel* is well correlated to the beginning of nuclear reactor construction in the FRG with a time lag of seven years. (Source: Marchetti, 1988.)

Also if one takes an individual journal, like *Der Spiegel*, one gets exactly the same result (*Figure 24.5*). The coverage is highly correlated with reactor construction start-ups both in the USA and the Federal Republic of Germany, if with many years delay time. Also accident waves are correlated with periodic press coverage. The Three Mile Island accident occurred when public attention, as measured by the periodical press, was at a maximum. This certainly does not establish a cause–effect relationship, but the time sequence proves invalid the reverse cause–effect, i.e., that the accident wave is the stimulus for press coverage.

The variety of subjects touched never ends, from the witch hunt of the Middle Ages, to the adoption of stamps in the Western hemisphere or the writing of papers on the greenhouse effects of CO_2. The same elementary mathematics neatly covers the fact.

At this point I have to draw some conclusions. As the economist Walt Rostow once said after listening to one of my presentations, *we may have struck a deeper level of truth*. The deeper level, as I see it now, is that our culture operates as the carrier of action paradigms at all levels of spatial and hierarchical integration. Humanity then operates as a gigantic quasifractal system, where the equation is always the diffusion equation but its parameters depend on the level of fractality.

If this is true, and everything we have done to date conveys that, a really deeper level of truth has been struck revealing the single and all-pervasive mechanism of the working of society. Studying primary energy substitution can lead very far indeed!

References

Marchetti, C., 1975, On strategies and fate, in W. Häfele *et al.*, *Second Status Report of the IIASA Project on Energy Systems 1975*, RR-76-1, International Institute for Applied Systems Analysis, Laxenburg, Austria.

Marchetti, C., 1980, Society as a learning system: Discovery, invention, and innovation cycles revisited, *Technological Forecasting & Social Change* **18**:267–282.

Marchetti, C., 1983, The automobile in a systems context, the past 80 years and the next 20 years, *Technological Forecasting & Social Change* **23**:3–23.

Marchetti, C., 1986, *Stable Rules in Social Behavior*, IBM conference Academia Brasileira de Ciências 1986, IBM Brasil.

Marchetti, C., 1988, *On Society and Nuclear Energy*, International Institute for Applied Systems Analysis, Laxenburg, Austria, mimeo.

Marchetti, C. and Nakićenović, N., 1979, *The Dynamics of Energy Systems and the Logistic Substitution Model*, RR-79-13, International Institute for Applied Systems Analysis, Laxenburg, Austria.

Mazur, A., 1984, Media influences on public attitudes toward nuclear power, in W.R. Freudenburg and E.A. Rosa, eds., *Public Reactions to Nuclear Power. Are There Critical Masses?* IAAS Selected Symposium 93, American Association for the Advancement of Science, Washington, DC, USA.

Mensch, G., 1975, *Das technologische Patt, Innovationen überwinden die Depression*, Umschau Verlag, Frankfurt, FRG.

Stewart, H.B., 1982, *Technology, Innovation, and Business Growth*, Nutevco, San Diego, CA, USA.

List of Contributors

Mr. Jesse H. Ausubel
Fellow in Science and Public Policy
The Rockefeller University, NR 403
1230 York Avenue
New York, NY 10021-65399
USA

Prof. Robert U. Ayres
Department of Engineering and
Public Policy
Carnegie-Mellon University
Pittsburgh, PA 15213-3890
USA

Prof. Frank M. Bass
School of Management
University of Texas
System Eugene McDermott
Dallas, Texas 75230
USA

Prof. Harvey Brooks
Science, Technology and Public
Policy Program
Harvard University
John F. Kennedy School of
Government
Cambridge, MA 02138
USA

Prof. Emilio Casetti
Department of Geography
The Ohio State University
103 Bricker Hall
Columbus, Ohio 43210-1361
USA

Mr. Alain Debecker
Digital Equipment Corporation
International (Europe)
12 avenue des Morgines
C.P. 176
CH-1213 Petit Lancy 1
Geneva
Switzerland

Drs. Paul Diederen
Maastricht Economic Research
Institute on Innovation and
Technology (MERIT)
Tongersestraat 49
P.O. Box 616
NL-6200 MD Maastricht
The Netherlands

Prof. Giovanni Dosi
Dipartimento di Scienze Economiche
Universita Degli Studi di Roma
via Nomentana, 41
I-00161 Rome
Italy

Dr. Ike Ezekoye
Westinghouse Research Laboratory
Pittsburgh, PA 15235
USA

Prof. C. Cindy Fan
Department of Geography
University of California
Los Angeles, CA 90024
USA

Prof. Dominique Foray
Ecole Centrale Paris
Laboratoire "Strategie & Technologie"
Grande Voie des Vignes
F-92295 Châtenay Malabry Cédex
France

Dr. Sergei Yu. Glaziev
Central Economic Mathematical
Institute (CEMI)
32, Krasikova Street
117418 Moscow
USSR

Prof. Ove Granstrand
Department of Industrial Management
and Economics
Chalmers University of Technology
S-41296 Göteborg
Sweden

Dr. Arnulf Grübler
IIASA
A-2361 Laxenburg
Austria

Dr. Jonny Hjelm
Department of History
Umeå University
S-90187 Umeå
Sweden

Dr. Dirk-Jan F. Kamann
Faculty of Economics
University of Groningen
P.O. Box 800
NL-9700 AV Groningen
The Netherlands

Dr. Yuri Kaniovsky
IIASA
A-2361 Laxenburg
Austria

Prof. Charlie Karlsson
Head, Department of Economics
University of Karlstad
P.O. Box 9501
S-65009 Karlstad
Sweden

Prof. Maryellen R. Kelley
School of Urban and Public Affairs
Carnegie-Mellon University
Pittsburgh, PA 15213-3890
USA

Drs. René Kemp
Maastricht Economic Research
Institute on Innovation and
Technology (MERIT)
Tongersestraat 49
P.O. Box 616
NL-6200 MD Maastricht
The Netherlands

Prof. Harold A. Linstone
Systems Science Ph.D. Program
Portland State University
Portland, Oregon 97207-0751
USA

Prof. Vijay Mahajan
Department of Marketing
Administration
Graduate School of Business
The University of Texas at Austin
Austin, Texas 78712-1176
USA

Dr. Cesare Marchetti
IIASA
A-2361 Laxenburg
Austria

Prof. George Modelski
Department of Political Science
University of Washington
Seattle, Washington 98195
USA

Dr. Theodore Modis
Digital Equipment Corporation
International (Europe)
12 avenue des Morgines
C.P. 176
CH-1213 Petit Lancy 1
Geneva
Switzerland

Prof. Eitan Muller
Leon Recanti Graduate School of
Business Administration
Tel-Aviv University
IL-69978, Tel Aviv
Israel

Prof. Joan Muysken
Maastricht Economic Research
Institute on Innovation and
Technology (MERIT)
Tongersestraat 49
P.O. Box 616
NL-6200 MD Maastricht
The Netherlands

Dr. Nebojša Nakićenović
IIASA
A-2361 Laxenburg
Austria

Prof. Peter Nijkamp
Department of Economics
Free University of Amsterdam
P.O. Box 7161
NL-1007 MC Amsterdam
The Netherlands

Prof. William Peirce
Department of Economics
Weatherhead School of Management
Case Western Reserve University
Cleveland, Ohio 44106
USA

Dr. Gardner Perry III
Science and Mathematics Department
North Seattle Community College
Seattle, Washington 98103
USA

Prof. Gerhard Rosegger
Frank Tracy Carlton
Professor of Economics
Weatherhead School of Management
Case Western Reserve University
Cleveland, Ohio 44106
USA

Dr. Gerald Silverberg
Maastricht Economic Research
Institute on Innovation and
Technology (MERIT)
Tongersestraat 49
P.O. Box 616
NL-6200 MD Maastricht
The Netherlands

Prof. John E. Tilton
Department of Mineral Economics
Colorado School of Mines
Golden, Colorado 80401
USA

Drs. G. Rombout de Wit
Maastricht Economic Research
Institute on Innovation and
Technology (MERIT)
Tongersestraat 49
P.O. Box 616
NL-6200 MD Maastricht
The Netherlands

International Conference on

Diffusion of Technologies and Social Behavior

14–16 June 1989

P R O G R A M

Wednesday, 14 June 1989

08:30 Registration
09:00 Welcoming Address, R. Pry, IIASA Director
09:15 Introduction to Conference, N. Nakićenović
09:30 Summary of Venice Diffusion Conference, G. Dosi

SESSION I: THEORIES OF INNOVATION DIFFUSION

10:00 **A. Temporal Diffusion**
Chairperson: T. Lee
Panel: A. Grübler, H. Linstone, V. Mahajan
H. Linstone, *The Need for Multiple Perspectives on Technology Diffusion: Insights and Lessons*
V. Mahajan, *New Product Diffusion Models in Marketing: Status and Outlook*
A. Grübler, *Diffusion: Long-term Patterns and Discontinuities*
11:00 **Discussion**
12:00 **Lunch**
13:45 **B. Spatial Diffusion**
Chairperson: T. Hägerstrand
Panel: L. Brown, K. Kamann, M. Sonis
L. Brown, *Overview of Spatial Diffusion Studies and Future Research Directions*
K. Kamann, *Technogenesis: Origins and Diffusion in a Turbulent Environment*
M. Sonis, *Spatio-Temporal Innovation Diffusion, Schumpeterian (Ecological) Competition and Dynamic Choice: A New Synthesis*
14:45 **Discussion**
15:30 Coffee

Wednesday, 14 June 1989
SESSION I: THEORIES OF INNOVATION DIFFUSION (*Continued*)

16:00 **C. Diffusion in Economics**
Chairperson: J. Tilton
Panel: R. Ayres, V. Faltsman, G. Dosi, G. Silverberg
G. Dosi, *The Research on Innovation Diffusion: An Assessment*
R. Ayres, *Competition and Complementarity in Diffusion*
V. Faltsman, *Scientific and Technological Progress in the USSR: Change of Priorities*
G. Silverberg, *Adoption and Diffusion of Technology as a Collective Evolutionary Process*
17:15 **Discussion**
18:30 Departure for Heurigen

Thursday, 15 June 1989
SESSION I: THEORIES OF INNOVATION DIFFUSION (*Continued*)

08:30 **D. Diffusion as a Social Process**
Chairperson: H. Brooks
Panel: C. Marchetti, G. Modelski
H. Brooks, *Introduction*
G. Modelski, *Democratization in Long-range Perspective*
C. Marchetti, *Stable Rules in Social Behavior*
09:15 **Discussion**
09:45 Coffee

SESSION II: CASE STUDIES

10:00 **A. Plenary Session**
Chairperson: R. Vacca
Panel: V. Livshits, N. Nakićenović, G. Rosegger, J. Tilton
J. Tilton, *Material Substitution: The Role of New Technology*
G. Rosegger, *Interfirm Cooperation in the Motor Vehicle Industry*
V. Livshits, *Problems of Technological Change in Soviet Infrastructure*
N. Nakićenović, *The Diffusion of Pervasive Systems*
11:30 **Discussion**
12:30 Lunch
14:30 **A. SESSION II** (*Continued*)
T. Lee, *Global Alliance and Technology Diffusion*
15:15 **B. Parallel Sessions**
Session B1: New Manufacturing and Information Technologies
Session B2: Diffusion Case Studies
Session B3: Case Study Methodology
18:30 Reception

Thursday, 15 June 1989, starting: 15:15
SESSION II: CASE STUDIES (Parallel Sessions)

B1. New Manufacturing and Information Technologies
Chairperson: R. Ayres

H. Brooks, *Diffusion of Programmable Automation: Why Small Firms Are Not Likely to Adopt It*
M. Kelley, *The Uneven Pattern of Diffusion of Programmable Automation Among US Manufacturers*
C. Edquist, *Empirical Differences Between OECD Countries in the Diffusion of New Product and Process Technologies*
H.-D. Haustein, *Diffusion of Automation in GDR's Industry: Statistical Analysis and Forecasting by DYNAMAT*
I. Tchijov, *CIM Diffusion Processes*
C. Karlsson, *The Adoption of Applications of Information Technology for Process Development*
T. Åstebro, *The Intra-firm Diffusion of Electronic Mail Systems*

B2. Diffusion Case Studies
Chairperson: G. Rosegger

E. Casetti (presented by L. Brown), *The Spatial Diffusion of AIDS*
P. Diederen, *The Diffusion of Technologies in Dutch Banking*
T. Modis, *Diffusion of Services on Aging Products*
D. Foray, *Morphological Analysis, Diffusion and Lock-out of Technologies*
S. Glazev, *Diffusion and New Socio-technical Paradigms*
O. Ullmann, *On the Diffusion of Old and New Solar Technologies*
J. Hjelm, *The Diffusion of Chain Saws in Swedish Forestry*

B3. Case Study Methodology
Chairperson: V. Mahajan

R. Vacca (with V. Franchina), *Logistic Curves: Construction, Fit & Uncertainty*
A. Medvedev, *Product and Process Innovation: In Search of Harmony*
A. Debecker, *Determination of the Uncertainties in S-curve Logistic Fits*
V. Zhianov, *Mathematical Models of Technology Diffusion*
Y. Kaniovsky, *Innovation Processes Modeled Based on a Generalized Urn Scheme*
S. Lobanov, *Assessment of Efficiency of Technological Change*
S. Mori, *A Frame of a Dynamic I-O Model Associated with R&D*

Friday, 16 June 1989

SESSION III: APPLICATIONS IN PRIVATE AND PUBLIC POLICY

08:30 **Policy and Research**
Chairperson: G. Silverberg
Panel: R. Ayres, B. Guile, W. Peirce, V. Rudashevski
R. Ayres, *Technometrics: The Problem Relevance (Are the Measures We Need, the Measures We Have?*
V. Rudashevski, *Organizational Problems of Technological Transfer*
W. Peirce, *Diffusion Policy in the "Single Europe" Institutional Structure and Industrial Prospects for the European Communities*

09:30 **Discussion**

10:00 Coffee and Software Demonstration by Digital Corporation

10:30 **Policy Applications**
Chairperson: F. Schmidt-Bleek
Panel: Å. Andersson, J.-E. Aubert, R. Pry, B. Sullivan
Å. Andersson, *The K–Society*
B. Sullivan, *Using Life Cycles Diffusion Analysis in Long-term Strategy Formulation*
F. Malik (presented by A. Eggler), *Fit of Diffusion and Substitution Analysis in the Concept of Strategic Management*
R. Pry, *Summary of Policy Issues*

12:00 **Discussion**

12:30 Cold Buffet

14:00 **SESSION IV: SUMMARY AND CONCLUSION OF MEETING**
Chairperson: J. Ausubel, *Introduction and Summary*
Panel: R. Ayres, G. Rosegger, V. Mahajan, *Key Questions Raised During Meeting*

15:00 **Discussion**

16:00 Closure of Meeting, N. Nakićenović

16:05 Informal Discussions and *vin d'honneur*

List of Participants

Åke Andersson
Institute for Future Studies
Sweden

Thomas Åstebro
Chalmers University of Technology
Sweden

Jean-Eric Aubert
OECD
France

Jesse H. Ausubel
The Rockefeller University
USA

Petr Aven
IIASA

Robert U. Ayres
Carnegie-Mellon University
USA

Luigi Bodda
IIASA

Ernest Braun
Institut für Sozio-ökonomische
Entwicklungsforschung und
Technikbewertung (ISET)
Austria

Harvey Brooks
John F. Kennedy School of
Government
Harvard University
USA

Lawrence Brown
Ohio State University
USA

Johann Cas
Institut für Sozio-ökonomische
Entwicklungsforschung und
Technikbewertung (ISET)
Austria

Marie Caron
IIASA

Rosalia Castellano
IIASA - YSSP

Alain Debecker
Digital Equipment Corporation
International (Europe)
Switzerland

Paul Diederen
Maastricht Economic Research
Institute on Innovation and
Technology (MERIT)
The Netherlands

Giovanni Dosi
Universita Degli Studi di Roma
Italy

Charles Edquist
University of Linköping
Sweden

Andreas Eggler
Management Zentrum St. Gallen
Switzerland

Vladimir Faltsman
Research Institute of Economics
USSR Academy of Sciences
USSR

Dominique Foray
Ecole Centrale Paris
Laboratoire "Strategie & Technologie"
France

Valerio Franchina
VECTOR
Italy

Sergei Yu. Glaziev
Central Economic Mathematical
Institute (CEMI)
USSR

Aldo Gotta
IIASA - YSSP

Arnulf Grübler
IIASA

Bruce Guile
National Academy of Engineering
USA

Torsten Hägerstrand
Royal University of Lund
Sweden

Heinz-Dieter Haustein
Institut für Sozialistische
Wirtschaftsführung
GDR

Simeon Hein
IIASA - YSSP

Jonny Hjelm
University of Umeå
Sweden

Dirk-Jan F. Kamann
University of Groningen
The Netherlands

Yuri Kaniovsky
IIASA

Ildar Karimov
IIASA - YSSP

Charlie Karlsson
University of Karlstad
Sweden

Maryellen R. Kelley
School of Urban and Public Affairs
Carnegie-Mellon University
USA

Ivan Kopernicky
Slovak Commission for Planning
and Investment
Czechoslovakia

Coen Kreuwels
IIASA - YSSP

Thomas H. Lee
Massachusetts Institute of Technology
USA

Tarmo Lemola
Technical Research Center of Finland
Finland

Harold A. Linstone
Systems Science Ph.D. Program
Portland State University
USA

Veniamin Lifshitz
All-Union Research Institute for
Systems Studies (VNIISI)
USSR

Serguei Lobanov
Central Economic and Mathematical Institute (CEMI)
USSR

Raimo Lovio
Finish National Fund for Research and Development (SITRA)
Finland

Lennart Lundmark
University of Umeå
Sweden

Vijay Mahajan
University of Texas at Austin
USA

Cesare Marchetti
IIASA

Thomas Marmefelt
IIASA - YSSP

Andrei G. Medvedev
Leningrad Institute for Engineering Economics
USSR

Hans Mieras
Directorate of Technology Policy
Ministry of Economic Affairs
The Netherlands

George Modelski
University of Washington
USA

Theodore Modis
Digital Equipment Corporation International (Europe)
Switzerland

Shunsuke Mori
Department of Industrial Administration
Science University of Tokyo
Japan

Nebojša Nakićenović
IIASA

Helga Nowotny
Institut für Wissenschaftstheorie und Wissenschaftsforschung
University of Vienna
Austria

Igor Očka
Research Institute of Technology and Economy in Mechanical Engineering (VUSTE)
Czechoslovakia

William Peirce
Department of Economics
Case Western Reserve University
USA

Paul Pissjak
Institut für Sozio-ökonomische Entwicklungsforschung und Technikbewertung (ISET)
Austria

Wolfgang Polt
IIASA - YSSP

Robert H. Pry
Director
IIASA

Jukka Ranta
Technical Research Center of Finland
Finland

Michele Rivard
IIASA - YSSP

Hans Holger Rogner
IIASA

Gerhard Rosegger
Department of Economics
Case Western Reserve University
USA

Vladimir Rudashevsky
All-Union Research Institute for
Systems Studies (VNIISI)
USSR

Uwe Schmidt
IIASA - YSSP

Friedrich Schmidt-Bleek
IIASA

Christoph Schneider
IIASA

Stanislav S. Shatalin
Institute of Economics and Fore-
casting of Technological Progress
USSR Academy of Sciences
USSR

Gerald Silverberg
Maastricht Economic Research
Institute on Innovation and
Technology (MERIT)
The Netherlands

Michael Sonis
Bar-Ilan University
Israel

Brian Sullivan
Futures Compendium
Government of Alberta
Canada

Iouri Tchijov (†)
IIASA

John E. Tilton
Department of Mineral Economics
Colorado School of Mines
USA

Gabrielle Tondl
IIASA

Oskar Ullmann
Ludwig Bölkow Stiftung
FRG

Roberto Vacca
Via Oddvone di Cluny 3
I-00153 Rome
Italy

Pentti Vuorinen
Finnish National Fund for
Research and Development (SITRA)
Finland

Jörg Walter
IIASA - YSSP

Michael Wingenfeld
IIASA - YSSP

Pekka Ylä-Anttila
The Research Institute of the
Finnish Economy (ETLA)
Finland

Vladimir Zhianov
Central Economic Mathematical
Institute (CEMI)
USSR

Leopold Ziegler
Bundesministerium für Wissenschaft
und Forschung
Austria

Note: IIASA - YSSP denotes participants in the Young Scientists' Summer Program.

THE INTERNATIONAL INSTITUTE FOR APPLIED SYSTEMS ANALYSIS

is a nongovernmental research institution, bringing together scientists from around the world to work on problems of common concern. Situated in Laxenburg, Austria, IIASA was founded in October 1972 by the academies of science and equivalent organizations of twelve countries. Its founders gave IIASA a unique position outside national, disciplinary, and institutional boundaries so that it might take the broadest possible view in pursuing its objectives:

To promote international cooperation in solving problems arising from social, economic, technological, and environmental change

To create a network of institutions in the national member organization countries and elsewhere for joint scientific research

To develop and formalize systems analysis and the sciences contributing to it, and promote the use of analytical techniques needed to evaluate and address complex problems

To inform policy advisors and decision makers about the potential application of the Institute's work to such problems

The Institute now has national member organizations in the following countries:

Austria
The Austrian Academy of Sciences

Bulgaria
The National Committee for Applied Systems Analysis and Management

Canada
The Canadian Committee for IIASA

Czech and Slovak Federal Republic
The Committee for IIASA of the Czech and Slovak Federal Republic

Finland
The Finnish Committee for IIASA

France
The French Association for the Development of Systems Analysis

Germany
Association for the Advancement of IIASA

Hungary
The Hungarian Committee for Applied Systems Analysis

Italy
The National Research Council (CNR) and the National Commission for Nuclear and Alternative Energy Sources (ENEA)

Japan
The Japan Committee for IIASA

Netherlands
The Netherlands Organization for Scientific Research (NWO)

Poland
The Polish Academy of Sciences

Sweden
The Swedish Council for Planning and Coordination of Research (FRN)

Union of Soviet Socialist Republics
The Academy of Sciences of the Union of Soviet Socialist Republics

United States of America
The American Academy of Arts and Sciences

MIX
Papier aus verantwortungsvollen Quellen
Paper from responsible sources
FSC® C105338

If you have any concerns about our products,
you can contact us on
ProductSafety@springernature.com

In case Publisher is established outside the EU,
the EU authorized representative is:
Springer Nature Customer Service Center GmbH
Europaplatz 3, 69115 Heidelberg, Germany

Printed by Libri Plureos GmbH
in Hamburg, Germany